Lecture Notes in Computer Scien

T0238512

Commenced Publication in 1973
Founding and Former Series Editors:
Gerhard Goos, Juris Hartmanis, and Jan van Leeuwen

Nelma Moreira   Rogério Reis (Eds.)

Implementation and Application of Automata

17th International Conference, CIAA 2012
Porto, Portugal, July 17-20, 2012
Proceedings

 Springer

Volume Editors

Nelma Moreira
Universidade do Porto
Faculdade de Ciências
Departamento de Ciência de Computadores
Rua do Campo Alegre 1021-1055
4169-007 Porto, Portugal
E-mail: nam@dcc.fc.up.pt

Rogério Reis
Universidade do Porto
Faculdade de Ciências
Departamento de Ciência de Computadores
Rua do Campo Alegre 1021-1055
4169-007 Porto, Portugal
E-mail: rvr@dcc.fc.up.pt

ISSN 0302-9743   e-ISSN 1611-3349
ISBN 978-3-642-31605-0   e-ISBN 978-3-642-31606-7
DOI 10.1007/978-3-642-31606-7
Springer Heidelberg Dordrecht London New York

Library of Congress Control Number: 2012940849

CR Subject Classification (1998): F.1.1-2, F.1, F.2, F.4, E.1, H.3

LNCS Sublibrary: SL 1 – Theoretical Computer Science and General Issues

Typesetting: Camera-ready by author, data conversion by Scientific Publishing Services, Chennai, India

Printed on acid-free paper

Springer is part of Springer Science+Business Media (www.springer.com)

Preface

The 17th International Conference on Implementation and Application of Automata (CIAA 2012) was held at the Faculdade de Ciências da Universidade do Porto, in Porto, Portugal during July 17–20, 2012.

The CIAA conference series is one of the major annual conferences for researchers, application developers, and users of automata-based systems. The topics of the conference include automata applications in, for example, formal verification methods, natural language processing, pattern matching, data storage and retrieval, and bioinformatics, as well as theoretical work on automata theory. Sixteen previous CIAA conferences took place in: Blois (2011), Winnipeg (2010), Sydney (2009), San Francisco (2008), Prague (2007), Taipei (2006), Nice (2005), Kingston (2004), Santa Barbara (2003), Tours (2002), Pretoria (2001), London Ontario (2000), Potsdam (WIA 1999), Rouen (WIA 1998), London Ontario (WIA 1997), London Ontario (WIA 1996).

For the editors, the CIAA was always synonymous with Sheng Yu's presence, commitment, and enthusiasm. He was in great part responsible for the idea of having the conference in Porto, and when he unexpectedly passed away, this year, it was naturally a great shock. The CIAA community will miss his profound knowledge, and his invaluable and innovative contribution to the area. We will miss also his friendship and immense kindness.

This volume of *Lecture Notes in Computer Science* contains the invited contributions and the accepted papers presented at CIAA 2012. The submission and refereeing process was supported by the EasyChair conference management system. The 53 papers submitted to CIAA 2012 were from 21 countries all over the world, including Algeria, Canada, China, Czech Republic, France, Germany, Greece, Iran, Italy, Japan, The Netherlands, Poland, Portugal, Russian Federation, Slovakia, South Africa, Spain, Thailand, Turkey, UK, and USA. Each submission was reviewed by at least three referees and discussed by the members of the Program Committee. A total of 21 regular papers and seven short papers were selected for presentation at the conference. There were five invited talks presented by Janusz Brzozowski, Paul Gastin, José Nuno Oliveira, Grzegorz Rozenberg, and Kai Salomaa. We warmly thank the invited speakers and all the authors of the submitted papers. Their efforts were the basis for the success of the conference.

We would like to thank all the members of the Program Committee and the external referees, for their work in evaluating the papers and their valuable comments that led to the selection of the contributed papers.

We thank Alfred Hofmann and Elke Werner from Springer for their help in making this volume available before the conference.

We are grateful to all the members of the Organizing Committee for their efforts in the preparation of the scientific sessions and social events. A special thank goes to José Pedro Rodrigues for his help in the graphic design for all the conference materials. Thanks also go to the staff, our colleagues, and the students Ivone Amorim, Rizó Isrfov, Eva Maia, Davide Nabais, and David Pereira, among others, of the Department of Computer Science of FCUP.

We want to thank the European Association for Theoretical Computer Science (EATCS) for the scientific sponsorship, and Universidade do Porto (UP), Universidade da Beira Interior (UBI), Centro de Matemática da Universidade do Porto (CMUP), Laboratório de Inteligência Artificial e Ciência de Computadores (LIACC), Câmara Municipal do Porto, Multicert, and iTech-ON for their kind financial support. Finally, we would like to thank all the participants of CIAA 2012. Looking forward to CIAA 2013 in Halifax, Nova Scotia, Canada.

July 2012   Nelma Moreira
                                                                 Rogério Reis

Organization

CIAA 2012 was organized by Faculdade de Ciências da Universidade do Porto with the support of Universidade da Beira Interior.

Invited Speakers

Janusz Brzozowski	University of Waterloo, Canada
Paul Gastin	ENS de Cachan, France
José Nuno Oliveira	Universidade do Minho, Portugal
Grzegorz Rozenberg	Leiden Center for Natural Computing, The Netherlands
Kai Salomaa	Queen's University, Kingston, Canada

Program Committee

Marie-Pierre Béal	Université Paris Est, France
Béatrice Bouchou-Markhoff	Université François Rabelais, Tours, France
Patricia Bouyer	ENS de Cachan, France
Cezar Câmpeanu	University of Prince Edward Island, Canada
Pascal Caron	Université de Rouen, France
Jean-Marc Champarnaud	Université de Rouen, France
Jan Daciuk	Gdańsk University of Technology, Poland
Michael Domaratzki	Manitoba University, Canada
Yo-Sub Han	Yonsei University, South Korea
Tero Harju	University of Turku, Finland
Markus Holzer	Justus-Liebig-Universität Giessen, Germany
Oscar Ibarra	University of California, Santa Barbara, USA
Masami Ito	Kyoto Sangyo University, Japan
Joost-Pieter Katoen	RWTH Aachen University, Germany
Stavros Konstantinidis	University of Halifax, Canada
Andreas Maletti	University of Stuttgart, Germany
Sebastian Maneth	University of New South Wales, Australia
Denis Maurel	Université Francois Rabelais, Tours, France
Ian Mcquillan	University of Saskatoon, Canada
Mehryar Mohri	New York University, USA
Nelma Moreira (co-chair)	Universidade do Porto, Portugal
Alexander Okhotin	University of Turku, Finland
Giovanni Pighizzini	Università degli Studi di Milano, Italy
Bala Ravikumar	Sonoma State University, USA
Rogério Reis (co-chair)	Universidade do Porto, Portugal
Kai Salomaa	Queen's University, Kingston, Canada

Colin Stirling	Edinbourgh University, UK
Mikhail Volkov	Ural State University, Russia
Bruce Watson	University of Pretoria, South Africa
Hsu-Chun Yen	National Taiwan University, Taiwan
Sheng Yu	University of Western Ontario, Canada

Additional Referees

Cyril Allauzen	Szymon Grabowski	Damien Nouvel
Rajeev Alur	Hermann Gruber	Scott Owens
Ivone Amorim	Peter Habermehl	Beatrice Palano
Nathalie Aubrun	Brent Heeringa	Xiaoxue Piao
Franziska Biegler	Lucian Ilie	Elena Pribavkina
Fabienne Braune	Christina Jansen	Daniel Quernheim
Sabine Broda	Hadrien Jeanne	Narad Rampersad
Arturo Carpi	Artur Jeż	Klaus Reinhardt
Jeong-Won Cha	Derrick Kourie	Pierre Rety
Alessandra Cherubini	Yoshiyuki Kunimochi	Michael Riley
Salimur Choudhury	Giuseppe Lami	Abiel Roche-Lima
Loek Cleophas	António Machiavelo	Emanuele Rodaro
Flavio D'Alessandro	Kalpana Mahalingam	Mathieu Sablik
Manfred Droste	Eva Maia	Shinnosuke Seki
Krystian Dudzinski	Katja Meckel	Tinus Strauss
Szilard Zsolt Fazekas	Ludovic Mignot	Jean-Marc Talbot
Francesca Fiorenzi	Benjamin Monmege	Marco Trubian
Mário Florido	František Mráz	Bianca Truthe
Marianne Flouret	Kim Nguyen	Sabrina von Styp
Nathalie Friburger	Thomas Noll	

Steering Committee

Jean-Marc Champarnaud	Université de Rouen, France
Oscar Ibarra	University of California, Santa Barbara, USA
Denis Maurel	Université François Rabelais, Tours, France
Kai Salomaa	Queen's University, Kingston, Canada
Sheng Yu	University of Western Ontario, Canada (Chair)

Organizing Committee

Sabine Broda	Universidade do Porto, Portugal
António Machiavelo	Universidade do Porto, Portugal
Nelma Moreira	Universidade do Porto, Portugal
Rogério Reis	Universidade do Porto, Portugal
Simão Melo de Sousa	Universidade da Beira Interior, Portugal

Sponsors

CMUP
Centro de **Matemática**
Universidade do Porto

FACULDADE DE CIÊNCIAS
UNIVERSIDADE DO PORTO

UNIVERSIDADE DA BEIRA INTERIOR
Covilhã | Portugal

PORTO
SOCIAL
FUNDAÇÃO

PORTO
Cidade de Ciência

PORTO
Câmara Municipal

Table of Contents

Invited Talks

Regular Papers

Short Papers

In Memoriam Sheng Yu

Yuan Gao[1] and Kai Salomaa[2]

[1] Department of Computer Science, The University of Western Ontario, London, Ontario N6A 5B7, Canada
ygao72@csd.uwo.ca
[2] School of Computing, Queen's University, Kingston, Ontario K7L 3N6, Canada
ksalomaa@cs.queensu.ca

Professor Sheng Yu passed away unexpectedly in London, Canada on January 23, 2012, one day before his 62nd birthday. Sheng was one of the world leading theoretical computer scientists. His strong commitment to excellence in scholarship has touched everyone who has worked or studied with him. This includes a large segment of the "CIAA community" and, in particular, the authors of the current article. Sheng Yu's work will continue to influence and inspire automata theory research for a long time to come.

Sheng's undergraduate studies were delayed by China's cultural revolution, however, this did not prevent him from achieving a brilliant academic career. He completed his Ph.D. in 1986 in Waterloo, Canada, under the guidance of Karel Culik II. In the late 1980's Sheng taught for a few years at Kent State

N. Moreira and R. Reis (Eds.): CIAA 2012, LNCS 7381, pp. 1–4, 2012.
© Springer-Verlag Berlin Heidelberg 2012

University and after that has been with the Computer Science Department at the University of Western Ontario in London, Canada.

Sheng's research is reflected in more than 150 refereed publications and he has given invited plenary lectures at numerous international conferences. He collaborated with a wide range of scientists and has more than 70 co-authors in publications in a wide range of areas. Sheng supervised 9 Ph.D. students and numerous M.Sc. students. He was always remarkably dedicated to his students and willing to help at any time of the day. Many of Sheng's former students have leading positions in academia and industry.

Sheng is most widely known for his work on automata theory and formal languages, where his major contributions include giving a new classification of cellular automata, introducing synchronization expressions and languages and establishing their fundamental properties and, in particular, he made significant contributions and opened up new avenues of research in the area of state complexity of regular languages on which we will elaborate below. Sheng's comprehensive survey on regular languages [12] has become a standard reference in this area. Also, Sheng (together with co-authors) proved that the inclusion problem for pattern languages is unsolvable. The latter was a long-standing open problem and researchers working in the area had expected the opposite result.

In addition to his work on formal languages, Sheng carried out research on a broad front and made significant contributions in a wide range of areas, including object oriented programming and parallel processing. A fairly up to date list of his publications can be found in the special issue of *Theoretical Computer Science* [6] published in honor of Sheng's 60th birthday. Independently of what he was working on, Sheng always developed his own original ideas while paying careful attention to the work of others.

One of Sheng's major contributions to theoretical computer science is in the area of descriptional complexity of regular languages. In the early 1990's he initiated a systematic study of the state complexity of finite automata, which has during the last 20 years developed into a very active research area. At the University of Western Ontario Sheng was the leader of a group consisting of several graduate students, postdocs and other research collaborators.

Sheng's work was especially focused on aspects that connect automata theory to new applications in current computer science and on the implementation of automata. He introduced notions like cover automata, new variants of regular expressions, and approximation methods for state complexity. Just a few references on these topics include [1,2,3,7,8,9,14,15], and the reader can find in [6] a bibliography of Sheng's work. Sheng was one of the first automata theory researchers to effectively use software tools for state complexity lower bound constructions [13,16]. At the same time Sheng continued to work on deep abstract questions of automata theory and, recently, proved together with Arto Salomaa that determining the state complexity of combined language operations is undecidable [10]. Complementing the general undecidability result, the work of Sheng's state complexity research group has culminated in the determination of the precise worst-case state complexity of all combinations of two basic language operations [4].

In the year 2000 Sheng chaired at the University of Western Ontario the conference titled *A Half-Century of Automata Theory* with many famous early pioneers of the field as plenary speakers [11]. In his opening remarks to the conference, Sheng emphasized that the best days of automata theory still lay ahead of us and, indeed, in the last decade there has been much renewed interest in this area, as evidenced by the success of conferences like CIAA, DLT and DCFS. Again in 2010, the world's leading automata theory and formal languages researchers gathered in London, Ontario when Sheng chaired the *14th International Conference on Developments in Language Theory* [5].

In 1996, together with Derick Wood, Sheng founded the conference series *Implementation and Application of Automata, CIAA* (originally the meetings were called Workshop on Implementing Automata, WIA). The CIAA conference series has become a leading venue for research on new applications of automata theory and the conferences have attracted many new researchers to this area. During Sheng's tenure as CIAA steering committee chair starting from 2001, the conferences have been held on five different continents.

We conclude with a few personal remarks. Both of the authors have been working in Sheng's group for extended periods of time, as a Ph.D. student and a postdoc, respectively. We have benefited greatly from Sheng's research expertise and from being part of his group. We have always admired Sheng's insight and vision in science and he has been a role model in research for us. Deep inside us we keep a high esteem and gratitude for him. Personally Sheng was warm hearted, easy-going and a great friend for both of us. We will always remember him.

Sheng was accompanied for a large part of his life by his wife Lizhen. We extend our deep sympathy to Lizhen and to Sheng's family in China.

Yuan Gao and Kai Salomaa
London and Kingston, April 2012

Acknowledgement. We thank the CIAA 2012 organizers for hosting a special session in memory of Professor Sheng Yu. The photograph has been kindly provided by Rogério Reis.

References

1. Câmpeanu, C., Păun, A., Yu, S.: An efficient algorithm for constructing minimal cover automata for finite languages. Int. J. Found. Comput. Sci. 13(1), 83–97 (2002)
2. Câmpeanu, C., Santean, N., Yu, S.: Minimal cover-automata for finite languages. Theor. Comput. Sci. 267(1-2), 3–16 (2001)
3. Câmpeanu, C., Salomaa, K., Yu, S.: A formal study of practical regular expressions. Int. J. Found. Comput. Sci. 14(6), 1007–1018 (2003)
4. Cui, B., Gao, Y., Kari, L., Yu, S.: State complexity of combined operations with two basic operations. Theoret. Comput. Sci. (2012), doi:10.1016/j.tcs.2012.02.030

5. Gao, Y., Lu, H., Seki, S., Yu, S. (eds.): DLT 2010. LNCS, vol. 6224. Springer, Heidelberg (2010)
6. Ilie, L., Rozenberg, G., Salomaa, A., Salomaa, K. (eds.): Formal Languages and Applications: A Collection of Papers in Honor of Sheng Yu. Theoret. Comput. Sci., vol. 410(24-25) (2009)
7. Ilie, L., Yu, S.: Follow automata. Inf. Comput. 186(1), 140–162 (2003)
8. Ilie, L., Yu, S., Zhang, K.: Word complexity and repetitions in words. Int. J. Found. Comput. Sci. 15(1), 41–55 (2004)
9. Konstantinidis, S., Santean, N., Yu, S.: On implementing recognizable transductions. Int. J. Comput. Math. 87(2), 260–277 (2010)
10. Salomaa, A., Salomaa, K., Yu, S.: Undecidability of the State Complexity of Composed Regular Operations. In: Dediu, A.-H., Inenaga, S., Martín-Vide, C. (eds.) LATA 2011. LNCS, vol. 6638, pp. 489–498. Springer, Heidelberg (2011)
11. Salomaa, A., Wood, D., Yu, S. (eds.): A Half-Century of Automata Theory: Celebration and Inspiration. World Scientific (2001)
12. Yu, S.: Regular languages. In: Rozenberg, G., Salomaa, A. (eds.) Handbook of Formal Languages, vol. I, pp. 41–110. Springer (1997)
13. Yu, S.: Grail+: A symbolic computation environment for finite-state machines, regular expressions and finite languages (2002), http://www.csd.uwo.ca/Research/grail
14. Yu, S., Gao, Y.: State Complexity Research and Approximation. In: Mauri, G., Leporati, A. (eds.) DLT 2011. LNCS, vol. 6795, pp. 46–57. Springer, Heidelberg (2011)
15. Yu, S., Păun, A. (eds.): CIAA 2000. LNCS, vol. 2088. Springer, Heidelberg (2001)
16. Yu, S., Zhuang, Q., Salomaa, K.: The state complexity of some basic operations on regular languages. Theoret. Comput. Sci. 125, 315–328 (1994)

In Search of Most Complex Regular Languages*

Janusz Brzozowski

David R. Cheriton School of Computer Science, University of Waterloo,
Waterloo, ON, Canada N2L 3G1
`brzozo@uwaterloo.ca`

Abstract. Regular languages that are most complex under common
complexity measures are studied. In particular, certain ternary languages
$U_n(a, b, c)$, $n \geqslant 3$, over the alphabet $\{a, b, c\}$ are examined. It is proved
that the state complexity bounds that hold for arbitrary regular languages are also met by the languages $U_n(a, b, c)$ for union, intersection,
difference, symmetric difference, product (concatenation) and star. Maximal bounds are also met by $U_n(a, b, c)$ for the number of atoms, the quotient complexity of atoms, the size of the syntactic semigroup, reversal,
and 22 combined operations, 5 of which require slightly modified versions. The language $U_n(a, b, c, d)$ is an extension of $U_n(a, b, c)$, obtained
by adding an identity input to the minimal DFA of $U_n(a, b, c)$. The witness $U_n(a, b, c, d)$ and its modified versions work for 14 more combined
operations. Thus $U_n(a, b, c)$ and $U_n(a, b, c, d)$ appear to be universal witnesses for alphabets of size 3 and 4, respectively.

Keywords: combined operation, finite automaton, operation, regular
language, state complexity, syntactic semigroup, witness.

*I dedicate this work to the memory of Sheng Yu whose extensive research
on state complexity led to many questions studied in this paper.*

1 Introduction

State complexity is currently an active area of research in the theory of formal
languages; for references, see the surveys in [1,30] and the bibliography at the
end of this paper. The *state complexity of a regular language* [30] L over a finite
alphabet Σ is the number of states in the minimal (complete) deterministic
finite automaton (DFA) recognizing the language. An equivalent notion is that
of *quotient complexity* [1] of L, which is the number of distinct left quotients of
L, where the quotient of $L \subseteq \Sigma^*$ by a word $w \in \Sigma^*$ is the language $w^{-1}L =
\{x \in \Sigma^* \mid wx \in L\}$. This paper uses *complexity* for both of these equivalent
notions, and this term will not be used for any other property here.

The *(state/quotient) complexity of an operation* on regular languages is the
maximal complexity of the language resulting from the operation as a function of

---

* This work was supported by the Natural Sciences and Engineering Research Council
of Canada under grant No. OGP0000871.

N. Moreira and R. Reis (Eds.): CIAA 2012, LNCS 7381, pp. 5–24, 2012.

the complexities of the arguments. For example, for $K, L \subseteq \Sigma^*$, the complexity of the union $K \cup L$ is mn, if the complexities of K and L are m and n, respectively.

There are two parts to the process of establishing the complexity of an operation. First, one must find an *upper bound* on the complexity of the result of the operation by using quotient computations or automaton constructions. Second, one must find *witnesses* that meet this upper bound. One usually defines a sequence $(L_n \mid n \geqslant k)$ of languages, where k is some small positive integer. This sequence will be called a *stream*. The languages in a stream differ only in the parameter n. For example, one might study unary languages $(\{a^n\}^* \mid n \geqslant 1)$ that have zero a's modulo n. A unary operation then takes its argument from a stream $(L_n \mid n \geqslant k)$. For a binary operation, one adds as the second argument a stream $(K_n \mid n \geqslant k)$, usually different from the first. Also, the witness streams are normally different for different operations.

In this paper I pose the question: Is it possible to use the *same* stream for all the operations? In other words, is there a *universal witness*? The answer is "yes" for all of the common operations.

Section 2 describes common conditions that make a language difficult to handle, introduces the main witness stream $(U_n(a, b, c) \mid n \geqslant 3)$ (U for "universal"), and states the main theorem. Properties of a single language, unary operations, and binary operations are discussed in Sections 3–5, respectively. It is shown in Sections 6 and 7 that the bounds for several combined operations are also met by $U_n(a, b, c)$, or by other streams closely related to $U_n(a, b, c)$. Section 8 deals with combined operations that (seem to) require witnesses over four-letter alphabets. The witness $U_n(a, b, c)$ is then extended to $U_n(a, b, c, d)$, where d is an added identity input in the minimal DFA of $U_n(a, b, c)$.

If K and L are regular languages, let $K \cup L$, $K \cap L$, $K \setminus L$ and $K \oplus L$ be their union, intersection, difference, and symmetric difference, let L^R be the reverse of L, and let M be another regular language. Witnesses derived from $U_n(a, b, c)$ and $U_n(a, b, c, d)$ are presented for the following 36 combined operations:

$$K \cup L^R, \ K \cap L^R, \ K \setminus L^R, \ K \oplus L^R, \ L^R \setminus K,$$
$$K^R \cup L^R, \ K^R \cap L^R, \quad KL^R, \ K^R L, \quad (KL)^R, \ (L^R)^*,$$
$$K \cup L^*, \ K \cap L^*, \ K \setminus L^*, \ K \oplus L^*, \ L^* \setminus K,$$
$$K^* \cup L^*, \ K^* \cap L^*, \quad KL^*, \ K^* L, \quad (KL)^*, \ (K \cup L)^*,$$
$$K^2 \cup L^2, \ K^2 \cap L^2, \ K^2 \setminus L^2, \ K^2 \oplus L^2,$$
$$(KL) \cup M, \ (KL) \cap M, \ (KL) \setminus M, \ (KL) \oplus M, \ M \setminus (KL),$$
$$(K \cup L)M, \ (K \cap L)M, \ K(L \cup M), \ K(L \cap M) \ \text{and} \ K(L \setminus M).$$

Section 9 concludes the paper.

2   Conditions for the Complexity of Languages

If a language L_n is most difficult, what properties should it have? Below are some suggestions to help answer this question.

2.1 Properties of a Single Language

Properties that make a single language L_n difficult to handle are discussed first.

A0: The (state/quotient) complexity of $L_n \subseteq \Sigma^*$ should be n. It is assumed that the complexity of the language is fixed at some integer $n \geqslant 1$, and all the other properties are expressed in terms of n.

A1: The complexity of each quotient of L_n should be n. The complexity of each quotient is bounded from above by n, because the DFA $\mathcal{D} = (Q, \Sigma, \delta, q_0, F)$ that defines L_n also defines $w^{-1}L_n$ for any word $w \in \Sigma^*$, if its initial state is changed to $\delta(q_0, w)$. This requirement is easy to meet, since every strongly connected DFA defines a language satisfying this condition.

A2: The number of atoms of L_n should be 2^n. Atoms of regular languages were introduced in 2011 by Brzozowski and Tamm [4], and the theory was slightly modified in [5]. The newer model, which admits up to 2^n atoms, is used here.

An *atom* of a regular language with quotients $K_0, \ldots, K_{n-1}$ is a non-empty intersection of the form $\widetilde{K_0} \cap \cdots \cap \widetilde{K_{n-1}}$, where $\widetilde{K_i}$ is either K_i or $\overline{K_i}$, $\overline{K_i}$ being the complement of K_i with respect to Σ^*. Thus the number of atoms is bounded from above by 2^n, and it was proved in [5] that this bound is tight. Since every quotient of L_n (including L_n itself) is a union of atoms, the atoms of L_n are its basic building blocks. So it is reasonable that L_n should have the maximal number of atoms.

A3: The complexity of each atom of L_n should be maximal. It was shown in [5] that the complexity of the atoms with 0 or n complemented quotients is bounded from above by $2^n - 1$, and the complexity of any atom with r complemented quotients, where $1 \leqslant r \leqslant n - 1$, by

$$f(n,r) = 1 + \sum_{k=1}^{r} \sum_{h=k+1}^{n-r+k} C_h^n \cdot C_k^h,$$

where C_j^i is the binomial coefficient i *choose* j. It was also shown in [5] that these bounds are tight. It is reasonable to expect that the building blocks of a language should be as complex as possible.

A4: The syntactic semigroup of L_n should have cardinality n^n. The *Myhill congruence* [22] $\approx_L$ of $L \subseteq \Sigma^*$ is defined as follows: For $x, y \in \Sigma^*$,

$$x \approx_L y \text{ if and only if } uxv \in L \Leftrightarrow uyv \in L \text{ for all } u, v \in \Sigma^*.$$

The *syntactic semigroup* [20,24] of L is the quotient semigroup $\Sigma^+ / \approx_L$. It is isomorphic to the *semigroup of transformations* by non-empty words in the minimal DFA of L [20]. The semigroup of transformations is normally used to represent the syntactic semigroup.

Since there are n^n possible transformations of a set of n elements, n^n is an upper bound on the size of the syntactic semigroup of L_n. That the bound is tight follows from the 1935 theorem of Piccard [23] who proved that three transformations of a set of n elements are sufficient to generate all n^n transformations. Also in 1935, Eilenberg showed that fewer than three generators are not possible [28]. In the context of automata, it was first noted without proof by Maslov [19] in 1970 that n^n is a tight bound.

2.2 Unary Operations

B1: The complexity of the reverse of L_n should be 2^n. It follows from the 1959 subset construction of Rabin and Scott [25] that the upper bound is 2^n. It was first shown by Mirkin [21] in 1966 that this bound can be met.

B2: The complexity of the star of L_n should be $2^{n-1} + 2^{n-2}$. It was first noted without proof by Maslov [19] in 1970 that this is a tight upper bound. A proof was provided by Yu, Zhuang and Salomaa [29] in 1994.

2.3 Binary Operations

Two types of binary operations are examined next: boolean operations and product (concatenation or catenation). Four boolean operations union ($\cup$), symmetric difference ($\oplus$), intersection ($\cap$) and difference ($\setminus$) are considered; they are chosen because the complexity of every other binary boolean operation can be obtained from the complexities of these four. Denote by $K_m \circ L_n$ any one of these four operations.

C1: The complexity of $K_m \circ L_n$ should be mn. The upper bound for the boolean operations is mn, since $w^{-1}(K_m \circ L_n) = (w^{-1}K_m) \circ (w^{-1}L_n)$. That the bound is tight for union was noted without proof by Maslov [19] in 1970, and proved for both union and intersection by Yu, Zhuang and Salomaa [29] in 1994. Symmetric difference and difference were treated by Brzozowski [1] in 2010.

C2: The complexity of the product $K_m L_n$ should be $(m-1)2^n + 2^{n-1}$. Maslov [19] stated without proof in 1970 that this bound is tight, and Yu, Zhuang and Salomaa [29] provided a proof in 1994.

2.4 The Witness

The language stream that turns out to be the universal witness for all the operations listed above is defined as follows:

Definition 1. *For $n \geqslant 3$, let $\mathcal{U}_n = \mathcal{U}_n(a,b,c) = (Q, \Sigma, \delta, q_0, F)$, where $Q = \{0, \ldots, n-1\}$ is the set of states[1], $\Sigma = \{a,b,c\}$ is the alphabet, $q_0 = 0$ is the initial state, $F = \{n-1\}$ is the set of final states, $\delta(q, a) = q+1 \bmod n$, $\delta(0,b) = 1$, $\delta(1,b) = 0$, $\delta(q,b) = q$ for $q \notin \{0,1\}$, $\delta(n-1,c) = 0$, and $\delta(q,c) = q$ for $q \neq n-1$. Let $U_n = U_n(a,b,c)$ be the language accepted by $\mathcal{U}_n$.*

The structure of the DFA $\mathcal{U}_n(a,b,c)$ is shown in Fig. 1.

---

[1] Although Q, δ, and F depend on n, this dependence is not shown to keep the notation as simple as possible.

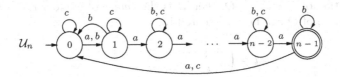

Fig. 1. DFA $\mathcal{U}_n$ of witness language U_n

A language $K \subseteq \Sigma^*$ is *permutationally equivalent* to a language $L \subseteq \Sigma^*$ if K can be obtained from L by permuting the letters of Σ. For example, let π be the permutation $\pi(a) = b$, $\pi(b) = c$ and $\pi(c) = a$; then $\pi(a(b^* \cup cc)) = b(c^* \cup aa)$. Similarly, let $\mathcal{K} = \mathcal{L}(\pi(a), \pi(b), \pi(c))$ be the DFA obtained from $\mathcal{L}(a, b, c)$ by changing the roles of the inputs according to permutation π. Then $\mathcal{K}$ is *permutationally equivalent* to $\mathcal{L}$. In such cases, K ($\mathcal{K}$) is essentially the same language (DFA) as L ($\mathcal{L}$), except that its inputs have been renamed. Obviously, if two languages are permutationally equivalent, then they have the same one-language complexity properties, and the same complexities of unary operations.

Specifically, for this paper let $\mathcal{U}_n(b, a, c)$ be the DFA obtained from $\mathcal{U}_n(a, b, c)$ by interchanging the roles of the inputs a and b. For some operations input c is not needed; then let $\mathcal{U}_n(a, b, \emptyset)$ be the DFA of Definition 1 restricted to inputs a and b, and let $U_n(a, b, \emptyset)$ be the language recognized by this binary DFA. Also, $\mathcal{U}_n(a, \emptyset, \emptyset)$ and $U_n(a, \emptyset, \emptyset)$ are $\mathcal{U}_n(a, b, c)$ and $U_n(a, b, c)$ restricted to a.

Theorem 1 (Main Theorem). *The stream $(U_n(a, b, c) \mid n \geqslant 3)$ meets conditions* **A0–A4**, **B1,B2** *and* **C2**, *whereas* **C1** *is met by two closely related streams $(U_m(a, b, c) \mid m \geqslant 3)$ and $(U_n(b, a, c) \mid n \geqslant 3)$. Moreover,*

- **A0** *and* **A1** *are met by $(U_n(a, \emptyset, \emptyset) \mid n \geqslant 3)$.*
- **B2** *is met by $(U_n(a, b, \emptyset) \mid n \geqslant 3)$.*
- **C1** *is met by $(U_m(a, b, \emptyset) \mid m \geqslant 3)$ and $U_n(b, a, \emptyset) \mid n \geqslant 3)$.*

3 Properties of a Single Language

Conditions **A0–A4** are now briefly discussed for the language U_n.

A0 Complexity of the Language: $U_n(a, \emptyset, \emptyset)$ has n quotients because DFA $\mathcal{U}_n(a, \emptyset, \emptyset)$ is minimal. This holds since state i accepts a^{n-1-i} and no other state accepts this word, for $0 \leqslant i \leqslant n - 1$; hence no two states are equivalent.

A1 Complexity of Quotients: Each quotient of $U_n(a, \emptyset, \emptyset)$ has complexity n, since DFA $\mathcal{U}_n(a, \emptyset, \emptyset)$ is strongly connected.

A2 Number of Atoms: It was proved in [5] that U_n has 2^n atoms. This is discussed further below, in connection with **B1 Reversal**.

A3 Complexity of Atoms: The bounds given in the previous section were derived in [5].

Some background is needed before the next property can be discussed. A *transformation* of a set $Q = \{0, \ldots, n - 1\}$ is a mapping of Q into itself [11]. If t is a transformation of Q and $i \in Q$, then it is the *image* of i under t. An arbitrary transformation of Q can be represented by

$$t = \begin{pmatrix} 0 & 1 & \cdots & n-2 & n-1 \\ i_0 & i_1 & \cdots & i_{n-2} & i_{n-1} \end{pmatrix},$$

where $i_k = kt$, $0 \leqslant k \leqslant n-1$, and $i_k \in Q$. The notation $t = [i_0, i_1, \ldots, i_{n-1}]$ is also used for the transformation t above.

A *permutation* of Q is a mapping of Q *onto* itself. For $2 \leqslant k \leqslant n$, a permutation t is a *cycle* of length k, if there exist pairwise different elements $i_1, \ldots, i_k$ such that $i_1 t = i_2, i_2 t = i_3, \ldots, i_{k-1} t = i_k$, and $i_k t = i_1$. A cycle is denoted by $(i_1, i_2, \ldots, i_k)$. A *transposition* is the cycle (i, j) of length 2 that interchanges i and j and does not affect any other elements. A *singular* transformation, denoted by $\binom{i}{j}$, has $it = j$ and $ht = h$ for all $h \neq i$. The *identity* transformation of Q is denoted by 1_Q.

The set of all permutations of n elements is isomorphic to the symmetric group of degree n and has $n!$ elements. The following result is due to Piccard [23]:

Theorem 2 (Permutations). *For $n \geqslant 3$, the set of all $n!$ permutations of the set $\{0, \ldots, n-1\}$ is generated by a cycle of length n and a transposition (i, j).*

The set of all transformations of a finite set Q is a semigroup under composition, in fact, a monoid $\mathcal{T}_Q$ of n^n elements. In 1935 Piccard [23] proved that three transformations of Q are sufficient to generate $\mathcal{T}_Q$. Dénes [10] studied more general generators; his formulation is used here:

Theorem 3 (Transformations). *For $n \geqslant 3$, the set of all n^n transformations of the set $\{0, \ldots, n-1\}$ is generated by a cycle of length n, a transposition (i, j), and a singular transformation $\binom{k}{\ell}$.*

Every word w in Σ^+ performs a transformation of the set of states of a DFA defined by $q \to \delta(q, w)$. The set of all such transformations is the semigroup of transformations also called the *transition semigroup* of the DFA [24].

A4 Cardinality of Syntactic Semigroup: By Theorem 3, the syntactic semigroup of $U_n(a, b, c)$ has cardinality n^n, since the transformations performed by inputs a, b and c generate all possible transformations of Q.

4 Unary Operations

B1 Reversal: In 1966 Mirkin [21] used a DFA very similar to $U_n(a, b, c)$ to meet the 2^n bound for reversal. It is defined by inputs $a : (0, 1, \ldots, n-1)$, $b : (0, n-2)$ and $c : \binom{0}{n-1}$, with initial state 0 and final state 0. The syntactic semigroup of the language of this DFA has size n^n. Another similar DFA with inputs $a : (0, 1, \ldots, n-1)$, $b : (0, 1)$ and $c : \binom{0}{n-1}$ and initial state 0 and final state 0 was used by Leiss [18] in 1981; the semigroup is also of size n^n.

Salomaa, Wood, and Yu [27] showed the following result:

Theorem 4 (Transformations and Reversal). *Let $\mathcal{D}$ be a minimal DFA with n states accepting a language L. If the transformation semigroup of $\mathcal{D}$ has n^n elements, then the quotient complexity of L^R is 2^n.*

From this and **A4** it follows that U_n^R has 2^n quotients. In view of the following result proved by Brzozowski and Tamm [5], U_n has 2^n atoms.

Theorem 5 (Atoms). *The number of atoms of a regular language L is equal to the complexity of L^R.*

There are also binary witnesses that reach the bound 2^n for L^R. For a detailed discussion see the recent paper by Jirásková and Šebej [17]. Their witness has input a that performs the cycles $(0, 1, 2)$ and $(3, 4, \ldots, n-1)$, and input b that has the cycles $(0, 1)$ and $(2, 3)$ and is an identity on the remaining states.

Is there a close relation between the size of the syntactic semigroup and the quotient complexity of reversal? Besides the result for regular languages in Theorem 4, there are other examples where the languages that have maximal syntactic semigroups also meet the maximal bound for reversal. This is the case for right ideals [6] and prefix-free languages [3]. For left and two-sided ideals [6] and for suffix-, bifix-, and factor-free languages [3] there are only conjectured upper bounds on the size of the syntactic semigroup, but the languages that meet these bounds also meet the maximal bounds for reversal.

The witness of Jirásková and Šebej [17] shows that it is possible for a language to reach the bound 2^n for reversal without having syntactic complexity of n^n. It is also possible for a language to have the maximal syntactic complexity for its class and not reach the bound for reversal. For example, it was shown by Brzozowski and Li [2] that the star-free language defined by the 3-state DFA with inputs $a : [0, 0, 1]$, $b : [1, 1, 2]$, $c : [0, 2, 2]$ and $d : [0, 1, 2]$ and final state 0 meets the maximal bound 10 for the size of the syntactic semigroup. But it does not meet the conjectured upper bound 7 for reversal. However, that bound is met if the final states are 0 and 2.

Does there always exists a language that meets both bounds?

B2 Star: The tightness of the bound for star is now proved. The language $(U_n(a, b, \emptyset))^*$ is accepted by the ε-NFA $\mathcal{S}_n = (Q_{\mathcal{S}}, \{a, b\}, \delta_{\mathcal{S}}, \{s\}, \{s, n-1\})$, where $Q_{\mathcal{S}} = Q \cup \{s\}$, $s \notin Q$, $\delta_{\mathcal{S}}(s, a) = \delta_{\mathcal{S}}(s, b) = \{1\}$, $\delta_{\mathcal{S}}(q, x) = \{\delta(q, x)\}$ for all $q \in Q$, $x \in \Sigma$, and $\delta_{\mathcal{S}}(n-1, \varepsilon) = \{0\}$. The ε-NFA $\mathcal{S}_4$ is shown in Figure 2.

Theorem 6 (Star). *For $n \geqslant 3$, the complexity of $(U_n(a, b, \emptyset))^*$ is $2^{n-1} + 2^{n-2}$.*

Proof. It will be proved that $\{s\}$, all 2^{n-1} subsets of Q containing 0, and all $2^{n-2} - 1$ non-empty subsets of $\{1, \ldots, n-2\}$ are reachable and pairwise distinguishable, giving the DFA of $(U_n(a, b, \emptyset))^*$ a total of $2^{n-1} + 2^{n-2}$ states.

Since s is the initial state, $\{s\}$ is reachable by ε, and $\{0\}$ by ab. It will be shown how to reach the remaining sets from $\{0\}$. Note that any subset containing $n-1$ must also contain 0.

First it is proved that all 2^{n-1} subsets of Q containing 0 are reachable. Since

$$\{0\} \xrightarrow{a^{n-1}} \{0, n-1\} \xrightarrow{a} \{0, 1\} \xrightarrow{(ab)^{i-1}} \{0, i\},$$

for $2 \leqslant i \leqslant n-2$, all two-element subsets of Q containing 0 are reachable.

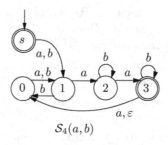

$S_4(a,b)$

Fig. 2. NFA for $(U_4(a,b,\emptyset))^*$

For $k \geqslant 2$, if any k-element set containing 0 can be reached, then so can be any $(k+1)$-element set containing 0 and $n-1$, for if $i_1 < i_2 < \cdots < i_k$, then

$$\{0, i_2 - i_1, \ldots, i_{k-1} - i_1, n - 1 - i_1\} \xrightarrow{a^{i_1}} \{0, i_1, i_2, \ldots, i_{k-1}, n - 1\}.$$

For $k \geqslant 3$, if any k-element set containing 0 and $n-1$ can be reached, then so can be any k-element set containing 0. This holds because

$$\{0, i_2 - i_1, \ldots, i_{k-1} - i_1, n - 1\} \xrightarrow{a(ab)^{i_1-1}} \{0, i_1, \ldots, i_{k-1}\}.$$

It follows now that all 2^{n-1} subsets of Q containing 0 are reachable. Since also

$$\{0, i_2 - i_1, \ldots, i_k - i_1\} \xrightarrow{a^{i_1-1}} \{i_1, i_2, \ldots, i_k\},$$

all the $2^{n-2} - 1$ non-empty subsets of $\{1, \ldots, n-2\}$ are reachable.

It remains to prove that all subsets are pairwise distinguishable. Set $\{s\}$ and any subset of Q containing $n-1$ differ from any subset of Q not containing $n-1$, because they accept the empty word. Also, $\{s\}$ differs from any subset of Q containing $n-1$, because the latter accepts b. Finally, if set P contains $0 \leqslant i < n-1$ but set R does not, then P accepts a^{n-1-i}, and R does not. $\square$

Since the required number of subsets can be reached by words in $\{a,b\}^*$, and they are pairwise distinguishable by words in $\{a,b\}^*$, it follows that the complexity of $(U_n(a,b,c))^*$ with the added input c is also $2^{n-1} + 2^{n-2}$.

Discussion: For $n = 1$, there are only two languages, $\emptyset$ and Σ^*. The complexity of $\emptyset^* = \varepsilon$ is 2, and that of $(\Sigma^*)^* = \Sigma^*$ is 1; the bound does not apply here.

For $n = 2$, the language of Definition 1 is well defined, but inputs a and b coincide. The star of U_2 has complexity 2 only; hence $U_2(a, \emptyset, c)$ is not most complex here. However, the bound $2^1 + 2^0 = 3$ is met by the language over $\{a,b\}$ of all the words with an odd number of a's [29].

5 Binary Operations

5.1 Boolean Operations

Since $K_n \cup K_n = K_n \cap K_n = K_n$, and $K_n \setminus K_n = K_n \oplus K_n = \emptyset$, two different languages have to be used to reach the bound mn if $m = n$. Figure 3 shows the DFA $\mathcal{U}_4(a, b, \emptyset)$ and the DFA $\mathcal{U}_5(b, a, \emptyset)$ permutationally equivalent to $\mathcal{U}_5(a, b, \emptyset)$. The direct product of $\mathcal{U}_4(a, b, \emptyset)$ and $\mathcal{U}_5(b, a, \emptyset)$ is in Figure 4.

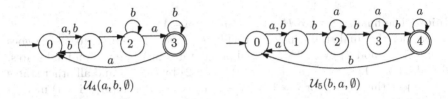

Fig. 3. DFA's of $U_4(a, b, \emptyset)$ and $U_5(b, a, \emptyset)$

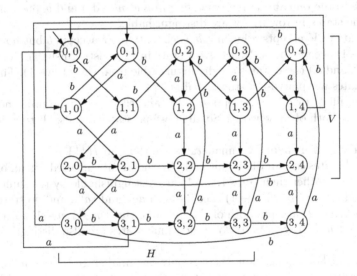

Fig. 4. Direct product of $\mathcal{U}_4(a, b, \emptyset)$ with $\mathcal{U}_5(b, a, \emptyset)$

Theorem 7 (Boolean Operations). *The complexity of $U_m(a, b, \emptyset) \circ U_n(b, a, \emptyset)$ is mn for $m, n \geqslant 3$.*

Proof. In the direct product, state $(0,0)$ is the initial state, and state $(1,1)$ is reached by a. From $(1,1)$, state $(2,0)$ is reached by a, and state $(0,2)$, by b. From $(0,2)$, state $(i,2)$ is reached by a^i; hence all the states in column 2 are reachable. From $(1,2)$, state $(0,3)$ is reached by b and the states in column 3 are reached by words in a^*. This repeats until state $(0, n-1)$ is reached by b from $(1, n-2)$, and other states in column $n - 1$ are then reached by words in a^*. Thus all the states in columns $2, \ldots, n - 1$ are reachable.

Next, $(1, 0)$ is reached from $(0, n - 1)$ by b, and $(0, 1)$ from $(1, 0)$ by b. The remaining states $(i, 0)$, $i \geqslant 2$, are reached from $(1, 1)$ by $a(ba)^{i-2}$ and states $(i, 1)$, by $(ab)^{i-1}$. Thus all the states in columns 0 and 1 are also reachable.

It remains to prove that all the states are pairwise distinguishable. Let H (for *horizontal*) be the set $H = \{(m - 1, 0), (m - 1, 1), \ldots, (m - 1, n - 2)\}$, and let V (for *vertical*) be $V = \{(0, n - 1), (1, n - 1), \ldots, (m - 2, n - 1)\}$. The boolean operations are now considered one by one.

Union: The final states are $H \cup V \cup \{(m - 1, n - 1)\}$.

Consider the final states in $V' = V \cup \{(m - 1, n - 1)\}$. First, $(0, n - 1)$ goes to $(m - 1, m \bmod 2)$ by ba^{m-2}, and all other states in V' go to non-final states. Second, $(1, n - 1)$ goes to $(m - 1, (m - 1) \bmod 2)$ by ba^{m-1}, and all other states in V' reject this word. For $i > 1$, $(i, n - 1)$ goes to $(m - 1, (m - 1 - i) \bmod 2)$ by ba^{m-1-i}, and all other states in V' reject this word. Thus all the states in column $n - 1$ are distinguishable.

Next, take the final states in $H' = H \cup \{(m - 1, n - 1)\}$. By an argument symmetric to the one above, interchanging m and n and a and b, one concludes that all the states in row $m - 1$ are distinguishable.

Each state in V accepts a but not b, each state in H accepts b but not a, and $(m - 1, n - 1)$ accepts both. Hence every state in V is distinguishable from every state in H, and all these states are distinguishable from $(m - 1, n - 1)$. Therefore all final states are pairwise distinguishable.

Any non-final state (i, j) accepts a^{m-1-i} and b^{n-1-j}, but no other non-final state accepts both of these words. So all non-final states are also distinguishable.

Symmetric Difference: The final states are those in $H \cup V$.

The final states are all distinguishable by the argument used for union. The non-final states other than $(m - 1, n - 1)$ are distinguishable by the same words as for union. State $(m - 1, n - 1)$ accepts both ab^n and ba^m, and no state other than $(m - 2, n - 2)$ accepts both of these words. But $(m - 1, n - 1)$ rejects aba, while $(m - 2, n - 2)$ accepts it. So all non-final states are also distinguishable.

Intersection: For intersection, there is only one final state $(m - 1, n - 1)$. The non-final states q and words w_q accepted only by those states are listed below:

1. $q = (0, j)$ with $n - 1 - j$ even, $w_q = b^{n-1-j}a^{m-1}$,
2. $q = (0, j)$ with $n - 1 - j$ odd, $w_q = b^{n-1-j}a^{m-2}$,
3. $q = (1, j)$ with $n - 1 - j$ even, $w_q = b^{n-1-j}a^{m-2}$,
4. $q = (1, j)$ with $n - 1 - j$ odd, $w_q = b^{n-1-j}a^{m-1}$,
5. for $i \geqslant 2$, $q = (i, j)$, $w_q = b^{n-1-j}a^{m-1-i}$.

Difference: For difference, the final states are H. State $(m - 1, j)$ rejects b^{n-1-j}, but other final states accept it. So all final states are distinguishable.

For non-final states $q = (i, j)$ and (h, l), other than $(m - 1, n - 1)$:

1. If $i = h$ and $j \neq l$, then q rejects w_q, while (h, l) accepts it, where w_q is defined as for intersection. Thus all the non-final states in the same row are distinguishable.
2. Two non-final states $q = (i, j)$ and (h, j) in the same column, with $j < n - 1$ and $i \neq h$, are distinguishable by a^{m-1-i}.
3. If $j = l = n - 1$ and $i \neq h$, then q accepts $a^{m-1-i}b$, while (h, l) rejects it.

Any non-final state (i, j) with $j < n - 1$ is distinguished from $(m - 1, n - 1)$ by a^{m-1-i}. State $(0, n - 1)$ accepts ba^{m-2}, while $(m - 1, n - 1)$ rejects it. Similarly, $(1, n - 1)$ accepts ba^{m-1}, while $(m - 1, n - 1)$ rejects it. For $2 \leqslant i \leqslant n - 2$, $(i, n - 1)$ accepts ba^{m-1-i}, but $(m - 1, n - 1)$ rejects it. □

Although it is impossible for the stream $(U_n(a, b, \emptyset), n \geqslant 3)$ to meet the bound for boolean operations when $m = n$, this stream is as complex as it could possibly be in view of the following:

Conjecture 1 $(K_m \circ L_n, m \neq n)$
If $m \neq n$, the complexity of $U_m(a, b, \emptyset) \circ U_n(a, b, \emptyset)$ is mn.
(Verified for $3 \leqslant m, n \leqslant 10$ and some higher values.)

Note about Conjectures: The 19 conjectures in this paper have 35 different claims. The proofs are not trivial; sometimes two such proofs constitute an entire paper, for example, in [7,9,16,26]. Because of the limitations of time, space, and the author's energy, the proofs are omitted, although some of the claims have been verified. The conjectures are supported by Grail and GAP computations.

5.2 Product

It is shown next that the complexity of the product of $U_m(a, b, c)$ with $U_n(a, b, c)$ reaches the maximal possible bound.

To avoid confusion of states, let $\mathcal{U}_m = \mathcal{U}_m(a, b, c) = (Q_m, \Sigma, \delta_m, q_0, \{q_{m-1}\})$, where $Q_m = \{q_0, \ldots, q_{m-1}\}$, and let $\mathcal{U}_n = \mathcal{U}_n(a, b, c)$, as in Definition 1. Define the ε-NFA $\mathcal{P} = (Q_m \cup Q_n, \Sigma, \delta_{\mathcal{P}}, \{q_0\}, \{n - 1\})$, where $\delta_{\mathcal{P}}(q, a) = \{\delta_m(q, a)\}$ if $q \in Q_m$, $a \in \Sigma$, $\delta_{\mathcal{P}}(q, a) = \{\delta_n(q, a)\}$ if $q \in Q_n$, $a \in \Sigma$, and $\delta_{\mathcal{P}}(q_{m-1}, \varepsilon) = \{0\}$. This ε-NFA accepts $U_m U_n$, and is illustrated in Figure 5 for $m = 4$ and $n = 5$.

Theorem 8 (Product). *For* $m, n \geqslant 2$, *the complexity of* $U_m(a, b, c) U_n(a, b, c)$ *is* $(m - 1)2^n + 2^{n-1}$.

Proof. It will be shown that all $(m - 1)2^n$ subsets of states of $\mathcal{P}$ of the form $\{q_i\} \cup S$, where $i < m - 1$ and S is any subset of Q_n, are reachable, as well as all 2^{n-1} subsets of the form $\{q_{m-1}, 0\} \cup S$, where S is any subset of $\{1, \ldots, n - 1\}$. *All the arithmetic below is modulo* n.

First, study how states of the form $\{q_0\} \cup S$ can be reached. Since $\{q_0\}$ is the initial set of states, it is reached by ε. Sets $\{q_i\}$ are reached from $\{q_0\}$ by a^i, for $i = 1, \ldots, m - 2$, and $\{q_{m-1}, 0\}$, by a^{m-1}.

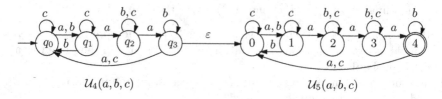

Fig. 5. ε-NFA $\mathcal{P}$ of $U_4(a,b,c)U_5(a,b,c)$

From $\{q_{m-1},0\}$, $\{q_0,0\}$ is reached by c, and $\{q_0,1\}$ by a. From $\{q_0,1\}$, $\{q_0,i\}$ is reached by $(ab)^{i-1}$, for $i = 2, \ldots, n-1$. Hence all the sets of the form $\{q_0\} \cup S$, where $|S| \leqslant 1$ are reachable.

Second, it will be shown that, if $\{q_{m-1},0\} \cup S$ can be reached for all sets $S \subseteq \{1, \ldots, n-1\}$ with $|S| = k \geqslant 0$, then $\{q_0\} \cup T$ can be reached for all $T = \{t_0, t_1, \ldots, t_k\} \subseteq \{0, \ldots, n-1\}$ with $0 \leqslant t_0 < t_1 < \cdots < t_k \leqslant n-1$. There are three cases to consider:

1. $t_0 = 0$: Use $\{q_{m-1},0,t_2 - t_1, \ldots, t_k - t_1, n-1\} \xrightarrow{a(ab)^{t_1-1}} \{q_0,t_1,t_2,\ldots,t_k,0\}$.
2. $t_0 = 1$: Use $\{q_{m-1},0,t_1 - 1, \ldots, t_k - 1\} \xrightarrow{a} \{q_0,1,t_1,\ldots,t_k\}$.
3. $t_0 > 1$: Use $\{q_{m-1},0,t_1-(t_0-1), \ldots, t_k-(t_0-1)\} \xrightarrow{bc(ab)^{t_0-1}} \{q_0,t_0,t_1,\ldots,t_k\}$.

Third, consider sets $\{q_{m-1},0\} \cup S$, $S \subseteq \{1, \ldots, n-1\}$. It has already been shown that $\{q_{m-1},0\}$ is reachable. Suppose that all the sets of the form $\{q_0\} \cup S$ with $|S| = k \geqslant 1$, $0 \notin S$ can be reached. Then to reach $\{q_{m-1},0,t_1,\ldots,t_k\}$ with $1 \leqslant t_1 < \cdots < t_k \leqslant n-1$, use $\{q_0,t_1 - (m-1),\ldots,t_k - (m-1)\} \xrightarrow{a^{m-1}} \{q_{m-1},0,t_1,\ldots,t_k\}$.

Finally, for $0 < i < m-1$, $\{q_i,t_1,\ldots,t_k\}$ is reached by a^i from $\{q_0,t_1 - i,\ldots,t_k - i\}$, where $0 \leqslant t_1 < \cdots < t_k \leqslant n-1$. Hence all the required states can be reached.

It will now be proved that all these subsets are pairwise distinguishable.

Consider $s = \{q_i\} \cup S$ and $t = \{q_j\} \cup T$, where $0 \leqslant i, j \leqslant m-1$ and $S \neq T$, $S, T \subseteq Q_n$. If k is in $S \oplus T$, then a^{n-1-k} distinguishes s and t.

Next suppose $s = \{q_i\} \cup S$ and $t = \{q_j\} \cup S$ with $i < j < m-1$. Applying $(ca)^{m-1-j}$ sends $t = \{q_j\} \cup S$ to $t' = \{q_{m-1},0\} \cup S'$ for some $S' \subseteq \{1,\ldots,n-1\}$, but sends $s = \{q_i\} \cup S$ to $s' = \{q_{i+m-1-j}\} \cup S'$, and this pair can be distinguished since the subsets of Q_n are different. If $i > 0$ and $j = m-1$, apply $(ca)^{m-1-i}$. Then $s = \{q_i\} \cup S$ is sent to $s' = \{q_{m-1},0\} \cup S'$, and $t = \{q_{m-1}\} \cup S$ is sent to $t' = \{q_k\} \cup S'$ for some $S' \subseteq \{1,\ldots,n-1\}$ and $k < m-1$.

This leaves the case where $i = 0$ and $j = m-1$. Then use ba to send $t = \{q_j\} \cup S$ to $t' = \{q_0\} \cup S'$ and $s = \{q_i\} \cup S$ to $s' = \{q_2\} \cup S'$. Now $(ca)^{m-3}$ can be applied to make the subsets of Q_n different.

Since all reachable sets are pairwise distinguishable, the bound is met.   □

Discussion. The restrictions of U_n to two letters do not meet the bound for product. This is a defect of U_n, since there exist binary witnesses for product. Maslov [19] used the DFA $\mathcal{K}_m$ with input a performing the cycle $(0, \ldots, m-1)$, with b being the identity, and a DFA $\mathcal{L}_n$ with a performing the transposition $(n-2, n-1)$, and b mapping i to $i+1$ for $i < n-1$, and $n-1$ to $n-1$. Yu, Zhuang and Salomaa [29] used ternary languages.

6   Combined Operations with $U_m(a,b,c)$ and $U_n(b,a,c)$

To simplify the notation, denote $U_n(b,a,c)$ by $\tilde{U}_n$.

Gao and Yu [15] studied the complexities of $K_m \cup L_n^R$ and $K_m \cap L_n^R$, and showed that they are both $m2^n - (m-1)$, and are met using a quaternary alphabet. Their results can be improved and extended as follows: (1) *ternary alphabets* suffice, (2) the *same language stream* can be used for K_m and L_n for both union and intersection, (2) the same language stream is also a witness for two *difference* operations and *symmetric difference*, and (4) the bound for symmetric difference is $m2^n$.

Conjecture 2 ($K_m \circ L_n^R$ and $L_n^R \setminus K_m$)
For $m, n \geqslant 3$, the complexities of $U_m \cup U_n^R$, $U_m \cap U_n^R$, $U_m \setminus U_n^R$, $U_n^R \setminus U_m$ are all $m2^n - (m-1)$, whereas that of $U_m \oplus U_n^R$ is $m2^n$. (Verified for $3 \leqslant m, n \leqslant 10$.)

It was shown in [12] by Gao, Kari, and Yu that the complexities of $K_m^R \cup L_n^R$ and $K_m^R \cap L_n^R$ are $(2^m - 1)(2^n - 1) + 1$ with witnesses over a six-letter alphabet. The bound can also be met by ternary languages:

Conjecture 3 ($K_m^R \cup L_n^R$ and $K_m^R \cap L_n^R$)
For $m, n \geqslant 3$, the complexities of $U_m^R \cup \tilde{U}_n^R$ and $U_m^R \cap \tilde{U}_n^R$ are $(2^m - 1)(2^n - 1) + 1$. (Verified for $3 \leqslant m, n \leqslant 7$.)

Incidentally, the same bound can be reached by these witnesses for difference.

It was shown in [9] by Cui, Gao, Kari, and Yu that the complexity of $K_m L_n^R$ is $(m-1)2^n + 2^{n-1} - (m-1)$ with ternary witnesses. Two permutationally equivalent witnesses also work:

Conjecture 4 ($K_m L_n^R$)
For $m, n \geqslant 3$, the complexity of $U_m U_n^R$ is $(m-1)2^n + 2^{n-1} - (m-1)$. (Verified for $3 \leqslant m \leqslant 7$ and $3 \leqslant n \leqslant 6$.)

If the complexities of K_m, L_n and M_p are m, n and p, Cui, Gau, Kari, and Yu [8] showed that the complexity of $(K_m L_n) \cap M_p$ is $((m-1)2^n + 2^{n-1})p$; here the complexity of the result is the composition of the complexities of product and intersection. They also showed that the same bound holds for $(K_m L_n) \cup M_p$. This can be generalized to all binary boolean operations:

Conjecture 5 ($(K_m L_n) \circ M_p$ and $M_p \setminus (K_m L_n)$)
The complexities of $(U_m U_n) \circ \tilde{U}_p$ and $\tilde{U}_p \setminus (U_m U_n)$ are all $((m-1)2^n + 2^{n-1})p$ for $m, n, p \geqslant 3$. (Verified for various values of m, n, and p.)

Cui, Gao, Kari, and Yu [8] also proved that the complexity of $(K_m \cap L_n)M_p$ is the composition of complexities of intersection and product, $(mn-1)2^p + 2^{p-1}$. They used quaternary witnesses, but there are ternary witnesses:

Conjecture 6 $((K_m \cap L_n)M_p)$
The complexity of $(U_m \cap \tilde{U}_n)U_p$ is $(mn-1)2^p + 2^{p-1}$ for $m, n, p \geqslant 3$. (Verified for various values of m, n, and p.)

In the case of $(K_m \cap L_n)\Sigma^*$, the languages $K_m = U_m(a, b, \emptyset)$ and $L_n = U_n(b, a, \emptyset)$ also reach the bound mn.

Gao and Yu [15] showed that the complexity of $K_m \cup L_n^*$ is $3m2^{n-2} - (m-1)$. These results are extended here to symmetric difference and to one difference operation. For the remaining boolean operations see Conjecture 9.

Conjecture 7 $(K_m \cup L_n^*, K_m \oplus L_n^*, \text{ and } L_n^* \setminus K_m)$
The complexities of the operations $U_m \cup \tilde{U}_n^*$, $U_m \oplus \tilde{U}_n^*$ and $\tilde{U}_n^* \setminus U_m$ are all $3m2^{n-2} - (m-1)$ for $m, n \geqslant 3$. (Verified for $3 \leqslant m, n \leqslant 10$.)

7   Combined Operations with "Dialects" of $U_n(a, b, c)$

For the combined operations in this section, the witness $U_n(a, b, c)$ no longer works. However, the class of witnesses can be extended beyond those permutationally equivalent to $U_n(a, b, c)$.

Definition 2. *A* dialect *of the language* $U_n(a, b, c)$ *is any ternary language* $V_n(a, b, c)$ *of complexity* n, *in which one input* $a : (0, \ldots, n-1)$ *performs a cyclic permutation of the* n *states in the minimal DFA of* V_n, *a second input* $b : (i, j)$ *performs a transposition of two states, and the third input is a singular transformation* $c : \binom{k}{\ell}$.

Clearly, a dialect $V_n(a, b, c)$ of $U_n(a, b, c)$ satisfies **A0** and **A1**. By Theorem 3, its syntactic semigroup is of size n^n, and so it satisfies **A4**. By Theorem 5, it satisfies **A2**. I conjecture that it satisfies **A3**. By Theorem 4, it satisfies **B1**. So it seems that dialects have many desirable single-language properties.

A Combined Operation with Binary Witnesses: In 2007 A. Salomaa, K. Salomaa, and S. Yu [26] showed that the complexity of $(K_m \cup L_n)^*$ is $2^{m+n-1} - (2^{m-1} + 2^{n-1} - 1)$ with ternary witnesses. Jirásková and Okhotin [16] used binary witnesses. It is shown below that permutationally equivalent binary dialects of $U_n(a, b, c)$ can also be used.

Let $\mathcal{S}_n = \mathcal{S}_n(a, b) = (Q, \Sigma, \delta_S, 0, \{0\})$, where $a : (0, 1, \ldots, n-1)$, and $b : \binom{0}{1}$. In this dialect, both the final state and the singular transformation have been changed. Let S_n be the language of $\mathcal{S}_n$, and let $\tilde{S}_n = S_n(b, a)$.

Theorem 9 $((K_m \cup L_n)^*)$
For $m, n \geqslant 3$, the complexity of $(S_m \cup \tilde{S}_n)^*$ is $2^{m+n-1} - (2^{m-1} + 2^{n-1} - 1)$.

Proof The proof follows closely that of [16]. The DFA's of languages $S_4(a, b)$ and $S_5(b, a)$ are shown in Fig. 6. If the dotted transitions are added, the resulting

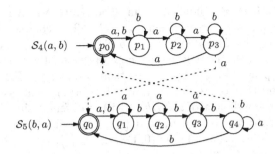

Fig. 6. DFA's of $S_4(a, b, \emptyset)$ and $S_5(b, a, \emptyset)$

NFA $\mathcal{N} = (Q_N, \Sigma, \delta_N, \{p_0, q_0\}, \{p_0, q_0\})$ in the general case accepts $(S_m \cup \tilde{S}_n)^*$. Note that all the reachable subsets of Q_N always contain at least one state from $S_m(a, b)$ and at least one state from $S_n(b, a)$. The reachable sets are (a) those consisting of a non-empty subset of $P \setminus \{p_0\}$ together with a non-empty subset of $Q \setminus \{q_0\}$ and (b), those consisting of $\{p_0, q_0\}$ together with any subset of $P \cup Q$. The number of reachable states is precisely the bound.

Any subset of cardinality 2 is reached as follows: $\{p_0, q_0\}$ is the initial state, and $\{p_i, q_j\}$, for $1 \leqslant i, j$, is reached by $ba^{i-1}b^{j-1}$. Next, use induction on the size of the set in several cases. To reach $\{p_{i_1}, \ldots, p_{i_k}, q_{j_1}, \ldots, q_{j_l}\}$, proceed as follows:

1. If $i_1 = j_1 = 0$ and ($i_2 > 1$ or $k = 1$), start with the set (of size $k + l - 1$) $\{p_{m-1}, p_{i_2-1} \cdots, p_{i_k-1}, q_{j_2}, \ldots, q_{j_l}\}$ and apply a.
2. If $i_1 = j_1 = 0$ and ($j_2 > 1$ or $l = 1$), this is symmetric to Case 1.
3. If $i_1 = j_1 = 0$ and $i_2 = j_2 = 1$, start with the set (of size $k + l - 1$) $\{p_{m-1}, p_0, p_{i_3-1}, \ldots, p_{i_k-1}, q_0, q_{j_3}, \ldots, q_{j_l}\}$ and apply a.
4. If $i_1 \geqslant 1$ and $j_1 \geqslant 1$, start with set S of size $k + l$, reachable by 1–3, where $S = \{p_0, p_{i_2-i_1}, \ldots, p_{i_k-i_1}, q_0, q_{j_2-(j_1-1)}, \ldots, q_{j_l-(j_1-1)}\}$, and apply $a^{i_1}b^{j_1-1}$.

Hence all the required states are reachable. Since only state p_i accepts a^{m-i}, and only p_j accepts b^{n-j}, all subsets are pairwise distinguishable. $\square$

Incidentally, $(S_m \oplus \tilde{S}_n)^*$ also reaches the bound $2^{m+n-1} - (2^{m-1} + 2^{n-1} - 1)$.

The star of the reverse was studied by Gao, Salomaa, and Yu [14], who showed that the complexity of this operation is 2^n with a ternary witness. Here, a dialect $\mathcal{U}_{\{0\},n}(a, b, c)$ of $\mathcal{U}_n(a, b, c)$ with final state changed to 0 can be used:

Conjecture 8 $((L_n^R)^*)$
For $n \geqslant 3$, the complexity of $(U_{\{0\},n}^R)^*$ is 2^n. (Verified for $3 \leqslant n \leqslant 7$.)

Gao and Yu [15] studied the intersection $K_m \cap L_n^*$; this result is extended here to a difference operation. The dialect which is the complement of $U_m(a, b, c)$ applies here.

Conjecture 9 $(K_m \cap L_n^*$ and $K_m \setminus L_n^*)$
For $m, n \geqslant 3$, the complexities of $\overline{U_m} \cap \tilde{U}_n^*$ and $\overline{U_m} \setminus \tilde{U}_n^*$ are both $3m2^{n-2} - (m-1)$. (Verified for $3 \leqslant m, n \leqslant 10$.)

The language $K_m L_n^*$ was studied by Cui, Gao, Kari, and Yu [9]. If the only final quotient of L_n is L_n itself, then $L_n = L_n^*$. The complexity of $K_m L_n^*$ is then that of $K_m L_n$; by Theorem 8, $U_m(a, b, c)$ and $U_n(a, b, c)$ meet this bound. Hence assume that there is at least one final quotient of L_n other than L_n. In that case, it was proved in [9] that the quotient complexity of $K_m L_n^*$ is at most $(3m - 1)2^{n-2}$, and that this bound is tight with ternary witnesses.

Here one can use a dialect of $U_n(a, b, c)$ with a different singular transformation. Let $\mathcal{T}_n = \mathcal{T}_n(a, b, c) = (Q, \Sigma, \delta_T, 0, \{n - 1\})$, where $Q = \{0, \ldots, n - 1\}$, $\Sigma = \{a, b, c\}$, $a : (0, 1, \ldots, n - 1)$, $b : (0, 1)$, and $c : \binom{1}{0}$. Let T_n be the language of $\mathcal{T}_n$ and let $\tilde{T}_n(a, b, c) = T_n(b, a, c)$.

Conjecture 10 $(K_m L_n^*)$
For $m, n \geqslant 3$, the complexity of $T_m \tilde{T}_n^*$ is $(3m - 1)2^{n-2}$.
(Verified for $3 \leqslant m, n \leqslant 6$.)

8 Witnesses Over Quaternary Alphabets

Operations that (appear to) require an alphabet of four letters are treated next.

8.1 Witnesses $U_n(a, b, c, d)$ and $\widehat{U}_n(a, b, c, d)$

Let $\mathcal{U}_n(a, b, c, d) = (Q, \Sigma, \delta_{\mathcal{U}}, 0, \{n - 1\})$, where $a : (0, 1, \ldots, n - 1)$, $b : (0, 1)$, $c : \binom{n-1}{0}$, and $d : 1_Q$. Thus $\mathcal{U}_n(a, b, c) = \mathcal{U}_n(a, b, c, \emptyset)$. Let $U_n(a, b, c, d)$ be the language of $\mathcal{U}_n(a, b, c, d)$. Also let $\widehat{U}_n(a, b, c, d) = U_n(d, c, b, a)$; then $\widehat{U}_n(a, b, c, d)$ and $U_n(a, b, c, d)$ are permutationally equivalent.

From now on, the symbols $\mathcal{U}_n$ and U_n stand for $\mathcal{U}_n(a, b, c, d)$ and $U_n(a, b, c, d)$, and the same applies to the versions with the "hat".

It was shown by Cui, Gao, Kari, and Yu [8] that quaternary witnesses meet the bound $3 \cdot 2^{m+n-2} - 2^n + 1$ for $(K_m L_n)^R$. Here U_n and $\widehat{U}_n$ also work:

Conjecture 11 $((K_m L_n)^R)$
For $m, n \geqslant 3$, the complexity of $(U_m \widehat{U}_n)^R$ is $3 \cdot 2^{m+n-2} - 2^n + 1$.
(Verified for $3 \leqslant m, n \leqslant 7$.)

It was shown in [8] by Cui, Gao, Kari, and Yu that quaternary witnesses meet the bound $5 \cdot 2^{m+n-3} - (2^{m-1} + 2^n - 1)$ for $K_m^* L_n$. Here, one can also use U_n and $\widehat{U}_n$:

Conjecture 12 $(K_m^* L_n)$
For $m, n \geqslant 3$, the complexity of $U_m^* \widehat{U}_n$ is $5 \cdot 2^{m+n-3} - (2^{m-1} + 2^n - 1)$.
(Verified for $3 \leqslant m, n \leqslant 7$.)

It was shown Cui, Gao, Kari, and Yu in [8] that quaternary witnesses meet the bound $mn2^p - (m + n - 1)2^{p-1}$ for $(K_m \cup L_n)M_p$. Here U_n and $\widehat{U}_n$ also work:

Conjecture 13 $((K_m \cup L_n)M_p)$
For $m, n, p \geqslant 3$, the complexity of $(U_m \cup \widehat{U}_n)U_p$ is $mn2^p - (m + n - 1)2^{p-1}$.
(Verified for some values of m, n and p.)

The definition of "dialect" is extended to four inputs: $V_n(a, b, c, d)$ is a *dialect* of $U_n(a, b, c, d)$ if $V_n(a, b, c, \emptyset)$ is a dialect of $U_n(a, b, c, \emptyset)$ and $d : 1_Q$.

8.2   Witnesses $V_n(a, b, c, d)$ and $\widehat{V}_n(a, b, c, d)$

Let $\mathcal{V}_n = \mathcal{V}_n(a, b, c, d) = (Q, \Sigma, \delta_V, 0, \{n - 1\})$, where $a : (0, 1, \ldots, n - 1)$, $b : (n - 2, n - 1)$, $c : \binom{n-1}{n-2}$, and $d : 1_Q$. Let $V_n = V_n(a, b, c, d)$ be the language of $\mathcal{V}_n$; so $V_n(a, b, c, d)$ is a dialect of $U_n(a, b, c, d)$. Also let $\widehat{\mathcal{V}}_n(a, b, c, d) = \mathcal{V}_n(d, c, b, a)$; then $\widehat{\mathcal{V}}_n$ and $\mathcal{V}_n$ are permutationally equivalent.

It was shown in [8] by Cui, Gao, Kari and Yu that quaternary witnesses meet the bound $3 \cdot 2^{m+n-2}$ for $K_m^R L_n$. Here V_n and $\widehat{V}_n$ can be used:

Conjecture 14 ($K_m^R L_n$)
The complexity of $V_m^R \widehat{V}_n$ is $3 \cdot 2^{m+n-2}$ for $m, n \geqslant 3$. (Verified for $3 \leqslant m, n \leqslant 7$.)

8.3   Witnesses $W_n(a, b, c, d)$ and $\widehat{W}_n(a, b, c, d)$

Let $\mathcal{W}_n = \mathcal{W}_n(a, b, c, d) = (Q, \Sigma, \delta_W, 0, \{n - 1\})$, where $a : (0, 1, \ldots, n - 1)$, $b : (n - 2, n - 1)$, $c : \binom{1}{0}$, and $d : 1_Q$. Let $W_n = W_n(a, b, c, d)$ be the language of $\mathcal{W}_n$; so $W_n(a, b, c, d)$ is a dialect of $U_n(a, b, c, d)$. Also let $\widehat{\mathcal{W}}_n(a, b, c, d) = \mathcal{W}_n(d, c, b, a)$; then $\widehat{\mathcal{W}}_n$ and $\mathcal{W}_n$ are permutationally equivalent.

It was shown in [13] by Gao, Kari and Yu that quaternary witnesses meet the bound $9 \cdot 2^{m+n-4} - (3 \cdot 2^{m-2} + 3 \cdot 2^{n-2} - 2)$ for $K_m^* \cup L_n^*$ and $K_m^* \cap L_n^*$. Here W_n and $\widehat{W}_n$ apply:

Conjecture 15 ($K_m^* \cup L_n^*$ and $K_m^* \cap L_n^*$)
The complexities of $W_m^* \cup \widehat{W}_n^*$ and $W_m^* \cap \widehat{W}_n^*$ are $9 \cdot 2^{m+n-4} - (3 \cdot 2^{m-2} + 3 \cdot 2^{n-2} - 2)$ for $m, n \geqslant 3$. (Verified for $3 \leqslant m, n \leqslant 7$.)

It was shown in [12] by Gao, Kari and Yu that quaternary witnesses meet the bound $[(m - 1)2^m + 2^{m-1}] \cdot [(n - 1)2^n + 2^{n-1}]$ for $K_m^2 \cup L_n^2$ and $K_m^2 \cap L_n^2$. This is extended here to all four boolean operations:

Conjecture 16 ($K_m^2 \circ L_n^2$)
The complexity of $W_m^2 \circ \widehat{W}_n^2$ is $[(m-1)2^m + 2^{m-1}] \cdot [(n-1)2^n + 2^{n-1}]$ for $m, n \geqslant 3$. (Verified for $3 \leqslant m, n \leqslant 5$.)

It was shown in [14] by Gao, Salomaa and Yu that quaternary witnesses meet the bound $2^{m+n-1} + 2^{m+n-4} - (2^{m-1} + 2^{n-1} - m - 1)$ for $(K_m L_n)^*$. Here W_n and $\widehat{W}_n$ can also be used:

Conjecture 17 ($(K_m L_n)^*$)
The complexity of $(W_m \widehat{W}_n)^*$ is $2^{m+n-1} + 2^{m+n-4} - (2^{m-1} + 2^{n-1} - m - 1)$ for $m, n \geqslant 3$. (Verified for $3 \leqslant m, n \leqslant 5$.)

It was shown in [7] by Cui, Gao, Salomaa and Yu that quinary witnesses meet the bound $(m - 1)(2^{n+p} - 2^n - 2^p + 2) + 2^{n+p-2}$ for $K_m(L_n \cup M_p)$. Here the alphabet size is reduced to 4:

Conjecture 18 $(K_m(L_n \cup M_p))$
The complexity of $W_m(W_n \cup \widehat{W}_p)$ is $(m-1)(2^{n+p} - 2^n - 2^p + 2) + 2^{n+p-2}$, for $m, n, p \geqslant 3$. (Verified for $3 \leqslant m, n, p \leqslant 4$ and several larger values.)

It was shown in [7] by Cui, Gao, Salomaa and Yu that quaternary witnesses meet the bound $(m-1)2^{np} + 2^{np-1}$ for $K_m(L_n \cap M_p)$. This result is extended here also to $K_m(L_n \setminus M_p)$:

Conjecture 19 $(K_m(L_n \cap M_p)$ and $K_m(L_n \setminus M_p))$
The complexities of $W_m(W_n \cap \widehat{W}_p)$ and $W_m(W_n \setminus \widehat{W}_p)$ are $(m-1)2^{np} + 2^{np-1}$ for $m, n, p \geqslant 3$. (Verified for $3 \leqslant m, n, p \leqslant 4$ and several larger values.)

9   Conclusions

It is clear that a witness over an alphabet of three or four letters cannot be a witness when a larger alphabet is required. Also, $U_n(a, b, c)$ and $U_n(a, b, c, d)$ cannot be witnesses in a proper subclass of regular languages, since they do not possess the special properties of that class. However, in several cases it was possible to use $U_n(a, b, c)$ by "embedding" it in larger witnesses. For example, the DFA of a right ideal—a language L_n satisfying $L_n \Sigma^* = L_n$—can be constructed as shown in Fig. 7 to meet the upper bound on the size of the syntactic semigroup [6]. Similar constructions have been used for left ideals, and two-sided ideals [6], and for prefix-free, suffix-free, bifix-free and factor-free languages [3].

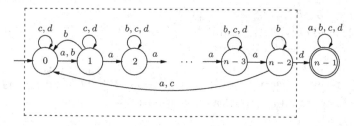

Fig. 7. Right ideal with n^{n-1} transformations

Although $U_n(a, b, c)$ and $U_n(a, b, c, d)$ succeed in all the cases I tried, they do have some shortcomings:

1. No binary language related to $U_n(a, b, c)$ seems to satisfy the reversal bound.
2. No binary languages related to $U_n(a, b, c)$ seem to satisfy the product bound.
3. Dialects of $U_n(a, b, c)$ and $U_n(a, b, c, d)$ had to be used for some operations.

In spite of these shortcomings, the results presented here strongly suggest that these witnesses ought to be considered when one is looking at new operations. The main remaining open questions are:

1. Does there exist a better ternary (quaternary) witness that would overcome the shortcomings listed above?
2. In general, does there exist a universal n-ary witness for operations that require witnesses over alphabets of n letters?

Finally, I remark that this paper must surely have a record number of conjectures! This would not have been possible without computer programs.

Acknowledgment. I am very grateful to Baiyu Li for helping me debug several proofs, for carrying out some computations, and for multiple proofreadings. I thank David Liu and Hellis Tamm for proofreading, and Lila Kari for providing references to work on combined operations.

References

1. Brzozowski, J.: Quotient complexity of regular languages. J. Autom. Lang. Comb. 15(1/2), 71–89 (2010)
2. Brzozowski, J., Li, B.: Syntactic complexities of some classes of star-free languages. In: Proceedings of the 14th International Workshop on Descriptional Complexity of Formal Systems (DCFS). LNCS. Springer, Heidelberg (to appear, 2012)
3. Brzozowski, J., Li, B., Ye, Y.: Syntactic complexity of prefix-, suffix-, bifix-, and factor-free regular languages. Theoret. Comput. Sci. (in press, 2012)
4. Brzozowski, J., Tamm, H.: Theory of Átomata. In: Mauri, G., Leporati, A. (eds.) DLT 2011. LNCS, vol. 6795, pp. 105–116. Springer, Heidelberg (2011)
5. Brzozowski, J., Tamm, H.: Quotient complexity of atoms of regular languages. In: Proceedings of the 16th International Conference on Developments in Language Theory (DLT). LNCS. Springer, Heidelberg (to appear, 2012)
6. Brzozowski, J., Ye, Y.: Syntactic Complexity of Ideal and Closed Languages. In: Mauri, G., Leporati, A. (eds.) DLT 2011. LNCS, vol. 6795, pp. 117–128. Springer, Heidelberg (2011)
7. Cui, B., Gao, Y., Kari, L., Yu, S.: State complexity of two combined operations: catenation-union and catenation-intersection. Int. J. Found. Comput. Sc. 22(8), 1797–1812 (2011)
8. Cui, B., Gao, Y., Kari, L., Yu, S.: State complexity of combined operations with two basic operations. Theoret. Comput. Sci. 437, 82–102 (2012)
9. Cui, B., Gao, Y., Kari, L., Yu, S.: State complexity of two combined operations: catenation-star and catenation-reversal. Int. J. Found. Comput. Sc. 23(1), 51–66 (2012)
10. Dénes, J.: On transformations, transformation semigroups and graphs. In: Erdös, P., Katona, G. (eds.) Theory of Graphs. Proceedings of the Colloquium on Graph Theory held at Tihany 1966, pp. 65–75. Akadémiai Kiado (1968)
11. Ganyushkin, O., Mazorchuk, V.: Classical Finite Transformation Semigroups: An Introduction. Springer (2009)
12. Gao, Y., Kari, L., Yu, S.: State complexity of union and intersection of square and reversal on k regular languages. Theoret. Comput. Sci. (in press, 2012)
13. Gao, Y., Kari, L., Yu, S.: State complexity of union and intersection of star on k regular languages. Theoret. Comput. Sci. 429, 98–107 (2012)

14. Gao, Y., Salomaa, K., Yu, S.: The state complexity of two combined operations: star of catenation and star of reversal. Fund. Inform. 83(1-2), 75–89 (2008)
15. Gao, Y., Yu, S.: State complexity of combined operations with union, intersection, star, and reversal. Fund. Inform. 116, 1–12 (2012)
16. Jirásková, G., Okhotin, A.: On the state complexity of star of union and star of intersection. Fund. Inform. 109, 1–18 (2011)
17. Jirásková, G., Šebej, J.: Note on Reversal of Binary Regular Languages. In: Holzer, M., Kutrib, M., Pighizzini, G. (eds.) DCFS 2011. LNCS, vol. 6808, pp. 212–221. Springer, Heidelberg (2011)
18. Leiss, E.: Succinct representation of regular languages by boolean automata. Theoret. Comput. Sci. 13, 323–330 (1981)
19. Maslov, A.N.: Estimates of the number of states of finite automata. Dokl. Akad. Nauk SSSR 194, 1266–1268 (1970) (Russian); English translation: Soviet Math. Dokl. 11, 1373–1375 (1970)
20. McNaughton, R., Papert, S.A.: Counter-Free Automata. M.I.T. Research Monographs, vol. 65. The MIT Press (1971)
21. Mirkin, B.G.: On dual automata. Kibernetika (Kiev) 2, 7–10 (1966) (Russian); English translation: Cybernetics 2, 6–9 (1966)
22. Myhill, J.: Finite automata and representation of events. Wright Air Development Center Technical Report 57–624 (1957)
23. Piccard, S.: Sur les fonctions définies dans les ensembles finis quelconques. Fund. Math. 24, 298–301 (1935)
24. Pin, J.E.: Syntactic semigroups. In: Handbook of Formal Languages. Word, Language, Grammar, vol. 1, pp. 679–746. Springer, New York (1997)
25. Rabin, M., Scott, D.: Finite automata and their decision problems. IBM J. Res. and Dev. 3, 114–129 (1959)
26. Salomaa, A., Salomaa, K., Yu, S.: State complexity of combined operations. Theoret. Comput. Sci. 383, 140–152 (2007)
27. Salomaa, A., Wood, D., Yu, S.: On the state complexity of reversals of regular languages. Theoret. Comput. Sci. 320, 315–329 (2004)
28. Sierpiński, W.: Sur les suites infinies de fonctions définies dans les ensembles quelconques. Fund. Math. 24, 209–212 (1935)
29. Yu, S., Zhuang, Q., Salomaa, K.: The state complexities of some basic operations on regular languages. Theoret. Comput. Sci. 125, 315–328 (1994)
30. Yu, S.: State complexity of regular languages. J. Autom. Lang. Comb. 6, 221–234 (2001)

A Formal Framework for Processes Inspired by the Functioning of Living Cells

Andrzej Ehrenfeucht[1] and Grzegorz Rozenberg[1,2]

[1] Department of Computer Science,
University of Colorado at Boulder,
Boulder, CO 80309, U.S.A.
[2] Leiden Institute of Advanced Computer Science (LIACS),
Leiden University,
Niels Bohrweg 1, 2300 RA Leiden, The Netherlands
`rozenber@liacs.nl`

Natural Computing is concerned with both human–designed computing inspired by nature and computing taking place in nature. The former research strand investigates computational techniques, models of computation and computational devices inspired by nature. The latter research strand investigates, in terms of information processing, processes taking place in nature.

"Standard" examples of the first research strand include evolutionary computation with paradigms inspired by Darwinian evolution of species, neural computation with paradigms inspired by the functioning of the brain, quantum computation with paradigms inspired by quantum mechanics, and molecular computation with paradigms inspired by molecular biology.

Representative examples of the second research strand are investigations into the computational nature of self-assembly, the computational nature of developmental processes, the computational nature of brain processes, the system biology approach to bionetworks where cellular processes are investigated in terms of communication and interaction, and the computational nature of biochemical reactions.

The second strand of research (which grows "explosively" now) underscores the fact that computer science is also the fundamental science of information processing, and as such a basic science for other scientific disciplines, such as, e.g., biology. This point of view is shared by both computer scientists and by scientists from natural sciences, e.g., by biologists.

We refer the reader to [8] and [10] for an insight into the fascinating, fast growing, and genuinely interdisciplinary area of natural computing.

Formal/computational understanding of the functioning of the living cell is one of the goals of natural computing – in the terminology presented above it belongs to the second strand of research of natural computing. In the framework of reaction systems one views the functioning of the living cell in terms of *interactions* between biochemical reactions. On the level of abstraction assumed by this framework the functioning of the living cell is determined by the pattern of interactions between biochemical reactions taking place in the living cell rather than by the specific tasks of each individual reaction. The second basic

N. Moreira and R. Reis (Eds.): CIAA 2012, LNCS 7381, pp. 25–27, 2012.

assumption of this framework is that the interactions between biochemical reactions are driven by two main mechanisms, facilitation and inhibition: reactions may facilitate or inhibit each other.

Reaction system is the central technical construct/model of the framework of reaction systems. Its formulation/definition follows the philosophy outlined above. The level of abstraction assumed for this model makes it a qualitative rather than a quantitative model. However, it still takes into account the basic bioenergietics of the living cell, and it takes into account the essential fact that the living cell is an open system and thus its behaviour is influenced by its environment.

Research topics in the framework of reaction systems are motivated either by biological considerations or by the need to understand the underlying computations. Examples of research topics are:

(1) Relationships to various models of computation such as finite transition systems ([2]), switching circuits ([2], boolean vector functions ([6]), and Petri nets ([9]).
(2) Formation of modules in biochemical/developmental processes ([5]).
(3) Causalities between entities in reaction systems ([1]).
(4) The role and nature of biochemical decay ([3]).
(5) The issue of time in processes of reaction systems ([7]).
(6) Adding quantitative/numerical parameters to (the qualitative model of) reaction systems ([4]).

We close this extended abstract by pointing out that reaction systems, which originated as a model for investigating the functioning of the living cell, are by now also attractive as a basic/novel model of computation.

References

1. Brijder, R., Ehrenfeucht, A., Rozenberg, G.: A note on Causalities in Reaction Systems. Electronic Communications of EASST 30 (2010)
2. Brijder, R., Ehrenfeucht, A., Rozenberg, G.: A Tour of Reaction Systems. International Journal of Foundations of Computer Science 22(7), 1499–1517 (2011)
3. Brijder, R., Ehrenfeucht, A., Rozenberg, G.: Reaction Systems with Duration. In: Kelemen, J., Kelemenová, A. (eds.) Computation, Cooperation, and Life. LNCS, vol. 6610, pp. 191–202. Springer, Heidelberg (2011)
4. Ehrenfeucht, A., Kleijn, J., Koutny, M., Rozenberg, G.: Qualitative and Quantitative Aspects of a Model for Processes Inspired by the Functioning of the Living Cell. In: Katz, E. (ed.) Biomolecular Information Processing. From Logic Systems to Smart Sensors and Actuators (to appear)
5. Ehrenfeucht, A., Rozenberg, G.: Events and Modules in Reaction Systems. Theoretical Computer Science 376, 3–16 (2007)
6. Ehrenfeucht, A., Rozenberg, G.: Reaction Systems. Fundamenta Informaticae 75, 263–280 (2007)
7. Ehrenfeucht, A., Rozenberg, G.: Introducing Time in Reaction Systems. Theoretical Computer Science 410, 310–322 (2009)

8. Kari, L., Rozenberg, G.: The Many Facets of Natural Computing. Communications of ACM 51, 72–83 (2008)
9. Kleijn, J., Koutny, M., Rozenberg, G.: Modelling Reaction Systems with Petri Nets. In: Proceedings of BIOPPN 2011, vol. 724, pp. 36–52. CEUR-WS (2011)
10. Rozenberg, G., Bäck, T., Kok, J.: Handbook of Natural Computing. Springer, Heidelberg (2012)

Adding Pebbles to Weighted Automata[*]

Paul Gastin and Benjamin Monmege

LSV, ENS Cachan, CNRS, Inria, France
`firstname.lastname@lsv.ens-cachan.fr`

Abstract. We extend weighted automata and weighted rational expressions with 2-way moves and (reusable) pebbles. We show with examples from natural language modeling and quantitative model-checking that weighted expressions and automata with pebbles are more expressive and allow much more natural and intuitive specifications than classical ones. We extend Kleene-Schützenberger theorem showing that weighted expressions and automata with pebbles have the same expressive power. We focus on an efficient translation from expressions to automata. We also prove that the evaluation problem for weighted automata can be done very efficiently if the number of (reusable) pebbles is low.

1 Introduction

Regular expressions have always been used to specify patterns. Popular because they propose a concise and intuitive way of denoting such patterns, they have also a long history in the formal language community. A seminal result, known as Kleene's theorem, establish that the (denotational) regular expressions have the same expressive power as the (operational) finite state automata. Efficient translation algorithms of regular expressions into finite automata are crucial since expressions are convenient to denote patterns and automata are amenable to efficient algorithms. Regular expressions and finite automata have been extended in several directions, e.g., tree (walking) automata, (regular) XPath, etc.

Nowadays, quantitative models and quantitative analysis are intensively studied, resulting in a revision of the foundation of computer science where classical yes/no answers are replaced by quantities such as probability, energy consumption, reliability, cost, etc. In the 60s, Schützenberger provided a generic way of turning qualitative into quantitative systems, starting the theory of weighted automata [31] (see [18,16,3] for recent books on this theory). Indeed, probabilistic automata and word transducers appear as instances of that framework, which found its way into numerous application areas such as natural language processing, speech recognition or digital image compression. Schützenberger proved the equivalence between weighted automata and weighted regular expressions, extending Kleene's theorem. Various translation algorithms can be extended from the Boolean framework to the weighted case, see [27,29] for surveys about these methods, and [22] which obtains Schützenberger's theorem as a corollary of Kleene's theorem.

---

[*] Supported by LIA INFORMEL.

N. Moreira and R. Reis (Eds.): CIAA 2012, LNCS 7381, pp. 28–51, 2012.
© Springer-Verlag Berlin Heidelberg 2012

In Sections 4 and 5, we extend weighted expressions and automata with 2-way moves and pebbles. There are several motivations for these extensions. First, as shown in Section 2 for applications in natural language processing and quantitative model-checking, 2-way moves and pebbles allow more natural and more concise descriptions of the quantitative expressions we need to evaluate. Second, in the weighted case, 2-way and pebbles do increase the expressive power as already observed in [8] in relation with weighted logics or in [26] in the probabilistic setting. This is indeed in contrast with the Boolean case where 2-way and pebbles do not add expressive power over words (see, e.g., [19]) even though they allow more succinct descriptions (see, e.g., [4]). Our work is also inspired by pebble tree-walking automata and in particular their links with powerful logics, XPath formalisms and caterpillar expressions on trees [17,10,6,30,5].

In Sections 6 and 7, we generalize Kleene and Schützenberger theorems to weighted expressions and automata with 2-way moves and pebbles. We establish their expressive power equivalence by providing effective translations in both directions. Showing how to transform an *operational* automaton into an equivalent *denotational* expression is indeed very interesting from a theoretical point of view, but is less useful in practice. On the other hand, we need highly efficient translations from the convenient denotational formalism of expressions to operational automata which, as stated above, are amenable to efficient algorithms. Efficiency is measured both wrt. the size of the resulting automaton and wrt. the space and time complexities of the translation. We show that, Glushkov's [20] or Berry-Sethi [2] translations, which are among the best ones in the Boolean case, can be extended to weighted expressions with 2-may moves and pebbles. The constructions for the rational operations (sum, product, star) can be adapted easily to cope with 2-way moves, even though the correctness proofs are more involved and require new theoretical grounds such as series over a partial monoid as explained in Section 4.1. The main novelty in Sections 6 and 7 is indeed the treatment of pebbles in the translations between expressions and automata.

To complete the picture, we study in Section 8 the evaluation problem of a weighted automaton with 2-way moves and *reusable* pebbles over a given word. The algorithm is polynomial in the size of the word, where the degree is 1 plus the number of reusable pebbles. We can even decrease the degree by 1 for *strongly layered* automata. This applies when we only have one reusable pebble, and we obtain an algorithm which is linear in the size of the input word. This is in particular the case for automata derived from weighted LTL.

The paper focuses on intuitive explanations and examples for a better understanding of weighted expressions with 2-way moves and pebbles, and of the translations between automata and expressions. Most proofs are omitted and will appear in a longer version.

2 Motivations

We give in this section two motivating examples for studying weighted expressions and automata with 2-way moves and pebbles.

2.1   Language Modeling

Since decades, weighted automata have been extensively used in Natural Language Processing (see [21]), in particular for automatic translation, speech recognition or transliteration. All these tasks have in common to split the problem into independent parts, certain directly related to the specific task and others related to the knowledge of the current language. For example, in the translation task from French sentences to English sentences, one splits the problem into first knowing translation of single words and then modeling English sentences (knowledge which is independent from the translation task). The second part, namely to know whether a sequence of words is a *good* English sentence, is known as *language modeling*. Often this knowledge is learned from a large corpus of English texts, and stored into a formal model, e.g., a weighted finite state automaton representing the probability distribution $\mathbb{P}$ of well-formed English sentences. The translation task is then resolved by first generating several English sentences from the original French one (due to ambiguity of the word-by-word translation task), and then choosing among this set of sentences the ones with highest probability.

One broadly used language model is the n-gram model, where the probability of a word in a sentence depends only on the previous $n - 1$ words: for example in a 1-gram model, only the individual word frequencies are relevant to generate well-formed English sentences, whereas in a 2-gram model, the probability of a word depends on the very same frequency distribution and also the previous word. To formally describe these models, and further study them, let us define them using regular expressions. Let D denote the dictionary of words in the language. Suppose we are given the conditional probability distributions $\mathbb{P}(u_n \mid u_1, \ldots, u_{n-1})$ in the n-gram model (with $u_i \in D$ for all i). The probability of a sentence $(u_i)_{1 \leq i \leq m} \in D^m$ can be given by the following weighted regular expression in a 1-gram model and a 3-gram model:

$$E_1 = \left(\sum_{u \in D} u\mathbb{P}(u) \right)^* \qquad E_3 = \rightarrow\rightarrow\left(\sum_{u,v,w \in D} \leftarrow\leftarrow uvw\mathbb{P}(w \mid u, v) \right)^*$$

where symbols $\rightarrow$ and $\leftarrow$ denote a right or left move, respectively, no matter what word it is reading. Expression E_1 is a classical weighted regular expression where the Kleene star iterates the computation of the inner expression, which here computes the probability of the current word u. Expression E_3 has the opportunity to move forward and backward: this allows to easily recover the context whereas a 1-way automaton would have to store the context in its states. Notice that expression E_3 is quite readable and intuitive. One could write an equivalent 1-way expression, but imagine how intricate it would be since positions would have to encode the context, i.e., the last two words. This is an important motivation for studying 2-way expressions and automata.

Actually, expression E_3 is not small since the sum hides the very big set D^3: for a dictionary of size 1 million, this seems already unpracticable. But in practice, a much smaller expression could be sufficient. First, for many words, the frequency distribution of the word w is a sufficiently good approximation of

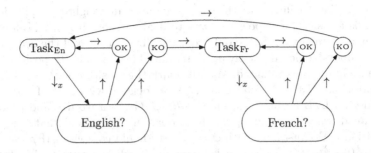

Fig. 1. Pebble automaton for the multi-language modeling task

the conditional probability $\mathbb{P}(w \mid u, v)$. Let us denote D_0 this set of words. For instance, the probability of observing the word **the** may not really depend on the previous words. Then, let D_1 be the set of words (disjoint from D_0) such that only the previous word is necessary to describe the probability. Finally, let D_2 be the rest of the dictionary. Now, we may replace expression E_3 by the following expression, whose size is much smaller if D_0 and D_1 contain enough words:

$$\left(\sum_{w \in D_0} w\mathbb{P}(w) + \sum_{w \in D_1, v \in D} {\leftarrow} vw\mathbb{P}(w \mid v) + \sum_{w \in D_2, u, v \in D} {\leftarrow}{\leftarrow} uvw\mathbb{P}(w \mid u, v) \right)^*$$

To motivate the introduction of pebbles, let us add internationalization, which means that the user has the ability to write/speak alternately in two or more languages, e.g., English and French. All tasks such as automatic translation or speech recognition are now more complex since there is no a priori knowledge of the current language of the speaker. Again, splitting the problem into independent parts, we have to know the probability distributions $\mathbb{P}_L$ for every involved language L, and, assuming a current language L, we should be able to solve the language processing task with a procedure Task_L. Then, before processing the next word, we start a computation which re-reads the current prefix of the text in order to compute using $\mathbb{P}_L$ the probability that the current language L is still valid. The next word is then processed with the current or the alternate language (see Figure 1). In order to compute the probability that the current language is still valid, we mark the current position with a pebble ($\downarrow_x$) and read the current prefix of the text with the automaton modeling the current language. Then we return to the marked position and lift the pebble ($\uparrow$) in order to resume the top level computation.

2.2   Weighted Linear Temporal Logic

Whereas weighted automata and weighted expressions have been extensively studied, logical formalisms adapted to the weighted case still need deeper understanding. This is especially true for weighted *linear* temporal logics [23], whereas weighted *branching* temporal logics have received more attention [13,12,25,7].

We would like to illustrate that using pebbles in weighted expressions or automata is a natural and powerful way to deal with nesting in LTL formulas. For this motivating example, we only consider finite words and the probabilistic setting. Temporal logics implicitly use a free variable to denote the position where the formula has to be evaluated. We will mark this position with a pebble, say x, in expressions $E_\varphi(x)$ or automata $\mathcal{A}_\varphi(x)$ associated with an LTL formula φ.

Consider an LTL formula $\mathsf{F}\varphi$, for *Finally* φ. Given a word u and a position i in the word, we want to compute the probability $\mathbb{P}(\mathsf{F}\varphi, u, i)$ that φ holds on u at position i. For instance, with $\varphi = \frac{1}{3}a$, we should compute $\mathbb{P}(\mathsf{F}\varphi, abba, 0) = \frac{1}{3} + \frac{2}{3}(0 + \frac{2}{3}(0 + \frac{2}{3}(\frac{1}{3} + 0)))$: either φ is satisfied immediately with probability $\frac{1}{3}$ or it is not (probability $\frac{2}{3}$) and (product) it should be satisfied later. More generally, we have

$$\mathbb{P}(\mathsf{F}\varphi, u, i) = \mathbb{P}(\varphi, u, i) + \mathbb{P}(\neg\varphi, u, i)\Big(\mathbb{P}(\varphi, u, i+1)$$
$$+ \mathbb{P}(\neg\varphi, u, i+1)\big(\mathbb{P}(\varphi, u, i+2) + \cdots\big)\Big).$$

For every LTL formula φ, we want to give an equivalent expression $E_\varphi(x)$ which evaluates to $\mathbb{P}(\varphi, u, i)$ over word u when pebble x marks position i. For *Finally* φ, we set

$$E_{\mathsf{F}\varphi}(x) = \triangleright? \to^* x? \big((y!E_{\neg\varphi}(y)) \to \big)^* (y!E_\varphi(y)) \to^* \triangleleft?$$
$$= \triangleright? \to^* x? \sum_{n \geq 0} \big((y!E_{\neg\varphi}(y)) \to \big)^n (y!E_\varphi(y)) \to^* \triangleleft?.$$

The expression starts at the beginning of the word ($\triangleright?$), moves right ($\to^*$) until it sees the marked position ($x?$), for each possible $n \geq 0$ it iterates n times the computation of $\neg\varphi$ with the current position marked with y ($y!E_{\neg\varphi}(y)$) and moving right between two computations, and it finally computes φ with $y!E_\varphi(y)$ before moving to the end of the word ($\to^* \triangleleft?$).

Similarly, for *Globally* φ ($\mathsf{G}\varphi$), we have $\mathbb{P}(\mathsf{G}\varphi, u, i) = \prod_{j \geq i} \mathbb{P}(\varphi, u, j)$, leading to the simpler expression

$$E_{\mathsf{G}\varphi}(x) = \triangleright? \to^* x? \big((y!E_\varphi(y)) \to \big)^* \triangleleft?.$$

Finally, based on the equivalence $\varphi \, \mathsf{U} \, \psi \equiv (\neg\psi \wedge \varphi) \, \mathsf{U} \, \psi$, the expression for the *Until* modality is

$$E_{\varphi\mathsf{U}\psi}(x) = \triangleright? \to^* x? \big((y!(E_{\neg\psi}(y) \leftarrow^* E_\varphi(y))) \to \big)^* (y!E_\psi(y)) \to^* \triangleleft?.$$

In terms of automata, let us assume that for every formula φ, there is an automaton $\mathcal{A}_\varphi$ with 2 designated terminal states $\{\mathrm{OK}, \mathrm{KO}\}$, such that runs ending in OK computes expression E_φ and those ending in KO computes expression $E_{\neg\varphi}$. We have depicted below automata for the modalities *Finally* and *Globally*.

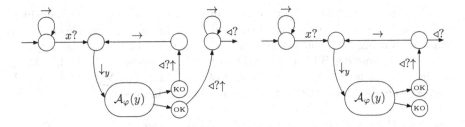

3 Preliminaries

Words. The set of non-empty words over a finite alphabet A is denoted A^+. We write $u = u_0 \cdots u_{n-1} \in A^+$ a non-empty word of length $|u| = n \geq 1$ with $u_i \in A$ for $0 \leq i < |u|$. The set of *positions* of u is $\mathrm{pos}(u) = \{0, 1, \ldots, |u|\}$. In particular, we include $|u|$ in $\mathrm{pos}(u)$ even though the last letter is on position $|u| - 1$.

Semirings. A semiring is a set $\mathbb{S}$ equipped with two binary internal operations denoted $+$ and $\times$, and two neutral elements 0 and 1 such that $(\mathbb{S}, +, 0)$ is a commutative monoid, $(\mathbb{S}, \times, 1)$ is a monoid, $\times$ distributes over $+$ and $0 \times s = s \times 0 = 0$ for every $s \in \mathbb{S}$. If the monoid $(\mathbb{S}, \times, 1)$ is commutative, the semiring itself is called commutative. See [15,27] for more discussions about semirings, especially complete and continuous ones, as we describe now.

A semiring $\mathbb{S}$ is *complete* if every family $(s_i)_{i \in I}$ of elements of $\mathbb{S}$ over an arbitrary indexed set I is summable to some element in $\mathbb{S}$ denoted $\sum_{i \in I} s_i$ and called *sum* of the family, such that the following conditions are satisfied:

- $\sum_{i \in \emptyset} s_i = 0$, $\sum_{i \in \{1\}} s_i = s_1$ and $\sum_{i \in \{1,2\}} s_i = s_1 + s_2$;
- if $I = \bigcup_{j \in J} I_j$ is a partition, $\sum_{j \in J} \left(\sum_{i \in I_j} s_i \right) = \sum_{i \in I} s_i$;
- $\left(\sum_{i \in I} s_i \right) \times \left(\sum_{j \in J} t_j \right) = \sum_{(i,j) \in I \times J} (s_i \times t_j)$.

Intuitively, this means that it is possible to define infinite sums that extends the binary addition and satisfies infinite versions of associativity and distributivity.

In a complete semiring, for every $s \in \mathbb{S}$, the element $s^* = \sum_{i \in \mathbb{N}} s^i$ exists (where s^i is defined recursively by $s^0 = 1$ and $s^{i+1} = s^i \times s$). Here are some examples of complete semirings.

- The Boolean semiring $(\{0, 1\}, \vee, \wedge, 0, 1)$ with $\sum$ defined as an infinite disjunction.
- $(\mathbb{R}_{\geq 0} \cup \{\infty\}, +, \times, 0, 1)$ with $\sum$ defined as usual for positive (not necessarily convergent) series: in particular, $s^* = \infty$ if $s \geq 1$ and $s^* = 1/(1-s)$ if $0 \leq s < 1$.
- $(\mathbb{N} \cup \{\infty\}, +, \times, 0, 1)$ as a complete subsemiring of the previous one.
- $(\mathbb{R} \cup \{-\infty\}, \min, +, -\infty, 0)$ with $\sum = \inf$ and $(\mathbb{R} \cup \{\infty\}, \max, +, \infty, 0)$ with $\sum = \sup$.
- Complete lattices such as $([0, 1], \min, \max, 0, 1)$.
- The semiring of languages over an alphabet A: $(2^{A^*}, \cup, +, \emptyset, \{\varepsilon\})$ with $\sum$ defined as (infinite) union.

In this paper, we consider *continuous* semirings which are complete semirings in which infinite sums can be approximated by finite partial sums. Formally, a complete semiring $\mathbb{S}$ is *continuous* if the relation $\leq$ defined over $\mathbb{S}$ by $a \leq b$ if $b = a + c$ for some $c \in \mathbb{S}$ is an order relation; and for every family $(s_i)_{i \in I}$ in $\mathbb{S}$, the sum $\sum_{i \in I} s_i$ is the least upper bound of the finite sums $\sum_{i \in J} s_i$ for $J \subseteq I$ finite. All the above complete semirings are also continuous.

Series and Polynomials. Let Z be a set. A series f over Z is a map $f \colon Z \to \mathbb{S}$. We denote by $\mathbb{S}\langle\langle Z \rangle\rangle$ the set of series over Z with coefficients in $\mathbb{S}$. The *support* of a series $f \in \mathbb{S}\langle\langle Z \rangle\rangle$ is the set $\{z \in Z \mid f(z) \neq 0\}$. A series with a finite support is called a *polynomial*. We denote by $\mathbb{S}\langle Z \rangle$ the set of polynomials over Z with coefficients in $\mathbb{S}$.

We can lift addition from $\mathbb{S}$ to $\mathbb{S}\langle\langle Z \rangle\rangle$ pointwise by $(f + g)(z) = f(z) + g(z)$ for all $z \in Z$. Then, $(\mathbb{S}\langle\langle Z \rangle\rangle, +, 0)$ is a commutative monoid where 0 is the series mapping every element $z \in Z$ to 0. If Z is a monoid and the semiring is complete, we can also define the (Cauchy) product of two series by $(f \times g)(z) = \sum_{z = xy} f(x)g(y)$ for all $z \in Z$. This sum may be infinite, but is well-defined since the semiring is complete. The Cauchy product is associative and admits as unit the characteristic function (denoted 1) of the neutral element of Z. Hence, $(\mathbb{S}\langle\langle Z \rangle\rangle, +, \times, 0, 1)$ is a semiring. When $\mathbb{S}$ is continuous, we can also lift infinite sums pointwise to $\mathbb{S}\langle\langle Z \rangle\rangle$ which becomes a continuous semiring.

4 Weighted Expressions with Pebbles

The syntax of our weighted expressions is carefully chosen so that an efficient translation to weighted automata can be obtained, essentially based on Glushkov's construction as we will see in Section 7. Formally, for a (continuous) semiring $\mathbb{S}$, an alphabet A and a set Peb of pebbles, the syntax is given by the grammar:

$$E ::= s \mid \varphi \mid \rightarrow \mid \leftarrow \mid x!E \mid E + E \mid E \cdot E \mid E^+$$
$$\varphi ::= a? \mid \triangleright? \mid \triangleleft? \mid x? \mid \neg\varphi \mid \varphi \wedge \varphi \mid \varphi \vee \varphi$$

with $s \in \mathbb{S}$, $a \in A$, $x \in$ Peb. We denote by Test the set of *test* formulas φ defined by the second line of the grammar above. For instance, one can check with $\triangleright?$ and $\triangleleft?$ whether we are at the beginning or at the end of the word. This is indeed useful since we have 2-way expressions. We denote by pebWE the set of weighted expressions with pebbles. Below, we give the intuitive meaning of our weighted expressions. We start without pebbles (i.e., without $x!E$). Then, we introduce pebbles. The formal semantics is given in Table 1.

Notice that from the *irreflexive* iteration $E^+ = \sum_{n>0} E^n$, we get also the classical Kleene star: $E^* \stackrel{\text{def}}{=} 1 + E^+$. Indeed, we also have $E^+ = E \cdot E^*$ but if we apply Glushkov's construction (blindly) to $E \cdot E^*$ we get an automaton with twice the number of states needed for E^+. This is basically why we prefer to have E^+ as a primary construct.

We have chosen to distinguish between *checking* the current position with some *test* φ and *moving* to the right or left position with $\rightarrow$ and $\leftarrow$. This is in the spirit of XPath in trees. This allows to write concise expressions, e.g., $E = \rightarrow^+ a? \leftarrow^+ b? \rightarrow^+ c? \leftarrow^+ d? \rightarrow^+$ to describe patterns consisting of an a having in its past a b, having in its future a c, having in its past a d. In the Boolean semantics, this expression defines words having this pattern. In the semiring $\mathbb{N}$ of natural numbers, the expression counts the number of occurrences of the pattern, e.g., $[\![E]\!](cabcdbadcbab) = 8$. Indeed we may write an equivalent 1-way expression for this pattern but it would be less concise and harder to decipher (see e.g., [4] for succinctness results in the Boolean case).

Let $u = u_0 u_1 \cdots u_{|u|-1} \in A^+$. A test φ will be evaluated at a position $i \in \text{pos}(u)$: $\triangleright?$ holds if $i = 0$, $\triangleleft?$ holds if $i = |u|$ and $a?$ holds if $i < |u|$ and $u_i = a$.

With the 2-way mechanism, a sub-expression such as $a? \leftarrow^+ b? \rightarrow^+ c? \leftarrow^+ d?$ may start from position i, end in position j and still visit the whole word. In order to inductively define the semantics of expressions, we assign to triples (u, i, j) a value $[\![E]\!](u, i, j) \in \mathbb{S}$.

It is also convenient to *check-and-move* so we introduce the macros $a \overset{\text{def}}{=} a? \rightarrow$ and $\bar{a} \overset{\text{def}}{=} a? \leftarrow$. Then, we can write $\rightarrow^* \texttt{blue} \leftarrow^+ \triangleright? \rightarrow^* \texttt{black} \rightarrow^*$ to define words having both $\texttt{blue}$ and $\texttt{black}$ as subwords. This allows to write classical (1-way) regular expressions such as $(ab)^+ aa$. In order to get the classical semantics for usual 1-way expressions, the evaluation of an expression on a whole word is defined as $[\![E]\!](u) = [\![E]\!](u, 0, |u|)$. For instance, $[\![a]\!](a) = [\![a?\rightarrow]\!](a, 0, 1) = 1$, $[\![\rightarrow^* a \rightarrow^*]\!](baaba) = [\![\rightarrow^* a? \rightarrow \rightarrow^*]\!](baaba, 0, 5) = 3$, and $[\![(2\rightarrow)^+]\!](u) = 2^{|u|}$.

Our 2-way expressions are uncomparable with expressions over the free group. Indeed, the expression $a\bar{a}b$ always evaluates to 0 in our setting, whereas over the free group it would evaluate to 1 on $b = a\bar{a}b$.

Notice that with the 2-way mechanism we may write $E = E_1 \triangleleft? \leftarrow^* \triangleright? E_2$ to compute the product (intersection in the boolean semantics) of the values computed by E_1 and E_2: $[\![E]\!](u) = [\![E_1]\!](u) \times [\![E_2]\!](u)$.

The 2-way mechanism together with iteration gives rise to infinite sums. This may be useful for probabilistic systems. For instance, in the *continuous* semiring $(\mathbb{R}_{\geq 0}^\infty, +, \times, 0, 1)$, consider the expression $E = (\neg\triangleleft?(s \rightarrow + (1 - s)\neg\triangleright?\leftarrow))^* \triangleleft?$ with $0 < s < 1$ some probability. Expression E describes a random walk[1] and it will be used again in Section 5. Let $F = \neg\triangleleft?(s \rightarrow + (1 - s)\neg\triangleright?\leftarrow)$ so that $E = F^* \triangleleft?$. Let u be a word of length $m \geq 2$. We can easily see that for all $i, j \in \text{pos}(u)$ and all $n > |j - i|$, the expression F^n computes a *positive* value on (u, i, j). Therefore, the expression F^* computes an infinite sum on (u, i, j). In the present case $(0 < s < 1)$, the series $\sum_{n \geq 0}[\![F^n]\!](u, i, j)$ converges and $[\![F^*]\!](u, i, j) \in \mathbb{R}_{\geq 0}$. On the other hand, for the expression $G = \neg\triangleleft?\rightarrow + \neg\triangleright?\leftarrow$, we can check that the series $\sum_{n \geq 0}[\![G^n]\!](u, i, j)$ diverges and we get $[\![G^*]\!](u, i, j) = \infty$. Since we are considering *complete* semirings, infinite sums exist and the semantics of an iteration E^* or E^+ is always well-defined.

---

[1] With $\alpha = \frac{1-s}{s}$, one can show that $[\![E]\!](u, 0, |u|) = \frac{1}{1 + \alpha + \ldots + \alpha^{|u|}}$.

Table 1. Semantics of weighted expressions

$$[\![s]\!](u, \sigma, i, j) = \begin{cases} s & \text{if } j = i \\ 0 & \text{otherwise} \end{cases} \qquad [\![\varphi]\!](u, \sigma, i, j) = \begin{cases} 1 & \text{if } j = i \land u, \sigma, i \models \varphi \\ 0 & \text{otherwise} \end{cases}$$

$$[\![\to]\!](u, \sigma, i, j) = \begin{cases} 1 & \text{if } j = i + 1 \\ 0 & \text{otherwise} \end{cases} \qquad [\![\leftarrow]\!](u, \sigma, i, j) = \begin{cases} 1 & \text{if } j = i - 1 \\ 0 & \text{otherwise} \end{cases}$$

$$[\![x!E]\!](u, \sigma, i, j) = \begin{cases} [\![E]\!](u, \sigma[x \mapsto i], 0, |u|) & \text{if } j = i < |u| \\ 0 & \text{otherwise} \end{cases}$$

$$[\![E + F]\!](u, \sigma, i, j) = [\![E]\!](u, \sigma, i, j) + [\![F]\!](u, \sigma, i, j)$$

$$[\![E \cdot F]\!](u, \sigma, i, j) = \sum_{k \in \text{pos}(u)} [\![E]\!](u, \sigma, i, k) \times [\![F]\!](u, \sigma, k, j)$$

$$[\![E^+]\!](u, \sigma, i, j) = \sum_{n > 0} [\![E^n]\!](u, \sigma, i, j)$$

We explain now the pebble mechanism used in our expressions. The construct $x!E$ marks with x the current position *in* u and evaluates E on the marked word, from beginning to end. Indeed, we can use $x?$ in E to test whether the current position is marked. For instance, consider

$$E = \to^+ a? \, x! \Big((\neg x?\to)^* \, b? \, (\neg x?\to)^+ \, c? \leftarrow^+ d? \to^+ \Big) \to^*$$

which is a variant of our first example. Here the pattern consists of an a for which the corresponding prefix contains a b, having in its future a c, having in its past a d. In particular, the c must be on the left of the current a which is marked with x. Hence, we get $[\![E]\!](cabcdbadcbab) = 4$.

As another example, on a word u, the expression $(x!((2\to)^+)\to)^+$ computes $2^{|u|^2}$ over the natural semiring[2]. Actually, the pebble is not tested in this expression: it is only used to restart the computation $|u|$ times.

We give now the formal semantics of tests and of pebWE. For each word $u \in A^+$, valuation $\sigma \colon \text{Peb} \to \text{pos}(u)$ and position $i \in \text{pos}(u)$, the semantics $u, \sigma, i \models \varphi$ of tests is defined inductively. The Boolean connectives are as usual. For the atoms, $\triangleright?$ holds if $i = 0$, $\triangleleft?$ holds if $i = |u|$, $a?$ holds if $i < |u|$ and $u_i = a$ and $x?$ holds if $\sigma(x) = i < |u|$ (the last position $|u|$ cannot be marked).

A *marked* word is a tuple (u, σ, i, j) where $u \in A^+$ is a word, $\sigma \colon \text{Peb} \to \text{pos}(u)$ is a valuation and $i, j \in \text{pos}(u)$ are positions. We denote by $\text{Mk}(A^+)$ the set of marked words (we will see below that it forms a partial monoid).

The semantics[3] of a pebWE E is a map $[\![E]\!] \colon \text{Mk}(A^+) \to \mathbb{S}$, i.e., a series over marked words: $[\![E]\!] \in \mathbb{S}\langle\!\langle \text{Mk}(A^+) \rangle\!\rangle$. It is is defined in Table 1. Note that,

---

[2] This function cannot be computed without pebble by a classical 1-way weighted expression. We can see this using Schützenberger's theorem since weighted automata only compute values in $2^{\mathcal{O}(|u|)}$.

[3] We may also define the semantics $[\![E]\!]_\mathcal{V}$ of an expression E using valuations over a subset $\mathcal{V} \subseteq \text{Peb}$, provided it contains the free pebbles of E.

since we are considering *complete* semirings, the infinite sum in the semantics of E^+ is always well-defined. If the expression has no free pebbles then we omit the valuation and simply write $[\![E]\!](u, i, j)$. For whole words we let $[\![E]\!](u, \sigma) = [\![E]\!](u, \sigma, 0, |u|)$ and $[\![E]\!](u) = [\![E]\!](u, 0, |u|)$ as explained above.

Notice that for tests φ_1 and φ_2, the expressions $\varphi_1 \wedge \varphi_2$ and $\varphi_1 \cdot \varphi_2$ are equivalent, but $\varphi_1 \vee \varphi_2$ and $\varphi_1 + \varphi_2$ are not equivalent in general. One can check that $\varphi_1 + \neg\varphi_1 \cdot \varphi_2$ is equivalent to the disjunction $\varphi_1 \vee \varphi_2$. Hence, conjunctions and disjunctions in tests are not necessary for the expressive power of pebWE and they could have been defined as macros.

Similar to the star-height of an expression, we define the *pebble*-depth:

$$\text{pebd}(s) = \text{pebd}(\varphi) = \text{pebd}(\leftarrow) = \text{pebd}(\rightarrow) = 0$$
$$\text{pebd}(E + F) = \text{pebd}(E \cdot F) = \max(\text{pebd}(E), \text{pebd}(F))$$
$$\text{pebd}(E^+) = \text{pebd}(E) \qquad\qquad \text{pebd}(x!E) = 1 + \text{pebd}(E).$$

4.1 Series Over a Partial Monoid

We show in this subsection that the set of marked words can be endowed with a *partial monoid* structure which allows to define a Cauchy product on series in $\mathbb{S}\langle\!\langle \text{Mk}(A^+) \rangle\!\rangle$. Since the sums can be lifted pointwise from $\mathbb{S}$ to series over $\mathbb{S}$, we show that $\mathbb{S}\langle\!\langle \text{Mk}(A^+) \rangle\!\rangle$ is actually a *continuous* semiring. Indeed, the semantics defined for sum, product and iteration of pebWE in Table 1 corresponds to sum, Cauchy product and star in the continuous semiring $\mathbb{S}\langle\!\langle \text{Mk}(A^+) \rangle\!\rangle$. This more formal view of the semantics of pebWE is especially useful for proofs, but since proofs are omitted in this paper this section may be skipped in a first reading.

Pebble weighted expressions and pebble weighted automata introduce two new difficulties. The first one comes from the 2-way navigation mechanism which prevents the computation of the behavior of an expression (or an automaton) using the concatenation of words in the underlying monoid, here the free monoid A^+. The second one comes indeed from pebbles which allow to restart the computation. To address both problems, we had to fix the word when defining the semantics and we no more use the monoid structure of A^+. Here, we define a *partial* monoid structure on the *marked words* and show how this allows us to reuse existing results from the classical theory of rational series.

A *partial* monoid is a triple $(Z, \cdot, Y)$ where Z is the set of elements, $\cdot : Z^2 \to Z$ is a *partially defined* associative concatenation[4] and $Y \subseteq Z$ is a set of *partial units* satisfying:

$$\forall x, z \in Z \quad \forall y \in Y \quad x \cdot y = z \implies x = z$$
$$\forall x, z \in Z \quad \forall y \in Y \quad y \cdot x = z \implies x = z$$
$$\forall z \in Z \quad \exists! y \in Y \quad y \cdot z = z$$
$$\forall z \in Z \quad \exists! y \in Y \quad z \cdot y = z.$$

---

[4] For all $x, y, z \in Z$, $(x \cdot y) \cdot z$ is defined iff $x \cdot (y \cdot z)$ is defined, and in this case $(x \cdot y) \cdot z = x \cdot (y \cdot z)$.

Indeed, a classical monoid is a partial monoid with the concatenation being totally defined and with the set of partial units being the singleton set consisting of the (real) unit.

We are especially interested in the partial monoid $(\mathrm{Mk}(A^+), \cdot, \mathrm{Unit}(A^+))$ of *marked words* over A^+ where

$$\mathrm{Mk}(A^+) = \{(u, \sigma, i, j) \mid u \in A^+, \ \sigma \colon \mathrm{Peb} \to \mathrm{pos}(u), \ i, j \in \mathrm{pos}(u)\}$$
$$\mathrm{Unit}(A^+) = \{(u, \sigma, i, i) \mid u \in A^+, \ \sigma \colon \mathrm{Peb} \to \mathrm{pos}(u), \ i \in \mathrm{pos}(u)\}$$

and the partial concatenation is defined for all $u \in A^+$, $\sigma \colon \mathrm{Peb} \to \mathrm{pos}(u)$ and $i, j, k \in \mathrm{pos}(u)$ by $(u, \sigma, i, k) \cdot (u, \sigma, k, j) = (u, \sigma, i, j)$ and it is undefined in all other cases. We can see that this partial concatenation is associative and that the above requirements for partial units are satisfied.

Note that a partial monoid needs not be graded and in particular, the partial monoid of marked words defined above is not graded. Hence, we cannot apply directly the theory of rational series over graded monoids as developed e.g. in [28]. Instead, we will use the theory of rational series over a *continuous* semiring $\mathbb{S}$ (see e.g., [27, III.5]). We first show that, even if the monoid Z, and more specifically $\mathrm{Mk}(A^+)$, is only partial, we can define (infinite) sums and (Cauchy) product on series over Z so that $\mathbb{S}\langle\langle Z \rangle\rangle$ forms a continuous semiring.

Let $\mathbb{S}$ be a continuous semiring and $(Z, \cdot, Y)$ be a partial monoid. As described in Section 3, infinite sums may be lifted from $\mathbb{S}$ to series in $\mathbb{S}\langle\langle Z \rangle\rangle$. We may also define the Cauchy product as usual. Note that, even though the concatenation in Z may be partially defined, the Cauchy product in $\mathbb{S}\langle\langle Z \rangle\rangle$ is always defined by $(f \times g)(z) = \sum_{x,y \in Z, z = x \cdot y} f(x) \times g(y)$ for $f, g \in \mathbb{S}\langle\langle Z \rangle\rangle$ and $z \in Z$. The sum ranges over all pairs (x, y) for which the concatenation is defined and such that $x \cdot y = z$. The sum may be finite or infinite but it must be nonempty since we have the left and right partial units for z. Finally, we let 1_Y be the characteristic function of the set Y of partial units of Z and we can easily check that it is a unit for the Cauchy product. Mimicking the proof for classical monoids, we can show the following.

Proposition 1. *If $\mathbb{S}$ is a continuous semiring and $(Z, \cdot, Y)$ is a partial monoid then the series $\mathbb{S}\langle\langle Z \rangle\rangle$ forms a continuous semiring $(\mathbb{S}\langle\langle Z \rangle\rangle, +, \times, 0, 1_Y)$.*

This allows to apply the theory of rational series over continuous semirings (see e.g., [27, III.5]). In particular, a star operation may be defined.

We can check that the semantics of pebWE in the continuous semiring $\mathbb{S}\langle\langle \mathrm{Mk}(A^+) \rangle\rangle$ as defined in Table 1 satisfies

$$[\![E + F]\!] = [\![E]\!] + [\![F]\!] \qquad\qquad [\![E^*]\!] = [\![E]\!]^*$$
$$[\![E \cdot F]\!] = [\![E]\!] \times [\![F]\!] \qquad\qquad [\![E^+]\!] = [\![E]\!]^+ .$$

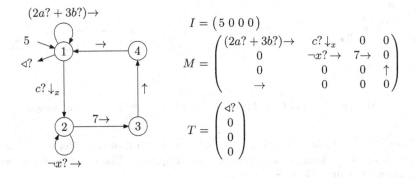

Fig. 2. A pebWA and its matrix representation

5   Weighted Automata with Pebbles

We fix a finite set Peb of pebbles and a (continuous) semiring $\mathbb{S}$. We denote by Move $= \{\leftarrow, \rightarrow, \uparrow\} \cup \{\downarrow_x \mid x \in \text{Peb}\}$ the set of possible moves of an automaton.

A *pebble weighted automaton* (pebWA) is a tuple $\mathcal{A} = (Q, A, I, M, T)$ with Q a finite set of states, A a finite alphabet, $I \in \mathbb{S}^Q$ a row vector assigning an initial weight to each state, $T \in \mathbb{S}\langle\text{Test}\rangle^Q$ a column vector assigning to each state a polynomial over tests, and $M \in (\mathbb{S}\langle\text{Test}\rangle\langle\text{Move}\rangle)^{Q \times Q}$ the transition matrix.

We explain first the semantics of a pebWA on the automaton $\mathcal{A}_1$ represented in Figure 2 with its matrix representation on the right.

Intuitively, we enter state 1 with weight 5. We can loop on state 1 if the current letter is either an a or a b, in which case we move right in the word. The weight of this loop is 2 or 3 depending on the current letter. If $\mathcal{A}_1$ reads letter c while being in state 1, then it drops pebble x and restarts at the beginning of the word in state 2. There, it moves right in the word, either staying in state 2 with weight 1 (provided the current position does not carry the pebble), or going to state 3 with weight 7. Once we reach state 3, we must lift the pebble and go to state 4. Then, we move right coming back to state 1.

An accepting run of $\mathcal{A}_1$ must start in state 1 and end in state 1. The weight of a run is the product of the weights of its transitions. Over the natural semiring $(\mathbb{N}^\infty, +, \times, 0, 1)$, each accepting run of $\mathcal{A}_1$ has weight $5 \times 2^{|u|_a} \times 3^{|u|_b} \times 7^{|u|_c}$. The non-deterministic choice in state 2 induces several runs. The semantics of the automaton is as usual the sum of the weights of all accepting runs. In our example,

$$[\![\mathcal{A}_1]\!](u) = 5 \times 2^{|u|_a} \times 3^{|u|_b} \times 7^{|u|_c} \times \prod_{\substack{i \in \text{pos}(u) \\ u_i = c}} (i+1).$$

Consider also the 2-way automaton $\mathcal{A}_2$ over the semiring $(\mathbb{R}_{\geq 0}^\infty, +, \times, 0, 1)$, with $0 < s < 1$. The matrix M of $\mathcal{A}_2$ admits as unique coefficient the polynomial $s \neg \triangleleft? \rightarrow + (1-s)(\neg \triangleright? \wedge \neg \triangleleft?) \leftarrow$, which, for clarity, we preferred to draw with

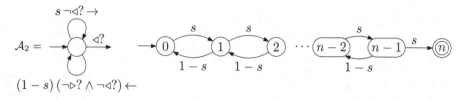

Fig. 3. Markov Chain obtained by synchronizing $\mathcal{A}_2$ with a word of length n

two loops in Figure 3. This is a compact and elegant way of representing a Markov chain describing a random walk, see Figure 3. The same example was described with a pebWE in Section 4.

As for expressions, we allow macros in M and T: for $a \in A$, we use $a \stackrel{\text{def}}{=} a? \cdot \rightarrow$ and $\bar{a} \stackrel{\text{def}}{=} a? \cdot \leftarrow$, for $d \in$ Move, we write d instead of $\mathtt{tt}? \cdot d$. For instance, the label of the loop on state 1 of $\mathcal{A}_1$ could be written $2a + 3b$.

For each $p, q \in Q$ and $d \in$ Move, we denote by $M_{p,q}^d \in \mathbb{S}\langle\text{Test}\rangle$ the coefficient of move d in $M_{p,q}$. For instance, $M_{1,2}^{\downarrow_x} = c?$ in $\mathcal{A}_1$. We collect these coefficients in matrices $M^d = (M_{p,q}^d) \in (\mathbb{S}\langle\text{Test}\rangle)^{Q \times Q}$.

We turn now to the formal definition of the semantics of pebWA. A *configuration* of $\mathcal{A}$ is a tuple (u, σ, q, i, π) with $u \in A^+$ a word, $\sigma\colon \text{Peb} \to \text{pos}(u)$ a valuation, $q \in Q$ the current state, $i \in \text{pos}(u)$ the current position, and $\pi \in (\text{Peb} \times \text{pos}(u))^*$ the stack of pebbles currently dropped. Since pebbles are reusable, the stack of pebbles may contain several occurrences of the same pebble dropped on different positions. In this case, only the last occurrence of each pebble is still visible for the automaton, older occurrences being hidden. This mechanism mimics the ability in pebWE to reuse the same pebble x in nested expressions $x!E$. We extract the visible pebbles from the stack π of dropped pebbles and the underlying valuation σ, hence defining a valuation σ_π by induction over π by $\sigma_\varepsilon = \sigma$ and $\sigma_{\pi(x,i)} = \sigma_\pi[x \mapsto i]$.

We define the semantics of pebWA in terms of a weighted transition system $\text{TS}(\mathcal{A})$ whose locations are the configurations of the automaton. The weight of $(u, \sigma, p, i, \pi) \rightsquigarrow (u, \sigma, q, j, \pi')$ is defined by

$$[\![M_{p,q}^{\rightarrow}]\!](u, \sigma_\pi, i, i) \quad \text{if } j = i+1 \text{ and } \pi' = \pi \tag{S1}$$

$$[\![M_{p,q}^{\leftarrow}]\!](u, \sigma_\pi, i, i) \quad \text{if } j = i-1 \text{ and } \pi' = \pi \tag{S2}$$

$$[\![M_{p,q}^{\downarrow_x}]\!](u, \sigma_\pi, i, i) \quad \text{if } j = 0, i < |u| \text{ and } \pi' = \pi(x, i) \tag{S3}$$

$$[\![M_{p,q}^{\uparrow}]\!](u, \sigma_\pi, i, i) \quad \text{if } \pi = \pi'(y, j) \text{ for some } y \in \text{Peb} \tag{S4}$$

where $[\![M_{p,q}^d]\!]$ is the semantics of $M_{p,q}^d \in \mathbb{S}\langle\text{Test}\rangle$, seen as a pebWE. Note from (S3) that a pebble cannot be dropped on position $|u|$ in agreement with the convention adopted for weighted expressions.

The set of transitions of $\mathrm{TS}(\mathcal{A})$ consists of those $(u, \sigma, p, i, \pi) \rightsquigarrow (u, \sigma, q, j, \pi')$ with a non-zero weight: hence $\mathrm{TS}(\mathcal{A})$ is a disjoint union of transition systems depending on the pair (u, σ) considered. A run of $\mathcal{A}$ is a path ρ in $\mathrm{TS}(\mathcal{A})$. Its weight is the product of the weights of its transitions from left to right.

Given a marked word $(u, \sigma, i, j) \in \mathrm{Mk}(A^+)$ and two states $p, q \in Q$, we define $[\![\mathcal{A}_{p,q}]\!](u, \sigma, i, j) = \sum_\rho \mathrm{weight}(\rho)$ where the sum ranges over all runs ρ from configuration $(u, \sigma, p, i, \varepsilon)$ to configuration $(u, \sigma, q, j, \varepsilon)$. This sum could be infinite, but is well defined since the semiring is complete. The semantics of $\mathcal{A}$ also use the initial and terminal weights:

$$[\![\mathcal{A}]\!](u, \sigma, i, j) = \sum_{p,q \in Q} I_p \times [\![\mathcal{A}_{p,q}]\!](u, \sigma, i, j) \times [\![T_q]\!](u, \sigma, j, j).$$

When reading the whole word, we simply write $[\![\mathcal{A}]\!](u, \sigma)$ for $[\![\mathcal{A}]\!](u, \sigma, 0, |u|)$. Note that we can compute the set of free pebbles of an automaton, i.e., the set of pebbles x that may be tested with x? before being dropped with $\downarrow_x$. If the automaton has no free pebble, then the underlying valuation σ is not necessary and we simply write $[\![\mathcal{A}]\!](u)$ for the semantics.

Layered automata. As observed in automaton $\mathcal{A}_1$, it is handy, if possible, to visualize a pebWA in terms of layers, where each layer contains subruns where no pebble is dropped or lifted. We will require in the following that there are a finite number of such layers: intuitively, this means that the depth of the current stack of pebbles is bounded by a fixed parameter K. Remark however that the stack may contain several occurrences of the same pebble. Also, due to the 2-way mechanism, runs may still be of unbounded size. More formally, we assume given a function $\ell: Q \to \{0, \ldots, K\}$ mapping each state to its layer. The top layer is K so $\ell(q)$ is the number of pebbles that can still be dropped on top of the stack. We want to start and end the computation at the top layer so we suppose that for all $q \in Q$, if $I_q \neq 0$ or $T_q \neq 0$ then $\ell(q) = K$. To maintain syntactically the condition along every possible run, we also suppose for all $p, q \in Q$ that if $M_{p,q}^\leftarrow \neq 0$ or $M_{p,q}^\rightarrow \neq 0$ then $\ell(q) = \ell(p)$; if $M_{p,q}^\uparrow \neq 0$ then $\ell(q) = \ell(p) + 1$; and for all $x \in \mathrm{Peb}$, if $M_{p,q}^{\downarrow_x} \neq 0$ then $\ell(q) = \ell(p) - 1$. An automaton $\mathcal{A}$ verifying these conditions will be called K-*layered* in the following. If we order states by decreasing layers, a 2-layered automaton $\mathcal{A} = (Q, A, I, M, T)$ is thus of the form

$$I = \left(\begin{array}{|c|c|c|} \hline I^{(2)} & 0 & 0 \\ \hline \end{array} \right), \quad M = \left(\begin{array}{|c|c|c|} \hline N^{(2)} & D^{(2)} & 0 \\ \hline L^{(1)} & N^{(1)} & D^{(1)} \\ \hline 0 & L^{(0)} & N^{(0)} \\ \hline \end{array} \right), \quad T = \left(\begin{array}{|c|} \hline T^{(2)} \\ \hline 0 \\ \hline 0 \\ \hline \end{array} \right) \quad (1)$$

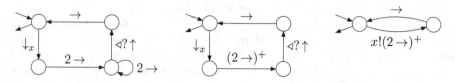

Fig. 4. A pebWA and two equivalent generalized pebWA

where entries in $N^{(i)}$ are in $\mathbb{S}\langle\text{Test}\rangle\langle\{\leftarrow, \rightarrow\}\rangle$, entries in $L^{(i)}$ are in $\mathbb{S}\langle\text{Test}\rangle\langle\{\uparrow\}\rangle$, and entries in $D^{(i)}$ are in $\mathbb{S}\langle\text{Test}\rangle\langle\{\downarrow_x \mid x \in \text{Peb}\}\rangle$. The entries of $I^{(2)}$ and $T^{(2)}$ are as usual in $\mathbb{S}$ and $\mathbb{S}\langle\text{Test}\rangle$ respectively.

6 From Automata to Expressions

In this section, we prove that every K-layered pebble weighted automaton admits an equivalent pebble weighted expression. We first show that, for 2-way weighted automata (or 0-layered pebWA), we can use the classical constructions, e.g., the state elimination method of Brzozowski and McCluskey [11], the procedure of McNaughton and Yamada [24] and the recursive algorithm [14]. We refer to the survey of Sakarovitch [29, Section 6.2] where these methods are presented and compared for 1-way weighted automata.

In the state elimination method, states are progressively suppressed and transitions are labeled with (weighted) rational expressions. To deal with pebbles, we will also eliminate the lower layers and subsume their computations with expressions of the form $x!E$. Therefore, it is convenient to consider automata allowing pebWE in the labels of transitions.

We first introduce these *generalized* pebWA. Then, we show how to compute pebWE equivalent to the behaviors of 0-layered generalized pebWA. Finally, we explain how to deal with drop and lift moves of K-layered automata.

6.1 Generalized Pebble Automata

We start with an example presented in Figure 4. The loop of the left automaton gives rise to the iteration $(2\rightarrow)^+$ on the middle automaton. Moreover, the drop/lift process has even been replaced with the $x!-$ feature of pebWE in the right automaton. This gives already the intuition of the construction of a pebWE equivalent to a pebWA. Note that the right automaton has a single layer whereas the left and middle ones have 2 layers.

Formally, a *generalized* pebWA (GpebWA) is a tuple $\mathcal{A} = (Q, A, I, M, T)$ with $I \in \mathbb{S}^Q$, $T \in \mathbb{S}\langle\text{Test}\rangle^Q$ and $M \in (\text{pebWE} + \mathbb{S}\langle\text{Test}\rangle\langle\text{Move} \setminus \{\leftarrow, \rightarrow\}\rangle)^{Q \times Q}$. Intuitively, the entries $M_{p,q}^\leftarrow \cdot \leftarrow + M_{p,q}^\rightarrow \cdot \rightarrow$ have been extended to arbitrary pebWE $M_{p,q}^{\text{pebWE}}$. The semantics of pebWA is easily extended to GpebWA. In fact, we only have to replace (S1-S2) by (G1-2) below:

$$[\![M_{p,q}^{\text{pebWE}}]\!](u, \sigma_\pi, i, j) \quad \text{if } \pi' = \pi \tag{G1-2}$$

The definition of K-layered automata can easily be extended to GpebWA. Layered automata are still of the form given in (1), the only difference being that the entries of matrices $N^{(i)}$ are now pebWE instead of simple polynomials in $\mathbb{S}\langle \text{Test}\rangle\langle\{\leftarrow, \rightarrow\}\rangle$. It is clear from the definition that every (K-layered) pebWA can be seen as a (K-layered) GpebWA.

6.2   Automata to Expressions: 0-layered Generalized pebWA

We deal in this section with GpebWA $\mathcal{A} = (Q, A, I, M, T)$ with no drop or lift transitions, i.e., 0-layered GpebWA where the entries of the transition matrix M are all pebWE.

Theorem 2. *Let $\mathcal{A} = (Q, A, I, M, T)$ be a 0-layered GpebWA. We can construct a matrix $\Phi(M) \in \text{pebWE}^{Q \times Q}$ which is equivalent to the automaton: for all $p, q \in Q$, we have $[\![\Phi(M)_{p,q}]\!] = [\![\mathcal{A}_{p,q}]\!]$, i.e., for all $(u, \sigma, i, j) \in \text{Mk}(A^+)$*

$$[\![\Phi(M)_{p,q}]\!](u, \sigma, i, j) = [\![\mathcal{A}_{p,q}]\!](u, \sigma, i, j).$$

Moreover, the entries of $\Phi(M)$ are in the rational closure[5] of the entries of M.

The matrix $\Phi(M)$ can be constructed from M using one of the classical algorithm, e.g., the recursive algorithm or McNaughton-Yamada algorithm. We can also apply the state elimination method or the system resolution method starting from any initial state p and final state q.

We quickly justify below the correctness of the construction based on the partial monoid structure of marked words introduced in Section 4.1. In a first reading, one may go directly to the next subsection.

Recall that $\mathbb{K} = \mathbb{S}\langle\!\langle \text{Mk}(A^+)\rangle\!\rangle$ is a continuous semiring by Proposition 1. For each finite set Q, the semiring of matrices $\mathbb{K}^{Q \times Q}$ is also continuous and, given a matrix H in $\mathbb{K}^{Q \times Q}$, the entries of the matrix $H^* = \sum_{n \geq 0} H^n \in \mathbb{K}^{Q \times Q}$ are in the rational closure of the entries of H. Moreover, H^* can be computed inductively: if $H = \begin{pmatrix} A & B \\ C & D \end{pmatrix}$ is a block decomposition, with A and D square matrices, then ([14])

$$H^* = \begin{pmatrix} (A + BD^*C)^* & A^*B(D + CA^*B)^* \\ D^*C(A + BD^*C)^* & (D + CA^*B)^* \end{pmatrix}. \tag{2}$$

We apply the above to $H = [\![M]\!] = ([\![M_{p,q}]\!])_{p,q \in Q} \in \mathbb{K}^{Q \times Q}$. As usual, the matrix M^n describes the paths of length n of the automaton $\mathcal{A}$ and the matrix $H^n = [\![M]\!]^n$ gives the semantics restricted to paths of length n. Summing over all paths, we obtain the full semantics:

$$[\![\mathcal{A}_{p,q}]\!] = \sum_{n \geq 0} ([\![M]\!]^n)_{p,q} = ([\![M]\!]^*)_{p,q}$$

Using recursively (2) on the matrix M, we obtain a matrix $\Phi(M)$ whose entries are in the rational closure of the entries of M and such that $[\![\Phi(M)]\!] = [\![M]\!]^*$.

---

[5] The rational closure is the closure under sum (+) concatenation ($\cdot$) and star (*).

6.3   Automata to Expressions: Generalized pebWA

We will now extend Theorem 2 to any K-layered GpebWA $\mathcal{A} = (Q, A, I, M, T)$. For $i \leq K$, we let $Q^{(i)} = \ell^{-1}(i)$ be the set of states in layer i.

Proposition 3. *Let $\mathcal{A} = (Q, A, I, M, T)$ be a 1-layered GpebWA. We can construct a 0-layered GpebWA $\mathcal{A}^{(1)} = (Q^{(1)}, A, I^{(1)}, M^{(1)}, T^{(1)})$ which is equivalent to $\mathcal{A}$: $[\![\mathcal{A}_{p,q}]\!] = [\![\mathcal{A}^{(1)}_{p,q}]\!]$ for all $p, q \in Q^{(1)}$.*

We use the layered decomposition given in Section 5 (1). To simplify the notation, we write $N = N^{(1)}$, $D = D^{(1)}$, $L = L^{(0)}$ and $P = N^{(0)}$ so that

$$
M = \begin{pmatrix} N & D \\ \hline L & P \end{pmatrix} = \begin{pmatrix} N & 0 \\ \hline 0 & 0 \end{pmatrix} + \begin{pmatrix} 0 & D \\ \hline L & P \end{pmatrix}.
$$

Let $p, q \in Q^{(1)}$ be in layer 1 and $p', q' \in Q^{(0)}$ be in layer 0. Then, D is a *drop*-matrix whose (p, p')-entry can be written $\sum_{x \in \text{Peb}} d^x_{p,p'} \cdot \downarrow_x$ with $d^x_{p,p'} \in \mathbb{S}\langle\text{Test}\rangle$. The (q', q)-entry of the *lift*-matrix L can be written $e_{q',q} \cdot \uparrow$ with $e_{q',q} \in \mathbb{S}\langle\text{Test}\rangle$. Now, P is a $Q^{(0)} \times Q^{(0)}$ matrix of pebWE and we may apply Theorem 2 in order to get a matrix $\Phi(P)$ of pebWE which is equivalent to the iteration of P: $[\![\Phi(P)]\!] = [\![P]\!]^*$. From (D, P, L), we define the $Q^{(1)} \times Q^{(1)}$ pebWE-matrix G by

$$
G_{p,q} = \sum_{p',q'} \sum_{x \in \text{Peb}} d^x_{p,p'} \cdot x! \big(\Phi(P)_{p',q'} \cdot e_{q',q} \cdot \to^*\big).
$$

The matrix G is also denoted $C(D, P, L)$ below. Note that the maximal pebble-depth of the entries of G is at most 1 plus the maximal pebble-depth of the entries of P since the construction $\Phi(P)$ does not increase the pebble-depth.

Lemma 4. *For all $p, q \in Q^{(1)}$ and all $(u, \sigma, i, j) \in \text{Mk}(A^+)$, we have*

$$
[\![G_{p,q}]\!](u, \sigma, i, j) = \sum_\rho \text{weight}(\rho)
$$

where the sum ranges over all runs ρ on (u, σ) from configuration (p, i, ε) to configuration (q, j, ε) and using only intermediate states in layer 0.

To conclude the proof of Proposition 3, we simply set

$$
M^{(1)} = N + G = N + C(D, P, L)
$$

and we can check using Lemma 4 that $[\![\mathcal{A}_{p,q}]\!] = [\![\mathcal{A}^{(1)}_{p,q}]\!]$ for all $p, q \in Q^{(1)}$.

Proposition 5. *Let $\mathcal{A} = (Q, A, I, M, T)$ be a K-layered GpebWA. We can construct a 0-layered GpebWA $\mathcal{A}^{(K)} = (Q^{(K)}, A, I^{(K)}, M^{(K)}, T^{(K)})$ which is equivalent to $\mathcal{A}$: $[\![\mathcal{A}_{p,q}]\!] = [\![\mathcal{A}^{(K)}_{p,q}]\!]$ for all $p, q \in Q^{(K)}$.*

We use again the notation of the layered decomposition. The proof is by induction on K. When $K = 0$ we simply have $\mathcal{A}^{(0)} = \mathcal{A}$, i.e., $M^{(0)} = N^{(0)}$. For $K > 0$, we set $M^{(K)} = N^{(K)} + C(D^{(K)}, M^{(K-1)}, L^{(K-1)})$ where the matrix $M^{(K-1)}$ is obtained by induction. The correctness follows from Proposition 3. From Theorem 2 and Proposition 5 we deduce:

Theorem 6. *Let $\mathcal{A} = (Q, A, I, M, T)$ be a K-layered GpebWA. The matrix $H = \Phi(M^{(K)})$ of pebWE satisfies $[\![H_{p,q}]\!] = [\![\mathcal{A}_{p,q}]\!]$ for all $p, q \in Q^{(K)}$. Therefore, the pebWE $E(\mathcal{A}) = I \times H \times T$ is equivalent to $\mathcal{A}$: $[\![E(\mathcal{A})]\!] = [\![\mathcal{A}]\!]$. Moreover, the pebble-depth of $E(\mathcal{A})$ is at most K if $\mathcal{A}$ is a K-layered pebWA.*

7 From Expressions to Automata

We describe in this section how to transform a weighted expression with pebbles to an equivalent weighted automaton with pebbles. Expressions are very convenient to denote in a rather clear and intuitive way the quantitative functions that we want to compute. On the other hand, automata are much more amenable to efficient algorithms, e.g., for evaluation as shown in Section 8. Hence, we need efficient translations from expressions to automata. Such translations have been well-studied both in the boolean and in the weighted (1-way) cases. Glushkov's translation (or Berry-Sethi) is acknowledged to be among the best ones. The good news is that this construction can be adapted to cope with 2-way moves and pebbles as we will show in this section. The construction is by structural induction on the expression.

Theorem 7. *For each pebWE E we can construct a layered pebWA $\mathcal{A}(E)$ such that $[\![\mathcal{A}(E)]\!] = [\![E]\!]$, i.e., for all $(u, \sigma, i, j) \in \mathrm{Mk}(A^+)$ we have*

$$[\![\mathcal{A}(E)]\!](u, \sigma, i, j) = [\![E]\!](u, \sigma, i, j).$$

Moreover, the number of layers in $\mathcal{A}(E)$ is the pebble-depth of E.

We define the *literal-length* $\ell\ell(E)$ of an expression as the number of occurrences of moves ($\leftarrow$ or $\rightarrow$) plus twice the number of occurrences of ! (in $x!-$). We will see that the number of states of $\mathcal{A}(E)$ will be $1 + \ell\ell(E)$. For a 2-way expression E of pebble-depth 0 (2-way-WE) the literal-length is simply the number of moves, which are the positions to be marked for Glushkov's construction.

For the rational operations ($+$, $\cdot$, *, and $^+$), we can still use the classical constructions even though we are working with pebWA. We recall these constructions below for the sake of completeness. The main novelty is indeed the treatment of pebbles.

We adopt the presentation of standard automata by Sakarovitch [29]. A standard automaton $\mathcal{A} = (Q, A, I, M, T)$ has a single initial state ι with (initial) weight 1, all other states have initial weight 0. Moreover, the initial state ι has no ingoing transition. We use both the graphical representation and the matrix representation of an automaton:

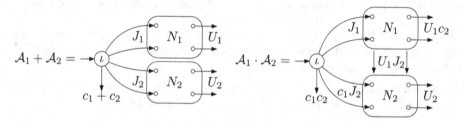

Since terminal weights allow polynomials over Test with the mapping $T\colon Q \to \mathbb{S}\langle\text{Test}\rangle$, we will be able to cope with expressions of the form $E \cdot \varphi?$ and $E \cdot s$ without adding unnecessary states. For $s \in \mathbb{S}$ and $\varphi \in \text{Test}$, we simply write s for $stt?$ and φ for 1φ, and also $\to$ for $1tt?\to$ and $\leftarrow$ for $1tt?\leftarrow$.

We start with atoms. Compared to the classical (1-way) translation, a slight difference is that we are using tests (φ) and moves ($\leftarrow,\to$) instead of letters ($a = a?\to$) for the atoms. The automata for the atoms are defined as

$$\mathcal{A}(s) = \longrightarrow \iota \xrightarrow{\ s\ } \qquad\qquad \mathcal{A}(\to) = \longrightarrow \iota \xrightarrow{\ \to\ } \bigcirc \xrightarrow{\ 1\ }$$

$$\mathcal{A}(\varphi) = \longrightarrow \iota \xrightarrow{\ \varphi\ } \qquad\qquad \mathcal{A}(\leftarrow) = \longrightarrow \iota \xrightarrow{\ \leftarrow\ } \bigcirc \xrightarrow{\ 1\ }$$

and we can easily see that they are equivalent to the corresponding atoms: if E is an atom then $[\![E]\!](u,\sigma,i,j) = [\![\mathcal{A}(E)]\!](u,\sigma,i,j)$ for all $(u,\sigma,i,j) \in \text{Mk}(A^+)$.

The constructions for sum and concatenation are as usual.

In the concatenation, we are overloading the product notation as follows. The product of two monomials $s_1\varphi_1$ and $s_2\varphi_2$ from $\mathbb{S}\langle\text{Test}\rangle$ should be understood as $(s_1 s_2)(\varphi_1 \wedge \varphi_2)$ to stay in $\mathbb{S}\langle\text{Test}\rangle$. Hence $c_1 c_2$ and the entries of $U_1 c_2$ are in $\mathbb{S}\langle\text{Test}\rangle$. Similarly, in $U_1 J_2$, the product of a monomial $s_1\varphi_1 \in \mathbb{S}\langle\text{Test}\rangle$ and a monomial $s_2\varphi_2 d$ (with $d \in \text{Move}$) is defined as $(s_1 s_2)(\varphi_1 \wedge \varphi_2)d$. Hence, the entries of the matrices $c_1 J_2$ and $U_1 J_2$ are in $\mathbb{S}\langle\text{Test}\rangle\langle\text{Move}\rangle$. The matrix representation is therefore:

$$\mathcal{A}_1 \cdot \mathcal{A}_2 = \begin{pmatrix} 1 & \boxed{0} & \boxed{0} \end{pmatrix} \begin{pmatrix} 0 & J_1 & c_1 J_2 \\ 0 & N_1 & U_1 J_2 \\ 0 & 0 & N_2 \end{pmatrix} \begin{pmatrix} c_1 c_2 \\ U_1 c_2 \\ U_2 \end{pmatrix}$$

For instance, the automaton for $2a = 2 \cdot a? \cdot \to$ is computed as follows:

$$\longrightarrow \iota \downarrow^{2} \quad \cdot \quad \longrightarrow \iota \downarrow^{a?} \quad \cdot \quad \longrightarrow \iota \xrightarrow{\to} \bigcirc \downarrow^{1} \quad = \quad \longrightarrow \iota \xrightarrow{\ 2a\ } \bigcirc \downarrow^{1}$$

Similarly, for the expression $E = (2a? + b?)\rightarrow(2b? + 3c?)$ we compute the concatenation of 3 automata as follows:

Finally, the star is also computed as usual with the following construction.

$$A^* = \begin{pmatrix} 1 & \boxed{} & 0 \end{pmatrix} \begin{pmatrix} 0 & \boxed{c^*J} \\ 0 & \boxed{N + Uc^*J} \end{pmatrix} \begin{pmatrix} \boxed{c^*} \\ \boxed{Uc^*} \end{pmatrix}$$

Notice that $c^* \in \mathbb{S}$ is well-defined since the semiring is complete. As for the concatenation, we can check that the entries of Uc^* are in $\mathbb{S}\langle\text{Test}\rangle$ and the entries of Uc^*J are in $\mathbb{S}\langle\text{Test}\rangle\langle\text{Move}\rangle$. The strict iteration A^+ is computed similarly by simply changing the final weight of ι to c^+ (note that $0^+ = 0$), but keeping the other occurrences of c^* in c^*J, Uc^*J and Uc^*.

For instance, for expression $E = \rightarrow^+ a?\leftarrow^+ b?\rightarrow^+ c?\leftarrow^+ d?\rightarrow^+$ introduced in Section 4, we can compute the automaton as follows:

Finally, we give the construction for $x!E$ which should drop the pebble on the current position, evaluate E from beginning to end ($\lhd?$) of the word and finally lift the pebble. From a standard automaton A equivalent to E, we construct the following standard automaton $x!A$:

$$x!A = \begin{pmatrix} 1 & 0 & 0 & \boxed{} & 0 \end{pmatrix} \begin{pmatrix} 0 & 0 & \downarrow_x & 0 \\ 0 & 0 & 0 & 0 \\ 0 & 0 & 0 & J \\ 0 & U\lhd?\uparrow & 0 & N \end{pmatrix} \begin{pmatrix} 0 \\ 1 \\ 0 \\ 0 \end{pmatrix}$$

The correctness of this construction follows easily from Proposition 3. Assume for simplicity that $\mathcal{A}$ is a 0-layered automaton, then $x!\mathcal{A}$ is a 1-layered automaton and it's layer decomposition is shown both in the graphical and matricial representations above. Let $Q = \{\iota, \tau, \iota'\} \uplus Q'$ be the set of states of $x!\mathcal{A}$ where Q' are the non-initial states of $\mathcal{A}$. Using the notation of Section 6.3, the drop-matrix D of $x!\mathcal{A}$ contains a single non-zero entry which is $D_{\iota,\iota'} = \downarrow_x$, the non-zero entries of the lift-matrix L are in the $Q' \times \tau$ column $U_\triangleleft?\uparrow$, and F is the transition matrix of $\mathcal{A}$. Therefore, the matrix $G = C(D, F, L)$ has a single non-zero entry which is

$$G_{\iota,\tau} = \sum_{q' \in Q'} x!\big((\Phi(F)_{\iota',q'} U_{q'})\triangleleft?\to^*\big) \equiv x!\Big(\sum_{q' \in Q'} \Phi(F)_{\iota',q'} U_{q'}\Big) \equiv x!E\,.$$

As last example, consider again expression E below used in Section 4:

$$E = \to^+ a?\, x!\Big((\neg x?\to)^* b?\, (\neg x?\to)^+ c? \leftarrow^+ d? \to^+\Big) \to^*\,.$$

The construction applied to E gives the following pebWA.

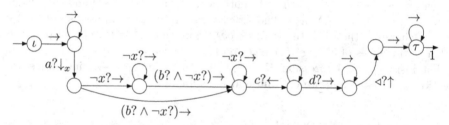

To close this section, we briefly discuss the complexity of our translation. Clearly, the number of states of the automaton $\mathcal{A}(E)$ is the literal-length $\ell\ell(E)$ of expression E. The time complexity is cubic in the length of E. It should be possible to get a quadratic algorithm by generalizing the notion of star normal form introduced in [9] for word languages or the algorithm presented in [1] for classical weighted expressions and automata.

8   Evaluation of Pebble Weighted Automata

In this section, we study the evaluation problem of a K-layered pebWA $\mathcal{A}$ with reusable pebbles: given a word u and a valuation $\sigma\colon \text{Peb} \to \text{pos}(u)$, compute $[\![\mathcal{A}]\!](u,\sigma)$. The challenge is important since, even if the word is fixed, the number of accepting runs may be infinite.

Let $\mathcal{A} = (Q, A, I, M, T)$ be a K-layered pebWA. As in Section 6.3, for $i \leq K$, we let $Q^{(i)} = \ell^{-1}(i)$ be the set of states in layer i.

Theorem 8. *Given a K-layered pebWA with p pebbles and a word $w \in A^+$, we can compute with $\mathcal{O}((K+1)|w|^{p+1})$ matrix operations (sum, product, iteration) the values $[\![\mathcal{A}_{p,q}]\!](w,\sigma)$ for all $p, q \in Q^{(K)}$ and valuations $\sigma\colon \text{Peb} \to \text{pos}(w)$.*

It is important to notice that the complexity only linearly depends on the number K of layers. The number of pebbles occurs in the exponent but since we allow reusable pebbles, this number may be much smaller than the number of layers. This is in the same vein as restricting the number of variable names, e.g., in first-order logic, without restricting the quantifier depth. Restricting the number of variable names often results in much lower complexity. For instance, the complexity of the evaluation (model-checking) problem of first-order logic over relational structures drops from PSPACE to PTIME when the number of variable names is bounded [32,33].

We have seen in Section 2.2 that weighted LTL formulas can be described with pebWE using two pebbles x and y. Actually, the same constructions are valid if we reuse pebble x instead of y. For instance, *until* may be described with

$$E_{\varphi \mathsf{U} \psi}(x) = \triangleright? {\to}^* x? \big((x!(E_{\neg \psi}(x) {\leftarrow}^* E_\varphi(x))) {\to} \big)^* (x! E_\psi(x)) {\to}^* \triangleleft? \,.$$

Therefore, any weighted LTL formula φ may be described with a pebWE E_φ using a single pebble x. The pebble-depth of E_φ being the nesting depth of modalities in φ. Using Theorem 7 we obtain a layered pebWA $\mathcal{A}_\varphi$ equivalent to E_φ. The number K of layers in $\mathcal{A}_\varphi$ is the pebble-depth of E_φ, i.e., the nesting depth of φ. Moreover, $\mathcal{A}_\varphi$ uses only one pebble. Theorem 8 yields an evaluation algorithms using $\mathcal{O}((K+1)|w|^2)$ matrix operations. We see below that there is an algorithm which is also linear in $|w|$.

We say that a K-layered pebWA $\mathcal{A} = (Q, A, I, M, T)$ is *strongly K-layered* if in each layer only a fixed pebble may be dropped: for all $i \leq K$, there is a pebble $x_i \in \mathrm{Peb}$ such that for all $q, q' \in Q$ and $x \in \mathrm{Peb}$, if $\ell(q) = i$ and $x \neq x_i$ then $M_{q,q'}^{\downarrow x} = 0$.

Theorem 9. *Given a strongly K-layered pebWA with p pebbles and a word $w \in A^+$, we can compute with $\mathcal{O}((K+1)|w|^{\max(1,p)})$ matrix operations (sum, product, iteration) the values $[\![\mathcal{A}_{p,q}]\!](w, \sigma)$ for all states $p, q \in Q^{(K)}$ and valuations $\sigma \colon \mathrm{Peb} \to \mathrm{pos}(w)$.*

Notice that if $p \leq 1$ then any K-layered pebWA is *strongly K-layered*. In this case, we get an evaluation algorithm using $\mathcal{O}((K+1)|w|)$ matrix operations. This is in particular the case for pebWA arising from weighted LTL formulas.

9   Discussion

To conclude, let us briefly mention some interesting topics that could be studied in the future. As already stated in Section 7, one should try to obtain a quadratic algorithm for the translation of pebWE to pebWA. Next, as in Section 8 for the evaluation problem, one should develop efficient algorithms for quantitative model-checking, emptiness, containment, etc.

We have no restriction over the syntax of expressions or automata. In particular, 2-way moves may give rise to unbounded loops which is why we considered continuous semirings. We believe that continuous semirings are suitable for most applications. But in case one needs to work without this hypothesis, it is possible to put restrictions on the syntax of expressions and automata in order to rule out unbounded loops and have a well-defined semantics in arbitrary semirings. For instance, one may restrict iterations to forward proper or backward proper expressions.

The correctness of our translations between pebWE and pebWA relies on the partial monoid structure of marked words, which does not use concatenation of words. We can also endow marked trees with such a partial monoid structure. Therefore, pebWE can be extended to trees with a semantics in the continuous semiring of series over marked trees. We obtain in this way a weighted extension of caterpillar expressions or Regular XPath. Similarly, one may define *tree-walking* pebWA. We believe that the translations presented in this paper also apply to pebWE over trees and tree-walking pebWA.

A more prospective problem is to replace the $x!-$ construction of pebWE with a *chop* product $E;F$ which evaluates E on the current prefix and F on the current suffix. We can easily simulate this relativization mechanism using a pebble to mark the current position. The converse is an interesting problem which needs to be investigated: is it possible to simulate pebbles with chop products?

Acknowledgements. The authors would like to thank Benedikt Bollig and Jacques Sakarovitch for helpful discussions.

References

1. Allauzen, C., Mohri, M.: A Unified Construction of the Glushkov, Follow, and Antimirov Automata. In: Královič, R., Urzyczyn, P. (eds.) MFCS 2006. LNCS, vol. 4162, pp. 110–121. Springer, Heidelberg (2006)
2. Berry, G., Sethi, R.: From regular expressions to deterministic automata. Theoretical Computer Science 48, 117–126 (1986)
3. Berstel, J., Reutenauer, C.: Noncommutative rational series with applications, Cambridge. Encyclopedia of Mathematics & Its Applications, vol. 137 (2011)
4. Birget, J.-C.: State-complexity of finite-state devices, state compressibility and incompressibility. Theory of Computing Systems 26, 237–269 (1993)
5. Bojańczyk, M.: Tree-Walking Automata. In: Martín-Vide, C., Otto, F., Fernau, H. (eds.) LATA 2008. LNCS, vol. 5196, pp. 1–2. Springer, Heidelberg (2008)
6. Bojańczyk, M., Samuelides, M., Schwentick, T., Segoufin, L.: Expressive Power of Pebble Automata. In: Bugliesi, M., Preneel, B., Sassone, V., Wegener, I. (eds.) ICALP 2006. LNCS, vol. 4051, pp. 157–168. Springer, Heidelberg (2006)
7. Bollig, B., Gastin, P.: Weighted versus Probabilistic Logics. In: Diekert, V., Nowotka, D. (eds.) DLT 2009. LNCS, vol. 5583, pp. 18–38. Springer, Heidelberg (2009)
8. Bollig, B., Gastin, P., Monmege, B., Zeitoun, M.: Pebble Weighted Automata and Transitive Closure Logics. In: Abramsky, S., Gavoille, C., Kirchner, C., Meyer auf der Heide, F., Spirakis, P.G. (eds.) ICALP 2010. LNCS, vol. 6199, pp. 587–598. Springer, Heidelberg (2010)
9. Brüggeman-Klein, A.: Regular expressions into finite automata. Theoretical Computer Science 120, 197–213 (1993)

10. Brüggeman-Klein, A., Wood, D.: Caterpillars: A context specification technique. Markup Languages 2(1), 81–106 (2000)
11. Brzozowski, J.A., McCluskey, E.J.: Signal flow graph techniques for sequential circuit state diagrams. IEEE Trans. on Electronic Computers 12(9), 67–76 (1963)
12. Buchholz, P., Kemper, P.: Model checking for a class of weighted automata. Discrete Event Dynamic Systems 20(1), 103–137 (2009)
13. Ciesinski, F., Größer, M.: On Probabilistic Computation Tree Logic. In: Baier, C., Haverkort, B.R., Hermanns, H., Katoen, J.-P., Siegle, M. (eds.) Validation of Stochastic Systems. LNCS, vol. 2925, pp. 147–188. Springer, Heidelberg (2004)
14. Conway, J.: Regular Algebra and Finite Machines. Chapman & Hall (1971)
15. Droste, M., Kuich, W.: Semirings and formal power series. In: Handbook of Weighted Automata [16], ch. 1, pp. 3–27
16. Droste, M., Kuich, W., Vogler, H.: Handbook of Weighted Automata. EATCS Monographs in Theoretical Computer Science. Springer (2009)
17. Engelfriet, J., Hoogeboom, H.J.: Tree-walking pebble automata. In: Jewels are Forever, pp. 72–83. Springer (1999)
18. Ésik, Z., Kuich, W.: Modern Automata Theory. Electronic book (2007), http://dmg.tuwien.ac.at/kuich
19. Globerman, N., Harel, D.: Complexity results for two-way and multi-pebble automata and their logics. Theoretical Computer Science 169, 161–184 (1996)
20. Glushkov, V.M.: The abstract theory of automata. Russian Math. Surveys 16, 1–53 (1961)
21. Knight, K., May, J.: Applications of weighted automata in natural language processing. In: Handbook of Weighted Automata [16], ch. 14, pp. 555–579
22. Kuske, D.: Schützenberger's theorem on formal power series follows from kleene's theorem. Theoretical Computer Science 401(1-3), 243–248 (2008)
23. Mandrali, E.: Weighted LTL with Discounting. In: Moreira, N., Reis, R. (eds.) CIAA 2012. LNCS, vol. 7381, pp. 353–360. Springer, Heidelberg (2012)
24. McNaughton, R., Yamada, H.: Regular expressions and state graphs for automata. IRE Trans. on Electronic Computers 9(1), 39–47 (1960)
25. Meinecke, I.: A Weighted μ-Calculus on Words. In: Diekert, V., Nowotka, D. (eds.) DLT 2009. LNCS, vol. 5583, pp. 384–395. Springer, Heidelberg (2009)
26. Ravikumar, B.: On some variations of two-way probabilistic finite automata models. Theoretical Computer Science 376(1-2), 127–136 (2007)
27. Sakarovitch, J.: Elements of Automata Theory. Cambridge University Press (2009)
28. Sakarovitch, J.: Rational and recognisable power series. In: Handbook of Weighted Automata [16], ch. 4, pp. 103–172
29. Sakarovitch, J.: Automata and expressions. In: AutoMathA Handbook (to appear, 2012)
30. Samuelides, M., Segoufin, L.: Complexity of Pebble Tree-Walking Automata. In: Csuhaj-Varjú, E., Ésik, Z. (eds.) FCT 2007. LNCS, vol. 4639, pp. 458–469. Springer, Heidelberg (2007)
31. Schützenberger, M.-P.: On the definition of a family of automata. Information and Control 4, 245–270 (1961)
32. Vardi, M.: The complexity of relational query languages. In: Proceedings of STOC 1982, pp. 137–146. ACM Press (1982)
33. Vardi, M.: On the complexity of bounded-variable queries. In: Proceedings of PODS 1995, pp. 266–276. ACM Press (1995)

Typed Linear Algebra for Weigthed (Probabilistic) Automata

José N. Oliveira

High Assurance Software Laboratory,
INESC TEC and University of Minho,
Braga, Portugal
jno@di.uminho.pt

Abstract. There is a need for a language able to reconcile the recent upsurge of interest in quantitative methods in the software sciences with logic and set theory that have been used for so many years in capturing the qualitative aspects of the same body of knowledge. Such a *lingua franca* should be typed, polymorphic, diagrammatic, calculational and easy to blend with traditional notation.

This paper puts forward *typed linear algebra* (LA) as a candidate notation for such a role. Typed LA emerges from regarding matrices as morphisms of suitable categories whereby traditional linear algebra is equipped with a type system.

In this paper we show typed LA at work in describing weighted (probabilistic) automata. Some attention is paid to the interface between the index-free language of matrix combinators and the corresponding index-wise notation, so as to blend with traditional set theoretic notation.

Keywords: Weighted automata, linear algebra, categories of matrices.

> "*Quantitative Formal Methods deals with systems whose behaviour of interest is more than the traditional Boolean "correct" or "incorrect" judgment. (...) The aim of the workshop was to create a new forum where current and novel theories and application areas of quantitative methods could be discussed, together with the verification techniques that might apply to them.*"
>
> Andova et al. [2]

1 Introduction

There is a trend towards *quantitative methods* in computing. Further to predicting that something "may happen", going quantitative should allow one to anticipate "*how often or costly it will happen*". Or, looking from the negative side of things, if something bad can take place one wishes to know how likely is it to occur.

As happened with other sciences in the past (eg. physics), computer science is in some sense becoming *probabilistic*. However, traditional notation for probabilities is too descriptive and not meant for proving and calculating software as we understand this activity today. Quoting Hehner [14]:

N. Moreira and R. Reis (Eds.): CIAA 2012, LNCS 7381, pp. 52–65, 2012.
© Springer-Verlag Berlin Heidelberg 2012

Perhaps a thousand years ago the philosophers of the time [might give] reasons why their answer is right. Now we don't argue; we formalize, calculate, and unformalize.

There has been work on tuning probabilistic notation and reasoning to software design. McIver and Morgan [22] develop a method for rigorous reasoning about probabilistic programs that includes a calculus which, in the Hoare style, operates at the level of the program text. At programming level, Erwig and Kollmansberger [11] give a collection of modules that make up a probabilistic functional programming library in Haskell based on the (finite) distribution monad. More recently, Gibbons and Hinze [13] have shown how to perform equational reasoning about programs that exploit both nondeterministic and probabilistic choice as part of a more ambitious plan to reason about effectful computations in general.

Sokolova [26] presents a coalgebraic analysis of probabilistic systems in a way that connects two main-stream research areas: coalgebraic reasoning and probabilistic modeling and verification. This work builds upon foundational work by Larsen and Skou [15] on probabilistic bisimulation. Broadening scope, recent work by Bonchi et al. [8] gives a coalgebraic perspective on so-called *linear weighted automata*, which generalize the probabilistic ones.

Weighted Automata. Weighted automata [9, 10, 8] are a generalisation of finite state, non-deterministic automata where each state transition, in addition to some input, involves a quantity indicative of the *weight* (expressing eg. cost or probability) of its execution. The minimal structure for expressing weights is a *semiring* $(\mathbb{S}; +, \times, 0, 1)$ where $(\mathbb{S}; +, 0)$ is a commutative monoid, $(\mathbb{S}; \times, 1)$ is a monoid, multiplication distributes over addition and 0 annihilates multiplication $(0 \times s = s \times 0 = 0)$.

Following [10], a weighted finite automaton $W = (A, Q; \lambda, \mu, \gamma)$ consists of an input alphabet A, a finite set of states Q and three functions: $\lambda, \gamma : Q \to \mathbb{S}$ are weight functions for entering and leaving a state, respectively, and $\mu : A \to \mathbb{S}^{Q \times Q}$ is such that $\mu(a)(p, q)$ indicates the cost of transition $p \xrightarrow{a} q$. Cost 0 means that there is no transition from p to q labelled a.

For $\mathbb{S}$ the Boolean algebra $\mathbb{B}$ of truth values, a weighted automaton becomes a (non-deterministic) labelled transition system (LTS), or non-deterministic finite-state automaton (FSA): $\mu(a) \in \mathbb{B}^{Q \times Q}$ is the state-transition relation associated to input a, λ is the set of initial states and γ the set of terminal states. For $\mathbb{S}$ the interval $[0, 1]$ of the real numbers $(\mathbb{R})$ W can be regarded as a *probabilistic automaton* under certain conditions [1]. Bonchi et al. [8] only consider μ and the output function γ. Their coalgebraic perspective twists the type of μ into $Q \to (\mathbb{S}^Q)^A$ and then amalgamates γ and μ into a coalgebra of functor $\mathsf{F}X = \mathbb{S} \times (\mathbb{S}^X)^A$.

State Transition Matrices. For each $a \in A$, $\mu(a) \in \mathbb{S}^{Q \times Q}$ can be regarded as a Q-indexed *matrix* expressing the cost of each state transition in which input a

---

[1] For a comprehensive analysis and taxonomy of probabilistic systems see eg. [26].

participates. In the same way, λ and γ can be regarded as Q-indexed vectors. It is therefore no wonder that the work on weighted automata often resorts to matrix terminology and operations such as matrix-matrix multiplication and matrix-vector multiplication. However, *linear algebra* (LA) is seldom assumed explicitly as the *central* notation and calculus — such reasoning takes place episodically, where convenient, conventional set theory doing the main job. This means that the main advantage of LA — the conciseness of blocked, index-free notation and its powerful algebra — is (partially) lost. There are, however, approaches in which LA is the main notational device, see eg. references [9, 28] which follow the tradition of Bloom et al. [7]. But such notation is *untyped* and therefore hard to combine with that of the relations, predicates and functions which are around.

Typed versus Untyped Mathematics. What does *(un)typed* mean in the previous sentence? It is a commonplace in mathematics to regard functions as special cases of relations (the deterministic, total ones) and relations as special cases of matrices (the Boolean ones, provided addition is trimmed to 1). Yet the three classes of object are treated in disparate ways, unrelatedly and with incompatible (if not contradictory) notation.

For instance, one writes $y = f(x)$ to define a function and $(x, y) \in Graph(f)$ — note how x and y swap position — to express the input/output pairs of the graph of function f, which is a relation. As far as typing is concerned, most people accept notation $f : A \to B$ for defining the signature of a function (as we have seen above) but only reluctantly will accept the same notation $R : A \to B$ to define the *type* of relation R, writing $R \subseteq A \times B$ instead. As far as matrices are concerned, writing $M : m \to n$ to declare the type of a matrix with m columns and n rows will look surprising — textbooks simply tell that M is of order $m \times n$ (or is it $n \times m$?), with loose typing rules. As for type checking, results are stated as *"valid only for matrices of the same order"* [1] and the like. Polymorphic functions are well-accepted. But telling that the identity matrix is as polymorphic as the identity function will sound odd to many people.

Relational mathematics [24] is a step forward towards conceptual unification between relations and matrices. But it is first and foremost *category theory* [20] which provides for successful unification, by regarding functions, relations and matrices as morphisms (arrows) of suitable categories. The category of functions is well known, that of relations less known and those of matrices by and large ignored.

In the sequel we will show how weighted automata can be described and reasoned about in the typed LA which emerges from regarding matrices as morphisms (rather than objects) of suitable categories, as pioneered by MacLane [20] and MacLane and Birkhoff [21]. This is part of a research line which started in [16] and whose aim is to provide evidence of the usefulness of changing notation (and reasoning style) and adopting *typed* LA as the lingua franca of quantitative methods in computer science.

2 Typed Linear Algebra

Computer scientists tend to regard matrices as rectangular shaped data structures implemented as bidimensional arrays, lists of lists and the like. Mathematicians tend to regard them as linear transforms, i.e. vector-to-vector operations. Yet matrices are abstract entities independent of either such views: they can be regarded as arrows of particular categories, whereby they become *typed*. This answers questions such as: what is the type of a matrix? What are their basic *constructors*? In what measure are these related to standard matrix operations and algebra?

By studying the categories of matrices of [20], the authors of [16] have identified typed, algebraically rich constructors aiming to repair the lack just mentioned. Backhouse [4] regards matrices as a way of compacting sets of equations into single equations which *is a tremendous improvement in concision that does not incur any loss of precision!* Reference [16] furthermore show how the very general concept of a *biproduct* [21] promotes individual values to blocks and value-level operations to block-level operations, in fact the great conceptual advantage offered by matrix notation.

Matrices as Arrows. A matrix M with n rows and m columns is a function which tells the value occupying each cell (r, c), for $1 \leq r \leq n$, $1 \leq c \leq m$. The type of such cell-values varies, but the minimal algebraic structure of semirings is required for matrix operations to make sense. Standard linear algebra operates over the richer structure of a *field* (further offering additive and multiplicative inverses) and the field of real numbers ($\mathbb{R}$) is often taken by default.

Interestingly, what is meant by the *type* of a matrix in the sequel does not bear a direct relationship to such algebraic structures: it rather provides (as in programming) a way of interfacing matrices with each other. The type of a matrix M with m columns and n rows will be denoted by the arrow $m \longrightarrow n$ between the number of columns and the number of rows. By writing $m \xrightarrow{M} n$ (or the equivalent $n \xleftarrow{M} m$) one declares matrix M and its type.

The most interesting matrix combinator is *composition*, commonly referred to as *matrix multiplication*. Denoting the (r, c)-th cell of a given matrix M by rMc [2], the (r, c)-th cell of composite matrix $M \cdot N$ is given by

$$r(M \cdot N)c = \langle \sum x :: (rMx) \times (xNc) \rangle \tag{1}$$

where $\times$ is the cell-level semiring multiplicative operation and $\sum$ is the finite iteration of its additive operation.

What is x in (1) and what is its range? This will be easy to answer by inspecting the types of both M and N:

$$n \xleftarrow{\quad M \quad} m \xleftarrow{\quad N \quad} k \tag{2}$$
$$\underbrace{\qquad\qquad\qquad}_{M \cdot N}$$

---

[2] Rather than the more conventional $M(r, c)$ — we will explain later why we propose a different notation.

Thus $1 \leq x \leq m$ and matrix multiplication can be abstracted by arrow composition.

For every n there is a matrix of type $n \longleftarrow n$ which is the unit of composition. This is nothing but the *identity matrix* of size n, indistinguishably denoted by $n \xleftarrow{id_n} n$ or $n \xleftarrow{1} n$. This is the diagonal of size n, that is [3], $r(id)c \triangleq r = c$ under the $\{0,1\}$ encoding of the Booleans:

$$id_n = \begin{pmatrix} 1 & 0 & \cdots & 0 \\ 0 & 1 & \cdots & 0 \\ \vdots & \vdots & \ddots & \vdots \\ 0 & 0 & \cdots & 1 \end{pmatrix} \qquad n \xleftarrow{\;id_n\;} n$$

Therefore,

$$id_n \cdot M \;=\; M \;=\; M \cdot id_m \qquad\qquad \begin{array}{c} m \xleftarrow{\;id_m\;} m \\[2pt] M \Big\downarrow \;\;\swarrow{\scriptstyle M}\;\; \Big\downarrow M \\[2pt] n \xleftarrow{\;id_n\;} n \end{array} \qquad (3)$$

where the subscripts m and n can be omitted wherever the underlying type diagrams are assumed.

Equipped with composition (2) and identity (3), matrices form a *category* whose *objects* are matrix dimensions and whose *morphisms* ($m \xleftarrow{M} n$ etc) are the matrices themselves [20, 21]. Strictly speaking, there is one such category per matrix cell-level algebra. Notation $Mat_{\mathbb{S}}$ will be used to denote such a category, parametric on semiring $\mathbb{S}$ or any other (richer) algebraic structure.

Vectors as Arrows. Vectors are special cases of matrices in which one of the dimensions is 1, for instance

$$v = \begin{pmatrix} v_1 \\ \vdots \\ v_m \end{pmatrix} \qquad \text{and} \qquad w = (w_1 \ldots w_n)$$

Column vector v is of type $m \longleftarrow 1$ (m rows, one column) and row vector w is of type $1 \longleftarrow n$ (one row, n columns). Our convention is that lowercase letters (eg. v, w) denote vectors and uppercase letters (eg. M, N) denote arbitrary matrices.

Converse of a Matrix. One of the kernel operations of linear algebra is *transposition*, whereby a given matrix changes shape by turning its rows into columns and vice-versa. Given matrix $n \xleftarrow{M} m$, notation $m \xleftarrow{M^\circ} n$ denotes its transpose, or converse. The following idempotence and contravariance laws hold:

---

[3] Notation $x \triangleq y$ means $x = y$ by definition.

$$(M^\circ)^\circ = M \tag{4}$$
$$(M \cdot N)^\circ = N^\circ \cdot M^\circ \tag{5}$$

Bilinearity. Given two matrices of the same type $n \xleftarrow{\ M,N\ } m$ it makes sense to add them up index-wise, leading to matrix $M + N$ where symbol $+$ promotes the underlying semiring additive operator to matrix-level. Likewise, additive unit cell value 0 is promoted to matrix 0 wholly filled with 0s, the unit of matrix addition and zero of matrix composition:

$$M + 0 = M = 0 + M \tag{6}$$
$$M \cdot 0 = 0 = 0 \cdot M \tag{7}$$

Composition is bilinear relative to $+$:

$$M \cdot (N + P) = M \cdot N + M \cdot C \tag{8}$$
$$(N + P) \cdot M = N \cdot M + P \cdot M \tag{9}$$

In the same way $M + N$ denotes the promotion of addition of matrix cells to matrix addition, the same promotion can take place with respect to the whole semiring algebra. For instance, cell value multiplication leads to matrix multiplication, denoted $M \times N$ or simply MN (for M and N of the same type), also known as the *Hadamard product*, which is commutative, associative and distributive over addition (ie. bilinear). Clearly,

$$M \times \top = \top \times M = M \tag{10}$$

where matrix $\top$ is of the same type as M and is wholly filled with 1s.

Type Generalization. Matrix types (the end points of arrows) can be generalized to arbitrary, denumerable sets since addition in $\mathbb{S}$ is commutative, that is, the summation of (1) can be evaluated in arbitrary order.

In fact, and as is standard in relational mathematics [24], objects in categories of matrices can be generalized from numeric dimensions ($n, m \in \mathbb{N}_0$) to arbitrary denumerable types (A, B), taking disjoint union $A + B$ for $m + n$, Cartesian product $A \times B$ for mn, unit type 1 for number 1, the empty set $\emptyset$ for 0, etc. Conversely, dimension n corresponds to the type made of the initial segment of the natural numbers up to n. Our convention is that lowercase letters (eg. n, m) denote the traditional dimension types (natural numbers), letting uppercase letters denote arbitrary other types.

3   Weighted Automata as $Mat_\mathbb{S}$ Arrows

Following [8], we consider in the sequel a simpler notion of weighted automaton $W = (Q, A; \mu, \gamma)$ which deals without the input weight function λ. This facilitates the comparison between the coalgebraic approach of [8] and our own and helps in staying with the binary matrix block combinators of [16], to be presented shortly. For this purpose, we assign the type $Q \longrightarrow 1$ to output function γ, which is therefore regarded as a row vector in $Mat_\mathbb{S}$. Concerning μ, it can either be regarded as a matrix of type $Q \times A \longrightarrow Q$ or of type $Q \longrightarrow Q \times A$, as these types are isomorphic in $Mat_\mathbb{S}$ [4]. We prefer the second (coalgebraic) alternative and therefore regard the following diagram as representation of weighted automaton $W = (Q, A; \mu, \gamma)$:

$$Q \times A \xleftarrow{\;\mu\;} Q \xrightarrow{\;\gamma\;} 1 \tag{11}$$

Clearly, both μ and γ can be packaged into a single coalgebra (matrix) of type $(Q \times A) + 1 \xleftarrow{\;W\;} Q$ and made of two blocks

$$W = \begin{bmatrix} \mu \\ \gamma \end{bmatrix} \tag{12}$$

provided we explain what the meaning of combinator $[-]$ is. This leads into matrix block notation and its algebra.

Block Notation. Two basic binary combinators are available for building matrices out of other matrices, say M and N:

- $[M|N]$ — M and N side by side (read $[M|N]$ as "M junc N")
- $\begin{bmatrix} M \\ N \end{bmatrix}$ — M on top of N (read $\begin{bmatrix} M \\ N \end{bmatrix}$ as "M split N").

That is, matrices are stacked either vertically ($\begin{bmatrix} M \\ N \end{bmatrix}$) or horizontally ($[M|N]$). Dimensions should agree, as shown in the diagram below, taken from [16], where m, n, p and t are types:

$$[M|N] = M \cdot \pi_1 + N \cdot \pi_2 \tag{13}$$

$$\begin{bmatrix} P \\ Q \end{bmatrix} = i_1 \cdot P + i_2 \cdot Q \tag{14}$$

The special matrices i_1, i_2, π_1 and π_2 are fragments of the identity matrix as given by the so-called *reflexion laws*,

$$[i_1 | i_2] = id$$
$$\begin{bmatrix} \pi_1 \\ \pi_2 \end{bmatrix} = id$$

---

[4] This follows from a self-adjunction in $Mat_\mathbb{S}$ which is studied in detail in [19]. The isomorphism reshapes matrices by reducing the number of columns by the same factor the number of rows increases, keeping the "rectangular area" and its information intact.

which play an important role in explaining the semantics of the two combinators. In brief, *junc* (13) and *split* (14) form a so-called *biproduct* [20]. The details of this, however, can be skipped for the purposes of this presentation, sufficing to be aware of the rich algebra of such combinators of which we single out two "fusion"-laws,

$$R \cdot [M|N] = [R \cdot M|R \cdot N] \tag{15}$$

$$\left[\frac{M}{N}\right] \cdot R = \left[\frac{M \cdot R}{N \cdot R}\right] \tag{16}$$

two structural equality laws,

$$[A|B] = [C|D] \equiv A = C \wedge B = D \tag{17}$$

$$\left[\frac{A}{B}\right] = \left[\frac{C}{D}\right] \equiv A = C \wedge B = D \tag{18}$$

and two absorption laws:

$$[A|B] \cdot (C \oplus D) = [A \cdot C|B \cdot D] \tag{19}$$

$$(C \oplus D) \cdot \left[\frac{A}{B}\right] = \left[\frac{C \cdot A}{D \cdot B}\right] \tag{20}$$

All these laws emerge as corollaries of the universal properties of biproducts. Mind the types: the laws are only valid for matrices which typecheck and types are obtained by unification, as explained in [16].

Weighted Automata as Matricial Coalgebras. As suggested by (12) above, weighted automaton W can be regarded as a coalgebra for $\mathcal{M}at_{\mathbb{S}}$ endofunctor $\mathsf{F}X = (X \otimes id) \oplus id$, where $\oplus$ and $\otimes$ are the so-called *direct sum* and *Kronecker* bifunctors. The former,

$$M \oplus N = [i_1 \cdot M|i_2 \cdot N]$$

is of type

$$
\begin{array}{ccc}
n & m & n+m \\
\Big\downarrow M & \Big\downarrow N & \Big\downarrow M \oplus N \\
k & j & k+j
\end{array}
$$

and the latter is of type

$$
\begin{array}{ccc}
n & m & n \times m \\
\Big\downarrow M & \Big\downarrow N & \Big\downarrow M \otimes N \\
k & j & k \times j
\end{array}
$$

Fusion laws

$$[M|N] \otimes C = [M \otimes C | N \otimes C]$$
$$\left[\frac{M}{N}\right] \otimes C = \left[\frac{M \otimes C}{N \otimes C}\right]$$

capture the meaning of Kronecker product block-wise. Index-wise, one has:

$$(y, x)(M \otimes N)(b, a) = (yMb) \times (xNa)$$

4   Weighted Automata Homomorphisms

A homomorphism between two weighted automata W and W' is a function h making the following Mat_S-diagram commute,

$$\begin{array}{ccc}
FQ & \xleftarrow{\;W\;} & Q \\
{\scriptstyle Fh}\downarrow & & \downarrow{\scriptstyle h} \\
FQ' & \xleftarrow{\;W'\;} & Q'
\end{array} \qquad\qquad (21)$$

for $FX = (X \otimes id) \oplus id$ (F-coalgebra homomorphism). The reader may wonder about how does h (a function) fit into a diagram of matrices. The explanation is easy: every function $A \xrightarrow{\;f\;} B$ can be represented in Mat_S by a matrix $[\![f]\!]$ of the same type defined by

$$b[\![f]\!]a \;\triangleq\; (b =_S f\, a)$$

where, in general, $y =_S x$ is the unit 1 of S if $y = x$ and 0 otherwise. Thus $[\![f]\!]$ is the matrix which represents the graph of f: there is a 1 in every entry of $[\![f]\!]$ addressed by $(f(a), a)$ and 0s everywhere else. As S is always implicit and all diagrams are drawn in Mat_S unless otherwise specified, subscript S in $=_S$ and the parentheses in $[\![f]\!]$ can be safely dropped.

Below we show how diagram (21) unfolds into the usual definition of weighted automata homomorphism [8], which is termed *functional* simulation in [9]. For this we will rely on typed, blocked linear algebra:

$$(Fh) \cdot W \;=\; W' \cdot h$$

$$\equiv \qquad \{ \text{ unfold } Fh \; ; \; W \text{ and } W' \text{ are splits defined by (12) } \}$$

$$((h \otimes id) \oplus id) \cdot \left[\frac{\mu}{\gamma}\right] \;=\; \left[\frac{\mu'}{\gamma'}\right] \cdot h$$

$$\equiv \qquad \{ \text{ absorption (20), identity (3) and fusion (16) } \}$$

$$\left[\frac{(h \otimes id) \cdot \mu}{\gamma}\right] \;=\; \left[\frac{\mu' \cdot h}{\gamma' \cdot h}\right]$$

$$\equiv \quad \{ \text{ equality (18) } \}$$

$$\begin{cases} (h \otimes id) \cdot \mu = \mu' \cdot h \\ \gamma = \gamma' \cdot h \end{cases} \tag{22}$$

The reader wishing to convert the equalities of (22) into index-wise formulas for cross-checking with other sources is invited to do so based on the following rules interfacing index-free and index-wise matrix notation, where N is an arbitrary matrix and f, g are functional matrices:

$$y(f \cdot N)x = \langle \sum z : y = f(z) : zNx \rangle \tag{23}$$

$$y(g^\circ \cdot N \cdot f)x = (g(y))N(f(x)) \tag{24}$$

These rules are expressed in the style of the Eindhoven quantifier calculus [3]. Their calculation (deferred to the appendix) provides evidence of the safe mix among matrix, predicate and function notation in typed LA.

We start by unfolding the first equality of (22):

$(h \otimes id) \cdot \mu = \mu' \cdot h$

$\equiv \quad \{ \text{ index-wise equality on matrices of type } Q' \times A \longleftarrow Q \ \}$

$(q', a)((h \otimes id) \cdot \mu)q = (q', a)(\mu' \cdot h)q$

$\equiv \quad \{ \text{ (24) on the right hand side, for } g, N, f := id, \mu', h \ \}$

$(q', a)((h \otimes id) \cdot \mu)q = (q', a)\mu'(h(q))$

$\equiv \quad \{ \text{ (23) for } f, N := h \otimes id, \mu \ \}$

$\langle \sum (p, b) : (q', a) = (h \otimes id)(p, b) : (p, b)\mu q \rangle = (q', a)\mu'(h(q))$

$\equiv \quad \{ \text{ since } (h \otimes id)(p, b) = (h(p), b); \text{ one-point rule [3] over } a = b \ \}$

$\langle \sum p : q' = h(p) : (p, a)\mu q \rangle = (q', a)\mu'(h(q))$

$\equiv \quad \{ \text{ liberally writing } p \overset{a}{\longleftarrow} q \text{ for the weight of the corresponding transition } \}$

$\langle \sum p : q' = h(p) : \ p \overset{a}{\longleftarrow} q \ \rangle = q' \overset{a}{\longleftarrow} h(q)$

In words: the weight associated to transition $q' \overset{a}{\longleftarrow} h(q)$ in the target automaton is the accumulation of the weights of all transitions $p \overset{a}{\longleftarrow} q$ in the source automaton for all p which h maps to q'.

Unfolding the other matrix equality in (22) is simpler: as γ, γ' are row vectors, we get, for all $q \in Q$, $1\gamma q = 1(\gamma' \cdot h)q$, since there is only one row. By (24) this becomes $1\gamma q = 1\gamma'(h(q))$, that is $\gamma(q) = \gamma'(h(q))$ once γ, γ' are regarded back as functions.

Summing up, both calculations show that weighted automata homomorphisms defined in a category of matrices coincide with those defined by Bonchi et al.

[8] in the category of sets. We regard this as just the beginning of a typed LA approach to weighted automata to be developed comprehensively in the near future.

5 Summary

This abstract addresses on-going work. Since the research presented in [16, 19], typed LA calculational techniques have been successfully applied to data mining [17] and probabilistic program calculation [23], the latter extending the algebra of programming of Bird and de Moor [6].

In the case of weighted automata, LA is a natural choice already identified by other researchers. Buchholz [9], for instance, praises matrix notation because it *allows an elegant and compact formulation of the theory*. Trčka [28] writes that *matrices (...) increase clarity and compactness, simplify proofs, make known results from linear algebra directly applicable* and also mentions their *didactic advantage*.

In broad terms, the approach put forward in this abstract proposes that LA be *typed* on the basis of a categorial approach in which index-free matrix terms form the main notation, diagrammatic representations and proofs included. That is to say, rather than accepting LA arguments embedded in ordinary set-theoretical reasoning, we propose that typed LA be regarded as a *lingua franca* for computing, the other approaches coming as suitable instantiations [5].

We should say we are not the first proposing this strategy. The acronym LAoP, for "linear algebra of programming" has been put forward already, albeit in a somewhat different setting, by Sernadas et al. [25], the key idea being *"to adopt linear algebra as the lingua franca of software verification"* [27]. Our contribution is the emphasis on LA polymorphic *types*. For this to work in practice, we believe the interfaces with standard logic, set theory and relation algebra should not be neglected. Schmidt [24] already relies on matrix notation for doing relation algebra. Our experiments eg. with the Eindhoven quantifier notation show that the interface between functions, relations, predicates and matrices is (at least pedagogically) relevant. The infix notation we adopt for matrix entries — yMx rather than $M(y, x)$ — intends to bridge with that commonly used for binary relations. For instance, $y \leq x$ is preferred to $\leq (y, x)$.

6 Current and Related Work

One of our targets is the *linear algebra of components* which, anticipated in [18], promises a quantitative expansion of the coalgebraic approach of Barbosa [5] on software components.

The work by Bonchi et al. [8] on a coalgebraic perspective on weighted automata promises a similar outcome but their use of linear algebra is on a different

---

[5] Even so general a framework as that of an *allegory* [12] arises from matrices whose data values form *locales*.

plan: triggered by the need to extend the powerset functor quantitatively, they introduce a *vector space* which weights (*quantifies*) multi-way state evolution. (In a sense, powersets become "metric".) Because this is carried in the category of sets, their coalgebras involve functor $W = \mathbb{K} \times (\mathbb{K}_\omega^-)^A$ over a field $\mathbb{K}$, where $\mathbb{K}_\omega^-$ is the so-called field valuation (exponential) functor. Our approach flattens such exponentials by changing category: the category of sets and functions gives room to the category of matrices built on top of $\mathbb{K}$. Thus $(\_)^A$ within sets becomes $(\_) \times A$ within matrices. In this way, weights no longer need to be taken explicitly into account, as the underlying matrix algebra circumspectly takes care of them.

Much remains to be done, in particular calling for the unification with related work. For instance, we would like to relate our ideas with those of Trčka [28], who presents a matrix approach to the notions of strong, weak and branching bisimulation ranging from labeled transition systems to Markov reward chains. This already is the aim of Buchholz [9], who targets at a *universal definition of bisimulation which can be applied to a wide class of model types such that the different forms of bisimulation can all be seen as specific cases*, helping to unify system analysis.

We believe matrix types will improve the approaches of both [9] and [28] in a significant way. But, above all, in its use of matrix categories our strategy is close to the iteration theory $Mat_{\mathcal{L}(X^*)}$ of Bloom et al. [7] whose morphisms are matrices with entries in the semiring of languages. We intend to investigate the relationship between both approaches in a thorough way.

Acknowledgements. The author is indebted to Nelma Moreira for her comments on an earlier draft of this extended abstract. This research was carried out in the context of the QAIS (Quantitative analysis of interacting systems: foundations and algorithms) project funded by the ERDF through the Programme COMPETE and by the Portuguese Government through FCT (Foundation for Science and Technology) contract PTDC/EIA-CCO/122240/2010.

Appendix

To calculate (23) we let $M := f$ in (1):

$$y(f \cdot N)x$$

$$= \quad \{ \text{ definition (1) } \}$$

$$\left\langle \sum z :: (y = f(z)) \times (zNx) \right\rangle$$

$$= \quad \{ \text{ rule (25) below } \}$$

$$\left\langle \sum z : y = f(z) : zNx \right\rangle$$

The rule used above,

$$\left\langle \sum x : p(x) : e(x) \right\rangle = \left\langle \sum x :: (p(x)) \times (e(x)) \right\rangle \tag{25}$$

is illustrative of the interface between predicate logic and the semiring algebra underneath: on the left hand side, $p(x)$ is a predicate expressing the range of a summation; on the right hand side it is encoded into $\mathbb{S}$: 1 if $p(x)$ holds, 0 otherwise. Since $0 \times s = 0$, all terms such that $p(x)$ doesn't hold boil down to 0 and don't affect the summation [6].

Similarly, for $M := g^\circ$ in (1):

$$y(g^\circ \cdot N)x$$

$$= \quad \{ \text{ definition (1) ; } y(g^\circ)z = z =_{\mathbb{S}} g(y) \}$$

$$\left\langle \sum z :: (z = g(y)) \times (zNx) \right\rangle$$

$$= \quad \{ \text{ rule (25) } \}$$

$$\left\langle \sum z : z = g(y) : zNx \right\rangle$$

$$= \quad \{ \text{ one-point rule [3] } \}$$

$$(g(y))Nx$$

Thus $y(g^\circ \cdot N)x = (g(y))Nx$. The calculation of $y(N \cdot f)x = yN(f(x))$ follows the same steps. Rule (24) puts these two equalities together.

References

1. Abadir, K., Magnus, J.: Matrix algebra. Econometric exercises, vol. 1. Cambridge University Press (2005)
2. Andova, S., McIver, A., D'Argenio, P.R., Cuijpers, P.J.L., Markovski, J., Morgan, C., Núñez, M. (eds.): Proceedings First Workshop on Quantitative Formal Methods: Theory and Applications. EPTCS, vol. 13 (2009)
3. Backhouse, R., Michaelis, D.: Exercises in Quantifier Manipulation. In: Uustalu, T. (ed.) MPC 2006. LNCS, vol. 4014, pp. 69–81. Springer, Heidelberg (2006)
4. Backhouse, R.: Mathematics of Program Construction, 608 pages. Univ. of Nottingham (2004), draft of book in preparation
5. Barbosa, L.: Towards a Calculus of State-based Software Components. Journal of Universal Computer Science 9(8), 891–909 (2003)
6. Bird, R., de Moor, O.: Algebra of Programming. Series in Computer Science. Prentice-Hall International (1997)
7. Bloom, S., Sabadini, N., Walters, R.: Matrices, machines and behaviors. Applied Categorical Structures 4(4), 343–360 (1996)
8. Bonchi, F., Bonsangue, M., Boreale, M., Rutten, J., Silva, A.: A coalgebraic perspective on linear weighted automata. Information and Computation 211, 77–105 (2012)
9. Buchholz, P.: Bisimulation relations for weighted automata. Theoretical Computer Science 393(1-3), 109–123 (2008)

---

[6] For $\mathbb{S}$ the Boolean semiring, $\sum$ is existential quantification, $\times$ is conjunction and equality (25) becomes an instance of the *trading rule* of existential quantification [3].

10. Droste, M., Gastin, P.: Weighted automata and weighted logics. In: Kuich, W., Vogler, H., Droste, M. (eds.) Handbook of Weighted Automata. EATCS Monographs in Theoretical Computer Science, ch. 5, pp. 175–211. Springer (2009)
11. Erwig, M., Kollmansberger, S.: Functional pearls: Probabilistic functional programming in Haskell. J. Funct. Program. 16, 21–34 (2006)
12. Freyd, P., Scedrov, A.: Categories, Allegories, Mathematical Library, vol. 39. North-Holland (1990)
13. Gibbons, J., Hinze, R.: Just do it: simple monadic equational reasoning. In: Proceedings of the 16th ACM SIGPLAN International Conference on Functional Programming, ICFP 2011, pp. 2–14. ACM, New York (2011)
14. Hehner, E.: A probability perspective. Formal Aspects of Computing 23, 391–419 (2011)
15. Larsen, K., Skou, A.: Bisimulation through probabilistic testing. Inf. Comput. 94(1), 1–28 (1991)
16. Macedo, H.D., Oliveira, J.N.: Matrices As Arrows! A Biproduct Approach to Typed Linear Algebra. In: Bolduc, C., Desharnais, J., Ktari, B. (eds.) MPC 2010. LNCS, vol. 6120, pp. 271–287. Springer, Heidelberg (2010)
17. Macedo, H.D., Oliveira, J.N.: Do the middle letters of "OLAP" stand for linear algebra ("LA")? Technical Report TR-HASLab:04:2011, INESC TEC and University of Minho, Gualtar Campus, Braga (2011)
18. Macedo, H.D., Oliveira, J.N.: Towards Linear Algebras of Components. In: Barbosa, L.S. (ed.) FACS 2010. LNCS, vol. 6921, pp. 300–303. Springer, Heidelberg (2010)
19. Macedo, H.D., Oliveira, J.N.: Typing linear algebra: A biproduct-oriented approach (2011) (accepted for publication in SCP)
20. MacLane, S.: Categories for the Working Mathematician. Springer, New-York (1971)
21. MacLane, S., Birkhoff, G.: Algebra. AMS Chelsea (1999)
22. McIver, A., Morgan, C.: Abstraction, Refinement and Proof For Probabilistic Systems. Monographs in Computer Science. Springer (2005)
23. Oliveira, J.: Towards a linear algebra of programming. Accepted for publication in Formal Aspects of Computing (2012)
24. Schmidt, G.: Relational Mathematics. Encyclopedia of Mathematics and its Applications, vol. 132. Cambridge University Press (November 2010)
25. Sernadas, A., Ramos, J., Mateus, P.: Linear algebra techniques for deciding the correctness of probabilistic programs with bounded resources. Tech. rep., SQIG - IT and IST - TU Lisbon, 1049-001 Lisboa, Portugal (2008), short paper presented at LPAR 2008, Doha, Qatar, November 22-27
26. Sokolova, A.: Coalgebraic Analysis of Probabilistic Systems. Ph.D. dissertation, Tech. Univ. Eindhoven, Eindhoven, The Netherlands (2005)
27. SQIG-Group: LAP: Linear algebra of bounded resources programs, iT & Tech. Univ. Lisbon (2011), http://sqig.math.ist.utl.pt/work/LAP
28. Trčka, N.: Strong, weak and branching bisimulation for transition systems and Markov reward chains: A unifying matrix approach. In: [2], pp. 55–65

A Pushdown Transducer Extension
for the OpenFst Library

Cyril Allauzen and Michael Riley

Google Research, 76 Ninth Avenue, New York, NY 10011, USA
{allauzen,riley}@google.com

Abstract. Pushdown automata are devices that can efficiently represent context-free languages, have natural weighted versions, and combine naturally with finite automata. We describe a pushdown transducer extension to OpenFst, a weighted finite-state transducer library. We present several weighted pushdown algorithms, some with clear finite-state analogues, describe their library usage and give some applications of these methods to recognition, parsing and translation.

1 Introduction

OpenFst is an open-source C++ software library for creating, combining, searching and optimizing finite-state transducers (FSTs) [4]. Weighted FSTs have many applications in speech and language processing, computational biology and other areas and the availability of flexible, large-scale algorithms libraries allows rapid experimentation and development [17]. However, there are problems that are not well-represented by finite automata such as aspects of natural language parsing or translation. In particular, a context-free representation may be better suited either because the language considered is not regular or is more compactly represented in a recursive manner.

In these cases, a common approach is to use a weighted context-free grammar as the representation. However, weighted *pushdown automata* offer an attractive alternative. As automata, they are more closely tied to computation and can share and mix with finite automata in a natural way [7]. Our goal here is to present several weighted pushdown algorithms, some with clear finite-state analogues, to describe their realization in a pushdown transducer extension to the OpenFst library and to give some applications of these methods and the library.

2 Definitions

Informally, pushdown transducers are finite-state transducers that have been augmented with a stack. Typically this is done by adding a stack alphabet and labeling each transition with a stack operation (a stack symbol to be pushed onto, popped or read from the stack) in additon to the usual input and output labels [1,6] and weight [12,20]. Our equivalent representation allows a transition to be labeled by a stack operation or regular input/output symbols but not both.

N. Moreira and R. Reis (Eds.): CIAA 2012, LNCS 7381, pp. 66–77, 2012.
© Springer-Verlag Berlin Heidelberg 2012

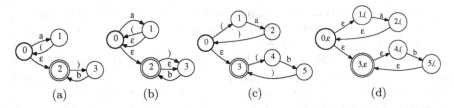

Fig. 1. PDA Examples: (a) Non-rational PDA A_1 accepting $\{a^n b^n | n \in \mathbb{N}\}$. (b) Rational (but not bounded-stack) PDA A_2 accepting $a^* b^*$. (c) Bounded-stack PDA A_3 accepting $a^* b^*$ and (d) its expansion A_4 as an FSA.

Stack operations are represented by pairs of open and close parentheses (pushing a symbol on and popping it from the stack). The advantage of this representation is that it is identical to the finite-state transducer representation except that certain symbols (the parentheses) have special semantics. As such, several finite-state algorithms either immediately generalize to this PDT representation or do so with minimal changes.

2.1 Dyck Languages

A (restricted) Dyck language consists of "well-formed" or "balanced" strings over a finite number of pairs of parentheses. Thus the string ([( ) ( )] { } [ ]) ( ) is in the Dyck language over three pairs of parentheses (following [6]).

More formally, let A and $\overline{A}$ be two finite alphabets such that there exists a bijection f from A to $\overline{A}$. Intuitively, f maps an opening parenthesis to its corresponding closing parenthesis. Let $\bar{a}$ denote $f(a)$ if $a \in A$ and $f^{-1}(a)$ if $a \in \overline{A}$. The *Dyck language* D_A over the alphabet $\widehat{A} = A \cup \overline{A}$ is then the language defined by the following context-free grammar: $S \to \epsilon$, $S \to SS$ and $S \to a S \bar{a}$ for all $a \in A$. We define the mapping $c_A : \widehat{A}^* \to \widehat{A}^*$ as follows. $c_A(x)$ is the string obtained by iteratively deleting from x all factors of the form $a\bar{a}$ with $a \in A$. Observe that $D_A = c_A^{-1}(\epsilon)$.

Let A and B be two finite alphabets such that $B \subseteq A$, we define the mapping $r_B : A^* \to B^*$ by $r_B(x_1 \ldots x_n) = y_1 \ldots y_n$ with $y_i = x_i$ if $x_i \in B$ and $y_i = \epsilon$ otherwise.

2.2 Pushdown Automata and Transducers

Formally, a *weighted pushdown transducer* (PDT) T over the tropical semiring $(\mathbb{R} \cup \{\infty\}, \min, +, \infty, 0)$ is a 9-tuple $(\Sigma, \Delta, \Pi, \overline{\Pi}, Q, E, I, F, \rho)$ where Σ and Δ are the finite input and output alphabets, Π and $\overline{\Pi}$ are the finite open and close parenthesis alphabets, Q is a finite set of states, $I \in Q$ the initial state, $F \subseteq Q$ the set of final states, $E \subseteq Q \times (\Sigma \cup \widehat{\Pi} \cup \{\epsilon\}) \times (\Delta \cup \widehat{\Pi} \cup \{\epsilon\}) \times (\mathbb{R} \cup \{\infty\}) \times Q$ a finite set of transitions, and $\rho : F \to \mathbb{R} \cup \{\infty\}$ the final weight function. Let $e = (p[e], i[e], o[e], w[e], n[e])$ denote a transition in E we require that if $i[e] \in \widehat{\Pi}$ or $o[e] \in \widehat{\Pi}$, then $i[e] = o[e]$. We define the *size* of T as $|T| = |Q| + |E|$.

A path π is a sequence of transitions $\pi = e_1 \ldots e_n$ such that $n[e_i] = p[e_{i+1}]$ for $1 \leq i < n$. We then define $p[\pi] = p[e_1]$, $n[\pi] = n[e_n]$, $i[\pi] = i[e_1] \cdots i[e_n]$, $o[\pi] = o[e_1] \cdots o[e_n]$ and $w[\pi] = w[e_1] + \ldots + w[e_n]$.

A path π is accepting if $p[\pi] = I$ and $n[\pi] \in F$. A path π is balanced if $r_{\widehat{\Pi}}(i[\pi]) \in D_{\Pi}$. A balanced path π accepts the pair $(x, y) \in \Sigma^* \times \Delta^*$ if it is a balanced accepting path such that $r_{\Sigma}(i[\pi]) = x$ and $r_{\Delta}(o[\pi]) = y$.

The *weight associated by* T to a pair of strings $(x, y) \in \Sigma^* \times \Delta^*$ is

$$T(x, y) = \min_{\pi \in P(x,y)} w[\pi] + \rho(n[\pi])$$

where $P(x, y)$ denotes the set of balanced paths accepting (x, y). A weighted transduction is recognizable by a weighted pushdown transducer iff it is algebraic [20] or equivalently iff it is recognizable by a weighted simple syntax-directed translation [1,14].

A *weighted pushdown automaton* (PDA) is a pushdown transducer where $i[e] = o[e]$ for all transition $e \in E$. A weighted language is recognizable by a weighted pushdown automaton iff it is context-free [1,12].

A pushdown transducer T has *bounded stack* if there exists $K \in \mathbb{N}$ such that for any path π from I such that $c_{\Pi}(r_{\widehat{\Pi}}(i[\pi])) \in \Pi^*$:

$$|c_{\Pi}(r_{\widehat{\Pi}}(i[\pi]))| \leq K. \tag{1}$$

If T has bounded stack, then it represents a rational transduction (see Section 4.1). Figure 1a-c gives examples of non-rational, rational and bounded-stack PDAs.

A pushdown transducer is *deterministic* if at any state with at least two outgoing transitions the input labels of the outgoing transitions are distinct and are either all input symbols (in Σ) or all close parentheses (in $\overline{\Pi}$).

A *weighted finite-state transducer or automaton* (FST or FSA) can be viewed as a PDT or PDA where the open and close parentheses alphabets are empty; see [16] for a stand-alone definition.

3   Implementation

The benefit of this definition of PDTs is that a PDT T can be represented as a pair of a FST specification, with input alphabet $\Sigma \cup \widehat{\Pi}$ and output alphabet $\Delta \cup \widehat{\Pi}$, and a parentheses mapping $f : \Pi \to \overline{\Pi}, a \mapsto \bar{a}$. This allows us to fully leverage the OpenFst library [4] for representing and manipulating the FST specifications of PDTs.

The PDA A_1 given in Figure 1a can be generated from the three text files given Figure 2. The `pda.txt` file is the textual description of the FSA specification of A_1 in the OpenFst format. The `symbols` file maps each symbol to an integer value used for the internal memory representation. Finally, the `parens` file describes the pair of open and close parentheses. The `fstcompile` binary command can be used to generate a binary file for the FSA specification of A_1:

```
fstcompile --acceptor --isymbols=symbols pda.txt > pda.fst
```

pda.txt	symbols	parens
0 1 a	eps 0	3 4
0 2 eps	a 1	
1 0 (	b 2	
2 3)	(3	
2	) 4	
3 2 b		

Fig. 2. Text files representing the PDA from Figure 1a

The pair of files (pda.fst, parens) is then the file representation of the PDA for the purposes of the library. For instance, the reverse of A_1 can then be computed by invoking the following command:

```
pdtreverse --pdt_parentheses=parens pda.fst > reverse-pda.fst
```

Using the C++ interface, a PDT is similarly represented by a pair consisting of an object of type StdFst and a vector<pair<int, int> > object representing the set of open and close parenthesis pairs. The following C++ code is equivalent to the command given above:

```
StdFst *pda = StdFst::Read("pda.fst");
vector<pair<int, int> > parens(1, make_pair(3,4));
StdVectorFst reverse_pda;
Reverse(*pda, parens, &reverse_pda);
```

Table 1 shows the operations available in the PDT library extension [2]. The shared file and memory representations for FSTs and PDTs allows some operations from the OpenFst library, such as Union or Invert for instance, to be applied to PDTs unmodified. Other operations can be implemented with minimal work by leveraging the corresponding FST operation. For instance, PDT reversal can be implemented by first calling the Reverse operation of OpenFst followed by replacing every occurence of a parenthesis $a \in \widehat{\Pi}$ by its matching parenthesis $\bar{a}$ in the resulting machine.

4 Algorithms

In this section, we present PDT algorithms that are not trivially derived from FST analogues. The algorithms that we chose were motivated by analogy to the finite automata or context-free grammar case, by their applications (see Section 5), and by their tractability.

4.1 Expansion

Given a bounded-stack PDT T, the *expansion* of T is the FST T' equivalent to T defined as follows.

A state in T' is a pair (q, z) where q is a state in T and $z \in \Pi^*$. A transition (q, a, b, w, q') in T results in a transition $((q, z), a', b', w, (q', z'))$ in T' only when one of the following conditions hold: (a) $a \in \Sigma \cup \{\epsilon\}$, $z' = z$, $a' = a$ and $b' = b$, (b)

Table 1. Algorithms for manipulating pushdown transducers and the corresponding binary commands

Operation	Algorithm	Section	Command
Union	FST alg.		fstunion
Concatenation	FST alg.*		fstconcat
Closure	FST alg.*		fstclosure
Reversal	trivial changes to FST alg.		pdtreverse
Inversion	FST alg.		fstinvert
Projection	FST alg.		fstproject
Expansion	PDT-specific alg.°	4.1	pdtexpand
Replacement	PDT-specific alg.	4.5	pdtreplace
Composition	non-trivial changes to FST alg.	4.2	pdtcompose
Determinization	FST alg. useful[†]		fstdeterminize
Epsilon removal	FST alg.		fstrmepsilon
Minimization	FST alg. useful[‡]		fstminimize
Shortest distance	PDT-specific alg.°	4.3	N/A
Shortest path	PDT-specific alg.°	4.3	pdtshortestpath
Pruned expansion	PDT-specific alg.°	4.4	pdtexpand
Pruning	PDT-specific alg. required	4.6	N/A
Connection	PDT-specific alg. required	4.6	N/A

[*]Assumes the presence of distinguished initial and final parentheses.
[°]Requires bounded-stack input.
[†]Reduces the redundancy but does not produce a deterministic PDT.
[‡]Reduces the size but does not perform PDT minimization.

$a \in \Pi$, $z' = za$, $a' = \epsilon$ and $b' = \epsilon$, or (c) $a \in \overline{\Pi}$, $z = z'\overline{a}$, $a' = \epsilon$ and $b' = \epsilon$. The initial state of T' is $I' = (I, \epsilon)$. A state (q, z) in T' is final iff q is final in T and $z = \epsilon$. We have $\rho'((q, \epsilon)) = \rho(q)$. The set of states of T' is the set of pairs (q, z) that can be reached from an initial state by transitions defined as above. The condition that T has bounded stack ensures that this set is finite (since it implies that for any such pair (q, z), $|z| \leq K$).

The complexity of the algorithm is linear in $O(|T'|) = O(e^{|T|})$. Figure 1d shows the result of the algorithm when applied to the PDA of Figure 1c.

4.2 Composition

The class of weighted pushdown transducers is closed under composition with weighted finite-state transducers [5,18]. Considering a pair (T_1, T_2) where one element is an FST and the other element a PDT and such that T_1 has input and output alphabets Σ and Δ and T_2 has input and output alphabets Δ and Γ, then there exists a PDT $T_1 \circ T_2$, the *composition* of T_1 and T_2, such that for all $(x, y) \in \Sigma^* \times \Gamma^*$: $(T_1 \circ T_2)(x, y) = \min_{z \in \Delta^*} (T_1(x, z) + T_2(z, y))$. We assume in the following that T_2 is an FST. We also assume that T_2 has no input-ϵ transitions. When T_2 has input-ϵ transitions, an epsilon filter [16,3] generalized to handle parentheses can be used.

SHORTESTDISTANCE(T)
1 **for each** $q \in Q$ and $a \in \Pi$ **do**
2    $B[q, a] \leftarrow \emptyset$
3   GETDISTANCE(T, I)
4 **return** $d[f, I]$

RELAX($q, s, w, \mathcal{S}$)
1 **if** $d[q, s] > w$ **then**
2    $d[q, s] \leftarrow w$
3   **if** $q \notin S$ **then**
4   ENQUEUE($\mathcal{S}, q$)

GETDISTANCE(T, s)
1 **for each** $q \in Q$ **do**
2    $d[q, s] \leftarrow \infty$
3 $d[s, s] \leftarrow 0$
4 $\mathcal{S}_s \leftarrow s$
5 **while** $\mathcal{S}_s \neq \emptyset$ **do**
6    $q \leftarrow$ HEAD($\mathcal{S}_s$)
7   DEQUEUE($\mathcal{S}_s$)
8   **for each** $e \in E[q]$ **do**
9   **if** $i[e] \in \Sigma \cup \{e\}$ **then**   ▷ $i[e]$ is a regular symbol
10   RELAX($n[e], s, d[q, s] + w[e], \mathcal{S}_s$)
11   **elseif** $i[e] \in \overline{\Pi}$ **then**   ▷ $i[e]$ is a close parenthesis
12       $B[s, \overline{i[e]}] \leftarrow B[s, \overline{i[e]}] \cup \{e\}$
13   **elseif** $i[e] \in \Pi$ **then**   ▷ $i[e]$ is an open parenthesis
14   **if** $d[n[e], n[e]]$ is undefined **then**
15   GETDISTANCE($T, n[e]$)
16   **for each** $e' \in B[n[e], i[e]]$ **do**
17         $w \leftarrow d[q, s] + w[e] + d[p[e'], n[e]] + w[e']$
18   RELAX($n[e'], s, w, \mathcal{S}_s$)

Fig. 3. PDT shortest distance algorithm. We assume that $F = \{f\}$ and $\rho(f) = 0$ to simplify the presentation

A state in $T = T_1 \circ T_2$ is a pair (q_1, q_2) where q_1 is a state of T_1 and q_2 a state of T_2. The initial state is $I = (I_1, I_2)$. Given a transition $e_1 = (q_1, a, b, w_1, q_1')$ in T_1, transitions out of (q_1, q_2) in T are obtained using the following rules.

If $b \in \Delta$, then e_1 can be matched with a transition (q_2, b, c, w_2, q_2') in T_2 resulting a transition $((q_1, q_2), a, c, w_1 + w_2, (q_1', q_2'))$ in T. If $b = \epsilon$, then e_1 is handled by staying in q_2 resulting in a transition $((q_1, q_2), a, \epsilon, w_1, (q_1', q_2))$. Finally, if $b = a \in \widehat{\Pi}$, e_1 is also handled by staying in q_2, resulting in a transition $((q_1, q_2), a, a, w_1, (q_1', q_2))$ in T.

A state (q_1, q_2) in T is final when both q_1 and q_2 are final, and then $\rho((q_1, q_2)) = \rho_1(q_1) + \rho_2(q_2)$. The complexity of the algorithm is $O(|T_1| |T_2|)$ in the worst case.

4.3   Shortest Distance and Shortest Path

A *shortest path* in a PDT T is a balanced accepting path with minimal weight and the *shortest distance* in T is the weight of such a path. We show that when T has bounded stack, the shortest distance and shortest path can be computed in $O(|T|^3 \log |T|)$ time (assuming T has no negative weights) and $O(|T|^2)$ space.

Given a state s in T with at least one incoming open parenthesis transition, we denote by C_s the set of states that can be reached from s by a balanced path. If s has several incoming open parenthesis transitions, a naive implementation might lead to the states in C_s being visited up to exponentially many times. The basic idea of the algorithm is to memoize the shortest distance from s to states in C_s. The pseudo-code is given in Figure 3.

GETDISTANCE(T, s) starts a new instance of the shortest-distance algorithm from s using the queue $\mathcal{S}_s$, initially containing s. While the queue is not empty, a state is dequeued and its outgoing transitions examined (line 5-9). Transitions labeled by non-parenthesis are treated as in Mohri [16] (line 9-10). When the considered transition e is labeled by a close parenthesis, all balancing incoming open parentheses in s labeled by $\overline{i[e]}$ are remembered by adding e to $B[s, \overline{i[e]}]$ (line 11-12). Finally, when e is labeled with an open parenthesis, if its destination has not already been visited, a new instance is started from $n[e]$ (line 14-15). The destination states of all transitions balancing e are then relaxed (line 16-18).

The space complexity of the algorithm is quadratic for two reasons. First, the number of non-infinite $d[q, s]$ is $|Q|^2$. Second, the space required for storing B is at most in $O(|E|^2)$ since for each open parenthesis transition e, the size of $|B[n[e], i[e]]|$ is $O(|E|)$ in the worst case. This last observation also implies that the accumulated number of transitions examined at line 16 is in $O(N|Q| |E|^2)$ in the worst case, where N denotes the maximal number of times a state is inserted in the queue for a given call of GETDISTANCE. Assuming the cost of a queue operation is $\Gamma(n)$ for a queue containing n elements, the worst-case time complexity of the algorithm can then be expressed as $O(N|T|^3 \Gamma(|T|))$. When T contains no negative weights, using a shortest-first queue discipline leads to a time complexity in $O(|T|^3 \log |T|)$. When all the C_s's are acyclic, using a topological order queue discipline leads to a $O(|T|^3)$ time complexity.

When T has been obtained by converting an RTN into a PDA (see Section 4.5), the polynomial dependency in $|T|$ becomes a linear dependency both for the time and space complexities. Indeed, for each q in T, there exists a unique s such that $d[q, s]$ is non-infinite. Moreover, for each open parenthesis transition e, there exists a unique close parenthesis transition e' such that $e' \in B[n[e], i[e]]$. When each component of the RTN is acyclic, the complexity of the algorithm is hence in $O(|T|)$ in time and space.

Similarly, when $T = T_1 \circ T_2$ and T_1 was obtained by converting an RTN into a PDA, the complexity becomes $O(N|T_1||T_2|^3 \Gamma(|T|))$ in time and $O(|T_1||T_2|^2)$ in space. This follows since for each (q_1, q_2) there exists a unique s_1 such that $d[(q_1, q_2), (s_1, s_2)]$ is non-infinite. Also, for each open parenthesis transition e, there exist at most $|T_2|$ close parenthesis transition e' such that $e' \in B[n[e], i[e]]$.

The algorithm can be modified (without changing the complexity) to compute the shortest path through T by keeping track of parent pointers.

4.4   Pruned Expansion

Given a bounded-stack PDT T, the *pruned expansion* of T with threshold β is an FST T'_β obtained by deleting from T' all states and transitions that belong to no accepting path π in T' such that $\lambda'(p[\pi]) + w[\pi] + \rho'(n[\pi]) \leq d + \beta$ where d is the shortest distance in T. A naive implementation consisting of fully expanding T and then applying the FST pruning algorithm would lead to a complexity in $O(|T'| \log |T'|) = O(e^{|T|}|T|)$.

Assuming that the reverse T^R of T is also bounded-stack, an algorithm whose complexity is in $O(|T| |T'_\beta| + |T|^3 \log |T|)$ can be obtained by first applying the shortest distance algorithm from the previous section to T^R and then using this to prune the expansion as it is generated. When invoking the `pdtexpand` command, the `--weight` flag can be used to specify the threshold β and trigger a pruned expansion of the input PDT.

4.5 Replacement

A *recursive transitive network* (RTN) R is specified by $(N, \Sigma, \Delta, (T_\nu)_{\nu \in N}, S)$ where N is an alphabet of nonterminals, Σ and Δ are the input and output alphabets, $(T_\nu)_{\nu \in N}$ is a family of FSTs with input alphabet $\Sigma \cup N$ and output alphabet Δ, and $S \in N$ is the root nonterminal.

A pair $(x, y) \in \Sigma^* \times \Delta^*$ is accepted by R if there exists an accepting path π in T_S such that recursively replacing any transition with input label $\nu \in N$ by an accepting path in T_ν leads to a path π^* with input x and output y. The weight associated by R is the minimum over all such π^* of $w[\pi^*] + \rho_S(n[\pi^*])$.

Given an RTN R, the *replacement* of R is the PDT T equivalent to R defined by the 10-tuple $(\Sigma, \Delta, \Pi, \overline{\Pi}, Q, E, I, F, \sigma, \rho)$ with $\Pi = Q = \bigcup_{\nu \in N} Q_\nu$, $I = I_S$, $F = F_S$, $\rho = \rho_S$, and $E = \bigcup_{\nu \in N} \bigcup_{e \in E_\nu} E^e$ where $E^e = \{e\}$ if $i[e] \notin N$ and otherwise $E^e = \{(p[e], n[e], \epsilon, w[e], I_\mu), (f, \overline{n[e]}, \epsilon, \rho_\mu(f), n[e]) | f \in F_\mu\}$ with $\mu = i[e] \in N$.

The complexity of the construction is in $O(|T|)$. If $|F_\nu| = 1$, then $|T| = O(\sum_{\nu \in N} |T_\nu|) = O(|R|)$. Creating a superfinal state for each T_ν would lead to a T whose size is always linear in the size of R.

4.6 Discussion

The PDT expansion algorithm can result in an FST that is not *trim*: it may contain useless states or transitions not on accepting paths. OpenFst provides the Connect operation that performs classical finite-automata trimming (using a depth-first search). By analogy, a PDT can be defined trim if each state and transition lies on a balanced, accepting path. Similarly, a PDT can be defined *pruned with threshold* β if each state and transition lies on a balanced, accepting path with weight $w \le d + \beta$ where d is the shortest distance in the PDT. In the future, we wish to add algorithms Connect to trim a bounded-stack PDT and Prune to prune a bounded-stack PDT within threshold β. Note these algorithms are different from the connected or pruned expansion of a PDT, since the results here, in general, are PDTs not FSTs.

5 Applications

5.1 Recognition

Suppose we have an acyclic weighted finite automaton L that represents the likelihood $Pr[x|s]$ of some observation x given a sentence $s \in L$. For example, x could be spoken or written words with $Pr[x|s]$ being acoustically or optically-derived likelihoods from an automatic speech recognition (ASR) or optical character recognition (OCR) system. Further, suppose we have a weighted context-free grammar G that represents the a priori probability $Pr[s]$ of each sentence in the grammar. We wish to compute the maximum a posteriori probability sentence, $\underset{s}{\operatorname{argmax}} Pr[x|s] Pr[s]$, given L and G.

To do so, we will first represent G as a pushdown automaton. A weighted context-free grammar (CFG) can be specified by (N, Σ, P, S) where N is an

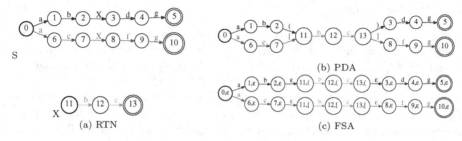

Fig. 4. Automata representations

alphabet of nonterminals, Σ is an alphabet of terminals, $P \subseteq N \times (N \cup \Sigma)^* \times (R \cup \{\infty\})$ are productions and S is the start symbol. A production (ν, α, w) is sometimes written as $\nu \to \alpha/w$.

To create a PDA that represents G, use each production (ν, α, w) to create the linear FSA $A_{\nu,\alpha,w}$ that accepts α with weight w. Then for each non-terminal ν, form the finite-state union $T_\nu = \cup_{(\nu,\alpha,w) \in P} A_{\nu,\alpha,w}$. Then $(N, \Sigma, \Sigma, (T_\nu)_{\nu \in N}, S)$ is an RTN R_G for which each accepting path π is in $1:1$ correspondence with a leftmost derivation of $i(\pi)$ in G [15]. Finally, use the construction in Section 4.5 to represent R_G as a PDA T_G.

For example, consider the context-free grammar: $S \to abXdg$, $S \to acXfg$ and $X \to bc$. Figure 4 shows several automata representations of this grammar. Figure 4a shows the RTN representation of this grammar with a 1:1 correspondence between each production in the CFG and each accepting path in the RTN components. Figure 4b shows the pushdown automaton representation generated from the RTN with the replacement algorithm of Section 4.5. Since this grammar's productions have no cyclic dependencies, the PDA has bounded stack and represents a regular language. Figure 4c shows the finite-state automaton representation of this grammar generated by the PDA using the expansion algorithm of Section 4.1.

For the probabilistic recognition example, we use negative log probabilities in the weighted finite automaton L and in the construction of the PDT T_G that represents CFG G. Then, the maximum a posteriori sentence can be found with $ShortestPath(L \cap T_G)$. With the command line operations, this becomes:

```
pdtcompose --pdt_parentheses=parens G.pda L.fsa |
pdtshortestpath --pdt_parentheses=parens > Map.fsa
```

since composition between acceptors is intersection.[1] The recognition has time complexity in $O(|L|^3 |T_G|)$ and space complexity in $O(|L|^2 |T_G|)$ since T_G has bounded stack and is derived from an RTN.

An advantage of the RTN, PDA, and FSA representations is that they can benefit from FSA epsilon removal, determinization and minimization algorithms applied to their components (for RTNs and PDAs) or their entirety (for FSAs). These steps could improve the time and space requirements of the recognition example.

---

[1] The compostion flag `--left_pdt=false` would be required if the arguments were exchanged.

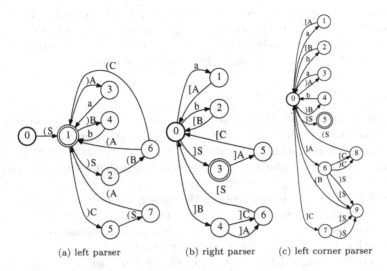

(a) left parser   (b) right parser   (c) left corner parser

Fig. 5. Different parsing strategies using PDTs

In a real-world example, this approach essentially is used to identify *voice action* queries in the Google Android speech platform. For example, a production could be $S \rightarrow$ send a message from X to Y where the non-terminals X and Y, for the sender and recipient, are rewritten as people's names. A match identifies a voice query as a messaging action.

5.2   Parsing

In the final example in the last section, we might not only wish to identify a messaging action in a voice query but also want to *parse* the input to find where the sender and recipient names are located. This is very similar to CFG recognition but with the output augmented with the parse bracketing. A classical approach is to augment the output tape of the PDT to include an index for each production [1]. We take another approach here: the parentheses are chosen to identify the production (or non-terminal) and the parentheses are retained in the shortest path output. With the command line operations, this is done with the flag --keep_parentheses. This does not increase the time or space complexity over recognition.

It has long been known that PDTs can be used to parse and that different parsing *strategies* can be achieved by compiling the CFG into different PDTs [1,13]. For example, the CFG: $S \rightarrow AB$, $S \rightarrow CB$, $C \rightarrow AS$, $A \rightarrow a$ and $B \rightarrow b$ can be left parsed ('top-down') by the PDT in Figure 5a, right parsed ('bottom-up') by the PDT in Figure 5b, and left-corner parsed by the PDT in Figure 5c [1]. Note an equivalent right parser can be obtained from the left parser by first reversing the right-hand side of the productions and then reversing the transducer.

The classical method to apply these parsers is equivalent to intersecting the PDT with the input string followed by the exponential expansion algorithm of Section 4.1. Lang [13] showed that the cubic tabular method of Earley can be naturally applied to PDTs; others give the weighted generalizations [21,19]. These approaches are closely related to intersecting the PDT with the input string followed by the shortest path algorithm of Section 4.3.

5.3   Translation

Hierarchical phrase-based translation, using a *synchronous context-free transla-tion grammar* (SCFG) G together with an n-gram target language model M, is a popular approach in machine translation [8]. The productions of the SCFG are of the form $S \rightarrow \langle uAvBw, xByAz \rangle$. This production says that $uAvBw$ translates to $xByAz$ where u, v, w, x, y, z are terminal strings and A and B are non-terminals that must be in $1 : 1$ correspondence in the source and target of the translation but not necessarily in the same order. If all the productions preserved this or-der, it would be possible to represent the translation grammar as a pushdown *transducer* but for a general SCFG this is not possible [1].

However, the result of the application of the input source string s to the probabilistic translation grammar G, which represents all possible translations of s by G, is compactly represented by a weighted RTN or PDA $T_{s,G}$ [11] [2]. It has bounded-stack, since the input s has already been applied to the SCFG.

Applying the n-gram language model M to $T_{s,G}$ and searching for the best resulting translation, typically the computationally expensive steps in transla-tion, becomes $ShortestPath(T_{s,G} \cap M)$. It has time complexity in $O(|T_{s,G}||M|^3)$ and space complexity in $O(|T_{s,G}||M|^2)$ since $T_{s,G}$ has bounded stack and is de-rived from an RTN. An alternative approach first expands $T_{s,G}$ to an FSA $F_{s,G}$ and then applies finite-state intersection and shortest path to give a time and space complexity of $O(|e^{|F_{s,G}|}|M|)$. Gonzalo, et al [11] give experimental results comparing these two approaches on a range of grammar and n-gram language model sizes in a large-scale English-Chinese translation system.

5.4   Discussion

For each of these tasks - recognition, parsing, or translation - real-world prob-lems might involve very large CFGs. In these cases, the cubic complexity of the shortest path algorithm may be prohibitive and *inadmissable* or *inexact* methods may be used that are not guaranteed to return the shortest path. One general approach is to prune away unpromising paths [8,10]. Another approach is to use a weaker, smaller grammar in a first pass, output a hypothesis set, and rescore that with the full grammar. For the latter method, the pruned expansion of Section 4.4 can be used to output the hypothesis sets.

Acknowledgments. We thank Mehryar Mohri for suggesting a PDT algorithms library and discussions and thank Bill Byrne, Adrià de Gispert and Gonzalo Igle-sias for working with us to adapt their pioneering automata approach for machine translation to PDTs along with their comprehensive evaluations of these methods.

---

[2] Another related representation, *hypergraphs*, are also often used for this purpose [11].

References

1. Aho, A.V., Ullman, J.D.: The Theory of Parsing, Translation and Compiling, vol. 1-2. Prentice-Hall (1972)
2. Allauzen, C., Riley, M.: Pushdown Transducers (2011), http://pdt.openfst.org
3. Allauzen, C., Riley, M., Schalkwyk, J.: Filters for Efficient Composition of Weighted Finite-State Transducers. In: Domaratzki, M., Salomaa, K. (eds.) CIAA 2010. LNCS, vol. 6482, pp. 28–38. Springer, Heidelberg (2011)
4. Allauzen, C., Riley, M., Schalkwyk, J., Skut, W., Mohri, M.: OpenFst: A General and Efficient Weighted Finite-State Transducer Library. In: Holub, J., Žďárek, J. (eds.) CIAA 2007. LNCS, vol. 4783, pp. 11–23. Springer, Heidelberg (2007), http://www.openfst.org
5. Bar-Hillel, Y., Perles, M., Shamir, E.: On formal properties of simple phrase structure grammars. In: Bar-Hillel, Y. (ed.) Language and Information: Selected Essays on their Theory and Application, pp. 116–150. Addison-Wesley (1964)
6. Berstel, J.: Transductions and Context-Free Languages. Teubner (1979)
7. Chen, S.F.: Designing a non-finite-state weighted transducer toolkit. Technical Report RC 24829, IBM Research Division (2009)
8. Chiang, D.: Hierarchical phrase-based translation. Computational Linguistics 33(2), 201–228 (2007)
9. Drosde, M., Kuick, W., Vogler, H. (eds.): Handbook of Weighted Automata. Springer (2009)
10. Hall, K., Johnson, M.: Language modeling using efficient best-first bottom-up parsing. In: Proceedings of ASRU (2003)
11. Iglesias, G., Allauzen, C., Byrne, W., de Gispert, A., Riley, M.: Hierarchical phrase-based translation representations. In: Proc. EMNLP, pp. 1373–1383 (2011)
12. Kuich, W., Salomaa, A.: Semirings, automata, languages. Springer (1986)
13. Lang, B.: Deterministic Techniques for Efficient Non-Deterministic Parsers. In: Loeckx, J. (ed.) ICALP 1974. LNCS, vol. 14, pp. 255–269. Springer, Heidelberg (1974)
14. Maryanski, F.J., Thomason, M.G.: Properties of stochastic syntax-directed translation schemata. International Journal of Computer and Information Sciences 8(2), 89–110 (1979)
15. Mohri, M.: Weighted grammar tools: the GRM library. In: Robustness in Language and Speech Technology, pp. 165–186. Kluwer (2001)
16. Mohri, M.: Weighted automata algorithms. In: Drosde, et al. [9], ch. 6, pp. 213–254
17. Mohri, M., Pereira, F.C.N., Riley, M.: Weighted finite-state transducers in speech recognition. Computer Speech and Language 16(1), 69–88 (2002)
18. Nederhof, M.-J., Satta, G.: Probabilistic parsing as intersection. In: Proceedings of 8th International Workshop on Parsing Technologies, pp. 137–148 (2003)
19. Nederhof, M.J., Satta, G.: Probabilistic parsing strategies. Journal of the ACM 53(3), 406–436 (2006)
20. Petre, I., Salomaa, A.: Algebraic systems and pushdown automata. In: Drosde, et al [9], ch. 7, pp. 257–289
21. Stolcke, A.: An efficient probabilistic context-free parsing algorithm that computes prefix probabilities. Computational Linguistics 21(2), 165–201 (1995)

Weak Inclusion for Recursive XML Types

Joshua Amavi, Jacques Chabin, and Pierre Réty

LIFO - Université d'Orléans, B.P. 6759, 45067 Orléans cedex 2, France
{joshua.amavi,jacques.chabin,pierre.rety}@univ-orleans.fr

Abstract. Considering that the *unranked* tree languages $L(G)$ and $L(G')$ are those defined by given *possibly-recursive XML types* G and G', this paper proposes a method to verify whether $L(G)$ is "approximatively" included in $L(G')$. The approximation consists in weakening the father-children relationships. Experimental results are discussed, showing the efficiency of our method in many situations.

Keywords: XML type, regular unranked-tree grammar, approximative inclusion.

1 Introduction

In database area, an important problem is schema evolution, particularly when considering XML types. XML is also used for exchanging data on the web. In this setting, we want to compare XML types in a loose way. To do it, we address the more general problem of approximative comparison of unranked-tree languages defined by regular grammars.

Example 1. Suppose an application where we want to replace an XML type G by a new type G' (*eg.*, a web service composition where a service replaces another, each of them being associated to its own XML message type). We want to analyse whether the XML messages supported by G' contains (in an approximate way) those supported by G. XML types are regular tree grammars where we just consider the structural part of the XML documents, disregarding data attached to leaves. Thus, to define leaves we consider rules of the form $A \rightarrow a[\epsilon]$.
Suppose that G and G' contain the following rules:

$\quad$ F $\rightarrow$ firstName[ϵ], L $\rightarrow$ lastName[ϵ] , T $\rightarrow$ title[ϵ] and Y $\rightarrow$ year[ϵ].

P defines a publication, and B is the bibliography.
In G : P $\rightarrow$ publi[$(F.L)^+.T.B^?$], B $\rightarrow$ biblio[P^+].
In G' : P $\rightarrow$ publi[$A^*.Pa$], A $\rightarrow$ author[$F.L$], Pa $\rightarrow$ paper[$T.Y.B^?$], B $\rightarrow$ biblio[P^+]

We want to know whether messages valid with respect to G can be accepted (in an approximate way) by G'. Notice that G accepts trees such as t in Figure 1 that are not valid with respect to schema G' but that represent the same kind of information G' deals with. Indeed, in G', the same information would be organized as the tree t' in Figure 1. $\qquad\square$

N. Moreira and R. Reis (Eds.): CIAA 2012, LNCS 7381, pp. 78–89, 2012.

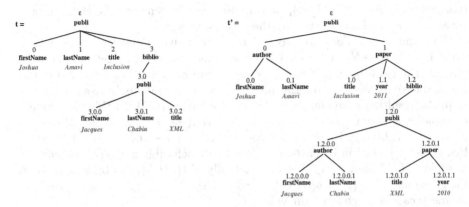

Fig. 1. Examples of trees t and t' valid with respect to G and G', respectively

The approximative criterion for comparing trees that is commonly used consists in weakening the father-children relationships (*i.e.*, they are implicitly reflected in the data tree as only ancestor-descendant). In this paper, we consider this criterion in the context of tree languages. We denote this relation *weak inclusion* to avoid confusion with the usual inclusion of languages (*i.e.*, the inclusion of a set of trees in another one).

Given two types G and G', we call $L(G)$ and $L(G')$ the sets (also called languages) of XML documents valid with respect to G and G', respectively. Our paper proposes a method for deciding whether $L(G)$ is *weakly included* in $L(G')$, in order to know if the substitution of G by G' can be envisaged. The unranked-tree language $L(G)$ is *weakly included* in $L(G')$ if for each tree $t \in L(G)$ there is a tree $t' \in L(G')$ such that t is weakly included in t'. Intuitively, t is weakly included in t' (denoted $t \lhd t'$) if we can obtain t by removing nodes from t' (a removed node is replaced by its children, if any). For instance, in Figure 1, t can be obtained by removing nodes *author*, *paper*, *year* from t', i.e. we have $t \lhd t'$.

To decide whether $L(G)$ is weakly included in $L(G')$, we consider the set of trees $WI(L(G')) = \{t \mid \exists t' \in L(G'), t \lhd t'\}$. Note that[1] $L(G)$ is weakly included in $L(G')$ iff $L(G) \subseteq WI(L(G'))$.

To compute $WI(L(G'))$, we have already proposed [3] a direct and simple approach using regular unranked-tree grammars (hedge grammars), assuming that $L(G')$ is bounded in depth, i.e. G' is not recursive. Given a hedge grammar G', the idea consists in replacing each non-terminal occurring in a right-hand-side of a production rule, by itself or its children. For example, if G' contains the rules $A \rightarrow a[B]$, $B \rightarrow b[C]$, $C \rightarrow c[\epsilon]$, we get the grammar $G'' = \{C \rightarrow c[\epsilon], B \rightarrow b[C|\epsilon], A \rightarrow a[B|(C|\epsilon)]\}$. This grammar generates $WI(L(G'))$, because in each regular expression, whenever we have a non-terminal X ($X \in \{A, B, C\}$)

---

[1] If $L(G)$ is weakly included in $L(G')$, for $t \in L(G)$ there exists $t' \in L(G')$ s.t. $t \lhd t'$. Then $t \in WI(L(G'))$, hence $L(G) \subseteq WI(L(G'))$.

    Conversely if $L(G) \subseteq WI(L(G'))$, for $t \in L(G)$ we have $t \in WI(L(G'))$. Thus there exists $t' \in L(G')$ s.t. $t \lhd t'$. Therefore $L(G)$ is weakly included in $L(G')$.

that generates x $(x \in \{a, b, c\})$, we can also directly generate the children of x (instead of x), and more generally the successors of x.

Unfortunately, it does not work if G' is recursive. Consider $G'_1 = \{A \rightarrow a[B.(A|\epsilon).C], B \rightarrow b[\epsilon], C \rightarrow c[\epsilon]\}$. If we work as previously, we get the rule $A \rightarrow a[B.(A|(B.(A|\epsilon).C)|\epsilon).C]$. However, a new occurrence of A has appeared in the rhs, and if we replace it again, this process does not terminate. If we stop at some step, the resulting grammar does not generate $WI(L(G'))$. In this very simple example, it is easy to see that $G''_1 = \{A \rightarrow a[B^*.(A|\epsilon).C^*], B \rightarrow b[\epsilon], C \rightarrow c[\epsilon]\}$ generates $WI(L(G'_1))$. But if we now consider $G'_2 = \{A \rightarrow a[B.(A.A \mid \epsilon).C], B \rightarrow b[\epsilon], C \rightarrow c[\epsilon]\}$, then in $WI(L(G'_2))$, a may be the left-sibling of b (which was not possible in $L(G'_2)$ nor in $WI(L(G'_1)))$. Actually, $WI(L(G'_2))$ can be generated by the grammar $G''_2 = \{A \rightarrow a[(A|B|C)^*], B \rightarrow b[\epsilon], C \rightarrow c[\epsilon]\}$. In other words, the recursive case is much more difficult.

In this paper, we address the general case: some symbols may be recursive, and some others may not. Given an arbitrary regular unranked-tree grammar G', we present a direct approach that computes a regular grammar, denoted $WI(G')$, that generates the language $WI(L(G'))$. To do it, terminal-symbols are divided into 3 categories: non-recursive, 1-recursive, 2-recursive. For $n > 2$, n-recursivity is equivalent to 2-recursivity. Surprisingly, the most difficult situation is 1-recursivity. We prove that our algorithm for computing $WI(G')$ is correct and complete. An implementation has been done in Java, and experiments are presented.

Consequently, for arbitrary regular grammars G and G', checking that $L(G)$ is weakly included into $L(G')$, i.e. checking that $L(G) \subseteq WI(L(G'))$, is equivalent to check that $L(G) \subseteq L(WI(G'))$, which is decidable since G and $WI(G')$ are regular grammars.

Paper Organisation: Section 2 gives some theoretical background. Section 3 presents how to compute $WI(G)$ for a given possibly-recursive regular grammar G, while Section 4 analyses some experimental results. Related work and other possible methods are addressed in Section 5. Due to the lack of space, missing proofs are given in [4].

2   Preliminaries

An XML document is an *unranked tree*, defined in the usual way as a mapping t from a set of positions $Pos(t)$ to an alphabet Σ. The set of the trees over Σ is denoted by T_Σ. For $v \in Pos(t)$, $t(v)$ is the label of t at the position v, and $t|_v$ denotes the sub-tree of t at position v. *Positions* are sequences of integers in $\mathbb{N}^*$ and $Pos(t)$ satisfies: $\forall u, i, j \, (j \geq 0, u.j \in Pos(t), 0 \leq i \leq j) \Rightarrow u.i \in Pos(t)$ (char ".". denotes the concatenation). The *size* of t (denoted $|t|$) is the cardinal of $Pos(t)$. As usual, ϵ denotes the empty sequence of integers, i.e. the root position. t, t' will denote trees.

Figure 1 illustrates trees with positions and labels: we have, for instance, $t(1) = lastName$ and $t'(1) = paper$. The sub-tree $t'|_{1.2}$ is the one whose root is *biblio*.

Definition 1. Position comparison: Let $p, q \in Pos(t)$. Position p is an *ancestor* of q (denoted $p < q$) if there is a non-empty sequence of integers r such that $q = p.r$. Position p is *to the left* of q (denoted $p \prec q$) if there are sequences of integers u, v, w, and $i, j \in \mathbb{N}$ such that $p = u.i.v$, $q = u.j.w$, and $i < j$. Position p is *parallel* to q (denoted $p \parallel q$) if $\neg(p < q) \wedge \neg(q < p)$.   □

Definition 2. Resulting tree after node deletion: For a tree t' and a non-empty position q of t', let us note $Rem_q(t') = t$ the tree t obtained from t' by removing the node at position q (a removed node is replaced by its children, if any). We have:

1. $t(\epsilon) = t'(\epsilon)$,
2. $\forall p \in Pos(t')$ such that $p < q$: $t(p) = t'(p)$,
3. $\forall p \in Pos(t')$ such that $p \prec q$: $t|_p = t'|_p$,
4. Let $q.0, q.1..., q.n \in Pos(t')$ be the positions of the children of position q, if q has no child, let $n = -1$. Now suppose $q = s.k$ where $s \in \mathbb{N}^*$ and $k \in \mathbb{N}$. We have:
   - $t|_{s.(k+n+i)} = t'|_{s.(k+i)}$ for all i such that $i > 0$ and $s.(k+i) \in Pos(t')$ (the siblings located to the right of q shift),
   - $t|_{s.(k+i)} = t'|_{s.k.i}$ for all i such that $0 \leq i \leq n$ (the children go up).   □

Definition 3. Weak Inclusion for Unranked Trees: The tree t is *weakly included in* t' (denoted $t \lhd t'$) if there exists a series of positions $q_1 \dots q_n$ such that $t = Rem_{q_n}(\cdots Rem_{q_1}(t'))$.   □

Example 2.
In Figure 1, $t = Rem_0(Rem_1(Rem_{1.1}(Rem_{1.2.0.0}(Rem_{1.2.0.1}(Rem_{1.2.0.1.1}(t'))))))$, then $t \lhd t'$. Notice that for each node of t, there is a node in t' with the same label, and this mapping preserves vertical order and left-right order. However the tree $t_1 = paper(biblio, year)$ is not weakly included in t' since *biblio* should appear to the right of *year*.   □

Definition 4. Regular Tree Grammar: A *regular tree grammar* (RTG) (also called *hedge grammar*) is a 4-tuple $G = (NT, \Sigma, S, P)$, where NT is a finite set of *non-terminal symbols*; Σ is a finite set of *terminal symbols*; S is a set of *start symbols*, where $S \subseteq NT$ and P is a finite set of *production rules* of the form $X \rightarrow a[R]$, where $X \in NT$, $a \in \Sigma$, and R is a regular expression over NT. We recall that the set of regular expressions over $NT = \{A_1, \dots, A_n\}$ is inductively defined by: $R ::= \epsilon \mid A_i \mid R|R \mid R.R \mid R^+ \mid R^* \mid R^? \mid (R)$ □

Grammar in Normal Form: As usual, in this paper, we only consider regular tree grammars such that (i) every non-terminal generates at least one tree containing only terminal symbols and (ii) distinct production rules have distinct left-hand-sides (*i.e.*, tree grammars *in normal form* [13]).

Thus, given an RTG $G = (NT, \Sigma, S, P)$, for each $A \in NT$ there exists in P exactly one rule of the form $A \rightarrow a[E]$, i.e. whose left-hand-side is A.   □

Example 3. The grammar $G_0 = (NT_0, \Sigma, S, P_0)$, where $NT_0 = \{X, A, B\}$, $\Sigma = \{f, a, c\}$, $S = \{X\}$, and $P_0 = \{X \rightarrow f[A.B], A \rightarrow a[\epsilon], B \rightarrow a[\epsilon], A \rightarrow c[\epsilon]\}$, is not in normal form. The conversion of G_0 into normal form gives the sets $NT_1 = \{X, A, B, C\}$ and $P_1 = \{X \rightarrow f[(A|C).B], A \rightarrow a[\epsilon], B \rightarrow a[\epsilon], C \rightarrow c[\epsilon]\}$.

Definition 5. Let $G = (NT, \Sigma, S, P)$ be an RTG (in normal form). Consider a non-terminal $A \in NT$, and let $A \rightarrow a[E]$ be the unique production of P whose left-hande-side is A.
$L^w(E)$ denotes the set of words (over non-terminals) generated by E.
  The set $L_G(A)$ of trees generated by A is defined recursively by:

$$L_G(A) = \{a(t_1, \ldots, t_n) \mid \exists u \in L^w(E), u = A_1 \ldots A_n, \forall i, t_i \in L_G(A_i)\}$$

The language $L(G)$ generated by G is: $L(G) = \{t \in T_\Sigma \mid \exists A \in S, t \in L_G(A)\}$.
A *nt-tree* is a tree whose labels are non-terminals. The set $L_G^{nt}(A)$ of nt-trees generated by A is defined recursively by:

$$L_G^{nt}(A) = \{A(t_1, \ldots, t_n) \mid \exists u \in L^w(E), u = A_1 \ldots A_n, \forall i (t_i \in L_G^{nt}(A_i) \lor t_i = A_i)\}$$

3 Weak Inclusion for Possibly-Recursive Tree Grammars

First, we need to compute the recursivity types of non-terminals. Intuitively, the non-terminal A of a grammar G is *2-recursive* if there exists $t \in L_G^{nt}(A)$ and A occurs in t at (at least) two non-empty positions $p, q \in Pos(t)$ s.t. $p \parallel q$. A is *1-recursive* if A is not 2-recursive, and A occurs in some $t \in L_G^{nt}(A)$ at a non-empty position. A is *not recursive*, if A is neither 2-recursive nor 1-recursive.

Example 4. Consider the grammar G of Example 1. P is 2-recursive since P may generate B, and B may generate the tree $biblio(P, P)$. B is also 2-recursive. On the other hand F, L, T, Y are not recursive. No non-terminal of G is 1-recursive.

Definition 6. Let $G = (NT, \Sigma, S, P)$ be an RTG in normal form. For a regular expression E, $NT(E)$ denotes the set of non-terminals occurring in E.
  - We define the relation $>$ over non-terminals by:
    $A > B$ if $\exists A \rightarrow a[E] \in P$ s.t. $B \in NT(E)$.
  - We define $>$ over multisets[2] of non-terminals, whose size is at most 2, by:
    - $\{A\} > \{B\}$ if $A > B$,
    - $\{A, B\} > \{C, D\}$ if $A = C$ and $B > D$,
    - $\{A\} > \{C, D\}$ if there exists a production $A \rightarrow a[E]$ in G and a word $u \in L(E)$ of the form $u = u_1 C u_2 D u_3$.

Remark 1. To check whether $\{A\} > \{C, D\}$, i.e. $\exists u \in L(E)$, $u = u_1 C u_2 D u_3$, we can use the recursive function "*in*" defined by $in(C, D, E) =$
- if $E = E_1 | E_2$, return $in(C, D, E_1) \lor in(C, D, E_2)$,
- if $E = E_1 . E_2$, return $(C \in NT(E_1) \land D \in NT(E_2)) \lor (C \in NT(E_2) \land D \in NT(E_1))$
  $\lor in(C, D, E_1) \lor in(C, D, E_2)$,

---

[2] Since we consider multisets, note that $\{A, B\} = \{B, A\}$ and $\{C, D\} = \{D, C\}$.

- if $E = E_1^*$ or $E = E_1^+$, return $(C \in NT(E_1)) \wedge (D \in NT(E_1))$,
- if $E = E_1^?$, return $in(C, D, E_1)$,
- if E is a non-terminal or $E = \epsilon$, return $false$.

This function terminates since recursive calls are always on regular expressions smaller than E. The runtime is $O(|E|)$, where $|E|$ is the size of E.

Definition 7. Let $>^+$ be the transitive closure of $>$

- The non-terminal A is 2-recursive iff $\{A\} >^+ \{A, A\}$.
- A is 1-recursive iff $A >^+ A$ and A is not 2-recursive.
- A is not recursive iff A is neither 2-recursive nor 1-recursive.

Remark 2. The transitive closure of $>$ can be computed using Warshall algorithm. If there are n non-terminals in G, there are $p = n + \frac{n.(n+1)}{2}$ multisets of size at most 2. Then a boolean matrix $p \times p$ can represent $>$, consequently the runtime for computing $>^+$ is $O(p^3) = O(n^6)$, which is polynomial.

Example 5. Using grammar G of Example 1, we have $\{P\} > \{B\} > \{P, P\}$. Therefore P is 2-recursive.
We have $\neg(\{F\} >^+ \{F, F\})$ and $\neg(F >^+ F)$, therefore F is not recursive.

Now, to define an RTG that generates $WI(L(G))$, we need additional notions.

Definition 8. Let $G = (NT, \Sigma, S, P)$ be an RTG in normal form.

- $\equiv$ is the relation over non-terminals defined by $A \equiv B$ if $A >^* B \wedge B >^* A$, where $>^*$ denotes the reflexive-transitive closure of $>$.
  Note that $\equiv$ is an equivalence relation, and if $A \equiv B$ then A and B have the same recursivity type. $\widehat{A}$ will denote the equivalence class of A.
- $Succ(A)$ is the set of non-terminals s.t. $Succ(A) = \{X \in NT \mid A >^* X\}$.
- For a set $Q = \{A_1, \ldots, A_n\}$ of non-terminals, $Succ(Q) = Succ(A_1) \cup \cdots \cup Succ(A_n)$.
- $Left(A)$ is the set of non-terminals defined by $Left(A) = \{X \in NT \mid \exists B, C \in \widehat{A}, \exists B \to b[E] \in P, \exists u \in L^w(E), u = u_1 X u_2 C u_3\}$.
- Similarly, $Right(A)$ is the set of non-terminals defined by $Right(A) = \{X \in NT \mid \exists B, C \in \widehat{A}, \exists B \to b[E] \in P, \exists u \in L^w(E), u = u_1 C u_2 X u_3\}$.
- $RE(A)$ is the regular expression E, assuming $A \to a[E]$ is the production rule of G whose left-hand-side is A.
- $\widehat{RE}(A) = RE(A)|RE(B_1)| \cdots |RE(B_n)$ where $\widehat{A} = \{A, B_1, \ldots, B_n\}$.

Example 6. With the grammar G' of Example 1, we have :

- $P \equiv B$, because $P > Pa > B$ and $B > P$.
- $\widehat{P} = \{P, Pa, B\}$.
- $Succ(A) = \{A, F, L\}$.
- $Left(P)$ is defined using non-terminals equivalent ($\equiv$) to P, i.e. B, P, Pa, and grammar G', which contains rules (among others):
  $B \to biblio[P^+]$, $P \to publi[A^*.Pa]$, $Pa \to paper[T.Y.B^?]$.

P^+ may generate $P.P$, therefore $Left(P) = \{P\} \cup \{A\} \cup \{T, Y\} = \{P, A, T, Y\}$.
- $RE(P) = A^*.Pa$
- $\widehat{RE}(P) = RE(P)|RE(Pa)|RE(B) = (A^*.Pa)|(T.Y.B^?)|P^+$.

Lemma 1. *Let A, B be non-terminals.*
- *If $A \equiv B$ then $Succ(A) = Succ(B)$, $Left(A) = Left(B)$, $Right(A) = Right(B)$, $\widehat{RE}(A) = \widehat{RE}(B)$.*
- *If A, B are not recursive, then $A \neq B$ implies $A \not\equiv B$, therefore $\widehat{A} = \{A\}$ and $\widehat{B} = \{B\}$, i.e. equivalence classes of non-recursive non-terminals are singletons.*

Proof. The first part is obvious.
$(A >^+ A) \implies (A >^+ A \wedge \neg(\{A\} >^+ \{A, A\})) \vee (\{A\} >^+ \{A, A\})$ which implies that A is (1 or 2)-recursive. Consequently, if A is not recursive, then $A \not>^+ A$. Now, if A, B are not recursive, $A \neq B$ and $A \equiv B$, then $A >^+ B \wedge B >^+ A$, therefore $A >^+ A$, which is impossible as shown above. $\qquad\square$

To take all cases into account, the following definition is a bit intricate. To give intuition, consider the following very simple situations:
- If the initial grammar G contains production rules $A \to a[B]$, $B \to b[\epsilon]$ (here A and B are not recursive), we replace these rules by $A \to a[B|\epsilon]$, $B \to b[\epsilon]$ to generate $WI(L(G))$. Intuitively, b may be generated or removed (replaced by ϵ). See Example 7 below for a more general situation.
- If G contains $A \to a[[(A.A.B)|\epsilon]$, $B \to b[\epsilon]$ (here A is 2-recursive and B is not recursive), we replace these rules by $A \to a[(A|B)^*]$, $B \to b[\epsilon]$ to generate $WI(L(G))$. Actually the regular expression $(A|B)^*$ generates all words composed of elements of $Succ(A)$. See Example 8 for more intuition.
- The 1-recursive case is more complicated and is illustrated by Examples 9 and 10.

Definition 9. For each non-terminal A, we recursively define a regular expression $Ch(A)$ (Ch for children). Here, any set of non-terminals, like $\{A_1, \ldots, A_n\}$, is also considered as being the regular expression $(A_1|\cdots|A_n)$.
- if A is 2-recursive, $Ch(A) = (Succ(A))^*$
- if A is 1-rec, $Ch(A) = (Succ(Left(A)))^*.Ch_{\widehat{A}}^{rex}(\widehat{RE}(A)).(Succ(Right(A)))^*$
- if A is not recursive, $Ch(A) = Ch_{\widehat{A}}^{rex}(RE(A))$

and $Ch_{\widehat{A}}^{rex}(E)$ is the regular expression obtained from E by replacing each non-terminal B occurring in E by $Ch_{\widehat{A}}(B)$, where

$$Ch_{\widehat{A}}(B) = \begin{cases} - \ \widehat{B}|\epsilon & \text{if } B \text{ is 1-recursive and } B \in \widehat{A} \\ - \ Ch(B) & \text{if } (B \text{ is 2 recursive) or } (B \text{ is 1-recursive and } B \notin \widehat{A}) \\ - \ B|Ch(B) & \text{if } B \text{ is not recursive} \end{cases}$$

By convention $Ch_{\widehat{A}}^{rex}(\epsilon) = \epsilon$ and $Ch(\epsilon) = \epsilon$.

Algorithm. *Input:* let $G = (NT, T, S, P)$ be a regular grammar in normal form. *Output:* grammar $G' = (NT, T, S, P')$ obtained from G by replacing each production $A \rightarrow a[E]$ of G by $A \rightarrow a[Ch(A)]$.

Theorem 1. *The computation of Ch always terminate, and $L(G') = WI(L(G))$.*

The proof is given in [4]. Let us now consider several examples to give more intuition about the algorithm and show various situations.

Example 7. Consider grammar $G = \{A \rightarrow a[B], B \rightarrow b[C], C \rightarrow c[\epsilon]\}$. A is the start symbol. Note that A, B, C are not recursive.
$Ch(C) = Ch_{\widehat{C}}^{rex}(RE(C)) = Ch_{\widehat{C}}^{rex}(\epsilon) = \epsilon.$
$Ch(B) = Ch_{\widehat{B}}^{rex}(RE(B)) = Ch_{\widehat{B}}^{rex}(C) = Ch_{\widehat{B}}(C) = C| Ch(C) = C|\epsilon.$
$Ch(A) = Ch_{\widehat{A}}^{rex}(RE(A)) = Ch_{\widehat{A}}^{rex}(B) = Ch_{\widehat{A}}(B) = B| Ch(B) = B|(C|\epsilon).$
Thus, we get the grammar $G' = \{A \rightarrow a[B|(C|\epsilon)], B \rightarrow b[C|\epsilon], C \rightarrow c[\epsilon]\}$ that generates $WI(L(G))$ indeed. In this particular case, where no non-terminal is recursive, we get the same grammar as in our previous work [3], though the algorithm was formalized in a different way.

Example 8. Consider grammar G that contains the rules:
$$A \rightarrow a[((C.A.A^?)|F|\epsilon], \ C \rightarrow c[D], \ D \rightarrow d[\epsilon], \ F \rightarrow f[\epsilon]$$

A is 2-recursive; C, D, F are not recursive. $Ch(D) = Ch(F) = \epsilon$. $Ch(C) = D|\epsilon$. $Succ(A) = \{A, C, D, F\}$. Considered as a regular expression, $Succ(A) = A|C|D|F$. Therefore $Ch(A) = (A|C|D|F)^*$. We get the grammar G':

$$A \rightarrow a[(A|C|D|F)^*], \ C \rightarrow c[D|\epsilon], \ D \rightarrow d[\epsilon], \ F \rightarrow f[\epsilon]$$

The tree t below is generated by G. By removing underlined symbols, we get $t' \lhd t$, and t' is generated by G' indeed. Note that a is a left-sibling of c in t', which is impossible in a tree generated by G.

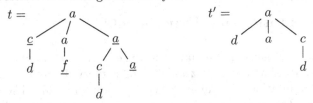

Example 9. Consider grammar G that contains the rules (A is the start symbol):
$$A \rightarrow a[(B.C.A^?.H)|F], \ B \rightarrow b[\epsilon], \ C \rightarrow c[D], \ D \rightarrow d[\epsilon], \ H \rightarrow h[\epsilon], \ F \rightarrow f[\epsilon]$$

A is 1-recursive; B, C, D, H, F are not recursive. $Ch(B) = Ch(D) = Ch(H) = Ch(F) = \epsilon$. $Ch(C) = D|\epsilon$. $\widehat{A} = \{A\}$, then $\widehat{RE}(A) = RE(A) = (B.C.A^?.H)|F$.
$Ch_{\widehat{A}}^{rex}(\widehat{RE}(A)) = Ch_{\widehat{A}}^{rex}((B.C.A^?.H)|F) = (Ch_{\widehat{A}}(B).Ch_{\widehat{A}}(C).Ch_{\widehat{A}}(A)^?.Ch_{\widehat{A}}(H))|Ch_{\widehat{A}}(F)$
$= ((B| Ch(B)).(C| Ch(C)).(A|\epsilon)^?.(H| Ch(H)))|(F| Ch(F))$
$= ((B|\epsilon).(C|(D|\epsilon)).(A|\epsilon)^?.(H|\epsilon))|(F|\epsilon)$, simplified into $(B^?.(C|D|\epsilon).A^?.H^?)|F^?$.
$Left(A) = \{B, C\}$, then $Succ(Left(A)) = \{B, C, D\}$.
$Right(A) = \{H\}$, then $Succ(Right(A)) = \{H\}$.

Considered as regular expressions (instead of sets), $Succ(Left(A)) = B|C|D$ and $Succ(Right(A)) = H$.

Therefore $Ch(A) = (Succ(Left(A)))^*.Ch_{\hat{A}}^{rex}(\widehat{RE}(A)).(Succ(Right(A)))^* = (B|C|D)^*.[(B^?.(C|D|\epsilon).A^?.H^?)|F^?].H^*$, which could be simplified into $(B|C|D)^*.(A^?|F^?).H^*$; we get the grammar G':

$$A \to a[(B|C|D)^*.(A^?|F^?).H^*],\; B \to b[\epsilon],\; C \to c[D|\epsilon],\; D \to d[\epsilon],\; H \to h[\epsilon],\; F \to f[\epsilon]$$

The tree t below is generated by G. By removing underlined symbols, we get $t' \lhd t$, and t' is generated by G' indeed. Note that c is a left-sibling of b in t', which is impossible for a tree generated by the initial grammar G. On the other hand, b, c, d are necessarily to the left of h in t'.

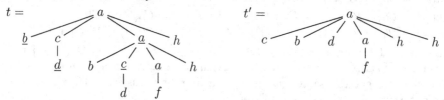

Example 10. The previous example does not show the role of equivalence classes. Consider $G = \{A \to a[B^?],\; B \to b[A]\}$. A and B are 1-recursive.

$A \equiv B$ then $\hat{A} = \hat{B} = \{A, B\}$. $Left(A) = Left(B) = Right(A) = Right(B) = \emptyset$.

Therefore $Ch(A) = Ch_{\hat{A}}^{rex}(\widehat{RE}(A)) = Ch_{\hat{A}}^{rex}(B^?|A) = (Ch_{\hat{A}}(B))^?|(Ch_{\hat{A}}(A)) = (\hat{B}|\epsilon)^?|(\hat{A}|\epsilon) = (A|B|\epsilon)^?|(A|B|\epsilon)$. Note that $\hat{A}$ and $\hat{B}$ have been replaced by $A|B$, which is needed as shown by trees t and t' below. $Ch(A)$ can be simplified into $A|B|\epsilon$. Since $A \equiv B$, $Ch(B) = Ch(A)$.

Then we get the grammar $G' = \{A \to a[A|B|\epsilon],\; B \to b[A|B|\epsilon]\}$.

The tree $t = a(b(a))$ is generated by G. By removing b, we get $t' = a(a)$ which is generated by G' indeed.

4   Implementation and Experiments

Our prototype is implemented in Java and the experiments are done on an Intel Quad Core i3-2310M with 2.10GHz and 8GB of memory. The only step that takes time is the computation of recursivity types of non-terminals. The difficulty is for deciding whether a recursive non-terminal is 2- or 1-recursive. To do it, we have implemented two algorithms: one using Warshall algorithm for computing $>^+$, whose runtime is $O(n^6)$ where n is the number of non-terminals[3], and another based on comparison of cycles in a graph representing relation $>$ (over non-terminals, not over multisets). In the worst case, the runtime of the second algorithm is at least exponential, since all cycles should be detected. Actually, the runtime of the first algorithm depends on the number n of non-terminals, whereas the runtime of the second one depends on the number of cycles in the graph.

---

[3] Since grammars are in normal form, n is also the number of production rules.

In Table 1, #1-*rec* denotes the number of 1-recursive non-terminals (idem for #0-*rec* and #2-*rec*), #Cycles is the number of cycles, and $|G|$ (resp. $|WI(G)|$) denotes the sum of the sizes of the regular expressions[4] occurring in the initial grammar G (resp. in the resulting grammar $WI(G)$). Results in lines 1 to 4 concern synthetic DTDs, while those in lines 5 to 6 correspond to real DTDs. The experiments show: if $n < 50$, the Warshall-based algorithm takes less than 6 seconds. Most often, the cycle-based algorithm runs faster than the Warshall-based algorithm. An example with $n = 111$ (line 3) took 7 minutes with the first algorithm, and was immediate with the second one. When the number of cycles is less than 100, the second algorithm is immediate, even if the runtime in the worst case is bad.

Now, consider the DTD (line 5) specifying the linguistic annotations of named entities performed within the National Corpus of Polish project [2, page 22]. After transforming this DTD into a grammar, we get 17 rules and some non-terminals are recursive. Both algorithms are immediate (few rules and few cycles). The example of line 6 specifies XHTML DTD[5] (with $n = 85$ and #Cycles = 9620). The Warshall-based algorithm and the cycle-based algorithm respond in 2 minutes.

Table 1. Runtimes in seconds for the Warshall-based and the Cycle-comparison algorithms

	Unranked grammars				Runtime (s)		Sizes	
	#0-*rec*	#1-*rec*	#2-*rec*	#Cycles	Warshall	Cycle-compar.	$|G|$	$\|WI(G)\|$
1	9	2	38	410	5.48	0.82	183	1900
2	34	4	12	16	5.51	0.08	126	1317
3	78	12	21	30	445	0.2	293	4590
4	8	2	16	788	0.38	1.51	276	397
5	14	0	2	1	0.08	0.01	30	76
6	30	0	55	9620	136.63	113.91	1879	22963

5   Related Work and Discussion

The (weak) tree inclusion problem was first studied in [12], and improved in [5,7,15]. Our proposal differs from these approaches because we consider the weak inclusion with respect to *tree languages* (and not only with respect to trees). Testing precise inclusion of XML types is considered in [6,8,9,14]. In [14], the authors study the complexity, identifying efficient cases. In [6] a polynomial algorithm for checking whether $L(A) \subseteq L(D)$ is given, where A is an automaton for unranked trees and D is a deterministic DTD.

In this paper, given a regular unranked-tree grammar G (hedge grammar), we have presented a direct method to compute a grammar G' that generates the set of trees (denoted $WI(L(G))$) weakly included in trees generated by G.

---

[4] The size of a regular expression E is the number of non-terminal occurrences in E.
[5] http://www.w3.org/TR/xhtml1/dtds.html

In [1], we have computed G' by transforming unranked-tree languages into binary-tree ones, using first-child next-sibling encoding. Then the weak-inclusion relation $\triangleleft$ is expressed by a context-free synchronized ranked-tree language, and using join and projection, we get G_1. By transforming G_1 into an unranked-tree grammar, we get G'. This method·is complex, and gives complex grammars.

Another way to compute G' could be the following. For each rule $A \to a[E]$ in G we add the *collapsing rule* $A \to E$. The resulting grammar G_1 generates $WI(L(G))$ indeed, but is not a hedge grammar: it is called *extended grammar*[6] in [11], and can be transformed into a context-free hedge grammar G_2 (without collapsing rules). Each hedge H of G_2 is a context-free word language over non-terminals defined by a word grammar, and if we consider its closure by sub-word[7], we get a regular word language H' defined by a regular expression [10]. Let G' be the grammar obtained from G_2 by transforming every hedge in this way. Then G' is a regular hedge grammar, and the language generated by G' satisfies $L(G') = L(G_2) \cup L_2$ where $L_2 \subseteq WI(L(G_2))$ (because of sub-word closure of hedges). Moreover $L(G_2) = L(G_1) = WI(L(G))$. Then $L_2 \subseteq WI(L(G_2)) = WI(WI(L(G))) = WI(L(G))$. Therefore $L(G') = WI(L(G)) \cup L_2 = WI(L(G))$.

References

1. Amavi, J.: Comparaison des langages d'arbres pour la substitution de services web (in French). Tech. Rep. RR-2010-13, LIFO, Université d'Orléans (2010)
2. Amavi, J., Bouchou, B., Savary, A.: On correcting XML documents with respect to a schema. Tech. Rep. 301, LI, Université de Tours (2012)
3. Amavi, J., Chabin, J., Halfeld Ferrari, M., Réty, P.: Weak Inclusion for XML Types. In: Bouchou-Markhoff, B., Caron, P., Champarnaud, J.-M., Maurel, D. (eds.) CIAA 2011. LNCS, vol. 6807, pp. 30–41. Springer, Heidelberg (2011)
4. Amavi, J., Chabin, J., Réty, P.: Weak inclusion for recursive XML types (full version). Tech. Rep. RR-2012-02, LIFO, Université d'Orléans (2012), http://www.univ-orleans.fr/lifo/prodsci/rapports/RR/RR2012/ RR-2012-02.pdf
5. Bille, P., Li Gørtz, I.: The Tree Inclusion Problem: In Optimal Space and Faster. In: Caires, L., Italiano, G.F., Monteiro, L., Palamidessi, C., Yung, M. (eds.) ICALP 2005. LNCS, vol. 3580, pp. 66–77. Springer, Heidelberg (2005)
6. Champavère, J., Gilleron, R., Lemay, A., Niehren, J.: Efficient Inclusion Checking for Deterministic Tree Automata and DTDs. In: Martín-Vide, C., Otto, F., Fernau, H. (eds.) LATA 2008. LNCS, vol. 5196, pp. 184–195. Springer, Heidelberg (2008)
7. Chen, Y., Shi, Y., Chen, Y.: Tree inclusion algorithm, signatures and evaluation of path-oriented queries. In: Symp. on Applied Computing, pp. 1020–1025 (2006)
8. Colazzo, D., Ghelli, G., Pardini, L., Sartiani, C.: Linear inclusion for XML regular expression types. In: Proceedings of the 18th ACM Conference on Information and Knowledge Management, CIKM, pp. 137–146. ACM Digital Library (2009)
9. Colazzo, D., Ghelli, G., Sartiani, C.: Efficient asymmetric inclusion between regular expression types. In: Proceeding of International Conference of Database Theory, ICDT, pp. 174–182. ACM Digital Library (2009)

---

[6] In [11], they consider automata, but by reversing arrows, we can get grammars.

[7] A sub-word of a word w is obtained by removing symbols from w. For example, *abeg* is a sub-word of *abcdefgh*.

10. Courcelle, B.: On constructing obstruction sets of words. Bulletin of the EATCS 44, 178–185 (1991)
11. Jacquemard, F., Rusinowitch, M.: Closure of Hedge-Automata Languages by Hedge Rewriting. In: Voronkov, A. (ed.) RTA 2008. LNCS, vol. 5117, pp. 157–171. Springer, Heidelberg (2008)
12. Kilpeläinen, P., Mannila, H.: Ordered and unordered tree inclusion. SIAM J. Comput. 24(2), 340–356 (1995)
13. Mani, M., Lee, D.: XML to Relational Conversion Using Theory of Regular Tree Grammars. In: Bressan, S., Chaudhri, A.B., Li Lee, M., Yu, J.X., Lacroix, Z. (eds.) EEXTT and DIWeb 2002. LNCS, vol. 2590, pp. 81–103. Springer, Heidelberg (2003)
14. Martens, W., Neven, F., Schwentick, T.: Complexity of Decision Problems for Simple Regular Expressions. In: Fiala, J., Koubek, V., Kratochvíl, J. (eds.) MFCS 2004. LNCS, vol. 3153, pp. 889–900. Springer, Heidelberg (2004)
15. Richter, T.: A New Algorithm for the Ordered Tree Inclusion Problem. In: Hein, J., Apostolico, A. (eds.) CPM 1997. LNCS, vol. 1264, pp. 150–166. Springer, Heidelberg (1997)

Synchronizing Automata
on Quasi-Eulerian Digraph*

Mikhail V. Berlinkov

Institute of Mathematics and Computer Science,
Ural Federal University 620083 Ekaterinburg, Russia
m.berlinkov@gmail.com

Abstract. We describe a new version of the so-called extension method that was used to prove quadratic upper bounds on the minimum length of reset words for various important classes of synchronizing automata. Our approach is formulated in terms of Markov chains; it is in a sense dual to the usual extension method and improves on a recent result by Jungers. As an application, we obtain a quadratic upper bound on the minimum length of reset words for a generalization of Eulerian automata.

1 Synchronizing Automata and the Černý Conjecture

Suppose $\mathscr{A}$ is a complete deterministic finite automaton whose input alphabet is A and whose state set is Q. The automaton $\mathscr{A}$ is called *synchronizing* if there exists a word $w \in A^*$ whose action *resets* $\mathscr{A}$, that is, w leaves the automaton in one particular state no matter at which state in Q it is applied: $q.w = q'.w$ for all $q, q' \in Q$. Any such word w is called *reset* (or *synchronizing*) for the automaton. The minimum length of reset words for a given automaton $\mathscr{A}$ is called the *reset length* of $\mathscr{A}$ and is denoted by $\mathfrak{C}(\mathscr{A})$.

Synchronizing automata serve as transparent and natural models of error-resistant systems in many applications (coding theory, robotics, testing of reactive systems) and also reveal interesting connections with symbolic dynamics and other parts of mathematics. For a brief introduction to the theory of synchronizing automata we refer the reader to the recent survey [15]. Here we discuss one of the main problems in this theory: proving an upper bound of magnitude $O(n^2)$ for the reset length of n-state synchronizing automata.

In 1964 Černý [3] constructed for each $n > 1$ a synchronizing automaton $\mathscr{C}_n$ with n states and 2 input letters whose reset length is $(n-1)^2$. The automaton $\mathscr{C}_4$ is shown in Fig. 1(left). Soon after that he conjectured that those automata represent the worst possible case, thus formulating the following hypothesis:

Conjecture 1 (Černý). Each synchronizing automaton $\mathscr{A}$ with n states has a reset word of length at most $(n-1)^2$, i.e. $\mathfrak{C}(\mathscr{A}) \leq (n-1)^2$.

---

* Supported by the Russian Foundation for Basic Research, grant 10-01-00793, and by the Presidential Program for young researchers, grant MK-266.2012.1.

N. Moreira and R. Reis (Eds.): CIAA 2012, LNCS 7381, pp. 90–100, 2012.

By now this simply looking conjecture is arguably the most longstanding open problem in the combinatorial theory of finite automata. Moreover, the best upper bound for the reset length of n-state synchronizing automata known so far is equal to $\frac{n^3-n}{6}$ and so is cubic[1] in n. This bound is due to Pin [11] and is based upon a combinatorial theorem conjectured by Pin and then proved by Frankl [6]. Since the Černý conjecture claims a quadratic value, it is of certain importance to prove quadratic upper bounds for some classes of synchronizing automata.

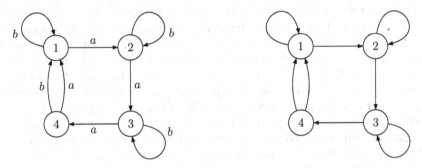

Fig. 1. The automaton $\mathscr{C}_4$ and its underlying graph

Several results of this sort have been obtained via a so-called extension method. In this method one constructs a reset word 'backwards', starting with a letter a such that the preimage of a certain state under the action of a is a non-singleton set P_1, then looking for a word w_1 such that the preimage of P_1 under the action of w_1 is a P_2 with $|P_2| > |P_1|$, then looking for a word w_2 such that the preimage of P_2 under the action of w_2 is a P_3 with $|P_3| > |P_2|$, and so on. One keeps climbing this way until one reaches the set Q of all states, and the resulting reset word is then $w_d \cdots w_1 a$ for some $d \leq |Q| - 2$. A crucial problem here is how to bound from above the lengths of the extending words $w_1, w_2, \ldots, w_d$. For some classes of synchronizing automata including circular automata [4], Eulerian automata [8] and one-cluster automata with prime length cycle [12], it is possible to bound these lengths from above by $|Q|$ and this suffices to prove the Černý conjecture for these classes. However, it is known [2] that in general no upper bound of the form $c|Q|$ with $c < 2$ may hold.

Recently Jungers [7] has used ideas of linear programming to develop an approach which is in a sense dual to the standard form of the extension method outlined in the preceding paragraph. In Section 3 we present a simplified and improved version of Jungers's result. Our proofs do not use linear programming but instead rely on a few basic facts from the theory of Markov chains. In Section 4 we apply our extension method to a class of synchronizing automata obtaining a quadratic upper bound on the reset length of automata in this class. Section 2 contains some preliminaries.

---

[1] A slightly better though still cubic upper bound $\frac{n(7n^2+6n-16)}{48}$ has been claimed in [14] but the published proof of this result contains an unclear part.

2   Exponents of Primitive Matrices vs. Reset Lengths

A real matrix is called *non-negative* (*positive*) if all its entries are non-negative (positive). A non-negative square matrix $M = (M_{i,j})$ is *primitive* if for some positive integer m the matrix M^m is positive. The least number m with this property is called the *exponent* of M and is denoted by $\exp(M)$. We also define the *weak exponent* of a matrix M (denoted $\mathrm{wexp}(M)$) as the minimum number m such that M^m has a positive row. Notice that the (weak) exponent depends only on the set $\mathrm{Supp}(M) = \{(i,j) \mid M_{i,j} > 0\}$ of indices of the positive entries and does not depend on values of these entries. In other words, if we assign to each primitive $n \times n$-matrix M its directed graph $G(M)$ with the vertex set $\{1, 2, \ldots, n\}$ and the edge set $\mathrm{Supp}(M)$, we conclude that the (weak) exponent depends only on $G(M)$ (or on its adjacency matrix which is a 0-1 matrix). Observe that for a 0-1 matrix M, the entry $M_{i,j}^t$ of the matrix M^t is equal to the number of directed paths of length exactly t from i to j in the graph $G(M)$. Therefore, in terms of the graph $G(M)$, the exponent of M is the least number m such that between every two vertices of $G(M)$ there is a directed path of length exactly m. In particular, the graph of a primitive matrix is strongly connected.

A primitive matrix M is said to be a *Wielandt matrix* if the graph $G(M)$ is isomorphic to the graph of the matrix

$$\begin{pmatrix} 0 & 1 & 0 & \ldots & 0 & 0 \\ 0 & 0 & 1 & \ldots & 0 & 0 \\ & & \cdots\cdots\cdots & & \\ 0 & 0 & 0 & \ldots & 0 & 1 \\ 1 & 1 & 0 & \ldots & 0 & 0 \end{pmatrix}.$$

The following proposition collects some basic properties of primitive matrices.

Proposition 1. *Let M be a primitive $n \times n$-matrix. Then:*

1. $\mathrm{wexp}(M) \le \exp(M) \le \mathrm{wexp}(M) + n - 1$;
2. $\exp(M) \le (n-1)^2 + 1$ *and the equality holds only for Wielandt matrices;*
3. $\mathrm{wexp}(M) \le (n-1)^2$.

Proof. The first inequality in claim 1 holds true by the definitions. The second inequality can be easily proved in terms of the graph $G(M)$. Indeed, let i be the index of a positive row in $M^{\mathrm{wexp}(M)}$. Then in $G(M)$ there is a path of length $d(i,t) \le n - 1$ from the vertex i to any fixed vertex t. Further, for each vertex s there is a path to some vertex q of length $n - 1 - d(i,t)$, and by the definition of the weak exponent, there is a path of length $\mathrm{wexp}(M)$ from q to i. Thus, between any two vertices s and t, there is a path $s \to q \to i \to t$ of length

$$n - 1 - d(i,t) + \mathrm{wexp}(M) + d(i,t) = \mathrm{wexp}(M) + n - 1,$$

and the claim is proved.

The inequality in claim 2 was discovered by Wielandt [16]. The second statement in this claim is due to Dulmage and Mendelsohn [5].

Claim 3 follows from claim 2 combined with the fact that the equality $\exp(M) = (n-1)^2 + 1$ only occurs when M is a Wielandt matrix (claim 2) in which case it is easy to see that $\mathrm{wexp}(M) = n^2 - 3n + 3 \leq (n-1)^2$.

In the rest of the paper, we assume that $\mathscr{A}$ is a synchronizing n-state automaton whose state set is Q and whose input alphabet is $A = \{a_1, a_2, \ldots, a_k\}$. We also assume that $n, k > 1$. The *underlying graph* $\mathrm{UG}(\mathscr{A})$ of $\mathscr{A}$ is the directed multigraph obtained from $\mathscr{A}$ by removing all edge labels. Fig. 1(right) shows the underlying graph of the Černý automaton $\mathscr{C}_4$. We freely transfer graph-theoretic terminology from underlying graphs of automata to automata themselves; in particular, we speak about strongly connected automata, Eulerian automata etc. Further we assume that the automaton $\mathscr{A}$ under consideration is strongly connected because the problem of finding a quadratic upper bound for the reset length of n-state synchronizing automata can be easily reduced to this case (see [10] for example).

Now we consider relations between primitive matrices and synchronizing automata. The proof of the following proposition can be found in [1] but we reproduce it here for the reader's convenience.

Proposition 2. *Let M denote the adjacency matrix of the underlying graph of the automaton $\mathscr{A}$. Then*

1. *M is a primitive matrix;*
2. *$\mathrm{wexp}(M) \leq \mathfrak{C}(\mathscr{A})$.*

Proof. Since $\mathscr{A}$ is a synchronizing automaton, there exists a reset word w of length $\ell = \mathfrak{C}(\mathscr{A})$ which takes all the states of the automaton to some state i. This means that the i-th row of M^ℓ is positive, so claim 2 is proved. Since $\mathscr{A}$ is strongly connected, claim 1 holds true as well.

It follows from the above propositions that the weak exponent of the underlying graph of $\mathscr{A}$ is at most $(n-1)^2$. This means that there are (unlabeled) directed paths of equal length $\ell \leq (n-1)^2$ from every state of the automaton $\mathscr{A}$ to some particular state. The Černý conjecture asserts additionally that such paths can be chosen to be labeled by some fixed word. It seems that this additional requirement should increase significantly the minimum length of such paths. Indeed, for many synchronizing automata the reset length by far exceeds the weak exponent of their underlying graphs. For instance, if synchronizing automata contains a loop then its weak exponent is at most $n - 1$ but its reset length can be equal to $(n-1)^2$ (the automata $\mathscr{C}_n$ in the Černý series can serve as examples here). However, in order to prove the Černý conjecture we only need a bound in the worst case, and in [1] strong connections between distributions of reset lengths of synchronizing automata and of exponents of primitive graphs have been observed. Thus, we believe that a deeper study of these connections may be useful.

3   Markov Chains and a New Extension Method

The aim of this paper is to obtain upper bounds on reset lengths by utilizing their connection with exponents of primitive graphs. For this, we associate a natural linear structure with the automaton $\mathscr{A} = (Q, A)$. Let $\mathbb{R}^n$ stand for the real n-dimensional linear space of column vectors. We assume that $Q = \{1, 2, \ldots, n\}$ and then assign to each subset $K \subseteq Q$ its *characteristic vector* $[K] \in \mathbb{R}^n$ defined as follows: the i-th entry of $[K]$ is 1 if $i \in K$, otherwise the entry is 0. For $q \in Q$ we write $[q]$ instead of $[\{q\}]$ to simplify notation.

For each word $w \in A^*$, the action of w on Q gives rise to a linear transformation of $\mathbb{R}^n$; we denote by $[w]$ the matrix of this transformations in the standard basis $[1], \ldots, [n]$ of $\mathbb{R}^n$. For instance, if $\mathscr{A} = \mathscr{C}_4$, that is, the automaton in Fig. 1(left), then $[ba] = \begin{pmatrix} 0 & 0 & 0 & 0 \\ 1 & 0 & 0 & 1 \\ 0 & 1 & 0 & 0 \\ 0 & 0 & 1 & 0 \end{pmatrix}$. Clearly, the matrix $[w]$ has exactly one non-zero entry in each column and this entry is equal to 1, in particular, $[w]$ is *column stochastic* (that is, each column sum is equal to 1). Observe that if $w = uv$, then $[w] = [v][u]$.

For $K \subseteq Q$ and $v \in A^*$ we denote by $K.v^{-1}$ the preimage of the subset K under the action of the word v, that is,

$$K.v^{-1} = \{q \mid q.v \in K\}.$$

One can easily check that $[K.v^{-1}] = [v]^T[K]$, where $[v]^T$ stands for the usual transpose of the matrix $[v]$. Recall that a word w is a reset word for $\mathscr{A}$ if and only if $q.w^{-1} = Q$ for some state q. Thus, in the language of linear algebra, we can rewrite the fact that w is a reset word as $[w]^T[q] = [Q]$.

For vectors $g_1, g_2 \in \mathbb{R}^n$, we denote their usual inner product by (g_1, g_2). Denote by $\mathbb{R}^n_+$ set of all positive vectors from $\mathbb{R}^n$. Let $p \in \mathbb{R}^n_+$ be a positive *stochastic* vector, that is, a positive vector whose entries sum up to 1. Then w is a reset word if and only if $([q.w^{-1}], p) = ([w]^T[q], p) = 1$. This condition is clearly necessary; its sufficiency follows from the fact that $[w]^T[q]$ is a 0-1 vector and from the condition that p is positive.

Recall that we have assumed that $A = \{a_1, a_2, \ldots, a_k\}$. Each positive stochastic vector $\pi \in \mathbb{R}^k_+$ defines a probability distribution on A in which the probability $p(a_j)$ is defined as the j-th entry of the vector π. For a word $v \in A^*$, let $\ell(v)$ denote the length of v and let $v(i)$ stand for the i-th letter of v, $i = 1, \ldots, \ell(v)$. We define the probability of a word v under the distribution π as

$$p(v) = \prod_{i=1}^{\ell(v)} p(v(i)).$$

Consider a process in which an agent randomly walks on the underlying graph of $\mathscr{A}$, choosing for each move an edge labeled a_i with probability $p(a_i)$. This is a Markov chain and the matrix $S(\mathscr{A}, \pi) = \sum_{i=1}^{k} p(a_i)[a_i]$ is called the *transition matrix* of this Markov chain. Observe that $\mathrm{Supp}(S(\mathscr{A}, \pi)) = \mathrm{Supp}(M)$ where M is the adjacency matrix of $\mathrm{UG}(\mathscr{A})$. By Proposition 2 we conclude that the matrix $S(\mathscr{A}, \pi)$ is primitive. Also, $S(\mathscr{A}, \pi)$ is easily seen to be column stochastic.

The following proposition summarizes a few properties of Markov chains that we need. They are, of course, well-known but we provide their proofs as we do not assume the reader's acquaintance with the theory of Markov chains. We denote by 1_n the uniform stochastic vector in $\mathbb{R}^n$, that is, the vector with all entries equal to $\frac{1}{n}$.

Proposition 3. *Let S be a column stochastic $n \times n$ primitive matrix. Then:*

1. 1_n *is a left eigenvector of S corresponding to the eigenvalue 1, that is,* $S^T 1_n = 1_n$;
2. *there exists a unique stationary distribution $\alpha \in \mathbb{R}_+^n$, that is, a positive stochastic vector satisfying $S\alpha = \alpha$;*
3. *1 is a unique eigenvalue of S with maximum absolute value and the corresponding eigenspace is one-dimensional.*

Proof. Since S is a column stochastic matrix, we have $S^T 1_n = 1_n$. Thus 1 is an eigenvalue of S and the corresponding left eigenvector 1_n is positive. Since S also is primitive, by the Perron-Frobenius theorem [9, Section 8.3] 1 is a unique eigenvalue of S with maximum absolute value and there is also a unique (up to a positive scalar) right positive eigenvector, that is, a vector α satisfying $S\alpha = \alpha$. Clearly, α can be chosen to be stochastic. Also by the Perron-Frobenius theorem the left and the right eigenspaces corresponding to the eigenvalue 1 are one-dimensional and equal to the linear spans $\langle 1_n \rangle$ and $\langle \alpha \rangle$ of 1_n and α respectively.

As discussed in Section 1, one of the most fruitful ways for finding quadratic upper bounds on the reset lengths of synchronizing automata is the extension method. Using the language of linear algebra, we can reformulate the method as follows. We choose some state q and find a letter a and a finite sequence of words $w_1, w_2, \ldots, w_d$ such that

$$\frac{1}{n} = ([q], 1_n) < ([a]^T[q], 1_n) < ([w_1 a]^T[q], 1_n) < \cdots$$
$$\cdots < ([w_d \cdots w_2 w_1 a]^T[q], 1_n) = 1. \quad (1)$$

It is clear that such a sequence can be constructed for any synchronizing automaton and that $d \le n - 2$ because each inner product in the sequence exceeds the previous one by at least $\frac{1}{n}$. Thus a quadratic upper bound on the reset length will be established as soon as one manages to prove that the lengths of $w_1, w_2, \ldots, w_d$ in (1) can be bounded by a linear (in n) function. For instance, if our automaton $\mathscr{A}$ is such that one can prove that $\ell(w_i) \le n$ for all $i = 1, \ldots, d$, then it can be easily shown that the Černý conjecture holds true for $\mathscr{A}$.

However, it is shown in [2] that there is a series of synchronizing automata with n states for which $\ell(w_i)$ cannot be bounded by cn for any $c < 2$ and for any sequence (1). This means that for some proper subset $X \subset Q$, the inequality $([v]^T[X], 1_n) \le ([X], 1_n)$ holds true for each word v of length at most cn. Therefore the Černý conjecture cannot be always achieved in this way. Thus, we have to change something. Jungers [7] has suggested an interesting idea that in our notation can be described as follows: one should substitute the uniform stochastic vector 1_n by an adaptive positive stochastic vector p which can depend on both the automaton $\mathscr{A} = (Q, A)$ and the proper subset $X \subset Q$ but has the property that there exists a word v of length at most n such that $([v]^T[X], p) > ([X], p)$. Jungers has explored this idea using techniques from linear programming and has proved that positive stochastic vectors with desired properties indeed exist for every synchronizing automaton and every proper subset of its states.

Our main contribution is the following result which shows that for every synchronizing automaton $\mathscr{A} = (Q, A)$ there is a positive stochastic vector p such that for each $X \subset Q$ there exists a word v of length at most n satisfying $([v]^T[X], p) > ([X], p)$. Thus, given a synchronizing automaton $\mathscr{A}$, a single vector serves all proper subsets of Q. Moreover, it turns out that the stationary distribution of any random walk on $\mathscr{A}$ satisfies the desired property.

For a positive integer r, denote by A^r the set of all words over A of length r.

Theorem 1. Let $\mathscr{A} = (Q, A)$ be a synchronizing automaton with $|Q| = n$ and $A = \{a_1, a_2, \ldots, a_k\}$. Let a stochastic vector $\pi \in \mathbb{R}_+^k$ define a probability distribution on A, and let $\alpha \in \mathbb{R}_+^n$ be the stationary distribution of the Markov chain with the transition matrix $S(\mathscr{A}, \pi) = \sum_{i=1}^{k} p(a_i)[a_i]$. Take a vector $x \in \mathbb{R}^n$ with $(x, \alpha) = 0$ and let $v \in A^*$ be a word of minimum length such that $([v]^T x, \alpha) > 0$. Then

1. $\sum_{u \in A^r} p(u)([u]^T x, \alpha) = 0$ for every positive integer r;
2. if $u \in A^*$ is a word with $\ell(u) < \ell(v)$ then $([u]^T x, \alpha) = 0$;
3. $\ell(v) \leq \dim\langle [u]\alpha \mid \ell(u) \leq n - 1 \rangle - 1 \leq n - 1$.

Proof. Denote $S = S(\mathscr{A}, \pi)$. Since $S\alpha = \alpha$, we have $S^r \alpha = \alpha$ for every positive integer r. It easy to see that

$$S^r = \sum_{u \in A^r} p(u)[u],$$

whence $\sum_{u \in A^r} p(u)[u]\alpha = \alpha$. Multiplying through by the vector x, we obtain

$$\sum_{u \in A^r} p(u)([u]^T x, \alpha) = \sum_{u \in A^r} p(u)([u]\alpha, x) = \left(\sum_{u \in A^r} p(u)[u]\alpha, x \right) = (\alpha, x) = 0.$$

This proves claim 1.

The equality in claim 1 immediately implies that if $([u]^T x, \alpha) \neq 0$ for some word u, then there exists a word w such that $\ell(w) = \ell(u)$ and $([w]^T x, \alpha) > 0$. Thus, claim 2 follows from the choice of v as a word of minimum length with $([v]^T x, \alpha) > 0$.

To prove claim 3, suppose that $\ell(v) \geq \dim\langle [u]\alpha \mid \ell(u) \leq n - 1 \rangle$. Then claim 2 implies that $([u]^T x, \alpha) = (x, [u]\alpha) = 0$ for every word u such that $\ell(u) < \dim\langle [u]\alpha \mid \ell(u) \leq n - 1 \rangle$. For $i \in \{1, 2, \ldots, n\}$, let U_i be the subspace spanned by all vectors $[u]\alpha$ with $\ell(u) \leq i - 1$. Then the chain

$$\langle \alpha \rangle = U_1 \subseteq U_2 \subseteq \cdots \subseteq U_n = \langle [u]\alpha \mid \ell(u) \leq n - 1 \rangle$$

of non-zero subspaces in the n-dimensional space $\mathbb{R}^n$ must become constant at some $j \leq \dim(U_n) \leq \ell(v)$, i.e.

$$U_1 \subset U_2 \subset \cdots \subset U_j = U_{j+1} = \cdots = U_n.$$

Observe that this implies that the subspace U_j is invariant with respect to all transformations induced by the letters in A, whence, in particular, $[v]\alpha$ belongs to U_j. Since $(x, [u]\alpha) = 0$ for every u with $\ell(u) \leq \dim(U_n) = \dim(U_j)$, we conclude that $(x, g) = 0$ for each $g \in U_{\dim(U_j)}$. By the choice of j we have $\dim(U_j) \geq j$, whence $U_j \subseteq U_{dim(U_j)}$. So $(x, g) = 0$ for each $g \in U_j$. As mentioned, $[v]\alpha$ belongs to U_j, hence $(x, [v]\alpha) = ([v]^T x, \alpha) = 0$, and this contradicts the condition $([v]^T x, \alpha) > 0$.

4   Quasi-Eulerian Automata

In view of Theorem 1, the lengths of all words w_i in the sequence

$$([q], \alpha) < ([w_1]^T[q], \alpha) < ([w_1 w_2]^T[q], \alpha) < \cdots < ([w_d \cdots w_2 w_1]^T[q], \alpha) = 1, \quad (2)$$

where α is the stationary distribution of the Markov chain with the transition matrix $S(\mathscr{A}, \pi)$, are bounded by $n - 1$. Unfortunately, here we encounter a 'complementary' difficulty: in general, it is quite hard to estimate the length of such a sequence because the increment on each step of (2) may be less than $\frac{1}{n}$. However, Theorem 1 can be applied for some classes of synchronizing automata. To start with, we register the following observation.

Corollary 1. *Let α be the stationary distribution of the Markov chain with the transition matrix $S(\mathscr{A}, \pi)$ for some synchronizing automaton $\mathscr{A} = (Q, A)$ and some probability distribution π on A. If L is the least common multiple of the denominators of the entries of α, then $\mathfrak{C}(\mathscr{A}) \leq 1 + (n-1)(L-2)$.*

Proof. Observe that if $x_1, x_2 \in \mathbb{R}^n$ are 0-1 vectors and $(x_2, \alpha) > (x_1, \alpha)$, then $(x_2, \alpha) \geq (x_1, \alpha) + \frac{1}{L}$. Since $\mathscr{A}$ is synchronizing, there exists a state $q \in Q$ and a letter $a \in A$ such that $|q.a^{-1}| > 1$. Set $w_1 = a$, then

$$([w_1]^T[q], \alpha) \geq ([q], \alpha) + \frac{1}{L} \geq \frac{2}{L}.$$

Suppose $([w_1]^T[q], \alpha) < 1$. Set $x_1 = [w_1]^T[q] - \|[w_1]^T[q]\|1_n$, then $(x_1, \alpha) = 0$, and let w_2 be a word of minimum length with $([w_2]^T x_1, \alpha) > 0$. Such a word w_2 exists because for any reset word u, we have

$$([u]^T x_1, \alpha) = ([Q] - \|[w_1]^T[q]\|1_n, \alpha) = 1 - \frac{1}{n}\|[w_1]^T[q]\| > 0$$

since $\|[w_1]^T[q]\| < n$ in view of $([w_1]^T[q], \alpha) < 1$. By Theorem 1, $\ell(w_2) \leq n - 1$ and

$$([w_2]^T[w_1]^T[q], \alpha) \geq (\|[w_1]^T[q]\|1_n, \alpha) + \frac{1}{L} \geq \frac{3}{L}.$$

Repeating this argument, we build a reset word $w_d w_{d-1} \cdots w_1$ where $\ell(w_1) = 1$ and $\ell(w_i) \leq n - 1$ for $i = 2, \ldots, d$. Since we start from $\frac{2}{L}$ and add at least $\frac{1}{L}$ on each step, we have $d - 1 \leq \frac{(1 - \frac{2}{L})}{\frac{1}{L}} = L - 2$. Hence

$$\mathfrak{C}(\mathscr{A}) \leq 1 + (n - 1)(d - 1) \leq 1 + (n - 1)(L - 2).$$

An automaton is said to be *Eulerian* if its underlying graph is Eulerian, or equivalently, if it is strongly connected and the in-degree and the out-degree of every vertex are equal (and hence they both are equal to the alphabet size). It is clear that $\mathscr{A}$ is Eulerian if and only if the matrix $S(\mathscr{A}, 1_n)$ is doubly stochastic. Following [13] we say that an automaton is $\mathscr{A} = (Q, A)$ is pseudo-Eulerian if we can find a probability distribution π on A such that the matrix $S(\mathscr{A}, \pi)$ is doubly stochastic.

Corollary 2. *If a synchronizing automaton $\mathscr{A}$ is pseudo-Eulerian, then*

$$\mathfrak{C}(\mathscr{A}) \leq 1 + (n - 1)(n - 2).$$

Proof. Choose a probability distribution π on A such that $S(\mathscr{A}, \pi)$ is row stochastic. Then the stationary distribution of the corresponding Markov chain is equal to 1_n, and Corollary 1 applies.

The bound of Corollary 2 was found by Kari [8] for Eulerian synchronizing automata and extended to pseudo-Eulerian automata by Steinberg [13]. Corollary 1 can be considered as a generalization of these results.

Proposition 4. *Let α be the stationary distribution of the Markov chain with the transition matrix $S(\mathscr{A}, \pi)$ for some synchronizing automaton $\mathscr{A} = (Q, A)$ and some probability distribution π on A. Suppose that for some positive integer c the vector α has $n - c$ equal entries. Then $\mathfrak{C}(\mathscr{A}) \leq 2^c(n - c + 1)(n - 1)$.*

Proof. Without any loss we may assume that $\alpha = (r_1, r_2, \ldots, r_c, r, r, \ldots, r)^T$. For $K \subset Q$, let K_i be the i-th entry of the vector $[K]$ and let f be the number of indices i such that $c < i \leq n$ and $K_i = 1$. Then $([K], \alpha) = \sum_{i=1}^{c} K_i r_i + fr$ whence this value is determined by the vector $f(K) = (K_1, K_2, \ldots, K_c, f)$, where $f \in \{0, \ldots, n-c\}$. Hence there are at most $2^c(n - c + 1)$ possible different values of $([K], \alpha)$ and the length d of any sequence of the form (2) does not exceed $2^c(n - c + 1)$. By Theorem 1, we can choose the words w_i in (2) such that $\ell(w_i) \leq n - 1$ for all $i = 1, \ldots, d$ whence the length of the reset word $w_d \cdots w_1$ does not exceed $2^c(n - c + 1)(n - 1)$.

As an application of Proposition 4, we prove a quadratic upper bound on the reset length for a new class of synchronizing automata. We call an automaton $\mathscr{A} = (Q, A)$ *quasi-Eulerian* with respect to a positive integer c if it has a 'pseudo-Eulerian part' $E_c \subset Q$ containing $n - c$ states only one of which can have incoming edges from the set $Q \setminus E_c$. If s is this 'entering' state, the condition on the component E_c amounts to saying that for some probability distribution π on A, the rows of the matrix $S(\mathscr{A}, \pi)$ that correspond to the states from $E_c \setminus \{s\}$ sum up to 1.

Theorem 2. *If a synchronizing automaton $\mathscr{A}$ is quasi-Eulerian with respect to a positive integer c, then $\mathfrak{C}(\mathscr{A}) \leq 2^c(n - c + 1)(n - 1)$.*

Proof. Choose a probability distribution π on A such that the rows of the matrix $S(\mathscr{A}, \pi)$ corresponding to the states from $E_c \setminus \{s\}$ sum up to 1. By Theorem 1 the stationary distribution α of the Markov chain with the transition matrix $S(\mathscr{A}, \pi)$ is a unique positive solution of the equation $(S - E)x = 0$. It is easy to show that all entries of α that correspond to the states from E_c are equal, whence Proposition 4 applies to α.

It is fairly easy to exhibit examples of quasi-Eulerian synchronizing automata. For instance, it can be shown that the automata $\mathscr{C}_n$ from the Černý series are quasi-Eulerian for $c = 1$. We hope that the ideas suggested in this paper may be used to attack the general problem of mastering a quadratic upper bound for the reset length of synchronizing automata.

Acknowledgement. The author thanks the anonymous referees for their useful remarks and suggestions.

References

1. Ananichev, D., Gusev, V., Volkov, M.: Slowly Synchronizing Automata and Digraphs. In: Hliněný, P., Kučera, A. (eds.) MFCS 2010. LNCS, vol. 6281, pp. 55–65. Springer, Heidelberg (2010)
2. Berlinkov, M.: On a conjecture by Carpi and D'Alessandro. Int. J. Found. Comput. Sci. 22(7), 1565–1576 (2011)
3. Černý, J.: Poznámka k homogénnym eksperimentom s konečnými automatami. Matematicko-fyzikalny Časopis Slovensk. Akad. Vied 14(3), 208–216 (1964) (in Slovak)
4. Dubuc, L.: Sur les automates circulaires et la conjecture de Černý. RAIRO Inform. Théor. Appl. 32, 21–34 (1998) (in French)
5. Dulmage, A.L., Mendelsohn, N.S.: Gaps in the exponent set of primitive matrices. Ill. J. Math. 8, 642–656 (1964)
6. Frankl, P.: An extremal problem for two families of sets. Eur. J. Comb. 3, 125–127 (1982)
7. Jungers, M.: The synchronizing probability function of an automaton. SIAM J. Discrete Math. 26, 177–192 (2011)
8. Kari, J.: Synchronizing finite automata on Eulerian digraphs. Theoret. Comput. Sci. 295, 223–232 (2003)
9. Meyer, C.D.: Matrix Analysis and Applied Linear Algebra. SIAM, Philadelphia (2000)
10. Pin, J.-E.: Le problème de la synchronization et la conjecture de Cerny, Thèse de 3ème cycle. Université de Paris 6 (1978)
11. Pin, J.-E.: On two combinatorial problems arising from automata theory. Ann. Discrete Math. 17, 535–548 (1983)
12. Steinberg, B.: The Černý conjecture for one-cluster automata with prime length cycle. Theoret. Comput. Sci. 412(39), 5487–5491 (2011)

13. Steinberg, B.: The averaging trick and the Černý conjecture. Int. J. Found. Comput. Sci. 22(7), 1697–1706 (2011)
14. Trahtman, A.N.: Modifying the Upper Bound on the Length of Minimal Synchronizing Word. In: Owe, O., Steffen, M., Telle, J.A. (eds.) FCT 2011. LNCS, vol. 6914, pp. 173–180. Springer, Heidelberg (2011)
15. Volkov, M.V.: Synchronizing Automata and the Černý Conjecture. In: Martín-Vide, C., Otto, F., Fernau, H. (eds.) LATA 2008. LNCS, vol. 5196, pp. 11–27. Springer, Heidelberg (2008)
16. Wielandt, H.: Unzerlegbare, nicht negative Matrizen. Math. Z 52, 642–648 (1950) (in German)

Cellular Automata on Regular Rooted Trees

Tullio Ceccherini-Silberstein[1], Michel Coornaert[2], Francesca Fiorenzi[3],
and Zoran Šunić[4]

[1] Dipartimento di Ingegneria, Università del Sannio
C.so Garibaldi 107, 82100 Benevento, Italy
`tceccher@mat.uniroma3.it`
[2] Institut de Recherche Mathématique Avancée, UMR 7501, Univ. Strasbourg
7 rue René Descartes, 67084 Strasbourg Cedex, France
`coornaert@math.unistra.fr`
[3] Laboratoire de Recherche en Informatique, UMR 8623, Bât 650 Univ. Paris-Sud 11
91405 Orsay Cedex France 91405 Orsay, France
`fiorenzi@lri.fr`
[4] Department of Mathematics, Texas A&M University
MS-3368, College Station, TX 77843-3368, USA
`sunic@math.tamu.edu`

Abstract. We study cellular automata on regular rooted trees. This includes the characterization of sofic tree shifts in terms of unrestricted Rabin automata and the decidability of the surjectivity problem for cellular automata between sofic tree shifts.

Keywords: Free monoid, sofic tree shift, unrestricted Rabin automaton, finite tree automaton, cellular automaton, surjectivity problem.

1 Introduction

In this paper, we study cellular automata between subshifts of A^{Σ^*} (also called tree shifts), where A is a finite nonempty set and Σ^* is a finitely generated free monoid identified with the $|\Sigma|$-regular rooted tree. We investigate, in particular, the decidability of the surjectivity problem for these cellular automata.

Amoroso and Patt [1] proved that the surjectivity and the injectivity problems have a positive answer in the one-dimensional case (i.e. when Σ^* is replaced by $\mathbb{Z}$ or $\mathbb{N}$). On the other hand, Kari [8] proved that these problems fail to be decidable in dimension $d \geq 2$ (i.e. for cellular automata defined on $A^{\mathbb{Z}^d}$). There are more general algorithms to decide the surjectivity of a cellular automaton on a finitely generated free monoid (which are deducible combining the results of Rabin [11], Muller and Schupp [10] and Thatcher and Wright [12]), but they are not practical. We have worked out the details in a limited setting of interest, namely when we start with an unrestricted Rabin automaton.

Tree shifts have been extensively studied by Aubrun and Béal in [2], [3] and [4]. In the present work we use a slightly different (but equivalent) setting.

A tree shift is said to be of finite type if it can be described as the set of configurations avoiding a finite number of forbidden patterns. Sofic tree shifts are defined as the images of tree shifts of finite type under cellular automata. In the

N. Moreira and R. Reis (Eds.): CIAA 2012, LNCS 7381, pp. 101–112, 2012.
© Springer-Verlag Berlin Heidelberg 2012

one-dimensional case, a sofic subshift of $A^{\mathbb{N}}$ may be characterized as the set of all right-infinite words accepted by some finite-state automaton. In our setting, we use the notion of an unrestricted Rabin automaton (see [11], [12], [6]), as well as a related notion of acceptance, in order to provide the analogous characterization of sofic tree shifts.

Let us now illustrate our decidability results. It is easy to decide the emptiness of a sofic tree shift accepted by a given unrestricted Rabin automaton. An idea to decide the surjectivity of a cellular automaton $\tau\colon A^{\Sigma^*} \to A^{\Sigma^*}$ could be to establish the emptiness of the tree language $A^{\Sigma^*} \setminus \tau(A^{\Sigma^*})$. But given a nontrivial tree shift, its complement always fails to be a subshift (nevertheless, Rabin theory guarantees that this tree language is still recognizable by a general Rabin automaton). In order to avoid this obstacle, we introduce the set of full-tree-patterns of a tree shift which is a finite tree language that characterizes the shift. Moreover, the full-tree-patterns of a sofic tree shift are recognizable by a suitably defined finite-tree automaton.

We prove that the recognizable sets of full-tree-patterns form a class which is closed under complementation and for which the emptiness problem is decidable. This allows us to find algorithms establishing both the surjectivity of a cellular automaton $\tau\colon A^{\Sigma^*} \to A^{\Sigma^*}$ and the equality of two sofic-tree shifts presented by unrestricted Rabin automata. With this latter result we can provide a general algorithm establishing the surjectivity of a cellular automaton defined between sofic tree shifts.

In this paper we just detail the decision procedures we mentioned above. The proofs of the results leading to these algorithms are omitted.

2 Definitions and Background

In the sequel, we denote by Σ and A two nonempty finite sets. In particular, the set A is called *alphabet* and its elements are called *labels* or *colors*.

2.1 The Free Monoid Σ^*

For $n \in \mathbb{N}$, we denote by Σ^n the set of all *words* $w = \sigma_1\sigma_2\cdots\sigma_n$ of *length* n (where $\sigma_i \in \Sigma$ for $i = 1, 2, \ldots, n$) over Σ. In particular $\varepsilon \in \Sigma^0$ indicates the only word of length 0 called the *empty word*. For $n \geq 1$, we denote by Δ_n the set $\bigcup_{i=0}^{n-1} \Sigma^i$ (that is, the set of all words of length $\leq n - 1$).

The *concatenation* of two words $w = \sigma_1\sigma_2\cdots\sigma_n \in \Sigma^n$ and $w' = \sigma'_1\sigma'_2\cdots\sigma'_m \in \Sigma^m$ is the word $ww' = \sigma_1\sigma_2\cdots\sigma_n\sigma'_1\sigma'_2\cdots\sigma'_m \in \Sigma^{m+n}$. Then the set $\Sigma^* = \bigcup_{n\in\mathbb{N}} \Sigma^n$, equipped with the multiplication given by concatenation, is a monoid with identity element the empty word ε. It is called the *free monoid* over Σ.

From the graph theoretical point of view, Σ^* is the vertex set of the $|\Sigma|$-regular rooted tree. The empty word ε is its root and, for every vertex $w \in \Sigma^*$, the vertices $w\sigma \in \Sigma^*$ (with $\sigma \in \Sigma$) are called the *children* of w. Each vertex is connected by an edge to each of its children.

2.2   Configurations and Tree Shifts

We denote by A^{Σ^*} the set of all maps $f \colon \Sigma^* \to A$. It is called the *space of configurations* of Σ^* over the alphabet A. When equipped with the *prodiscrete topology* (that is, with the product topology where each factor A of $A^{\Sigma^*} = \prod_{w \in \Sigma^*} A$ is endowed with the discrete topology), the configuration space is a compact, totally disconnected, metrizable space. Also, the free monoid Σ^* has a right action on A^{Σ^*} defined as follows: for every $w \in \Sigma^*$ and $f \in A^{\Sigma^*}$ the configuration $f^w \in A^{\Sigma^*}$ is defined by setting $f^w(w') = f(ww')$ for all $w' \in \Sigma^*$. This action, called the *shift action*, is continuous with respect to the prodiscrete topology.

Recall that a neighborhood basis of a configuration $f \in A^{\Sigma^*}$ is given by the sets $\mathcal{N}(f, n) = \{g \in A^{\Sigma^*} : g|_{\Delta_n} = f|_{\Delta_n}\}$ where $n \geq 1$ (as usual, for $M \subset \Sigma^*$, we denote by $f|_M$ the restriction of f to M).

A subset $X \subset A^{\Sigma^*}$ is called a *subshift* (or *tree shift*, or simply *shift*) provided that X is closed (with respect to the prodiscrete topology) and *shift-invariant* (that is, $f^w \in X$ for all $f \in X$ and $w \in \Sigma^*$).

2.3   Forbidden Blocks and Shifts of Finite Type

Let $M \subset \Sigma^*$ be a finite set. A *pattern* is a map $p \colon M \to A$. The set M is called the *support* of p and it is denoted by $\operatorname{supp}(p)$. We denote by A^M the set of all patterns with support M. A *block* is a pattern $p \colon \Delta_n \to A$. The integer n is called the *size* of the block. The set of all blocks is denoted by $\mathcal{B}(A^{\Sigma^*})$.

If X is a subset of A^{Σ^*} and $M \subset \Sigma^*$ is finite, the set of patterns $\{f|_M : f \in X\}$ is denoted by X_M. For $n \geq 1$, the notation X_n is an abbreviation for X_{Δ_n} (that is, the set of all blocks of size n which are restrictions to Δ_n of some configuration in X). We denote by $\mathcal{B}(X)$ the set of all blocks of X (that is, $\mathcal{B}(X) = \bigcup_{n \geq 1} X_n$).

Given a block $p \in \mathcal{B}(A^{\Sigma^*})$ and a configuration $f \in A^{\Sigma^*}$, we say that p *appears* in f if there exists $w \in \Sigma^*$ such that $(f^w)|_{\operatorname{supp}(p)} = p$. If p does not appear in f, we say that f *avoids* p. Let $\mathcal{F}$ be a set of blocks. We denote by $\mathsf{X}(\mathcal{F})$ the set of all configurations in A^{Σ^*} avoiding each block of $\mathcal{F}$, in symbols $\mathsf{X}(\mathcal{F}) = \{f \in A^{\Sigma^*} : (f^w)|_{\Delta_n} \notin \mathcal{F}, \text{ for all } w \in \Sigma^* \text{ and } n \geq 1\}$.

In analogy with the one-dimensional case (see for example [9, Theorem 6.1.21]), we have the following combinatorial characterization of subshifts: *a subset $X \subset A^{\Sigma^*}$ is a subshift if and only if there exists a set $\mathcal{F} \subset \mathcal{B}(A^{\Sigma^*})$ of blocks such that $X = \mathsf{X}(\mathcal{F})$.*

Let $X \subset A^{\Sigma^*}$ be a subshift. A set $\mathcal{F}$ of blocks as above is called a *defining set of forbidden blocks* for X. A subshift is *of finite type* if it admits a finite defining set of forbidden blocks.

Remark 1. We can always suppose that the forbidden blocks of a defining set of a given subshift of finite type all have the same support. This motivates the following definition: a shift of finite type has *memory* n if it admits a defining set of forbidden blocks of size n. Notice that a shift with memory n, also has memory m for all $m \geq n$.

Example 1 (Monochromatic children). Since $\Delta_2 = \{\varepsilon\} \cup \Sigma$, we can identify A^{Δ_2} with $A \times A^{\Sigma}$. Consider the set of blocks

$$\mathcal{F} = \left\{(a, (a_\sigma)_{\sigma \in \Sigma}) \in A \times A^{\Sigma} : a_\sigma \neq a_{\sigma'} \text{ for some } \sigma, \sigma' \in \Sigma\right\}.$$

The tree shift $\mathsf{X}(\mathcal{F}) \subset A^{\Sigma^*}$ is of finite type and exactly consists of those config-urations for which every vertex in Σ^* has monochromatic children. If $|\Sigma| = 2$ and $A = \{0, 1\}$ an example of a configuration in $\mathsf{X}(\mathcal{F})$ is given in Figure 1.

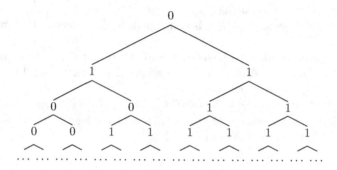

Fig. 1. A configuration of the tree shift presented in Example 1

2.4 Cellular Automata and Sofic Tree Shifts

Let $X \subset A^{\Sigma^*}$ be a tree shift. A map $\tau\colon X \to A^{\Sigma^*}$ is called a *cellular automaton* if it satisfies the following condition: there exists a finite subset $M \subset \Sigma^*$ and a map $\mu\colon A^M \to A$ such that $\tau(f)(w) = \mu((f^w)|_M)$ for all $f \in X$ and $w \in \Sigma^*$. The set M is called a *memory set* for τ and μ is the associated *local defining map*. We assume in the sequel (without loss of generality), that a memory set has the form $M = \Delta_n$, for a suitable $n \geq 1$.

The Curtis-Hedlund-Lyndon theorem gives a topologically characterization of cellular automata: *a map $\tau\colon X \to A^{\Sigma^*}$ is a cellular automaton if and only if it commutes with the shift action (that is, $(\tau(f))^w = \tau(f^w)$ for all $f \in X$ and $w \in \Sigma^*$), and is continuous (with respect to the prodiscrete topology on X).* For a proof in the one-dimensional case, see [9, Theorem 6.2.9]. See also [5, Theorem 1.8.1] and [7], for a more general setting. It immediately follows that the image of a tree shift under a cellular automaton is still a tree shift.

Remark 2. In the definition of a cellular automaton we have assumed that the alphabet of the shift X is the same as the alphabet of its image $\tau(X)$. In this assumption there is no loss of generality because if $\tau\colon X \to B^{\Sigma^*}$, one can always consider X as a subshifts of $(A \cup B)^{\Sigma^*}$. Classically, a cellular automaton is also a selfmapping $\tau\colon X \to X$. By dropping this hypothesis, we deal with a more general notion that, in the one-dimensional case, corresponds to that of *sliding block code* as defined in [9].

A subshift $X \subset A^{\Sigma^*}$ is called *sofic* provided there exist a subshift of finite type $Y \subset A^{\Sigma^*}$ and a cellular automaton $\tau\colon Y \to A^{\Sigma^*}$ such that $X = \tau(Y)$.

Remark 3. Every subshift of finite type is sofic but there are examples of sofic subshifts which are not of finite type (see [9, Example 2.1.5, Example 2.1.9]).

3   Unrestricted Rabin Graphs and Automata

An *unrestricted Rabin graph*, is a 4-tuple $\mathcal{G} = (S, \Sigma, A, \mathcal{T})$, where S is a nonempty set, called the set of *states* (or *vertices*) of $\mathcal{G}$ and $\mathcal{T}$ is a subset of $S \times A \times S^\Sigma$ whose elements are called *transition bundles*. When the state set S is finite $\mathcal{G} = (S, \Sigma, A, \mathcal{T})$ is called an *unrestricted Rabin automaton*.

Given a transition bundle $t = (s; a; (s_\sigma)_{\sigma \in \Sigma}) \in \mathcal{T}$ we denote by $\mathbf{i}(t) := s \in S$ its *initial state*, by $\lambda(t) := a \in A$ its *label*, by $\mathbf{t}(t) := (s_\sigma)_{\sigma \in \Sigma} \in S^\Sigma$ its *terminal sequence* and by $\mathbf{t}_\sigma(t) := s_\sigma \in S$ its σ-*terminal state*. A *bundle loop on* $s \in S$ is a transition bundle $t \in \mathcal{T}$ such that $\mathbf{i}(t) = \mathbf{t}_\sigma(t) = s$ for all $\sigma \in \Sigma$.

An unrestricted Rabin graph $\mathcal{G} = (S, \Sigma, A, \mathcal{T})$ is said to be *essential* provided that for each state $s \in S$ there is a transition bundle starting at s.

Definition 1 (Unrestricted Rabin graph of a configuration). The *unrestricted Rabin graph of a configuration* $f \in A^{\Sigma^*}$ is defined by $\mathcal{G}_f = (\Sigma^*, \Sigma, A, \mathcal{T}_f)$ where $\mathcal{T}_f = \{(w; f(w); (w\sigma)_{\sigma \in \Sigma}) : w \in \Sigma^*\}$.

Definition 2 (Homomorphism). A *homomorphism* from $\mathcal{G}_1 = (S_1, \Sigma, A, \mathcal{T}_1)$ to $\mathcal{G}_2 = (S_2, \Sigma, A, \mathcal{T}_2)$ is a map $\alpha \colon S_1 \to S_2$ such that $(\alpha(s); a; (\alpha(s_\sigma))_{\sigma \in \Sigma}) \in \mathcal{T}_2$ for all $(s; a; (s_\sigma)_{\sigma \in \Sigma}) \in \mathcal{T}_1$. By abuse of language/notation, we also denote by $\alpha \colon \mathcal{G}_1 \to \mathcal{G}_2$ such a homomorphism.

Definition 3 (Acceptance). Let $\mathcal{A} = (S, \Sigma, A, \mathcal{T})$ be an unrestricted Rabin automaton. We say that a configuration $f \in A^{\Sigma^*}$ is *accepted* (or *recognized*) by $\mathcal{A}$, if there exists a homomorphism $\alpha \colon \mathcal{G}_f \to \mathcal{A}$. In this case, we say that f *is accepted by* $\mathcal{A}$ *via* α. We denote by $\mathsf{X}_\mathcal{A}$ the set consisting of all those configurations $f \in A^{\Sigma^*}$ accepted by $\mathcal{A}$. An unrestricted Rabin automaton $\mathcal{A}$ is called a *presentation* for $X \subset A^{\Sigma^*}$ provided that $X = \mathsf{X}_\mathcal{A}$.

Remark 4. In the sequel, we shall always consider essential unrestricted Rabin automata. This is not restrictive since, by recursively removing all states that are source of no transition bundles, we can transform any unrestricted Rabin automaton $\mathcal{A}$ into an essential one $\mathcal{A}'$ which accepts the same subset, i.e. such that $\mathsf{X}_\mathcal{A} = \mathsf{X}_{\mathcal{A}'}$.

Remark 5. Explicitly, a configuration $f \in A^{\Sigma^*}$ is accepted by an unrestricted Rabin automaton $\mathcal{A} = (S, \Sigma, A, \mathcal{T})$ if there exists a map $\alpha \colon \Sigma^* \to S$ such that $(\alpha(w); f(w); (\alpha(w\sigma))_{\sigma \in \Sigma}) \in \mathcal{T}$ for all $w \in \Sigma^*$.

3.1   Graphical Representation

Let $|\Sigma| = k$. We identify Σ with the set $\{0, 1, \ldots, k - 1\}$. Hence, a transition bundle of an unrestricted Rabin automaton $\mathcal{A} = (S, \Sigma, A, \mathcal{T})$ is a $(k + 2)$-tuple

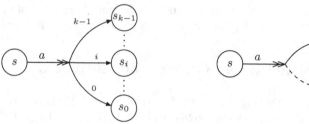

(a) A general labeled transition bundle.

(b) A labeled transition bundle of an unrestricted Rabin automaton in which $\Sigma = \{0, 1\}$.

Fig. 2. Representations of a transition bundle

$t = (s; a; s_0, \ldots, s_{k-1})$ and it can be visualized as in Figure 2(a). If $|\Sigma| = 2$ and $(s; a; s_0, s_1)$ is a transition bundle, we represent the edge from s to s_0 by a broken line and the edge from s to s_1 by a full line. This makes unnecessary to label the corresponding edges by 0 and 1, respectively (see Figure 2(b)).

Example 2. Consider the unrestricted Rabin automaton $\mathcal{A} = (A, \Sigma, A, \mathcal{T})$ where the bundle set is given by

$$\mathcal{T} = \{(a; a; (a_\sigma)_{\sigma \in \Sigma}) \in A \times A \times A^\Sigma : a_\sigma = a_{\sigma'} \text{ for all } \sigma, \sigma' \in \Sigma\}.$$

We then have that $\mathsf{X}_{\mathcal{A}}$ is the tree shift described in Example 1. If $|\Sigma| = 2$ and $A = \{0, 1\}$ the corresponding automaton is represented in Figure 3.

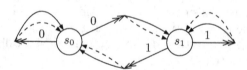

Fig. 3. The unrestricted Rabin automaton accepting the tree shift of Example 1

3.2   Unrestricted Rabin Automata and Sofic Shifts

Proposition 1. *Let $\mathcal{A} = (S, \Sigma, A, \mathcal{T})$ be an unrestricted Rabin automaton. Then $\mathsf{X}_{\mathcal{A}}$ is a sofic tree shift. Actually, up to a suitable extension of the alphabet A, there is an effective procedure to construct a tree shift of finite type $Y \subset A^{\Sigma^*}$ and a cellular automaton $\tau \colon Y \rightarrow A^{\Sigma^*}$ such that $\mathsf{X}_{\mathcal{A}} = \tau(Y)$.*

Let $M \subset \Sigma^*$ be a nonempty subset and $p \in A^M$ a pattern with support M. Given a word $w \in \Sigma^*$ we set $wM = \{wm : m \in M\} \subset \Sigma^*$ and denote by $^w p \in A^{wM}$ the pattern with support wM defined by $(^w p)(wm) = p(m)$ for all $m \in M$.

Definition 4 (Unrestricted Rabin automaton associated with a cellular automaton). Let $X \subset A^{\Sigma^*}$ be a tree shift of finite type and let $\tau \colon X \to A^{\Sigma^*}$ be a cellular automaton. Let $M = \Delta_n \subset \Sigma^*$ be a memory set for τ such that $n \geq 2$ and denote by $\mu \colon A^M \to A$ the corresponding local defining map. Fix $M' = \Delta_{n-1}$. The *unrestricted Rabin automaton* $\mathcal{A}(\tau, M, X)$ *associated with* τ is defined by $\mathcal{A}(\tau, M, X) = (X_{M'}, \Sigma, A, \mathcal{T})$, where $\mathcal{T} \subset X_{M'} \times A \times (X_{M'})^{\Sigma}$ consists of the bundles $(p; b; (p_\sigma)_{\sigma \in \Sigma})$ such that (*i.*) $p|_{\sigma M' \cap M'}$ equals $({}^\sigma p_\sigma)|_{\sigma M' \cap M'}$ for all $\sigma \in \Sigma$ (that is, $p(\sigma m) = p_\sigma(m)$ whenever $\sigma m \in \sigma M' \cap M'$); (*ii.*) the block $\bar{p} \colon M \to A$ coinciding with p on M' and with ${}^\sigma p_\sigma$ on $\sigma M'$ belongs to X_M for all $\sigma \in \Sigma$ (such a block $\bar{p} \in X_M$ is denoted by $\overline{(p; (p_\sigma)_{\sigma \in \Sigma})}$); (*iii.*) $b = \mu \left(\overline{(p; (p_\sigma)_{\sigma \in \Sigma})} \right)$.

A transition bundle of $\mathcal{A}(\tau, M, X)$ is illustrated in Figure 4 for $|\Sigma| = 2$.

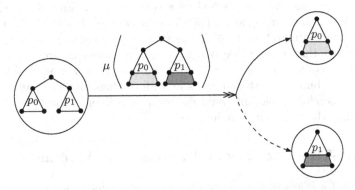

Fig. 4. A transition bundle of $\mathcal{A}(\tau, M, X)$ when $|\Sigma| = 2$

Proposition 2. *Let* $X \subset A^{\Sigma^*}$ *be a tree shift of finite type with memory* $n-1$ *and* $\tau \colon X \to A^{\Sigma^*}$ *be a cellular automaton with memory set* Δ_n. *Then* $X_{\mathcal{A}(\tau, \Delta_n, X)} = \tau(X)$.

Remark 6. Conditions on Δ_n in Proposition 2 are not restrictive. Proposition 2 says that $\mathcal{A}(\tau, \Delta_n, X)$ is a presentation of $\tau(X)$. In fact, we can actually show how to construct a pre-image of a configuration in $X_{\mathcal{A}(\tau, \Delta_n, X)}$. This leads in particular to a presentation of X as well.

Proposition 1 and Proposition 2 imply the following result. The bottom-up version of it has been proved by Béal and Aubrun in [4].

Corollary 1. *A tree shift* $X \subset A^{\Sigma^*}$ *is sofic if and only if it is accepted by some unrestricted Rabin automaton.*

3.3 Deterministic and Co-deterministic Presentations

An unrestricted Rabin automaton $\mathcal{A} = (S, \Sigma, A, \mathcal{T})$ is *deterministic* if, for each state $s \in S$, the transition bundles starting at s carry different labels. Analogously, $\mathcal{A}$ is *co-deterministic* if, for each sequence $\mathbf{s} \in S^{\Sigma}$, the transition bundles terminating at $\mathbf{s}$ (if there are any) carry different labels.

As stated below, for each unrestricted Rabin automaton $\mathcal{A}$ there exists a co-deterministic unrestricted Rabin automaton accepting the same shift.

Theorem 1 (Subset construction). *Let $\mathcal{A} = (S, \Sigma, A, \mathcal{T})$ be an unrestricted Rabin automaton. There exists a co-deterministic unrestricted Rabin automaton $\mathcal{A}_{\mathrm{cod}}$ such that $\mathsf{X}_{\mathcal{A}} = \mathsf{X}_{\mathcal{A}_{\mathrm{cod}}}$.*

The statement of the above theorem fails to hold, in general, for deterministic unrestricted Rabin automata, as shown in the following counterexample.

Example 3 (A sofic shift not admitting a deterministic presentation). Consider the tree shift X presented in Example 1. A non-deterministic presentation of X is given in Example 2. Suppose that X admits a deterministic presentation $\mathcal{A} = (S, \Sigma, A, \mathcal{T})$. First observe that, in this case, each *accessible* state (that is, each state that can be reached by a transition bundle), admits exactly one transition bundle starting at it. Thus for every accessible state $s \in S$ there exists exactly one configuration $f_s \in X$ accepted by a homomorphism $\alpha_s : \Sigma^* \to S$ *starting* at s, that is, such that $\alpha_s(\varepsilon) = s$. This implies that any state determines at most $|A|$ configurations (indeed, for a state s that is not accessible, there are at most $|A|$ bundles that start at s and all of these bundles end in accessible states). Therefore $\mathcal{A}$ accepts only finitely many different configurations, which contradicts the fact that X is infinite.

4   Full-Tree-Patterns and Finite-Tree Automata

Recall that a *k-ary rooted tree* is a rooted tree in which each vertex has at most k children. A *leaf* is a vertex without children. A *full k-ary rooted tree* is a rooted tree in which every vertex other than the leaves has k children. Hence Σ^* is the full k-ary rooted tree with no leaves, where $k = |\Sigma|$. A *subtree* of Σ^* is a connected subgraph of Σ^*. We shall always suppose that a subtree of Σ^* contains the root ε. If $T \subset \Sigma^*$ is a subtree and $w \in T$, we denote by $\Sigma_T(w)$ the set $\{\sigma \in \Sigma : w\sigma \in T\}$. Hence $w \in T$ is a leaf if and only if $\Sigma_T(w) = \varnothing$.

Given a subtree T, we denote by T^+ the subtree $T \cup \{w\sigma : w \in T, \sigma \in \Sigma\}$. Notice that T^+ is always a full subtree. If T is a full subtree, then T^+ is obtained by adding all the k children of each leaf in T.

Notice that for each $n \geq 1$ the set Δ_n is a full subtree whose leaves are the elements in Σ^{n-1}. Moreover, $\Delta_n^+ = \Delta_{n+1}$.

Finite full subtrees correspond to finite and complete prefix codes in [2].

A pattern defined on a finite full subtree T is called *full-tree-pattern*. The set of all full-tree-patterns is denoted by $\mathfrak{T}(A^{\Sigma^*})$. Given a shift $X \subset A^{\Sigma^*}$, we denote by $\mathfrak{T}(X)$ the set of all full-tree-patterns of X (that is, $\mathfrak{T}(X) = \bigcup_{T \subset \Sigma^*} X_T$, where the union ranges over all finite full subtrees T of Σ^*).

Definition 5 (Sub-bundle). Let $\mathcal{A} = (S, \Sigma, A, \mathcal{T})$ be an unrestricted Rabin automaton. Let $M \subset \Sigma$ be a subset. A tuple $(s; a; (s_\sigma)_{\sigma \in M}) \in S \times A \times S^M$ is called a *sub-bundle* of a transition bundle $(\bar{s}; \bar{a}; (\bar{s}_\sigma)_{\sigma \in \Sigma}) \in \mathcal{T}$ provided $s = \bar{s}$, $a = \bar{a}$, and $s_\sigma = \bar{s}_\sigma$ for each $\sigma \in M$.

Definition 6. Let $\mathcal{A} = (S, \Sigma, A, \mathcal{T})$ be an unrestricted Rabin automaton. Let $T \subset \Sigma^*$ be a subtree and let $f \colon T \to A$ be a map. One says that f is *accepted by* $\mathcal{A}$ if there exists a map $\alpha \colon T \to S$ such that, for each $w \in T$, $(\alpha(w); f(w); (\alpha(w\sigma))_{\sigma \in \Sigma_T(w)})$ is a sub-bundle of some $t \in \mathcal{T}$. In this case we say that f is accepted by $\mathcal{A}$ *via* α.

Note that, for a leaf $w \in T$, this latter acceptance condition reduces to saying that there exists a transition bundle starting at $\alpha(w)$ with label $f(w)$ (in fact, α is not defined on $w\sigma$ for any $\sigma \in \Sigma$).

Proposition 3. *Let $\mathcal{A} = (S, \Sigma, A, \mathcal{T})$ be an unrestricted Rabin automaton. Let $T \subset \Sigma^*$ be a subtree and suppose that $f \in A^T$ is accepted by $\mathcal{A}$. Then there exists a configuration $\bar{f} \in \mathsf{X}_\mathcal{A}$ such that $f = \bar{f}|_T$.*

We have the following characterization of acceptance which immediately results from Definition 6.

Proposition 4. *Let $\mathcal{A} = (S, \Sigma, A, \mathcal{T})$ be an unrestricted Rabin automaton. Let $T \subset \Sigma^*$ be a finite full subtree. A full-tree-pattern $p \in A^T$ is accepted by $\mathcal{A}$ if and only if there exists a map $\alpha \colon T^+ \to S$ such that $(\alpha(w); p(w); (\alpha(w\sigma))_{\sigma \in \Sigma}) \in \mathcal{T}$ for each $w \in T$.*

By abuse of language, if this acceptance condition holds and there is no ambiguity, we say that the full-tree-pattern p *is accepted by* $\mathcal{A}$ *via* α. Obviously, Proposition 4 applies whenever $T = \Delta_n$ for some $n \geq 1$ (recall that in this case $T^+ = \Delta_{n+1}$).

The following result follows from Proposition 3.

Corollary 2. *Let $\mathcal{A} = (S, \Sigma, A, \mathcal{T})$ be an unrestricted Rabin automaton. Let $p \in \mathfrak{T}(A^{\Sigma^*})$ be a full-tree-pattern. Then $p \in \mathfrak{T}(\mathsf{X}_\mathcal{A})$ if and only if p is accepted by $\mathcal{A}$.*

Remark 7. The blocks of a subshift determine the subshift. In fact, given two subshifts $X, Y \subset A^{\Sigma^*}$, we have $X = \mathsf{X}(\mathcal{B}(A^{\Sigma^*}) \setminus \mathcal{B}(X))$ so that $X = Y$ if and only if $\mathcal{B}(X) = \mathcal{B}(Y)$. This fact obviously generalizes to full-tree-patters: $X = Y$ if and only if $\mathfrak{T}(X) = \mathfrak{T}(Y)$.

4.1   Finite-Tree Automata

A *finite-tree automaton* is an unrestricted Rabin automaton $\mathcal{A} = (S, \Sigma, A, \mathcal{T})$ for which a subset $\mathcal{I} \subset S$ of *initial states* and a state $F \in S$, called *final state*, are specified. We shall denote it by $\mathcal{A}(\mathcal{I}, F)$. We say that a full-tree-pattern $p \in A^T$ is *accepted by* $\mathcal{A}(\mathcal{I}, F)$ if there exists a map $\alpha \colon T^+ \to S$ such that (*i.*) p is accepted by $\mathcal{A}$ via α (see Proposition 4); (*ii.*) $\alpha(\varepsilon) \in \mathcal{I}$; (*iii.*) $\alpha(w) = F$ if $w \in T^+ \setminus T$. We denote by $\mathfrak{T}(\mathcal{A}(\mathcal{I}, F))$ the set of all full-tree-patterns accepted by $\mathcal{A}(\mathcal{I}, F)$. A set of full-tree-patterns is called *recognizable* if it is of the form $\mathfrak{T}(\mathcal{A}(\mathcal{I}, F))$, for some finite-tree automaton $\mathcal{A}(\mathcal{I}, F)$. A finite-tree automaton $\mathcal{A}(\mathcal{I}, F)$ is *co-deterministic* if the unrestricted Rabin automaton $\mathcal{A}$ is co-deterministic.

Remark 8. As explained in Section 4, we only consider essential unrestricted Rabin automata. As far as finite-tree automata are concerned, we relax this assumption: each non final state is the source of some transition bundle, but no condition is required for the final state.

An unrestricted Rabin automaton $\mathcal{A} = (S, \Sigma, A, \mathcal{T})$ is called *co-complete* if for each $\mathbf{s} \in S^\Sigma$ and $a \in A$, there exists a transition bundle in $\mathcal{T}$ labeled by a and ending at $\mathbf{s}$. A finite-tree automaton $\mathcal{A}(\mathcal{I}, F)$ is *co-complete* if the unrestricted Rabin automaton $\mathcal{A}$ is co-complete.

A slight adaptation in the proof of Theorem 1 leads to the following result.

Theorem 2. *Let $\mathcal{A}$ be an unrestricted Rabin automaton. Then there is an effective procedure to construct a co-deterministic finite-tree automaton $\mathcal{A}_{\mathrm{cod}}(\mathcal{I}, F)$ such that $\mathfrak{T}(\mathsf{X}_\mathcal{A}) = \mathfrak{T}(\mathcal{A}_{\mathrm{cod}}(\mathcal{I}, F))$.*

The recognizable sets of full-tree-patterns form a class which is closed under complementation, as stated in the following theorem.

Theorem 3. *Let $\mathcal{A}(\mathcal{I}, F)$ be a co-deterministic finite-tree automaton. Then there exists a co-complete and co-deterministic finite-tree automaton $\mathcal{A}_\complement(\mathcal{I}_\complement, F_\complement)$ such that $\mathfrak{T}(A^{\Sigma^*}) \setminus \mathfrak{T}(\mathcal{A}(\mathcal{I}, F)) = \mathfrak{T}(\mathcal{A}_\complement(\mathcal{I}_\complement, F_\complement))$.*

Corollary 3. *Let $\mathcal{A}$ be an unrestricted Rabin automaton. Then there is an effective procedure to construct a co-complete and co-deterministic finite-tree automaton $\mathcal{A}_\complement(I, F)$ (with a single initial state) which accepts the complement of the set of all full-tree-patterns accepted by $\mathcal{A}$, in formulæ, $\mathfrak{T}(\mathcal{A}_\complement(I, F)) = \mathfrak{T}(A^{\Sigma^*}) \setminus \mathfrak{T}(\mathsf{X}_\mathcal{A})$.*

Corollary 4. *Let $\mathcal{A}$ be an unrestricted Rabin automaton. Let $\mathcal{A}_\complement(I, F)$ be as in Corollary 3. Then $\mathsf{X}_\mathcal{A} = A^{\Sigma^*}$ if and only if $\mathfrak{T}(\mathcal{A}_\complement(I, F)) = \varnothing$.*

The Emptiness Problem for Finite-Tree Automata. The emptiness problem for an unrestricted Rabin automaton is trivial (every nonempty essential automaton accepts at least a configuration), but this argument does not apply to the case of finite-tree automata. In this section we present an effective procedure to establish the emptiness of recognizable set of full-tree-patterns.

First, we define the *height of a finite subtree* $T \subset \Sigma^*$ as the minimal $n \in \mathbb{N}$ such that $T \subset \Delta_n$. The *height of a full-tree-pattern* $p \in A^T$ is the height of the (finite full) subtree T.

Let $\mathcal{A}(\mathcal{I}, F)$ be a finite-tree-automaton and let us show that there is an algorithm which establishes whether or not $\mathfrak{T}(\mathcal{A}(\mathcal{I}, F)) = \varnothing$. Observe that $\mathfrak{T}(\mathcal{A}(\mathcal{I}, F))$ is nonempty if and only if it contains a pattern of height $\leq |S|$, where S is the state set of $\mathcal{A}$ (we do not prove this fact in detail). Since there are finitely many full-tree-patterns of height $\leq |S|$ one can effectively check whether or not they are accepted by $\mathcal{A}(\mathcal{I}, F)$.

Since in principle we have to check all possible maps $\alpha \colon \Delta_{|S|+1} \to S$, this algorithm has exponential complexity in the size of S.

An Algorithm Establishing Whether Two Sofic Shifts Coincide. The *join* of $\mathcal{A}_1 = (S_1, \Sigma, A, \mathcal{T}_1)$ and $\mathcal{A}_2 = (S_2, \Sigma, A, \mathcal{T}_2)$ is the unrestricted Rabin automaton $\mathcal{A}_1 * \mathcal{A}_2 = (S_1 \times S_2, \Sigma, A, \mathcal{T}_\times)$ where $((s_1, s_2); a; (s'_\sigma, s''_\sigma)_{\sigma \in \Sigma}) \in \mathcal{T}_\times$ if and only if $(s_1; a; (s'_\sigma)_{\sigma \in \Sigma}) \in \mathcal{T}_1$ and $(s_2; a; (s''_\sigma)_{\sigma \in \Sigma}) \in \mathcal{T}_2$. Notice that $\mathsf{X}_{\mathcal{A}_1 * \mathcal{A}_2} = \mathsf{X}_{\mathcal{A}_1} \cap \mathsf{X}_{\mathcal{A}_2}$. Moreover, $\mathcal{A}_1 * \mathcal{A}_2$ is co-complete (respectively, co-deterministic), if $\mathcal{A}_1$ and $\mathcal{A}_2$ are co-complete (resp., co-deterministic).

We are now in position to describe our algorithm: let $\mathcal{A}_1 = (S_1, \Sigma, A, \mathcal{T}_1)$ and $\mathcal{A}_2 = (S_2, \Sigma, A, \mathcal{T}_2)$ be two unrestricted Rabin automata. Note that, by Remark 7, it suffices to establish whether or not

$$\mathfrak{T}(\mathsf{X}_{\mathcal{A}_1}) \setminus \mathfrak{T}(\mathsf{X}_{\mathcal{A}_2}) = \varnothing = \mathfrak{T}(\mathsf{X}_{\mathcal{A}_2}) \setminus \mathfrak{T}(\mathsf{X}_{\mathcal{A}_1}). \tag{1}$$

First construct the co-complete and co-deterministic finite-tree automata $\mathcal{A}'_1(I_1, F_1)$ and $\mathcal{A}'_2(I_2, F_2)$ as in Corollary 3, associated with $\mathcal{A}_1$ and $\mathcal{A}_2$, respectively. Consider the finite-tree automaton $(\mathcal{A}'_1 * \mathcal{A}'_2)(I_1, F)$, where $I_1 = (S_1 \setminus \{I_1\}) \times \{I_2\}$ and $F = (F_1, F_2)$. It can be seen that $\mathfrak{T}((\mathcal{A}'_1 * \mathcal{A}'_2)(I_1, F)) = \mathfrak{T}(\mathsf{X}_{\mathcal{A}_1}) \setminus \mathfrak{T}(\mathsf{X}_{\mathcal{A}_2})$. Analogously, by defining $I_2 = \{I_1\} \times (S_1 \setminus \{I_2\})$ one has $\mathfrak{T}((\mathcal{A}'_1 * \mathcal{A}'_2)(I_2, F)) = \mathfrak{T}(\mathsf{X}_{\mathcal{A}_2}) \setminus \mathfrak{T}(\mathsf{X}_{\mathcal{A}_1})$.

Thus (1) holds if and only if $\mathfrak{T}((\mathcal{A}'_1 * \mathcal{A}'_2)(I_1, F)) \bigcup \mathfrak{T}((\mathcal{A}'_1 * \mathcal{A}'_2)(I_2, F)) = \varnothing$. An effective procedure to establish this latter equality is then provided by the solution to the emptiness problem.

Remark 9. The algorithm above has exponential complexity in the maximal size of the state sets of the unrestricted Rabin automata. A different procedure can be applied to the class of *irreducible* unrestricted Rabin automata by using a minimization process. Actually, in [3] it is shown that there exists a canonical minimal co-deterministic presentation of an irreducible sofic tree shift. Thus another possible decision algorithm consists in computing the minimal presentations of the two shifts and checking whether they coincide or not. In this case Theorem 1 is needed while the procedure for the emptiness problem is not required. Hence this algorithm has in general an exponential complexity. The complexity can be reduced to be polynomial by only considering the class of co-deterministic irreducible tree shifts.

An Algorithm Establishing the Surjectivity of Cellular Automata. Observe first that giving a sofic shift $X \subset A^{\Sigma^*}$ corresponds, equivalently to giving a shift of finite type $Z \subset A^{\Sigma^*}$ and a surjective cellular automaton $\tau' : Z \to X$, or an unrestricted Rabin automaton $\mathcal{A}$ such that $X = \mathsf{X}_\mathcal{A}$. Propositions 1 and 2 provide an effective procedure to switch from one representation to the other.

Let $X, Y \subset A^{\Sigma^*}$ be two sofic shifts and $\tau : X \to Y$ a cellular automaton. Let us show that it is decidable whether τ is surjective or not. Let $Z \subset A^{\Sigma^*}$ and $\tau' : Z \to X$ as above. Now the cellular automaton $\tau : X \to Y$ is surjective if and only if the composite cellular automaton $\tau \circ \tau' : Z \to Y$ is surjective. Let $n \in \mathbb{N}$ be large enough so that the cellular automaton $\tau \circ \tau'$ has memory set Δ_n and that $n - 1$ is the memory of Z. By Proposition 2, the unrestricted Rabin automaton $\mathcal{A}(\tau \circ \tau', \Delta_n, Z)$ having state set Z_{n-1} is a presentation of $\tau(X)$. Then, it suffices to apply the algorithm in previous section to establish whether $Y = \tau(X)$.

Remark 10. If $X = Y = A^{\Sigma^*}$, then the algorithm becomes much simpler. Indeed, it can be proved by virtue of Corollary 4, Corollary 3 and by using the emptiness algorithm.

References

1. Amoroso, S., Patt, Y.N.: Decision procedures for surjectivity and injectivity of parallel maps for tessellation structures. J. Comput. System Sci. 6, 448–464 (1972)
2. Aubrun, N.: Dynamique symbolique des systèmes 2D et des arbres infinis. Ph.D. thesis, Universitè Paris-Est (2011)
3. Aubrun, N., Béal, M.-P.: Sofic and Almost of Finite Type Tree-Shifts. In: Ablayev, F., Mayr, E.W. (eds.) CSR 2010. LNCS, vol. 6072, pp. 12–24. Springer, Heidelberg (2010)
4. Aubrun, N., Béal, M.P.: Sofic tree-shifts (to appear, 2012)
5. Ceccherini-Silberstein, T., Coornaert, M.: Cellular Automata and Groups. Springer Monographs in Mathematics, Berlin (2010)
6. Comon, H., Dauchet, M., Gilleron, R., Löding, C., Jacquemard, F., Lugiez, D., Tison, S., Tommasi, M.: Tree automata techniques and applications (2007), http://www.grappa.univ-lille3.fr/tata
7. Fiorenzi, F.: Cellular automata and strongly irreducible shifts of finite type. Theoret. Comput. Sci. 299(1-3), 477–493 (2003)
8. Kari, J.: Reversibility and surjectivity problems of cellular automata. J. Comput. System Sci. 48(1), 149–182 (1994)
9. Lind, D.A., Marcus, B.H.: An Introduction to Symbolic Dynamics and Coding. Cambridge University Press, Cambridge (1995)
10. Muller, D.E., Schupp, P.E.: The theory of ends, pushdown automata, and second-order logic. Theoret. Comput. Sci. 37(1), 51–75 (1985)
11. Rabin, M.O.: Decidability of second-order theories and automata on infinite trees. Trans. Am. Math. Soc. 141, 1–35 (1969)
12. Thatcher, J.W., Wright, J.B.: Generalized finite automata theory with an application to a decision problem of second-order logic. Math. Syst. Theory

Strict Local Testability
with Consensus Equals Regularity

Stefano Crespi Reghizzi and Pierluigi L. San Pietro

Politecnico di Milano - DEI, Milano I-20133
{stefano.crespireghizzi,pierluigi.sanpietro}@polimi.it

Abstract. A recent language definition device named consensual is based on agreement between similar words. Considering, say, a regular set of words over a bipartite alphabet made by pairs of unmarked/marked letters, the match relation specifies when such words agree. Therefore a regular set (the "base") over the bipartite alphabet specifies another language over the unmarked alphabet, called the consensual language. A word is in the consensual language if a set of corresponding matching words is in the base. From previous results, the family of consensual languages based on regular sets have an NLOGSPACE word problem, include non-semilinear languages, and are incomparable with the context-free (CF) ones; moreover the size of a consensual specification can be in a logarithmic ratio with respect to a NFA for the same language. We study the consensual languages that are produced by other language families: the Strictly Locally Testable of McNaughton and Papert and the context-free/sensitive ones. Using a recent generalization of Medvedev's homomorphic characterization of regular languages, we prove that regular languages are exactly the consensual languages based on strictly locally testable sets, a result that hints at a novel parallel decomposition of finite automata into locally testable components. The consensual family based on context-free sets strictly includes the CF family, while the consensual and the base families collapse together if the context-sensitive languages are chosen instead of the CF.

Keywords: formal languages, strict local testability, local language, non-counting, consensual language, counter machine, sliding-window, regular language, Medvedev theorem, homomorphic characterization, context-free, context-sensitive.

1 Introduction

A recently introduced language definition model, named consensual [5,6], is based on agreement or consensus between similar words. Consider a set of words from a, say, regular language over a bipartite alphabet qualified as *internal*, made by pairs of unmarked/marked terminals. A so called match relation specifies when two or more words over the internal alphabet agree: for that, the words must coincide when the marks are ignored, and in each position exactly one character is marked. This model was loosely inspired by current ideas about language

N. Moreira and R. Reis (Eds.): CIAA 2012, LNCS 7381, pp. 113–124, 2012.
© Springer-Verlag Berlin Heidelberg 2012

processing in the brain (see e.g. [3,11]) by simultaneous processes that interact and re-enforce each other.

Thus a, say, regular set – the *base* – over the internal alphabet specifies another language over the unmarked alphabet, called the *consensual language*: a word is in the consensual language if a corresponding set of matching words is in the base.

The family of consensual languages based on regular sets is known to be in NLOGSPACE, to include non-semilinear languages, and to be incomparable with the family of context-free (CF) languages; moreover, a DFA recognizing the base language needed to specify a regular language L can be exponentially smaller than the minimal NFA of L.

Here we proceed in the study of consensual languages that are produced using various language families as base: the subregular families of Local, Strictly Locally Testable ($\mathcal{SLT}$), and the context-free/sensitive languages. Using a recent generalization [7] of the old Medvedev's homomorphic characterization [12,10] of regular languages, we prove the main result, that regular languages are exactly the consensual languages based on $\mathcal{SLT}$; this result hints to an unusual parallel decomposition of finite automata into elementary sliding-window component machines.

In general, the consensual family obtained taking the base language from a given family always include the latter; for CF bases we find that the inclusion is strict (as for regular bases), whereas for context-sensitive bases the inclusion becomes identity.

The paper starts with the basic definitions and relevant properties of consensual languages (Sect. 2), then it reports new results on the use of the subregular families as base (Sect. 3), and, lastly, it considers the use of context-free/sensitive bases (Sect. 4). The conclusion mentions directions for continuation.

2   First Definitions and Properties

The terminal alphabet is denoted by Σ and the empty word by ϵ. For a word x, the length is denoted by $|x|$ and the i-th letter by $x(i)$, $1 \leq i \leq |x|$. A deterministic finite automaton (DFA) is specified as $A = (\Delta, Q, \delta, q_0, F)$ where: Δ is a finite alphabet; Q is a finite set of states; $\delta : Q \times \Delta \to Q$ is the state-transition function (or graph), always assumed to be *total*; q_0 is the initial state, and $F \subseteq Q$ is the set of final states.

The family of *strictly locally testable* languages of McNaughton and Papert [9], is next defined following mainly [4]. For simplicity we deal only with ϵ-free languages. For every word $w \in \Sigma^+$, for all $k \geq 2$, let $i_k(w)$ and $t_k(w)$ denote the prefix and, resp., the suffix of w of length k if $|w| \geq k$, or w itself if $|w| < k$. Let $f_k(w)$ denote the set of factors of w of length k. Extend i_k, t_k, f_k to languages as usual.

Definition 1. *A language L is k-strictly locally testable, shortly k-slt, if there exist finite sets $I_{k-1}, T_{k-1} \subseteq \Sigma^{k-1}$ and $F_k \subseteq \Sigma^k$ such that, for every $x \in \Sigma^k \cdot \Sigma^*$, the following condition holds: $x \in L \iff i_{k-1}(x) \in I_{k-1} \wedge t_{k-1}(x) \in T_{k-1} \wedge f_k(x) \subseteq F_k$.*

A language is strictly locally testable *(slt) if it is k-slt for some k, to be called the* width.

Words shorter than $k-1$ are ignored, but can be separately listed, if needed. The family $\mathcal{SLT}$ of slt languages is strictly included in the family $\mathcal{REG}$ of regular languages, and is a strict hierarchy ordered by the width value. The family obtained closing $\mathcal{SLT}$ by Boolean operations and concatenation is known as *Non-Counting* or star-free $\mathcal{NC}$ [9].

The above definition of k-slt language is equivalent to the one in [1], which is different from the one in [9]. The two definitions, however, produce the same family $\mathcal{SLT}$. Value $k = 2$ yields the well known family of *local languages* $\mathcal{LOC}$ (e.g., [12,2]).

Consensual languages. Here we formalize a simple mechanism to express agreement between strings belonging to a regular language, by means of an elementary letter by letter matching. Consider a bipartite alphabet $\widetilde{\Sigma}$, made by pairs of unmarked/marked characters of Σ and formalize the agreement by a k-ary relation, called match, that is satisfied by a set of k equally long strings if, in each position, exactly one word has an unmarked letter and the other strings have the same letter but marked. In our metaphor we view such strings as providing mutual consensus on the validity of the corresponding unmarked string. This justifies the name "consensual" proposed for the new family, which strictly includes the regular one. Formal definitions follow; more details are available in [6].

Let $\underline{\Sigma}$ be the alphabet obtained by *marking* each letter $a \in \Sigma$ as $\underline{a}$. The union $\Sigma \cup \underline{\Sigma}$ is named the *internal* alphabet (because its use is restricted to the technical device of consensual definitions) and denoted by $\widetilde{\Sigma}$.

We also use the marking function, denoted by $\_ : \widetilde{\Sigma} \to \Sigma$, defined as $\underline{x} = x$ if $x \in \underline{\Sigma}$, and as $\underline{x} = \underline{a}$ if $x = a \in \Sigma$; the function is then naturally extended from letters to words.

The notion that two or more words over the internal alphabet are in agreement is formalized next.

Definition 2. *The partial, symmetrical, and associative binary operator, called*

match, $@ : \widetilde{\Sigma} \times \widetilde{\Sigma} \to \widetilde{\Sigma}$, *is, for all* $a \in \Sigma$:
$$\begin{cases} a@\underline{a} = \underline{a}@a = a \; ; \\ \underline{a}@\underline{a} = \underline{a} \quad\quad\quad ; \\ undefined, \quad\quad in\ every\ other\ case. \end{cases}$$

The operator is naturally extended to words of equal length, by positing $\epsilon @ \epsilon = \epsilon$ *and, for all* $w, w' \in \widetilde{\Sigma}^*$, *with* $|w| = |w'|$, *and for all* $a, b \in \widetilde{\Sigma}$, $aw \; @ \; bw' = (a@b)(w@w')$.

In words, the match is undefined if $|w| \neq |w'|$, or in the case that, in some position i, $w(i)@w'(i)$ is undefined, which happens when both characters are in Σ, when both are in $\underline{\Sigma}$ and differ, and when either one is marked but is not the marked copy of the other.

For instance, $\underline{aabb} \; @ \; \underline{aabb} = \underline{aabb}$ while $\underline{aabb} \; @ \; \underline{aabb}$ is undefined.

Given $m > 0$ words $w_1, \ldots, w_m \in \widetilde{\Sigma}^*$, if $w = w_1@w_2@\ldots@w_m$ is defined, then it is called the *match* of $w_1, w_2, \ldots, w_m$, further qualified as *strong* if $w \in \Sigma^*$ or as *weak* otherwise. We also write $@\{w_1, w_2, \ldots, w_m\}$. The cardinality m is called the *degree* of the match. By Def. 2, if w is a strong match, for each position $1 \le i \le n$, exactly one word, say w_k, is unmarked, i.e., $w_k(i) \in \Sigma$ and $w_j(i) \in \underline{\Sigma}$ for all $j \ne k$. We say that word w_k *places* the letter into position i and the other words *consent* to it.

The match operator is extended to two (or more) languages $L', L'' \subseteq \widetilde{\Sigma}^*$ by means of $L'@L'' = \{w'@w'' \mid w' \in L', w'' \in L''\}$.

To define the consensual language we use the repeated application of the match to a language. Let $L^{1@} = L$, $L^{i@} = L@L^{(i-1)@}$, $i \ge 2$. (Notice that in general $L^{(i-1)@} \not\subseteq L^{i@}$).

Definition 3. *The* closure under match, *or* @-closure, *of a language $L \subseteq \widetilde{\Sigma}^*$ is $L^@ = \bigcup_{i \ge 1} L^{i@}$. Let $B \subseteq \widetilde{\Sigma}^+$ be a language over the internal alphabet. The* consensual language with base B *is defined as* $\mathcal{C}(B) = B^@ \cap \Sigma^*$.

Let $\mathcal{F}$ denote a language family; the family of consensual languages based on $\mathcal{F}$, *written* $\mathcal{C}_{\mathcal{F}}$, *is the collection of all languages* $\mathcal{C}(B)$ *with* $B \in \mathcal{F}$.

Therefore, a consensual language with base B includes all and only the strongly matches of the match closure.

To give an idea of the generative capacity of $\mathcal{C}_{\mathcal{REG}}$ we list some typical examples.

Example 1. The consensual definitions of the following languages, or of some variations thereof, are in [6]:

CF language:	$L_1 = \{a^n b^n \mid n \ge 1\}$
non-semilinear CS languages:	
series of identical unary numbers	$L_2 = \{a^n b a^n b a^n b \ldots a^n b \mid n > 0\}$
enumeration of unary numbers	$L_3 = \{baba^2 b \ldots ba^n b \mid n \ge 0\}$
enumeration of exponential unary numbers	$L_4 = \{aba^2\, ba^4 ba^8 b \ldots a^{2^m} b \mid m \ge 1\}$

We describe just two cases. Language $L_1 = \mathcal{C}(R)$, where $R = \underline{a}^* a \underline{a}^* \underline{b}^* b \underline{b}^*$. Clearly every word $a^n b^n$, $n \ge 1$, is a strong match of n words in R; each word places one letter a and one letter b and consents to any number of a's and b's.

Language L_4 is defined by $\mathcal{C}(B_1 \cup B_2)$, where:

$$B_1 = \underline{\Sigma}^* a \underline{a} (\underline{a}\, \underline{a})^* \underline{b}\, \underline{a}^* a \underline{a} \underline{a} (\underline{a}\, \underline{a})^* \underline{b} \underline{\Sigma}^*, \qquad B_2 = ab \underline{a} \underline{a} (\underline{b} \underline{a}^+)^* b (\underline{a} \underline{a})^+ b$$

We explain how each word $w = aba^2 b \ldots ba^{2^h} b \ldots ba^{2^m} b$, $m \ge 2$ is the match of many words in B_1 and one in B_2. Call the *h-th segment* of w each factor, bordered by two b's, of the form a^{2^h}, $0 \le h \le m$. B_1 ensures that in every h-th segment, $1 \le h \le m - 1$, for every a occurring in an *odd* position from the right edge of the segment, there are two occurrences of a in the $h + 1$-th segment in consecutive *even* positions (from the right edge). Hence, each segment is exactly twice as long as the previous segment. B_2 places every b, allows the correct "initialization" of the first and second segment, and the correct completion of the odd-positioned a's in the last segment, while ignoring other segments. Hence, $w \in \mathcal{C}(B_1 \cup B_2)$. The converse is analogous.

The reader will notice that the above sample of languages is not included in the families of $\mathcal{REG}$, $\mathcal{CF}$, and tree adjoining languages. On the other hand, the language of palindromes is not in $\mathcal{C}_{REG}$ [6], thus proving incomparability of $\mathcal{C}_{REG}$ and $\mathcal{CF}$.

Consider now the DFA recognizing the base language R and a word in $\mathcal{C}(R)$, which is the strong match of some words in R. The matching words correspond to as many DFA computations, to be next formalized by means of multisets of states: the multiplicity of a state in the multiset equals the number of computations that have reached that state. A multiset can be represented by multiplicity counters, (one counter per DFA state). Since the cardinality of the multiset is bounded by the length of the input word, using a binary encoding of counters, the word membership can be computed by a nondeterministic counter machine operating in logarithmic space, hence in deterministic polinomial time.

Clearly, if there is a bound i such that $R^{i@} = R^{(i+1)@}$, i.e., $R^@$ has *bounded degree*, the cardinality of the multiset is bounded and language $\mathcal{C}(R)$ is regular; but, of course, this is is not a necessary condition for regularity.

To define a transition relation for the counter machine, we need some notation for multisets.

Notation for Multisets. Given a set Q, in particular the set of states of a DFA, a *multiset* over Q is a total mapping $Z : Q \to \mathbb{N}$. The cardinality of multiset Z is $|Z| = \sum_{q \in Q} Z(q)$. For $q \in Q$, if $Z(q) > 0$ then we say that $q \in Z$ with *multiplicity* $Z(q)$. The notation to represent a finite multiset is similar to the one for a set, but representing the multiplicity of an element with an exponent or by repetitions of the element. For example, the multiset Z over $Q = \{p, q, r\}$, characterized by $Z(p) = 3, Z(q) = 0, Z(r) = 5$, is also represented by $\{p^3, r^5\}$ or $\{p, p, p, r, r, r, r, r\}$.

Given two multisets Z, Z' over Q, the *sum* $Z \uplus Z'$ and the *difference* $Z - Z'$ are the multisets specified by the following characteristic functions, for every $q \in Q$:

$$(Z \uplus Z')(q) = Z(q) + Z'(q), \quad (Z - Z')(q) = max(0, Z(q) - Z'(q))$$

If $f : Q \to \mathbb{N}^Q$ is a total mapping, associating each element $q \in Q$ with a multiset $f(q)$ and $Z : Q \to \mathbb{N}$ is a multiset $\{q_1, \ldots, q_m\}$, where $m = |Z|$ and the q_i's are not necessarily distinct, then let the *generalized sum* $\biguplus_{q \in Z} f(q)$ be $f(q_1) \uplus \cdots \uplus f(q_m)$.

Finally, for a multiset Z over Q, the *underlying set* is $[\![Z]\!] = \{q \in Q \mid Z(q) > 0\}$.

Given a DFA $A = (\widetilde{\Sigma}, Q, \delta, q_0, F)$, where δ is a total function, the function is naturally extended to a multiset Z over Q, positing $\delta(Z, a) = \biguplus_{q \in Z} \{\delta(q, a)\}$. From this we define a transition relation on multisets of states.

Definition 4. *The* consensual transition relation *of A, denoted by $\leadsto_A \subseteq \mathbb{N}^Q \times \Sigma \times \mathbb{N}^Q$, is defined, for $a \in \Sigma$ and for multisets Z, Z' over Q as:*

$$Z \overset{a}{\leadsto}_A Z' \text{ if } \exists q \in Z : \ Z' = \{\delta(q, a)\} \uplus \delta(Z - \{q\}, \underline{a}).$$

Relation $\overset{a}{\leadsto}_A$ can be extended as usual from a letter a to a word $w \in \Sigma^*$ via the inductive definition:

$$\begin{cases} Z \overset{\epsilon}{\leadsto}_A Z \\ Z \overset{wa}{\leadsto}_A Z'', \text{ if } \exists Z' \text{such that } Z \overset{w}{\leadsto}_A Z' \overset{a}{\leadsto}_A Z''. \end{cases}$$

It is evident that if $Z \overset{a}{\leadsto}_A Z'$ then $|Z| = |Z'|$, i.e., the cardinality does not change.

Two types of multisets have a special role: the *initial* multisets $\{(q_0)^k\}$, for every $k > 0$, and the *final* multisets Z such that $[\![Z]\!] \subseteq F$.

The following crisp definition of consensual languages is obtained.

Proposition 1. *[6] Let $R \subseteq \widetilde{\Sigma}^*$ and let $A = (\widetilde{\Sigma}, Q, \delta, q_0, F)$ be a DFA accepting R.*

Then $\mathcal{C}(R) = \{w \mid \exists k > 0 \text{ and a final multiset } Z \text{ such that } \{(q_0)^k\} \overset{w}{\leadsto}_A Z\}$. Moreover, $|Z| = k$.

Example 2. Consider language of Ex. 1: $L_1 = \{a^n b^n \mid n \geq 1\} = \mathcal{C}(R)$, where $R = \underline{a}^* a\underline{a}^* \underline{b}^* b\underline{b}^*$, is recognized by the DFA below. The consensual transition relation accepting $aaabbb$ is also shown:

$\{q_1, q_1, q_1\} \overset{a}{\leadsto}_A \{q_1, q_1, q_2\} \overset{a}{\leadsto}_A$
$\{q_1, q_2, q_2\} \overset{a}{\leadsto}_A \{q_2, q_2, q_2\} \overset{b}{\leadsto}_A$
$\{q_3, q_3, q_4\} \overset{b}{\leadsto}_A \{q_3, q_4, q_4\} \overset{b}{\leadsto}_A \{q_4, q_4, q_4\}$

From the fact that a word over Σ is the trivial strong match of itself, we have:

Proposition 2. *Monotonicity*
Every language family $\mathcal{F}$ is included in the consensual family $\mathcal{C}_{\mathcal{F}}$. For any language families $\mathcal{F}'$ and $\mathcal{F}''$ such that $\mathcal{F}' \subseteq \mathcal{F}''$ the inclusion $\mathcal{C}_{\mathcal{F}'} \subseteq \mathcal{C}_{\mathcal{F}''}$ holds.

We are going to refine the above inclusions for various language families.

3   Consensual Languages with Regular and Subregular Bases

As the consensual languages on regular bases can be non-regular, it is natural to investigate if any base smaller than $\mathcal{REG}$ would produce all and only the regular languages. Clearly the subfamily of finite languages does not deserve consideration, since it is immediate that $\mathcal{FIN}$ coincides with $\mathcal{C_{FIN}}$.

A very simple subfamily is $\mathcal{LOC}$, the local (i.e., 2-slt) languages, but a reasoning, to be postponed after the main result, shows that $\mathcal{C_{LOC}}$ is strictly included into $\mathcal{REG}$. We are going to prove a rather surprising property: that all and only the regular languages can be consensually defined by means of a slt base. For that we need a recent generalization [7] of Medvedev's theorem [12] for regular languages and some definitions.

Given a finite alphabet Δ, an *(alphabetic) homomorphism* is a mapping $\pi : \Delta \to \Sigma$. For a language $L' \subseteq \Delta^+$, its *(homomorphic) image* under π is the language $L = \{\pi(x) \mid x \in L'\}$.

Medvedev's theorem states that every regular language, called *source*, is the image of a 2-slt, i.e., local language whose alphabet size may be much larger than the one of the source. To talk precisely about the width of the slt language and about the ratio of the alphabet size of the slt and source languages, we need a definition.

Definition 5. *[7] For $k \geq 2, m \geq 1$, a language $L \subseteq \Sigma^+$ is (m, k)-homomorphic if there exist an alphabet Δ (called* local*) of cardinality m, a k-slt language $L' \subseteq B^+$, and a homomorphism $\pi : \Delta \to \Sigma$ such that $L = \pi(L')$.*

Clearly, if $L \subseteq \Sigma^+$ is k-slt then L is trivially $(|\Sigma|, k)$-homomorphic. Otherwise, a local alphabet larger than Σ is needed. For instance, the language $L = (aa)^+ \cup (bb)^+$ is not slt but the language $L' = (a'a)^+ \cup (b'b)^+$ is 2-slt. By defining $\pi : \{a, a', b, b'\} \to \{a, b\}$ as $\pi(a) = \pi(a') = a$, $\pi(b) = \pi(b') = b$, then $L = \pi(L')$ and hence L is $(4, 2)$-homomorphic. The alphabetic ratio of L' and L is $4/2 = 2$.

The traditional construction (e.g., in [12]) of a 2-slt language L' considers an NFA for L and uses the set of edges of the transition graph as local alphabet, i.e., up to $n^2 \cdot |\Sigma|$ elements. Hence we can restate Medvedev's property saying that every regular language on Σ is $(n^2 \cdot |\Sigma|, 2)$-homomorphic (the alphabetic ratio is n^2).

If one allows a width $k \geq 2$, it is possible to set a constant bound on the alphabetic ratio; however in general the ratio cannot be smaller than two [7].

Tayloring to the needs of consensual languages, we simplify the main result in [7] freezing the alphabetic ratio to value 2.

Theorem 1. *Given a finite alphabet Σ, if a regular language $L \subseteq \Sigma^+$ is accepted by a NFA with n states, then L is $(2 \cdot |\Sigma|, O(\lg n))$-homomorphic, i.e., there exists a slt language B with alphabet size $2 \cdot |\Sigma|$ and a homomorphism π such that $\pi(B) = L$.*

Moreover, the width of the slt language cannot be smaller than a logarithmic function of the size of the source language recognizer.

We are ready to state and prove the first part of our main result.

Lemma 1. *Every regular language $L \subseteq \Sigma^+$ can consensually be defined using a strictly locally testable base.*

Proof. We take as base the k-slt language $B \subseteq \tilde{\Sigma}^+$ resulting from the application of Th. 1 to L. Let $I_{k-1}, T_{k-1} \subseteq \tilde{\Sigma}^{k-1}$ and $F_k \subseteq \tilde{\Sigma}^k$ be the sets defining B (Def. 1).

Denote by $\hat{\ } : \tilde{\Sigma} \to \tilde{\Sigma}$ the mapping, named *complementary*, $\hat{c} = d \mid c@d \in \Sigma$, i.e., c and d make a strong match. Let $\hat{B} = \{y \mid \forall x \in B, 1 \leq i \leq |x| : y(i)@x(i) \in \Sigma\}$ be the "complementary" language.

Clearly, $\hat{B}$ too is a k-slt language defined by the sets $I'_{k-1}, T'_{k-1} \subseteq \tilde{\Sigma}^{k-1}$ and $F'_k \subseteq \tilde{\Sigma}^k$ obtained by respectively applying the complementary mapping to I_{k-1}, T_{k-1} and F_k.

The inclusion $\mathcal{C}\left(B \cup \hat{B}\right) \subseteq L$ follows from Th. 1, since $\pi(B) = \pi(\hat{B}) = L$.

To prove the converse inclusion $L \subseteq \mathcal{C}(B \cup \hat{B})$, observe that, by Th. 1, every word $x \in L$ is the homomorphic image of a word in B which, by construction, strongly matches with a word in $\hat{B}$.   □

The converse property is stated next.

Lemma 2. *Let B be a strictly locally testable language. Then $\mathcal{C}(B)$ is regular.*

Proof. Let B be defined by the sets $I_{k-1}, T_{k-1} \subseteq \tilde{\Sigma}^{k-1}$, and $F_k \subseteq \tilde{\Sigma}^k$. By [4], B is recognized by a DFA A such that its states are suffixes of length at most k of words in $\tilde{\Sigma}^+$. In particular, the initial state is denoted by ϵ. Formally, let $A = (\tilde{\Sigma}, Q, \delta, \epsilon, F)$ with $Q \subseteq \{\epsilon\} \cup F_k \cup I_{k-1}$, $F = \{q \in Q \mid t_{k-1}(q) \in T_{k-1}\}$. The transition relation δ can be defined so that for $w \in \tilde{\Sigma}^+$, $\delta(\epsilon, w) = t_k(w)$, i.e., states are just suffixes of length k of the input word.

Consider the consensual transition relation $\leadsto_A$ on multisets of Q.

For every $y \in \Sigma^*$, for every multiplicity $j \geq k + 1$, for every $Z : Q \to \mathbb{N}$, if $\{(\epsilon)^j\} \overset{y}{\leadsto}_A Z$ (i.e., Z is reached from initial multiset $\{(\epsilon)^j\}$), then we claim that there exists $\mathring{Z} \subseteq Q$ such that:

(I) $Z = \mathring{Z} \cup \{t_k(y)^{j-|\mathring{Z}|}\}$,

(II) if $y \neq \epsilon$, $@(\mathring{Z}) = t_k(y)$,

(III) $|\mathring{Z}| \leq k$, and if $|y| > 0$ then $|\mathring{Z}| > 0$;

Consider the set $Y = \{y' \in \tilde{\Sigma}^+ \mid y' = y, \delta(\epsilon, y') \in Z\}$. Since $\delta(\epsilon, y') = t_k(y')$, $Z = \biguplus_{y' \in Y}\{t_k(y')\}$. But clearly $@(Y) = y$, hence if $y', y'' \in Y$ and $y' \neq y''$, then $t_k(y')@t_k(y'')$ is defined. This means that $t_k(y') \neq t_k(y'')$, unless $t_k(y') = t_k(y'') = t_k(y)$. Hence, in Z no state, other than $t_k(y)$, may occur more than once. Let $\mathring{Z} = \{q \in Z \mid q \neq t_k(y)\}$.

Claim (I) follows, since by the above definition of $\mathring{Z}$, $Z = \mathring{Z} \cup \{t_k(y)^{|Z|-|\mathring{Z}|}\}$: from Prop. 1, width j is constant in a computation, i.e., $|Z| = j$.

Claim (II) follows since $@(\mathring{Z}) = @\{t_k(y') \mid y' \in Y\} = t_k(y)$.

Claim (III) also follows. When $y \neq \epsilon$, $|\mathring{Z}|$ is equal to the number of words in Y whose suffix of length k has at least one unmarked letter. But this number is at most k, since $@(Y)$ is defined, and it is at least one: there exists one word $y' \in Y$ ending with an unmarked symbol (since $@(Y) = y$), hence $t_k(y') \neq t_k(y)$, therefore $t_k(y') \in \mathring{Z}$.

We also claim that, for every $a \in \Sigma, y \in \Sigma^*$, for every $j > 0$, for every $Z_1, Z : Q \to \mathbb{N}$:

$$\text{if } \{(\epsilon)^j\} \overset{y}{\leadsto}_A Z_1 \overset{a}{\leadsto}_A Z \text{ then } -1 \leq |\mathring{Z}_1| - |\mathring{Z}| \leq 1. \tag{1}$$

By definition of $\leadsto_A$, there exists $q \in Z_1$, such that $Z = \{\delta(q, a)\} \uplus \delta(Z_1 - \{q\}, \underline{a})$.

Clearly, $\delta(q, a) \in \mathring{Z}$, since $\delta(q, a)$ cannot be in $t_k(y\underline{a})$. The only case where $|\mathring{Z}| > |\mathring{Z}_1|$ may occur when the above state is $q = t_k(y) : \delta(q, a) \in \mathring{Z}$, but $\delta(q, \underline{a}) = t_k(y\underline{a})$, hence $\mathring{Z}$ may have at most one state more than $\mathring{Z}_1$: $|\mathring{Z}_1| - |\mathring{Z}| \geq -1$. Also, there exists at most one state $p \in \mathring{Z}_1$ such that $\delta(p, \underline{a}) \notin \mathring{Z}$: by contradiction, if for some $r \in \mathring{Z}_1, r \neq p$, also $\delta(r, \underline{a}) \notin \mathring{Z}$ holds, then both $r, p \in \Sigma\Sigma^{k-1}$, hence $r@p$ is undefined (both p and r start with an unmarked symbol), which is impossible by Claim (I) above. Therefore, $|\mathring{Z}| \geq |\mathring{Z}_1| - 1$, which completes the proof of Claim (1).

To prove the statement of the lemma, we also claim that:

$$\forall y \in \Sigma^*, \forall j \geq k+1, \forall Z : Q \rightarrow \mathbb{N}, \text{ if } \{(\epsilon)^j\} \overset{y}{\leadsto}_A Z, \tag{2}$$

then there exists $Z' : Q \rightarrow \mathbb{N}$ such that $\{(\epsilon)^{k+1}\} \overset{y}{\leadsto}_A Z'$, with $[\![Z']\!] = [\![Z]\!]$.

The proof is by induction on $|y| \geq 0$. The base case $y = \epsilon$ is obvious, since in this case $\mathring{Z} = \emptyset$ and $Z' = \{(\epsilon)^{k+1}\}$. Assume $y = xa$, for $a \in \Sigma, x \in \Sigma^*$, with $\{(\epsilon)^j\} \overset{x}{\leadsto}_A Z_1 \overset{a}{\leadsto}_A Z$ for some $Z_1 : Q \rightarrow \mathbb{N}$. Since $Z_1 \overset{a}{\leadsto}_A Z$, $\mathring{Z}_1 \cup \{t_k(\underline{x})^{j-|\mathring{Z}_1|}\} \overset{a}{\leadsto}_A \mathring{Z} \cup \{t_k(\underline{xa})^{j-|\mathring{Z}|}\}$. Let $i = |\mathring{Z}_1| - |\mathring{Z}|$, with $-1 \leq i \leq 1$ by Claim (1). Hence, for every $h \geq 1$ the following property also holds:

$$\mathring{Z}_1 \cup \{t_k(\underline{x})^h\} \overset{a}{\leadsto}_A \mathring{Z} \cup \{t_k(\underline{xa})^{h+i}\}, \text{ with } h + i > 0. \tag{3}$$

The induction hypothesis holds for Z_1: there exists Z_1' such that $\{(\epsilon)^{k+1}\} \overset{x}{\leadsto}_A Z_1'$, with $[\![Z_1']\!] = [\![Z_1]\!]$. But $Z_1' = \mathring{Z}_1' \cup \{t_k(\underline{x})^{k+1-|\mathring{Z}_1'|}\}$, hence $\mathring{Z}_1' = \mathring{Z}_1$. Let $h = k + 1 - |\mathring{Z}_1| \geq 1$ and apply Property (3) above: $Z_1' = \mathring{Z}_1 \cup \{t_k(\underline{x})^h\} \overset{a}{\leadsto}_A \mathring{Z} \cup \{t_k(\underline{x})^{h+i}\}$. Hence, let $Z' = \mathring{Z} \cup \{t_k(\underline{x})^{k+1-|\mathring{Z}|}\}$: then, $[\![Z']\!] = [\![Z]\!]$ (since $\mathring{Z}_1' = \mathring{Z}_1$ and $t_k(\underline{x})$ is in both Z' and Z), and $\{(\epsilon)^{k+1}\} \overset{x}{\leadsto}_A Z_1' \overset{a}{\leadsto}_A Z'$, which is the induction hypothesis for Z.

From Claim (2) above, the statement of the lemma holds, since Z and Z' are both final or both non-final multisets, and Z' has size at most $k + 1$: $B^@ = B \cup B^{2@} \cup \cdots \cup B^{(k+1)@}$, which is regular since k is a constant. Hence, also $\mathcal{C}(B) = B^@ \cap \Sigma^+$ is regular. $\square$

Theorem 2. *The family of regular languages coincides with the family $\mathcal{C}_{SLT}$.*

We observe that, in general, the width k of the base language cannot be bounded, since in the worst case it logarithmically depends on the complexity of L, measured in terms of states of a minimal NFA of L [7]. In particular, from the case $k = 2$ it follows that local languages are not sufficient to generate all regular languages.

Corollary 1. *The family $\mathcal{C}_{LOC}$ is strictly included into $\mathcal{REG}$.*

We briefly examine the effect of taking as base a language from the non-counting family $\mathcal{NC}$. From the monotonicity property, it is $\mathcal{C}_{NC} \supseteq \mathcal{C}_{SLT}$, and from Th. 2 the inclusion is strict because the language $\{a^n b^n \mid n \geq 1\}$ is defined by the non-counting base shown in Ex. 2. Also non-context-free languages are included in the family, e.g., language L_2 of Ex. 1 may be defined by the non-counting base $(\underline{a}^* a\underline{a}^* \underline{b})^+ \cup (\underline{a}^+ \underline{b})^+$

4   Consensual Languages over Non-regular Bases

The notions of match function and consensual computation apply without change to any language families, in particular Chomsky's CF and context-sensitive (CS) classes. The two cases are now considered obtaining rather diverse properties.

 We first address the question of the language generative capacity of consensual computation on a CF base.

Proposition 3. *The family $\mathcal{C}_{CF}$ strictly includes both families $\mathcal{C}_{REG}$ and CF.*

Proof. The strict inclusion of $\mathcal{C}_{REG}$ into $\mathcal{C}_{CF}$ follows by monotonicity and since language $\{ucu^r \mid u \in \{a,b\}^+\} \in CF \subseteq \mathcal{C}_{CF}$ is not in $\mathcal{C}_{REG}$. The strict inclusion of the CF into $\mathcal{C}_{CF}$ follows by monotonicity and by the existence of non-context-free languages (e.g., Ex. 1) in $\mathcal{C}_{REG} \subseteq \mathcal{C}_{CF}$.

The situation of CS languages is entirely different.

Theorem 3. *Let B be a context-sensitive language on alphabet $\tilde{\Sigma}$. Then $C(B)$ is a context-sensitive language.*

Sketch of the proof. Consider a nondeterministic Turing machine M accepting B, with one input tape and $k > 0$ memory tapes. Since B is CS, M can be assumed to be linearly-bounded, i.e., given an input x the memory space is at most $O(|x|)$.

 Define another nondeterministic Turing machine M' with two more memory tapes, called tape $k + 1$ and tape $k + 2$, of alphabet $\tilde{\Sigma}$.

 Call INIT_h, $h = k + 1$ or $h = k + 2$, a state of M' such that M' copies the marked version $\underline{x}$ of the input x into the h-th tape. M' starts in INIT_{k+1} and then goes to state INIT_{k+2}. When finished, it enters a state called UNMARKING.

 In state UNMARKING, M' starts the following procedure: if all symbols of tape $k + 1$ are unmarked, then M' accepts. Otherwise, M' nondeterministically selects one or more tape positions corresponding to marked symbol on tape $k + 1$ and "unmarks" them on both tapes $k + 1$ and $k + 2$. M' then goes to state SIMULATE.

 In state SIMULATE, M' performs a simulation of M using the first k tapes of M' as memory tapes and the $k + 2$-th tape as input tape. If the simulation of M rejects, M' rejects too, otherwise M' re-enters state INIT_{k+2}.

 Given an input word x, M' always halts, either rejecting when in state SIMU-LATE or accepting when in state UNMARKING (when all symbols of tape $k+1$ are unmarked): each time M' enters UNMARKING state, it always "unmarks" at least one symbol on tape $k + 1$, and it never marks again any symbol of tape $k + 1$ after initialization, so it can enter UNMARKING at most $|x| + 1$ times. Clearly, M' accepts x if, and only if, there exist matching words in B whose strong match is x; M' guesses these words in the UNMARKING state. Also, M' is still linearly-bounded, since the two additional tapes only consume $2 \cdot |x|$ memory cells.   □

5 Conclusion

The consensual approach proposes a new way of looking at finite-state or other language devices as recognizers of sets of matching words, thus shifting the perspective from one word at a time to a parallel view of language recognition.

The characterization of the regular languages as consensual languages based on strictly locally testable sets is rather surprising and stimulating. It strengthens the idea that the use of multiple computations permits to simplify the devices used as individual computational units. Sciences such as linguistics, brain theory, and biology that make use of formal language models have an enduring interest for languages characterized by some form of local testability: for instance, [8] argues that certain DNA sequences are slt. It is hoped that our result will be of interest in such domains.

The other result, that context-sensitive languages do not gain power by consensual computation, sets an upper frontier on the base language families that are worth considering in a consensual setting.

Since the consensual model is fairly new, many questions are open for investigation. Concerning language family inclusion, we mention the questions $\mathcal{C}_{NC} \overset{?}{=} \mathcal{C}_{REG}$ and $\mathcal{C}_{CF} \overset{?}{=} \mathcal{C}_{CS}$. In addition there are open classical questions about minimality, properties of unary languages, decidability of equivalence, determinism of the counter (or multi-set) machine, as well as the study of closure properties beyond the basic ones considered in [6].

References

1. de Luca, A., Restivo, A.: A characterization of strictly locally testable languages and its applications to subsemigroups of a free semigroup. Information and Control 44(3), 300–319 (1980)
2. Berstel, J., Pin, J.-E.: Local languages and the Berry-Sethi algorithm. Theor. Comp. Sci. 155 (1996)
3. Braitenberg, V., Pulvermüller, F.: Entwurf einer neurologischen Theorie der Sprache. Naturwissenschaften 79, 103–117 (1992)
4. Caron, P.: Families of locally testable languages. Theor. Comp. Sci. 242(1-2), 361–376 (2000)
5. Crespi Reghizzi, S., San Pietro, P.: Consensual Definition of Languages by Regular Sets. In: Martín-Vide, C., Otto, F., Fernau, H. (eds.) LATA 2008. LNCS, vol. 5196, pp. 196–208. Springer, Heidelberg (2008), http://dx.doi.org/10.1007/978-3-540-88282-4_19
6. Crespi Reghizzi, S., San Pietro, P.: Consensual languages and matching finite-state computations. RAIRO - Theor. Inf. and Applic. 45(1), 77–97 (2011), http://dx.doi.org/10.1051/ita/2011012
7. Crespi Reghizzi, S., San Pietro, P.: From regular to strictly locally testable languages. In: Ambroz, P., Holub, S., Masáková, Z. (eds.) WORDS, Proc. 8th Int. Conf. Words 2011. EPTCS, vol. 63, pp. 103–111 (2011), http://dx.doi.org/10.4204/EPTCS.63
8. Head, T.: Formal language theory and DNA: an analysis of the generative capacity of specific recombinant behaviors. Bull. Math. Biology 49, 737–759 (1987)

9. McNaughton, R., Papert, S.: Counter-free Automata. MIT Press, Cambridge (1971)
10. Medvedev, Y.T.: On the class of events representable in a finite automaton. In: Moore, E.F. (ed.) Sequential Machines – Selected Papers (translated from Russian), pp. 215–227. Addison-Wesley, New York (1964)
11. Pulvermüller, F.: Sequence detectors as a basis of grammar in the brain. Theory in Biosciences 122, 87–104 (2003), http://dx.doi.org/10.1007/s12064-003-0039-6
12. Eilenberg, S.: Automata, Languages, and Machines. Academic Press (1974)

Nominal Automata for Resource Usage Control*

Pierpaolo Degano, Gian-Luigi Ferrari, and Gianluca Mezzetti

Dipartimento di Informatica — Universitá di Pisa
{degano,giangi,mezzetti}@di.unipi.it

Abstract. Two classes of nominal automata, namely Usage Automata (UAs) and Variable Finite Automata (VFAs) are considered to express resource control policies over program execution traces expressed by a nominal calculus (**Usages**). We first analyse closure properties of UAs, and then show UAs less expressive than VFAs. We finally carry over to VFAs the symbolic technique for model checking **Usages** against UAs, so making it possible to verify the compliance of a program with a larger class of security properties.

Introduction

Computational models based on finite alphabets seem insufficient to accurately describe programs that adapt their behaviour when plugged inside mutable operational environments, and that therefore offer a multiplicity of dynamic entities. Ubiquitous computing is an illustrative example of these phenomena. Since we cannot predict the actual identities of all the entities that programs may plug in, we can abstractly represent such mutable operational environments as issuing stimuli taken from infinite alphabets. The challenge is therefore developing structures to formally manage infinite alphabets.

In this paper, we exploit *nominal techniques* [14] to deal with these alphabets, the elements of which are called *urelements*. Urelements are atomic objects that are indistinguishable: we can always substitute one for another. The only thing that characterises an object made of urelements is its shape, rather than the actual urelements it is made of. There are many instances of nominal models in the real word. E.g. XML files may contain URL links coming from the web but, roughly, an XML Schema Definition can validate XML files ignoring the specific actual content of these links.

Nominal techniques have been fruitful considered in several fields. Nominal process calculi, namely calculi with dynamic name creations and name-passing, have been shown effective to deal with security and mobility [15]. Nominal automata, that recognise languages over an infinite alphabet, have been developed over the years [9,5,16,18,8,24,11,10]. Some of these formalisms, among which [9,8,7], operate on *data words*, i.e. strings of operations acting on data objects, e.g. reading a file or invoking a remote server. We refer to [22,20] for

---

* This work has been partially supported by IST-FP7-FET open-IP project ASCENS and Regione Autonoma Sardegna, L.R. 7/2007, project TESLA.

N. Moreira and R. Reis (Eds.): CIAA 2012, LNCS 7381, pp. 125–137, 2012.

a detailed survey and some comparisons. The motivating application of some nominal automata, for example Finite Memory Automata (FMA) [18] and Variable Finite Automata (VFAs) [16], is to express properties of XML and Datalog data. Also, HD-Automata [19] and Fresh Register Automata [24] can decide bisimulation properties of a finite control restriction of the π-calculus [21].

This paper builds over the nominal technique of [6]. There, a basic nominal process calculus, now called Usages, is proposed to abstractly represent the behaviour of programs that dynamically generate and operate over resources, through actions, so generating data words. Usages encompass full sequentialization, general recursive definitions, and a dynamic name generation operator, much like the π-calculus. Instead, Usages do not include name passing facilities.

Usage Automata, UAs for short, have been introduced [5,6] to specify and enforce security policies of a system, the behaviour of which is abstractly represented by the language of a usage. The security policies considered are actually safety properties, expressing that nothing bad will occur during an execution. To show that a usage U respects a policy φ, represented by a UA, [5] resorted to model checking, in spite of the possible infinite resources the usage U can generate. The verification of security is reduced to the emptiness problem of the intersection between a pushdown and a finite state automaton. The first, i.e. the model, comes from U, the second, i.e. the property, from the UA φ. Indeed, this is the classical automata based model checking technique by Vardi and Volper [25].

This paper aims at a further development of nominal models. In particular we show that classical automata techniques can contribute to assess and better evaluate the expressiveness and the exploitation of nominal models in practice. To this purpose we compare the expressivity of the Usages with other nominal models considered in the literature (Section 2). We will show Usages to have an expressivity never considered before.

Then we establish some closure properties of the UAs, by studying them as nominal automata from a language theoretic viewpoint (Section 2). In particular, we show that UAs are closed under intersection and union, but not under complement and Kleene star. When seen as policies, this result amounts to saying that we can impose two policies together or be happy if one out of two is obeyed. Instead, to consider secure only those traces that violate a given policy, one has to explicitly build a new UA, which is not guaranteed to exist at all.

Next we consider VFAs. We conservatively extend VFAs to deal with data words. We also show that this smooth extension of VFAs yields automata that are more expressive than UAs (in the spirit of the taxonomy of nominal automata developed in [22,20].

To conclude the paper, we express through VFAs a larger class of security policies on Usages. The crucial point is whether a usage U can be model checked against a VFA (Section 3). We face this problem by rephrasing the technique of [5] and by defining *symbolic* VFAs, that represent the languages of VFAs under collapsing. In this way, VFAs become standard Finite State Automata, thus making model checking possible.

We omit the proofs because of lack of space; they can be found in [12]. Occasionally, we will sketch our arguments to give some insights on our results.

1   Preliminaries

We recall the notion of Usages [6,2,5,3] and Usage Automata [5] and extend VFAs [16] to work on data words. These formalisms will be used to represent execution traces of programs that dynamically generate new resources, and to express properties of sets of traces.

We assume hereafter a finite set Act of actions that operate on a given infinite set Res of resources. The actions comprise a special action new that represent the generation of a fresh resource. An event is a pair $\alpha(r)$ with $\alpha \in$ Act and $r \in Res$, that represents the firing of the action α on the resource r. Then, an execution trace is a sequence of events, i.e. a *data word*. We only consider here Usages and Usage Automata with actions on a single resource and we refer the interested reader to [5] for the polyadic version.

The set of resources Res is partitioned into two subsets: Res_s and Res_d. Res_s is a finite set of static resources, typically the one that are hard-coded in the program. Instead, Res_d is a countable infinite set of dynamic resources, i.e. the urelements, that are dynamically created. Indeed, whenever a program can generate an execution trace $\alpha(a)\beta(d)\alpha(d)\alpha(d')\beta(a)$ with $a \in Res_s, d, d' \in Res_d$ it can also generate $\alpha(a)\beta(d')\alpha(d')\alpha(d)\beta(a)$, since they only differ on the urelements d, d' that appear in them. Indeed, d and d' are simply exchanged without confusing their identities.

1.1   Usages

Usages are a simple nominal calculus designed to abstractly represent the behaviour of programs that create and use resources dynamically [23,4]. In [4], e.g., these safe approximations are mechanically derived by a type and effect system from the expressions of a λ-calculus suitably extended to specify web services in a secure manner (e.g. featuring a call-by-contract primitive and security policies). For more details and examples the reader is referred to [6].

The syntax of Usages follows.

Definition 1.1 (Usages). *Let* Nam *be a countable set of names such that* Nam$\cap$ $Res = \emptyset$. *Usages are inductively defined as follows:*

$U, V ::=$	ϵ	*empty*
	h	*recursion variable*
	$\alpha(r)$	$\alpha(r) \in$ Act $\times$ $(Res \cup$ Nam$), \alpha \neq$ new
	$U \cdot V$	*sequence*
	$U + V$	*choice*
	$\mu h.U$	*recursion*
	$\nu n.U$	*resource creation,* $n \in$ Nam

Table 1. Operational semantics of **Usages**

$$\epsilon \cdot U, \mathcal{R} \xrightarrow{\epsilon} U, \mathcal{R} \qquad\qquad \alpha(r), \mathcal{R} \xrightarrow{\alpha(r)} \epsilon, \mathcal{R} \qquad\qquad \mu h.U, \mathcal{R} \xrightarrow{\epsilon} U\{\mu h.U/h\}, \mathcal{R}$$

$$\frac{U, \mathcal{R} \xrightarrow{\alpha(r)} U', \mathcal{R}'}{U \cdot V, \mathcal{R} \xrightarrow{\alpha(r)} U' \cdot V, \mathcal{R}'} \qquad \frac{U, \mathcal{R} \xrightarrow{\alpha(r)} U', \mathcal{R}'}{U + V, \mathcal{R} \xrightarrow{\alpha(r)} U', \mathcal{R}'} \qquad \frac{V, \mathcal{R} \xrightarrow{\alpha(r)} V', \mathcal{R}'}{U + V, \mathcal{R} \xrightarrow{\alpha(r)} V', \mathcal{R}'}$$

$$\frac{}{\nu n.U, \mathcal{R} \xrightarrow{\text{new}(r)} U\{r/n\}, \mathcal{R} \cup \{r\}} \quad \text{if } r \in Res_d \setminus \mathcal{R}$$

The operators of the calculus are similar to those of the π-calculus, but we have full sequentialization, general recursion and no parallel operator; μh and νn are binders, the first one on recursion variables, the second on names.

A usage is *closed* when it has no free names and no free variables; it is *initial* when it is closed and with no dynamic resources, i.e. it is never the case that a resource $r \in Res_d$ appear as parameter of an action.

The semantics of **Usages** is specified by the labelled transition system in Table 1. We associate with a usage the language consisting of all the prefixes of the traces labelling its computations. The configurations of the transition system are pairs $(U, \mathcal{R})$, where U is a usage and $\mathcal{R} \subseteq Res_d$ is the set of dynamic resources generated so far.

Definition 1.2 (Semantics of Usages). *Given a closed usage U let $[\![U]\!]$ be the set of traces $\eta = w_1 \ldots w_n (w_i \in (\mathsf{Act} \times Res) \cup \{\epsilon\}, 1 \le i \le n)$ such that:*

$$\exists U', \mathcal{R}'. \; U, \emptyset \xrightarrow{w_1} \cdots \xrightarrow{w_n} U', \mathcal{R}'$$

The following definition is technical and will be used in Section 3.

Definition 1.3 (Well formed traces). *A trace η is well-formed if it is never the case that:*
1. *$\eta = \eta' \text{new}(r)\eta''$ for some η', η'' with $r \in Res_s$ or*
2. *$\eta = \eta' \text{new}(r)\eta'' \text{new}(r)\eta'''$ for some $\eta', \eta'', \eta''', r$ or*
3. *$\eta = \eta' \alpha(r)\eta'' \text{new}(r)\eta'''$ for some $\eta', \eta'', \eta''', \alpha$*

1.2 Usage Automata

Usage Automata (UAs) [5] are nominal automata over data words. The motivating application of UAs is to express policies for controlling the usage of resources. Policies are regular sets of traces [17]. The UAs express security policies relying on the so-called default-accept approach thus recognizing unwanted traces. Before going into the formal definitions of UAs, consider the example in Fig. 1. The automaton describes the usage policy for opening, reading and writing files. Essentially, it amounts to saying that a file f must be opened before being used (f is a variable standing for a generic file). Starting from q_0, performing the action open on f brings the automaton to q_1, so allowing the file f to be read and written. Instead, an attempt of reading or writing a different

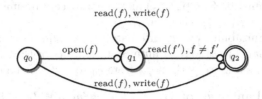

Fig. 1. An example of UA that describe the usage policy for opening, reading and writing files

file f' or an un-opened file brings to the offending state q_2. For instance, the string open(foo.txt) read(bar.txt) is offending, while open(foo.txt) read(foo.txt) is legal. We assume that a UA remains in the same state by recognising an action that does not match any of the labels of the outgoing edges. E.g. in Fig. 1, the self-loops in q_1 are redundant and are only displayed for readability.

To define UAs, it is convenient to assume a countable set of variables Var; from now onwards, let $V \subset$ Var. We start with a couple of auxiliary definitions.

Definition 1.4 (Substitution). *A substitution for V is a function $\sigma : V \rightarrow R, R \subseteq Res$.*

Hereafter a substitution σ is considered to be trivially extended on Res_s so that $\sigma(a) = a$ for all $a \in Res_s$. Hence, if $\sigma : V \rightarrow R$, the set R contains at least Res_s.

Below, we recall the syntax and the semantics of *guards*.

Definition 1.5 (Guards). *Given a set V of variables we inductively define the set G of guards on $Res_s \cup V$, ranged over by ζ, ζ', as follows:*

$$G_1, G_2 := true \mid \zeta = \zeta' \mid \neg G_1 \mid G_1 \wedge G_2$$

A given substitution $\sigma : V \rightarrow R$ satisfies a guard g, in symbols $\sigma \vDash g$, if and only if: ($g = true$) or ($g = (\zeta = \zeta')$ and $\sigma(\zeta) = \sigma(\zeta')$) or ($g = \neg g'$ and it is not the case that $\sigma \vDash g'$) or ($g = g' \wedge g''$ and $\sigma \vDash g'$ and $\sigma \vDash g''$).

Definition 1.6 (Usage Automata). *A Usage Automaton (UA) φ is $\langle S, Q, q_0, F, E \rangle$. The finite set $S \subseteq$ Act $\times$ ($Res_s \cup Var$) is its alphabet; Q is its finite set of states; q_0 its initial state; $F \subseteq Q$ the set of its final states; $E \subseteq Q \times S \times G \times Q$ is its finite set of edges with G set of guards on resources and variables in S.*

Given a UA φ, we will refer to the variables occurring in S with $Var(\varphi)$.

Definition 1.7 (Instantiation of UAs). *Let $\varphi = \langle S, Q, q_0, F, E \rangle$ be a UA and $\sigma : Var(\varphi) \rightarrow R$ be a substitution. The instantiation of φ under σ is the automaton $\varphi_\sigma = \langle R, Q, q_0, F, \delta_\sigma \rangle$, where $\delta_\sigma = X_\sigma \cup Comp_\sigma(X_\sigma)$ with*

$$X_\sigma = \{(q, \alpha(\sigma(v)), q') \mid (q, \alpha(v), g, q') \in E \text{ and } \sigma \vDash g\}$$
$$Comp_\sigma(X_\sigma) = \{(q, \alpha(r), q) \mid \alpha \in Act, r \in R \text{ and } \nexists q' \in Q.(q, \alpha(r), q') \in X_\sigma\}$$

Note that the completion $\text{Comp}_\sigma(X_\sigma)$ may possibly contain infinitely many self-loops of the form $(q, \alpha(r), q)$ when $r \in Res_d$.

Language recognizability by an automaton with infinitely many edges is defined much like that for standard Finite State Automata (FSA, for short): $\eta \in L(\varphi_\sigma)$ if there exists a finite path in φ_σ from q_0 to a $q' \in F$ labelled with η.

Definition 1.8 (Language of UAs). *The string $\eta \in L(\varphi)$ iff there exists a substitution $\sigma : Var(\varphi) \to R$ for some $R \subseteq Res$ such that $\eta \in L(\varphi_\sigma)$.*

1.3 Variable Finite Automata on Data Words

Now we conservatively extend Variable Finite Automata [16] to work over data words. To simplify notation we overload Act to also denote the actions of VFA.

Definition 1.9 (Variable Finite Automata). *The tuple $\mathcal{A} = \langle \text{Act}, \Omega, \Omega_s, X \cup \{y\}, A \rangle$ is a* Variable Finite Automaton (VFA), *where X is a finite set of variables;* Act *is a finite set of actions; and Ω is a possibly infinite alphabet with $\Omega_s \subseteq \Omega$ finite subset, $\Omega \cap X = \emptyset$. $A = \langle \Gamma, Q, q_0, F, \delta \rangle$ is a NFA with alphabet $\Gamma = \text{Act} \times (\Omega_s \cup X \cup \{y\})$ and $y \notin (\Omega \cup X)$ is a distinguished placeholder.*

Given a function $m : \Omega \to (\Omega_s \cup X \cup \{y\})$, let $m(\alpha(a)) = \alpha(m(a))$. When unambiguous, we will write $m(\eta)$ for m homomorphically applied to η.

Definition 1.10 (Language of VFAs). *A string $\eta \in (\text{Act} \times \Omega)^*$ is a legal instance of $w \in \Gamma^*$ and w is a witnessing pattern of η, if there exists a function $m : \Omega \to (\Omega_s \cup X \cup \{y\})$ such that $m(\eta) = w$ and m is a correspondence, i.e.*

1. $\forall a \in \Omega_s . m(a) = a$
2. $\forall x \in X$. if $(\exists a, b \in \Omega . m(a) = x$ and $m(b) = x)$ then $a = b$ and $a, b \notin \Omega_s$

A string $\eta \in L(\mathcal{A})$ iff there exists $w \in L(A)$ such that η is a legal instance of w.

Note that here we explicitly present the correspondence between strings and witnessing patterns as a function. Our definition is equivalent to that of [16] when actions are ignored.

Of course, we are interested in the behaviour of VFAs with infinite alphabets, typically when $\Omega = Res$, $\Omega_s = Res_s$, $\Omega \setminus \Omega_s = Res_d$.

2   Properties of Usages and UA

This section studies Usages and UAs from a formal languages point of view.

Usages: Usages are a process algebra whose semantics is given in term of set of traces. Hence, they can be also regarded as suitable grammar defining a language. Although we found in the literature no widely accepted notion of regular/context-free nominal language, we argue that the recursion operator μ of Usages makes the generated languages context-free and non-regular. This is justified by the fact that the language generated by $\nu n.\mu h.(\alpha(n) \cdot h \cdot \alpha(n) + hh + \epsilon)$ (miming balanced parenthesis) is not recognised by any of the following regular nominal automata: HD-Automata [19], Fresh Register Automata [24], Register Automata [18], VFAs and UAs. To the best of our knowledge, context-free nominal models have been studied only in [10], that introduces the class of *quasi context-free languages* (we refer the reader to Def. 1 of the original paper).

It turns out that the class of languages defined by Usages and that of quasi context-free languages have a non empty intersection, and that neither includes the other, as stated in Property 2.1 below. Our comparison takes care of the fact that quasi context-free languages are not defined on data words, i.e. they have no actions on resources, while UAs do. The strings generated by UAs belong to $(Act \times Res)^*$, with the additional constraint that an action on a dynamic resource r must be proceeded by $new(r)$. Instead, quasi context-free languages have no actions on resources. Thus, we will ignore actions, only considering the resources accessed when talking of Usages, so fixing Res to be the alphabet of both.

Property 2.1. There exists

1. a language generated by a usage U that is not quasi context-free;
2. a quasi context-free language that can be generated by no usage U.

For showing statement (1) above, consider the usage $U = \mu h.(\nu n.\alpha(n)) \cdot h$. As a matter of fact, there is no bound on the number of fresh resources that can occur in a string generated by U, while in a quasi context-free language the bound is given by the number of the registers, that are a characterizating feature of the models for generating/recognising these languages (for details, see [10]). The second statement holds because there is no usage U such that $[\![U]\!] = Res^*$, that is a quasi context-free language. We conjecture that this lack of expressivity of Usages comes from the absence of an explicit mechanism for disposing and reusing resources. In contrast, quasi context-free models can overwrite the content of a register, so forgetting that the resource previous contained therein already appear previously in the generated string.

Usage Automata: Logical connectives between policies have a counterpart as language operators. The complement of a language recognised by a UA φ is the set of traces that violates the negation of policy expressed by φ. The union/intersection of the languages recognised by two UAs φ, ψ is the set of traces violating the conjunction/disjunction of the two policies expressed by φ and ψ. Hence, closure properties are not only interesting from a theoretical viewpoint, but also deeply connected with the intended applications of UAs.

Theorem 2.1. *The set of languages accepted by UAs is closed under union and intersection. but it is not closed under complement and Kleene star.*

Below, we give an intuition on the proofs of non closedness, as they give some additional insight on the expressivity of UAs (see [12]). To see that UAs are not closed by complement, consider an UA that recognises the strings containing at least two occurrences of $\alpha(r)$ for some $r \in Res$. The complement of this

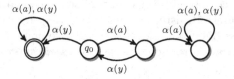

Fig. 2. A VFA the language of which is not accepted by any UA. The alphabet is $\Omega = Res_d \cup \{a\}$ with $\mathsf{Act} = \{\alpha\}$.

language contains those strings where all resources mentioned are pairwise distinct. This is a non-regular property, because an automaton should store all previously used resources. To show that UAs are not closed under Kleene star, consider a UA recognising the strings containing one $\alpha(a)$ followed by $\alpha(r)$ for some $r \in Res, r \neq a$. Now consider the Kleene star of the language $L(\varphi)$ and, by contradiction, let ψ be the UA recognising it. We can take arbitrary long strings $s \in L(\psi)$ of the form $s = \alpha(a)\alpha(d_1)\ldots\alpha(a)\alpha(d_n)$. Since variables in ψ are finite not all d_i can be bound to a variable, this implies that some self-loop labelled $\alpha(d_k)$ obtained by completion is traversed in recognising s. Then also $s' = \alpha(a)\alpha(d_1)\ldots\alpha(a)\ldots\alpha(a)\alpha(d_n)$ obtained from s removing $\alpha(d_k)$ is recognised by φ, this is a contradiction because $s' \notin L(\psi)^*$.

The last fact does not however reduce the power of UAs in expressing security policies. Consider a language of the form $L(\varphi)^*$ and let η be a trace in the semantics $[\![U]\!]$ of a usage U. For any $\eta \in L(\varphi)^*$ there exists also a prefix $\eta' \in L(\varphi)$ with $\eta' \in [\![U]\!]$ by Definition 1.2. This means that checking U against $L(\varphi)^*$ is the same as checking it against $L(\varphi)$. However, there exists a safety property that is not expressible by a UAs. Informally this safety policy requires to take a token $\alpha(a)$ before performing any other kind of action, while if two tokens are taken, any sequence of actions is then allowed. The above sketched safety property is expressed by the VFA in Fig. 2. In general, the following theorem holds.

Theorem 2.2. *UAs are less expressive than VFAs.*

It would also be interesting to formally compare the expressive power of some variants of UAs and of VFAs. We have a couple of preliminary results. First, consider the restriction of VFAs obtained by only permitting the distinguished placeholder y to occur in self-loops. We conjecture that this variant of VFAs has the same expressive power of UAs. Now, consider the extension of UAs with a wild-card, introduced in [3]. A wild-card can stand for any resource, and so it plays the role of the placeholder y in a VFA. Not surprisingly UAs extended in this way are just as expressive as VFAs.

3   Model Checking

As mentioned above, UAs have been introduced to specify and enforce security policies of systems, the behaviour of which is abstractly represented by the language of a usage U. The security policies considered are actually safety properties, expressing that nothing bad will occur during a computation, abstractly represented by the string η [17]. The approach taken in [5] follows the default-accept paradigm, i.e. only the unwanted behaviour is explicitly mentioned — this assumption justifies the way UAs are instantiated, and in particular the completion step made therein. Consequently, the language of φ is the set of *unwanted traces*, and an accepting state is considered offending. Then U respects the property φ, in symbols $U \vDash \varphi$, if and only if $\eta \in \llbracket U \rrbracket \Rightarrow \eta \notin L(\varphi)$.

To show that a usage U respects a policy φ, the authors of [5] resorted to model-checking, in spite of the possible infinite resources a usage can generate. This is done by carefully collapsing the verification to a well-known problem: the emptiness of the intersection between a pushdown and a finite state automata, that is decidable [13]. The first, i.e. the model, comes from U, the second, i.e. the property, from the UA φ. Indeed, this is the classical automata based model checking technique [25] with φ expressing unwanted traces, instead.

Since VFAs are more expressive than UA, the question arises whether the same checking technique of [5] can be used to model check a usage against more expressive policies, expressed by VFAs. The answer is positive and one can model check a usage U against a VFAs, defining policy compliance in the obvious way: $U \vDash \mathcal{A}$ if and only if $\eta \in \llbracket U \rrbracket \Rightarrow \eta \notin L(\mathcal{A})$.

To do this we introduce *symbolic* VFAs. Following [5], we let their alphabet be the finite set of witnesses $\mathsf{W} \subset \{\#_i\}_{i \in \mathbb{N}}$, where $\{\#_i\}_{i \in \mathbb{N}} \cap Res = \emptyset$. We also need a distinguished symbol $\_ \notin Res \cup \{\#_i\}_{i \in \mathbb{N}}$.

We recall from [5] the crucial notion of *collapsing* mapping, that is the link between Usages, VFAs and their symbolic counterparts. As a matter of fact, this is the technical machinery that deals with urelements, and it permits to abstract from their actual identity.

Definition 3.1 (Collapsing). *Given a finite set of witnesses* W, *a collapsing mapping* $\kappa : Res \to Res_s \cup \mathsf{W} \cup \{\_\}$ *of* $R \subset Res_d$ *onto* W *is a function such that:*

1. $\kappa(r \in Res_s) = r$ 2. $\kappa(R) = \mathsf{W}$ *and it is injective*   3. $\kappa(Res_d \setminus R) = \{\_\}$

We write $\kappa(\alpha(a))$ *for* $\alpha(\kappa(a))$ *and* $\kappa(\eta)$ *for the homomorphic extension of* κ *to* η.

The following property is very technical, and simplifies the procedure for model checking Usages against UAs and VFAs. Roughly, it states that well-formedness of traces can be checked by the so-called *unique-witness* automaton.

Property 3.1 (Unique-witness). Given a finite set of witnesses W and an initial usage U, there exists a *unique-witness FSA* N_W such that:

- $\eta \notin N_\mathsf{W} \Longrightarrow \forall \#_i \in \mathsf{W}$. there is a single $new(\#_i)$ in η
- $\eta \in \llbracket U \rrbracket \Longrightarrow \eta \notin L(N_\mathsf{W})$

By exploiting the construction given in [5], we can now associate with a usage U a symbolic pushdown automaton $\mathsf{B}_\mathsf{W}(U)$, the language of which is denoted by $L(\mathsf{B}_\mathsf{W}(U))$. The following theorem puts together some results proved in [5].

Theorem 3.1. *Given an initial usage* U, *there exist a finite set* W *of witnesses and a pushdown automaton* $B_W(U)$ *on the finite alphabet* $Act \times (Res_s \cup W \cup \{\_\})$ *such that:*

- *Given a collapsing* κ *such that* $\kappa(Res_d) \subseteq W \cup \{\_\}$ *then:*
  $\forall \eta.\ \eta \in [\![U]\!] \Rightarrow \kappa(\eta) \in L(B_W(U))$
- *Given a collapsing* κ *such that* $\kappa(Res_d) \supseteq W$, *then:*
  $\forall \eta'.\ (\eta' \in L(B_W(U)) \wedge \eta' \notin N_W) \Rightarrow (\exists \eta.\ \eta \in [\![U]\!] \wedge \eta' = \kappa(\eta))$

Definition 3.2 (Symbolic VFAs). *Let* $\mathcal{A} = \langle Act, Res, Res_s, X \cup \{y\}, A \rangle$ *be a VFA. Given a finite set of witnesses* W, *let* $Res_W = Res_s \cup W \cup \{\_\}$.
The symbolic VFA *on* W *is* $\mathcal{A}_W = \langle Act, Res_W, Res_s, X \cup \{y\}, A \rangle$. *Language recognition for symbolic VFAs additionally requires the correspondence* m *to be such that* $m(\_) = y$.

The following theorem makes clear the links between the language of a VFA and that of its symbolic automaton.

Theorem 3.2. *Let* $\mathcal{A} = \langle Act, Res, Res_s, X \cup \{y\}, A \rangle$ *be a VFA, and let* W *be a set of witnesses such that* $|W| = |X|$, $\mathcal{A}_W$ *as in Definition 3.2 and let* K *be the set of the collapsing* κ *such that* $\kappa(Res_d) = W \cup \{\_\}$, *then:*

- $\forall \eta.(\eta \in \mathcal{A} \Rightarrow \exists \kappa \in K.\kappa(\eta) \in \mathcal{A}_W)$
- $\forall \kappa \in K, \eta.(\kappa(\eta) \in \mathcal{A}_W \Rightarrow \eta \in \mathcal{A})$

We carry over VFAs the notions of substitution and instantiation, which transforms a VFA into a *Finite* State Automaton. The language recognised by any VFA can then be represented under collapsing by a finite class of its instantiations.

Definition 3.3 (Instantiation of VFAs). *Let* $\mathcal{A}_W = \langle Act, Res_W, Res_s, X \cup \{y\}, A \rangle$ *be a symbolic VFA with* $A = \langle \Gamma, Q, q_0, F, \delta \rangle$, $\Gamma = Act \times (Res_s \cup X \cup \{y\})$.
Given a function $\overline{m} : X \cup Res_s \to Res_s \cup W$ *it is a substitution for* $\mathcal{A}$ *if it is the identity on* Res_s *and it is injective on* X.
Given a substitution $\overline{m}$ *the instantiation of* $\mathcal{A}$ *is* $\mathcal{A}^{\overline{m}} = \langle Res_W, Q, q_0, F, \delta^* \rangle$, *where*

$$\delta^* = \{(q, \alpha(\overline{m}(v)), q') \mid (q, \alpha(v), q') \in \delta, v \neq y\} \cup$$
$$\{(q, \alpha(d), q') \mid (q, \alpha(y), q') \in \delta, d \in (Res_W \setminus (Res_s \cup Image(\overline{m}))\}$$

Note that, by the finiteness of W, $\mathcal{A}^{\overline{m}}$ is a standard FSA on a finite alphabet.

Theorem 3.3. *Let* $\mathcal{A} = \langle Act, Res_W, Res_s, X \cup \{y\}, A \rangle$ *be a symbolic VFA:*

$$\eta \in L(\mathcal{A}) \Leftrightarrow \exists\ substitution\ \overline{m}.\eta \in L(\mathcal{A}^{\overline{m}})$$

To simplify the technical development, we find convenient to resort to the well-known *weak-until* operator $A\ \mathfrak{W}\ B$ between automata, meaning that A holds until B holds or B always holds. We refer to a standard book on model checking, e.g. [1], or to [5] for more details.

Theorem 3.4 (Model checking). *Let U be an initial usage on the resources $Res = Res_d \cup Res_s$; let $\mathcal{A} = \langle Act, Res, Res_s, X \cup \{y\}, A \rangle$ be a VFA; and let W be a set of witnesses such that $|\mathsf{W}| = |X|$. Then $U \models \mathcal{A}$ if and only if:*
$$\forall \; substitution \; \overline{m} : X \cup Res_s \to Res_s \cup \mathsf{W}. \; L(\mathsf{B_W}(\mathsf{U})) \cap L(\mathcal{A}_\mathsf{W}^{\overline{m}} \; \mathfrak{W} \; N_\mathsf{W}) = \emptyset$$

This theorem gives us the means for an efficient model checking procedure. Given a substitution $\overline{m}$, it is indeed decidable to check whether $L(\mathsf{B_W}) \cap L(\mathcal{A}_\mathsf{W}^{\overline{m}} \; \mathfrak{W} \; N_\mathsf{W}) = \emptyset$ and there are finitely many substitutions $\overline{m}$, because Res_s, X and W are finite.

We can then re-use the model checker LocUsT [2] for verifying Usages against VFAs. As for complexity issues, we can restate the theorem established in [5] for VFAs. The proof is mostly the same with only minor changes regarding the number of instantiations of VFAs.

Theorem 3.5. *The worst-case asymptotic behaviour of model-checking an usage U against an automaton φ with n variables is $\mathcal{O}(|U|^{|n|+1})$.*

Conclusions

We have first studied two classes of nominal automata, namely Usage Automata (UAs) [5] and Variable Finite Automata (VFAs) [16], aiming at using them to express resource control policies. We analysed closure properties of the languages recognized by UAs, and showed that the expressive power of UAs is weaker that the one of VFAs. Then, we considered Usages [5], a nominal process calculus for modelling the (abstract) behaviour of programs with dynamic creation of resources. The class of languages defined by Usages is neither included nor contains the class of quasi context-free languages [10].

We slightly extended the symbolic technique of [5], that is based on collapsing and that reduces the two nominal automata mentioned above to standard Finite State Automata. Also the execution traces of a nominal calculus can be collapsed to traces of standard pushdown automata. This enables us to model check the compliance of execution traces against a property expressed in terms of a VFA. Indeed the collapsing above brings back us to the classical problem of verifying the emptiness of the intersection between a pushdown and a finite state automaton. Our results guarantee the correctness and the completeness of our proposal.

We plan to study whether the symbolic technique used here can be extended and applied to other classes of nominal automata (e.g. Finite Memory Automata) and to more expressive nominal process calculi to specify systems (e.g. with resources garbaging). It would be also interesting to further develop the taxonomy about nominal automata by placing UAs and VFAs into the expressivity hierarchy of [22]. A co-algebraic presentation of these automata could help, especially for investigating their relation with functors on nominal sets possibly with fusion of names. The operational approach to express properties based on nominal automata is deeply connected with the logical approach. It would then be important to exactly relate the expressive power of different kinds of nominal automata with that of various logics, e.g. EMSO [8,7] or LTL [1].

Acknowledgments. We would like to thank the anonymous referees for their comments that greatly helped us to improve the quality of our paper.

References

1. Baier, C., Katoen, J.P.: Principles of model checking. MIT Press (2008)
2. Bartoletti, M., Zunino, R.: LocUsT: a tool for checking usage policies. Tech. Rep. TR08-07, University of Pisa (2008)
3. Bartoletti, M., Costa, G., Degano, P., Martinelli, F., Zunino, R.: Securing Java with local policies. Journal of Object Technology 8(4), 5–32 (2009)
4. Bartoletti, M., Degano, P., Ferrari, G.L.: Planning and verifying service composition. Journal of Computer Security 17(5), 799–837 (2009)
5. Bartoletti, M., Degano, P., Ferrari, G.L., Zunino, R.: Model Checking Usage Policies. In: Kaklamanis, C., Nielson, F. (eds.) TGC 2008. LNCS, vol. 5474, pp. 19–35. Springer, Heidelberg (2009); Extended version to appear in Math. Stuct. Comp. Sci.
6. Bartoletti, M., Degano, P., Ferrari, G.L., Zunino, R.: Local policies for resource usage analysis. ACM Trans. Program. Lang. Syst. 31(6) (2009)
7. Benedikt, M., Ley, C., Puppis, G.: Automata vs. Logics on Data Words. In: Dawar, A., Veith, H. (eds.) CSL 2010. LNCS, vol. 6247, pp. 110–124. Springer, Heidelberg (2010)
8. Bollig, B.: An Automaton over Data Words That Captures EMSO Logic. In: Katoen, J.-P., König, B. (eds.) CONCUR 2011. LNCS, vol. 6901, pp. 171–186. Springer, Heidelberg (2011)
9. Bouyer, P.: A logical characterization of data languages. Information Processing Letters 84(2), 75–85 (2002)
10. Cheng, E.Y.C., Kaminski, M.: Context-free languages over infinite alphabets. Acta Inf. 35(3), 245–267 (1998)
11. Ciancia, V., Tuosto, E.: A novel class of automata for languages on infinite alphabets. Tech. rep., CS-09-003, University of Leicester, UK (2009)
12. Degano, P., Mezzetti, G., Ferrari, G.L.: Nominal models and resource usage control. Tech. Rep. TR-11-09, Dipartimento di Informatica, Università di Pisa (2011)
13. Esparza, J.: On the Decidability of Model Checking for Several μ-calculi and Petri Nets. In: Tison, S. (ed.) CAAP 1994. LNCS, vol. 787, pp. 115–129. Springer, Heidelberg (1994)
14. Gabbay, M.J., Pitts, A.M.: A new approach to abstract syntax with variable binding. Formal Aspects of Computing 13(3), 341–363 (2002)
15. Gordon, A.D.: Notes on Nominal Calculi for Security and Mobility. In: Focardi, R., Gorrieri, R. (eds.) FOSAD 2000. LNCS, vol. 2171, pp. 262–330. Springer, Heidelberg (2001)
16. Grumberg, O., Kupferman, O., Sheinvald, S.: Variable Automata over Infinite Alphabets. In: Dediu, A.-H., Fernau, H., Martín-Vide, C. (eds.) LATA 2010. LNCS, vol. 6031, pp. 561–572. Springer, Heidelberg (2010)
17. Hamlen, K.W., Morrisett, J.G., Schneider, F.B.: Computability classes for enforcement mechanisms. ACM Trans. on Programming Languages and Systems 28(1), 175–205 (2006)
18. Kaminski, M., Francez, N.: Finite-memory automata. Theoretical Computer Science 134(2), 329–363 (1994)

19. Montanari, U., Pistore, M.: π-Calculus, Structured Coalgebras and Minimal HD-Automata. In: Nielsen, M., Rovan, B. (eds.) MFCS 2000. LNCS, vol. 1893, pp. 569–578. Springer, Heidelberg (2000)

20. Neven, F., Schwentick, T., Vianu, V.: Towards Regular Languages over Infinite Alphabets. In: Sgall, J., Pultr, A., Kolman, P. (eds.) MFCS 2001. LNCS, vol. 2136, pp. 560–572. Springer, Heidelberg (2001)

21. Sangiorgi, D., Walker, D.: The Pi-Calculus - a theory of mobile processes. Cambridge University Press (2001)

22. Segoufin, L.: Automata and Logics for Words and Trees over an Infinite Alphabet. In: Ésik, Z. (ed.) CSL 2006. LNCS, vol. 4207, pp. 41–57. Springer, Heidelberg (2006)

23. Skalka, C., Smith, S., Horn, D.V.: Types and trace effects of higher order programs. Journal of Functional Programming 18(2), 179–249 (2008)

24. Tzevelekos, N.: Fresh-register automata. ACM SIGPLAN Notices 46(1), 295–306 (2011)

25. Vardi, M.Y., Wolper, P.: An automata-theoretic approach to automatic program verification (preliminary report). In: LICS, pp. 332–344. IEEE Computer Society (1986)

Weighted Nested Word Automata
and Logics over Strong Bimonoids*

Manfred Droste[1] and Bundit Pibaljommee[2]

[1] Institut für Informatik, Universität Leipzig
D-04009 Leipzig, Germany
droste@informatik.uni-leipzig.de
[2] Department of Mathematics, Faculty of Science, Khon Kaen University,
Khon Kaen 40002, Thailand
banpib@kku.ac.th

Abstract. Nested words have been introduced by Alur and Madhusudan as a model for e.g. recursive programs or XML documents and have received much recent interest. In this paper, we investigate a quantitative automaton model and a quantitative logic for nested words. The behavior resp. the semantics map nested words to weights taken from a strong bimonoid. Strong bimonoids can be viewed as semirings without requiring the distributivity assumption which was essential in the classical theory of formal power series; strong bimonoids include e.g. all bounded lattices and many other structures from multi-valued logics. Our main results show that weighted nested word automata and suitable weighted MSO logics are expressively equivalent. This extends the classical Büchi-Elgot result from words to a weighted setting for nested words.

1 Introduction

Nested words capture models with both a natural sequencing of positions and a hierarchical nested matching of these positions, as, e.g., in executions of recursively structured programs or in XML documents. They were introduced by Alur and Madhusudan [2] and are receiving much interest in the community. The interplay between automata on nested words, logical descriptions of their properties, and the accepted languages of nested words has been intensively studied (cf. [2] for a survey). It is the goal of this paper to develop for quantitative properties a weighted model of nested word automata and their logics.

Weighted automata on words were introduced and investigated already by Schützenberger [24]. Assuming for calculations that the weights form the algebraic structure of a semiring (e.g., like the natural numbers $(\mathbb{N}, +, \cdot, 0, 1)$), semiring-weighted automata soon developed a rich theory, as demonstrated by the books [12,25,15,5] and the recent handbook [9]. A suitable weighted MSO logic with the same expressive power as weighted word automata was developed in Droste and Gastin [7,8]. For nested words, a semiring-weighted automaton and an equivalent weighted logic were developed by Mathissen [19].

* This work has been supported by DFG-NRCT.

N. Moreira and R. Reis (Eds.): CIAA 2012, LNCS 7381, pp. 138–148, 2012.
© Springer-Verlag Berlin Heidelberg 2012

Logics with values in non-distributive lattices (unlike semirings, where multiplication distributes over addition) were proposed already by Birkhoff and von Neumann [3] as the logic of quantum mechanics. Recently, quantum automata and quantum logics in non-distributive lattices were investigated in [17,22,23,26,27]. For multi-valued model checking, non-distributive De Morgan algebras were employed in [18]. Recently, weighted automata and expressively equivalent weighted logics with values in strong bimonoids were studied in [11]. Strong bimonoids can be viewed as semirings without requiring the distributivity properties, and they comprise all bounded lattices.

In this paper, we will investigate weighted nested word automata and suitable weighted logics where the weights are taken from a strong bimonoid. We will prove an equivalence result for the expressive powers of the automata resp. the logics. For the nested word automata, we assume that their transitions are equipped with weights indicating, e.g., their use of resources or their probability of success. The behavior of such an automaton is a function from Σ^+ to the strong bimonoid $\mathbb{K}$. The classical model is obtained by letting $\mathbb{K}$ be the Boolean semiring $\{0,1\}$. Our weighted MSO-logic consists of all usual monadic second order formulas for nested words and all elements from the bimonoid $\mathbb{K}$ as constants. In order to define the semantics of negations of formulas, we assume that the bimonoid is equipped with a complement function interchanging 0 and 1. Such non-standard operations for the semantics of the logical connectives often occur in multi-valued logics, like Gödel logics, Lukasiewicz logics, Post logics, cf. [13]. Then the semantics of weighted sentences is again a function from Σ^+ to $\mathbb{K}$. In our first main result, we show that for arbitrary strong bimonoids with complement functions, weighted nested word automata and a suitably restricted weighted existential MSO logic have the same expressive power. This logic is (fortunately) essentially the same as the one in the original work of [7,8] for semiring-weighted logics over words. Our result extends the result of [19] who considered commutative semirings. In his proofs, Mathissen employed a transfer result from nested words to other structures like trees, which heavily employs both the commutativity and the distributivity assumptions of the underlying semiring. Since in our setting commutativity and distributivity typically fail, we have to give different proofs, using direct automata-theoretic constructions. In comparison to the setting of [11] for words, we have to deal with the more complicated structure of nested words and, correspondingly, the different transitions of nested word automata in a weighted setting.

In our second main result, we show that weighted nested word automata are expressively equivalent to the full weighted EMSO logic (without restriction), if the operations of the underlying strong bimonoid satisfy a local finiteness condition. For instance, the unit interval $[0, 1]$ with Lukasiewicz t-norm and t-conorm forms a non-distributive strong bimonoid satisfying these local finiteness assumptions. Our assumptions are also satisfied by all bounded lattices; therefore our equivalence result applies to all nested word automata and logics with values in bounded lattices (with complement function). Since all our constructions are effective, we obtain that e.g. the lattice-valued implication problem is decidable.

2 Strong Bimonoids

In this section, we recall the notion and examples of strong bimonoids. A *strong bimonoid* is a structure $\mathbb{K} = (K, +, \cdot, 0, 1)$ such that $(K, +, 0)$ is a commutative monoid, $(K, \cdot, 1)$ is a monoid and $k \cdot 0 = 0 = 0 \cdot k$ for every $k \in K$. We say that $\mathbb{K}$ is *distributive*, if it satisfies $(a + b) \cdot c = a \cdot c + b \cdot c$ and $c \cdot (a + b) = c \cdot a + c \cdot b$ for every $a, b, c \in K$. Then a *semiring* is a distributive strong bimonoid.

Example 1. (cf. [10])

1. The *Boolean semiring* is the semiring $(\mathbb{B}, \vee, \wedge, 0, 1)$ with $\mathbb{B} = \{0, 1\}$, and disjunction and conjunction as operations.
2. Bounded lattices (lattices containing a greatest element 1 and a smallest element 0) are strong bimonoids. As is well-known, there are large classes of lattices that are not distributive (cf. [14]).
3. The algebra $([0, 1], \oplus, \cdot, 0, 1)$ with the usual multiplication $\cdot$ of real numbers and either $a \oplus b = a + b - a \cdot b$ (*algebraic sum*) or $a \oplus b = \min\{a + b, 1\}$ (*bounded sum*), cf. [13], is a strong bimonoid.
4. Let X be an alphabet with $|X| \geq 2$. Consider the strong bimonoid $(X^* \cup \{\infty\}, \wedge, \cdot, \infty, \varepsilon)$ where $\wedge$ is the longest common prefix operation, $\cdot$ is the usual concatenation of words, and ∞ is a new element acting as zero. This bimonoid occurs in investigations for natural language processing, see [20].
5. Let $\mathcal{N}^* = (\mathbb{N}_\infty \cup \{0^*, 1^*\}, \oplus, \odot, 0^*, 1^*)$ where $\mathbb{N}_\infty = \mathbb{N} \cup \{\infty\}$, the operation $\oplus$ and $\odot$ are *both* the usual addition operation on $\mathbb{N}_\infty$, 0^* acts as zero, 1^* as one and $1^* \oplus x = x = x \oplus 1^*$ if $x \neq 0^*$. Then $\mathcal{N}^*$ is a strong bimonoid ([11]) but not a semiring.

The strong bimonoid $(K, +, \cdot, 0, 1)$ is called *bi-locally finite* if each finitely generated submonoid of both $(K, +, 0)$ and $(K, \cdot, 1)$ is finite. Clearly, any bounded lattice $(L, \vee, \wedge, 0, 1)$ is a bi-locally finite strong bimonoid. There are further examples:

Example 2. The strong bimonoid $(\{0\} \cup [\lambda, 1], \oplus, \odot, 0, 1)$ for $\lambda < \frac{1}{2}$, with $a \oplus b = \min\{a + b, 1\}$ and $a \odot b = a \cdot b$ if $a \cdot b \geq \lambda$, and 0 if $a \cdot b < \lambda$, is bi-locally finite.

In all of this paper, $\mathbb{K}$ will be a strong bimonoid.

3 Nested Words and Weighted Nested Word Automata

In this section we recall the notion of nested words and nested word automata introduced by Alur and Madhusudan [2], and we extend the notion of weighted nested word automata over an alphabet and a semiring introduced by Mathissen [19] to strong bimonoids. Let Δ be a finite alphabet and let Δ^+ be the free semigroup of finite non-empty words over Δ. If $w = a_1 \ldots a_n \in \Delta^+$ with $a_1, \ldots, a_n \in \Delta$, then we denote by $|w| = n$ the length of w and we let $\mathrm{dom}(w) = \{1, \ldots, n\}$. A *nested word* of length n over Δ is a pair (w, ν) consisting of a word $w \in \Delta^+$ with $|w| = n$ and a binary *nesting relation* ν on $\mathrm{dom}(w)$ such that for all $1 \leq i, j \leq n$:

(1) if $\nu(i,j)$ then $i < j$;
(2) if $\nu(i,j)$ and $\nu(i,j')$ then $j = j'$, and if $\nu(i,j)$ and $\nu(i',j)$ then $i = i'$;
(3) if $\nu(i,j)$ and $\nu(i',j')$ and $i < i'$ then either $j < i'$ or $j' < j$.

If $\nu(i,j)$, i is a *call position* and j is a *return position*. All other positions are called *internal positions*. We denote by $NW(\Delta)$ the collection of all nested words over Δ. Any subset of $NW(\Delta)$ is called a *language of nested words*.

Definition 1. *([2]) A nested word automaton (NWA) over Δ is a quadruple $\mathcal{A} = (Q, Q_{in}, \delta, Q_f)$ consisting of*

(1) a finite set Q of states,
(2) a set of initial states $Q_{in} \subseteq Q$,
(3) a set of final states $Q_f \subseteq Q$,
(4) a set of transition $\delta = (\delta_{call}, \delta_{int}, \delta_{ret})$, where
  - *$\delta_{call} \subseteq Q \times \Delta \times Q$ is a transition relation for call positions,*
  - *$\delta_{int} \subseteq Q \times \Delta \times Q$ is a transition relation for internal positions, and*
  - *$\delta_{ret} \subseteq Q \times Q \times \Delta \times Q$ is a transition relation for return positions.*

A *run* r of the automaton $\mathcal{A}$ over a nested word $nw = (a_1 \ldots a_n, \nu)$ is a sequence $(q_0, q_1, \ldots, q_n)$ on Q such that $q_0 \in Q_{in}$ and for each $1 \leq i \leq k$,

if i is a call position of ν, then $(q_{i-1}, a_i, q_i) \in \delta_{call}$,
if i is a internal position of ν, then $(q_{i-1}, a_i, q_i) \in \delta_{int}$ and
if i is a return position of ν with call-predecessor j, then $(q_{i-1}, q_{j-1}, a_i, q_i) \in \delta_{ret}$.

The automaton $\mathcal{A}$ *accepts* the nested word nw if it has a run $(q_0, \ldots, q_n)$ such that $q_n \in Q_f$. The language $\mathcal{L}(\mathcal{A})$ is the set of nested words accepted by $\mathcal{A}$.

A language L of nested words over Δ is *regular* if there exists a nested word automaton $\mathcal{A}$ over Δ such that $L = \mathcal{L}(\mathcal{A})$. As shown in [2], the class of regular languages of nested words is closed under set theoretical operations like union, intersection, and complement.

Next we define weighted nested word automata.

Definition 2. *Let $\mathbb{K}$ be a strong bimonoid. A weighted nested word automaton (WNWA for short) is a quadruple $\mathcal{A} = (Q, \iota, \delta, \kappa)$ where $\delta = (\delta_{call}, \delta_{int}, \delta_{ret})$ such that*

(1) Q is a finite set of states,
(2) $\delta_{call}, \delta_{int} : Q \times \Delta \times Q \to \mathbb{K}$ are the call resp. internal transition functions,
(3) $\delta_{ret} : Q \times Q \times \Delta \times Q \to \mathbb{K}$ is the return transition function, and
(4) $\iota, \kappa : Q \to \mathbb{K}$ are the initial resp. final distributions.

A *run* of $\mathcal{A}$ on $nw = (a_1 \ldots a_n, \nu)$ is a sequence $r = (q_0, \ldots, q_n) \in Q^{n+1}$ of states. The weight of r for nw at position $1 \leq i \leq n$ is given by

$$\mathrm{wt}_{\mathcal{A}}(r, nw, i) = \begin{cases} \delta_{call}(q_{i-1}, a_i, q_i) & \text{if } \nu(i,j) \text{ for some } i < j \le n \\ \delta_{int}(q_{i-1}, a_i, q_i) & \text{if } i \text{ is an internal position} \\ \delta_{ret}(q_{i-1}, q_{j-1}, a_i, q_i) & \text{if } \nu(j,i) \text{ for some } 1 \le j < i. \end{cases}$$

Now, the weight of r for nw is $\mathrm{wt}_{\mathcal{A}}(r, nw) = \prod_{1 \le i \le n} \mathrm{wt}_{\mathcal{A}}(r, nw, i)$.

The *behavior* $\|\mathcal{A}\| : NW(\Delta) \to \mathbb{K}$ of $\mathcal{A}$ is defined by

$$(\|\mathcal{A}\|, nw) = \sum_{r \in Q^{n+1}} \iota(q_0) \cdot \mathrm{wt}_{\mathcal{A}}(r, nw) \cdot \kappa(q_n).$$

For $\mathbb{K} = \mathbb{B}$ the Boolean semiring, we can consider the functions $\delta_{call}, \delta_{int}$ as subsets of $Q \times \Delta \times Q$ and δ_{ret} as a subset of $Q \times Q \times \Delta \times Q$ and we obtain nested word automata as described in Definition 1. Conversely, nested word automata can be considered as weighted nested word automata over the Boolean semiring. Therefore, weighted nested word automata form a generalization of nested word automata.

Example 3. 1. The following is a nondeterministic version of a similar example of Mathissen ([19]). Consider the randomized recursive pseudo-procedure **bar** where **flip(YY')** means flipping a fair coin Y and an unfair coin Y' at the same time. Let the probability of getting the head (H) and the tail (T) of the fair coin be $\frac{1}{2}$. For the unfair coin, let the probability of getting the head (H') be $\frac{2}{3}$ and getting the tail (T') be $\frac{1}{3}$. Consider the alphabet $\Delta = \{r, w, b, call, ret\}$ of atomic events which stand for read, write, beep, call, and return. Now an execution of bar could be as follow.

Case 1. read(x), flip the coins and see HT', call to recursively **bar**, read(x), flip the coins and see HH' or TT', beep, flip the coins and see HT' or TH', return from the recursive call, flip the coin and see HH' or TT', write(x), flip the coins and see HT' or TH', exit the program.

Or **Case 2.** read(x), flip the coins and see TH', call recursively **bar**, read(x),

```
proc bar( ){
   read(x);
   flip(YY');
   if(YY'==HH'||YY'=TT') {
         beep;
      else if(YY'==HT')
         bar( );
      else
         bar( );}
   flip(YY');while(YY'==HH'||
YY'==TT')
         write(x);
         flip(YY');
   exit;}
```

flip the coins and see HH' or TT', **beep**, flip the coins and see HT' or TH', return from the recursive call, flip the coins and see HH' or TT', write(x), flip the coins and see HT' or TH', exit the program. Then the nested word $7w = r.call.r.b.ret.w.ret$ with nesting relation $\nu = \{(2,5)\}$ models this execution of **bar** where ν encodes the recursive call of **bar**.

The probability of the word $7w$ is the probability of the execution of Case 1 plus the probability of the execution of Case 2 which equals $(1 \cdot \frac{1}{6} \cdot 1 \cdot \frac{3}{6} \cdot \frac{3}{6} \cdot \frac{3}{6} \cdot \frac{3}{6}) + (1 \cdot \frac{2}{6} \cdot 1 \cdot \frac{3}{6} \cdot \frac{3}{6} \cdot \frac{3}{6} \cdot \frac{3}{6}) = \frac{1}{32}$. Now we model **bar** using a nondeterministic nested weighted automata over $\Delta = \{r, w, b, call, ret\}$ and a strong bimonoid. We use the strong bimonoid $([0,1], \oplus, \cdot, 0, 1)$ where the operation $\oplus$ is the bounded sum of Example 1(3). The WNWA has five states $\{q_0, \ldots, q_4\}$. The non-zero initial and final weights and non-zero transition weights are defined as follows.

$$\iota(q_0) = \iota(q_1) = \kappa(q_4) = 1, \qquad \delta_{int}(q_0, r, q_2) = 1, \qquad \delta_{int}(q_1, r, q_2) = 1,$$
$$\delta_{call}(q_2, call, q_0) = 1/6, \quad \delta_{call}(q_2, call, q_1) = 2/6, \qquad \delta_{int}(q_2, b, q_3) = 3/6,$$
$$\delta_{int}(q_3, w, q_3) = 3/6, \quad \delta_{int}(q_3, ret, q_4) = 3/6, \quad \delta_{ret}(q_3, q_2, ret, q_3) = 3/6.$$

Then the automaton assigns the probability $\frac{1}{32}$ to the nested word $7w$.

2. Alur et al. [1] point out how a procedural computer execution program may be represented by finite or infinite nested words. The total computation time of a nondeterministic program can be modeled as the behavior of a WNWA over the strong bimonoid $\mathcal{N}^*$ of Example 1(5).

3. In algebraic path problems we count only paths with desired properties. This leads to strong bimonoids which are not distributive and hence not a semiring. For a range of examples, we refer the reader to [16], section 3.2.

A function $S : NW(\Delta) \to \mathbb{K}$ is called a *nested word series* or *series* (over Δ and $\mathbb{K}$). We call $S : NW(\Delta) \to \mathbb{K}$ *regular* if $S = \|\mathcal{A}\|$ for some WNWA $\mathcal{A}$. The series S is a *regular step function* if S assumes only finitely many values and for each $k \in \mathbb{K}$, $S^{-1}(k) = \{nw \in NW(\Delta) \mid S(nw) = k\}$ is regular. The set of all series over Δ and $\mathbb{K}$ is denoted by $\mathbb{K}\langle\!\langle NW(\Delta)\rangle\!\rangle$.

Let $L \subseteq NW(\Delta)$. The *characteristic function* $\mathbb{1}_L \in \mathbb{K}\langle\!\langle NW(\Delta)\rangle\!\rangle$ of L is defined by $\mathbb{1}_L(nw) = 1$ if $nw \in L$ and $\mathbb{1}_L(nw) = 0$ otherwise. Then L is regular iff the characteristic series $\mathbb{1}_L : NW(\Delta) \to \mathbb{B}$ is regular.

Lemma 1. *Any regular step function $S : NW(\Delta) \to \mathbb{K}$ is regular.*

For the proof of the following partial converse, we can adjust the proof of Theorem 11 of [10] as well as several other results from the theory of weighted automata over words to the nested word setting.

Theorem 1. *Let $\mathbb{K}$ be a bi-locally finite strong bimonoid and let $S : NW(\Delta) \to \mathbb{K}$ be regular. Then S is a regular step function.*

4   Weighted Logics on Nested Words

Let Δ be an alphabet. The formulas of the *monadic second-order logic of nested words* are given by the syntax:

$$\varphi ::= x = y \mid \mathrm{Lab}_a(x) \mid x \leq y \mid \nu(x, y) \mid x \in X \mid \varphi \vee \varphi \mid \neg\varphi \mid \exists x.\varphi \mid \exists X.\varphi$$

where $a \in \Delta$, x, y are first-order variables and X is a second-order variable. We denote by MSO the set of all monadic second orders formulas over Δ.

Let $\varphi \in$ MSO and let $\mathrm{Free}(\varphi)$ be the set of all free variables of φ. Let $\mathcal{V}$ be a finite set of first-order and second-order variables such that $\mathrm{Free}(\varphi) \subseteq \mathcal{V}$. A $(\mathcal{V}, nw)$-assignment γ is a function mapping first-order variables in $\mathcal{V}$ to elements of $\mathrm{dom}(nw)$ and second-order variables in $\mathcal{V}$ to subsets of $\mathrm{dom}(nw)$. If x is a first-order variable and $i \in \mathrm{dom}(nw)$, then $\gamma[x \to i]$ is the $(\mathcal{V} \cup \{x\}, nw)$-assignment which assigns x to i and equals γ on $\mathcal{V} \setminus \{x\}$. Similarly, $\gamma[X \rightarrow I]$ is defined for $I \subseteq \text{dom}(nw)$. The other definitions of the semantics are as usual. We write $(nw, \gamma) \models \varphi$ if φ holds in nw under the assignment γ.

We encode pairs (nw, γ) with $(\mathcal{V}, nw)$-assignment γ as nested words $((w, \nu), \sigma)$ as usual (cf. [7,8]) over the extended alphabet $\Delta_{\mathcal{V}} = \Delta \times \{0, 1\}^{\mathcal{V}}$ and with the same nesting relation as nw. We call $((w, \nu), \sigma)$ valid, if σ arises from a $(\mathcal{V}, nw)$-assignment. Clearly the language $N_{\mathcal{V}} = \{(nw, \sigma) \in NW(\Delta_{\mathcal{V}}) \mid (w, \sigma) \text{ is valid}\}$ is regular. If $\varphi \in \text{MSO}$ with $\text{Free}(\varphi) \subseteq \mathcal{V}$, we let $\mathcal{L}_{\mathcal{V}}(\varphi) = \{(nw, \sigma) \in N_{\mathcal{V}} \mid (nw, \sigma) \models \varphi\}$, the language defined by φ. We abbreviate and $\mathcal{L}(\varphi) = \mathcal{L}_{\text{Free}(\varphi)}(\varphi)$. Note that in case that φ is a *sentence*, i.e., $\text{Free}(\varphi) = \emptyset$, we consider $\mathcal{L}(\varphi)$ as a subset of $NW(\Delta)$.

Let $Z \subseteq \text{MSO}$. A language $L \subseteq NW(\Delta)$ is Z-definable if $L = \mathcal{L}(\varphi)$ for a sentence $\varphi \in Z$. The set EMSO comprises all MSO formulas φ of the form $\exists X_1 \ldots \exists X_m.\psi$, where ψ contains only quantification over first-order variables.

Theorem 2. (Alur and Madhusudan [2]) *A nested word language $L \in NW(\Delta)$ is regular iff L is MSO-definable iff L is EMSO-definable.*

Now we introduce our weighted monadic second order logics on nested words over strong bimonoids. We follow the approach of [7] for strings and semirings, cf. [19] for nested words and semirings and [11] for strings and strong bimonoids. We assume that the strong bimonoid $\mathbb{K}$ is equipped with a function $\bar{\ } : \mathbb{K} \rightarrow \mathbb{K}$ satisfying $\bar{0} = 1$ and $\bar{1} = 0$. We will regard $\bar{\ }$ as "complement function" and use it to define the semantics of the negation of formulas. Note that any strong bimonoid can be equipped with such a function by letting $\bar{0} = 1, \bar{1} = 0$, and defining $\bar{a}$ for $a \in K \setminus \{0, 1\}$ arbitrarily. In all of this section, let $\mathbb{K} = (K, +, \cdot, \bar{\ }, 0, 1)$ be a strong bimonoid with complement function. The set $\text{MSO}(\mathbb{K})$ of *weighted* MSO *formulas* over $\mathbb{K}$ and Δ is given by the following grammar:

$$\varphi ::= k \mid x = y \mid \text{Lab}_a(x) \mid x \leq y \mid \nu(x, y) \mid x \in X \mid \neg\varphi$$
$$\mid \varphi \vee \varphi \mid \varphi \wedge \varphi \mid \exists x.\varphi \mid \exists X.\varphi \mid \forall x.\varphi \mid \forall X.\varphi$$

where $k \in \mathbb{K}$ and $a \in \Delta$. Let $\varphi \in \text{MSO}(\mathbb{K})$ and $\text{Free}(\varphi) \subseteq \mathcal{V}$. The weighted semantics $[\![\varphi]\!]_{\mathcal{V}}$ of φ is a function from $NW(\Delta_{\mathcal{V}})$ to $\mathbb{K}$. Let $(nw, \sigma) \in NW(\Delta_{\mathcal{V}})$. If (nw, σ) is not valid, we put $[\![\varphi]\!]_{\mathcal{V}}(nw, \sigma) = 0$. If (nw, σ) with $nw = (a_1 \ldots a_n, \nu)$ is valid, we define $[\![\varphi]\!]_{\mathcal{V}}(nw, \sigma) \in \mathbb{K}$ inductively as in Table 1. For the products over $\text{dom}(nw)$, we follow the natural order of $\text{dom}(nw)$ resp. the lexicographic order on the power set of $\text{dom}(nw)$.

In the following, we shortly write $[\![\varphi]\!]$ for $[\![\varphi]\!]_{\text{Free}(\varphi)}$. Similarly to the string case (Proposition 3.3 of [7]) we can show that for every finite set $\mathcal{V}$ of variables containing $\text{Free}(\varphi)$ the semantics $[\![\varphi]\!]_{\mathcal{V}}$ and $[\![\varphi]\!]$ are consistent with each other, i.e., $([\![\varphi]\!]_{\mathcal{V}}, (nw, \sigma)) = ([\![\varphi]\!], (nw, \sigma_{|\text{Free}(\varphi)}))$ for every valid $(nw, \sigma) \in NW(\Delta_{\mathcal{V}})$.

We call a formula $\varphi \in \text{MSO}(\mathbb{K})$ *Boolean* ([4]) if it does not contain constants $k \in \mathbb{K} \setminus \{0, 1\}$ and does not use disjunction or existential quantifications. Clearly, every Boolean formula φ can be considered as a classical (unweighted) MSO-formula defining the language of $\mathcal{L}(\varphi)$. Then $[\![\varphi]\!] = \mathbb{1}_{\mathcal{L}(\varphi)}$. Clearly, Boolean formulas capture the full power of classical MSO logic. In particular, for every regular language $L \subseteq NW(\Delta)$ there is a Boolean sentence φ such that $\mathbb{1}_L = [\![\varphi]\!]$.

Table 1. MSO($\mathbb{K}$)-semantics

$$[\![k]\!]_\nu(nw,\sigma) \quad = k \text{ for all } k \in \mathbb{K} \qquad [\![x=y]\!]_\nu(nw,\sigma) = \begin{cases} 1, & \text{if } \sigma(x) = \sigma(y) \\ 0, & \text{otherwise,} \end{cases}$$

$$[\![\text{Lab}_a(x)]\!]_\nu(nw,\sigma) = \begin{cases} 1, & \text{if } a_{\sigma(x)} = a \\ 0, & \text{otherwise,} \end{cases} \qquad [\![x \le y]\!]_\nu(nw,\sigma) = \begin{cases} 1, & \text{if } \sigma(x) \le \sigma(y) \\ 0, & \text{otherwise,} \end{cases}$$

$$[\![\nu(x,y)]\!]_\nu(nw,\sigma) = \begin{cases} 1, & \text{if } \nu(\sigma(x),\sigma(y)) \\ 0, & \text{otherwise,} \end{cases} \quad [\![x \in X]\!]_\nu(nw,\sigma) = \begin{cases} 1, & \text{if } \sigma(x) \in \sigma(X) \\ 0, & \text{otherwise,} \end{cases}$$

$$[\![\neg\varphi]\!]_\nu(nw,\sigma) = \overline{[\![\varphi]\!]_\nu(nw,\sigma)}, \quad [\![\varphi \vee \psi]\!]_\nu(nw,\sigma) = [\![\varphi]\!]_\nu(nw,\sigma) + [\![\psi]\!]_\nu(nw,\sigma),$$

$$[\![\varphi \wedge \psi]\!]_\nu(nw,\sigma) = [\![\varphi]\!]_\nu(nw,\sigma) \cdot [\![\psi]\!]_\nu(nw,\sigma),$$

$$[\![\exists x.\varphi]\!]_\nu(nw,\sigma) = \sum_{i \in \text{dom}(nw)} [\![\varphi]\!]_{\nu \cup \{x\}}(nw,\sigma[x \to i]),$$

$$[\![\exists X.\varphi]\!]_\nu(nw,\sigma) = \sum_{I \subseteq \text{dom}(nw)} [\![\varphi]\!]_{\nu \cup \{X\}}(nw,\sigma[X \to I]),$$

$$[\![\forall x.\varphi]\!]_\nu(nw,\sigma) = \prod_{i \in \text{dom}(nw)} [\![\varphi]\!]_{\nu \cup \{x\}}(nw,\sigma[x \to i]),$$

$$[\![\forall X.\varphi]\!]_\nu(nw,\sigma) = \prod_{I \subseteq \text{dom}(nw)} [\![\varphi]\!]_{\nu \cup \{X\}}(nw,\sigma[X \to I]).$$

Example 4. Consider the strong bimonoid $\mathcal{N}^*$ of Example 1(5). Let $nw \in NW(\Delta)$. Define $\varphi = \exists x \exists y.(\nu(x,y) \wedge \forall z.((x < z \wedge z < y) \to ((\forall w.\neg\nu(z,w)) \wedge 1)))$ where $\varphi \to k$ abbreviates $\neg\varphi \vee (\varphi \wedge k)$. Then $[\![\varphi]\!](nw)$ is the sum of the lengths of open nested intervals in nw without further nesting inside.

We call a formula $\varphi \in$ MSO($\mathbb{K}$) *almost unambiguous* if it is constructed from constants k $(k \in \mathbb{K})$ and Boolean formulas, using disjunction, conjunction, and negation. A formula $\varphi \in$ MSO($\mathbb{K}$) is called *syntactically restricted*, if it satisfies the following conditions:

1. Whenever φ contains a conjunction $\psi \wedge \psi'$ as subformula but not in the scope of a universal first order quantifier, then either ψ and ψ' are almost unambiguous, or ψ or ψ' is Boolean.
2. Whenever φ contains $\forall X.\psi$ as a subformula, then ψ is a Boolean.
3. Whenever φ contains $\forall x.\psi$ or $\neg\psi$ as a subformula, then ψ is almost unambiguous.

We let srMSO($\mathbb{K}$) comprise all syntactically restricted formulas of MSO($\mathbb{K}$), and EMSO($\mathbb{K}$) all MSO($\mathbb{K}$)-formulas φ of the form $\varphi = \exists X_1 \ldots \exists X_n \cdot \psi$ such that ψ contains only first order quantifications. We put srEMSO($\mathbb{K}$) = srMSO($\mathbb{K}$) $\cap$ EMSO($\mathbb{K}$). The goal of this section is to prove the following two results.

Theorem 3. *Let $\mathbb{K}$ be any strong bimonoid with complement function, and let $S : NW(\Delta) \to \mathbb{K}$ be a series. Then the following are equivalent:*

(1) *S is regular.*
(2) *S is srMSO($\mathbb{K}$)-definable.*
(3) *S is srEMSO($\mathbb{K}$)-definable.*

Theorem 4. *Let $\mathbb{K}$ be any bi-locally finite strong bimonoid with complement function, and let $S : NW(\Delta) \to \mathbb{K}$ be a series. The following are equivalent:*

(1) *S is regular.*
(2) *S is EMSO($\mathbb{K}, \Delta$)-definable.*

For the proofs of the implications (2) $\to$ (1) of Theorem 3 resp. 4 we will proceed by induction over the structure of the formula, using the following lemma. Its proof involves automata constructions for sums, products, projections, and inverse projections of regular nested word series.

Lemma 2. *Let $\varphi, \psi \in$ MSO($\mathbb{K}$).*

(a) *If φ is almost unambiguous, then $[\![\varphi]\!]$ is a regular step function.*
(b) *If $[\![\varphi]\!]$ is a regular step function, then so is $[\![\neg\varphi]\!]$.*
(c) *If $[\![\varphi]\!]$ and $[\![\psi]\!]$ are regular step functions, then $[\![\varphi \vee \psi]\!]$ and $[\![\varphi \wedge \psi]\!]$ are also regular step functions.*
(d) *Let $[\![\varphi]\!]$ and $[\![\psi]\!]$ be regular. Then $[\![\varphi \vee \psi]\!]$ is regular. If φ or ψ is Boolean, then $[\![\varphi \wedge \psi]\!]$ is regular.*
(e) *If $[\![\varphi]\!]$ is regular, then $[\![\exists x.\varphi]\!]$ and $[\![\exists X.\varphi]\!]$ are regular.*
(f) *If $[\![\varphi]\!]$ is a regular step function, then $[\![\forall x.\varphi]\!]$ is regular.*

Proof of Theorems 3 and 4, (1) $\Rightarrow$ (3) *and* (1) $\Rightarrow$ (2): We construct a WNWA $\mathcal{A}$ over $\mathbb{K}$ and Δ with initial and final weights in $\{0, 1\}$, then we explicitly describe an srEMSO($\mathbb{K}$) sentence φ such that $\|\mathcal{A}\| = [\![\varphi]\!]$.
(2) $\Rightarrow$ (1): In case of Theorem 3, we show for any $\varphi \in$ srMSO($\mathbb{K}, \Delta$) that $[\![\varphi]\!]$ is regular by induction over the structure of φ. If φ is almost unambiguous, then by Lemmas 2 and 1, $[\![\varphi]\!]$ is regular. For the closure under disjunctions, the permitted conjunctions and existential quantifications, we apply Lemma 2. If φ contains $\forall X.\psi$ as a subformula, then ψ and hence also $\forall X.\psi$ is also Boolean, so $[\![\forall X.\psi]\!]$ is regular. Finally, if φ contains $\forall x.\psi$ or $\neg\psi$ as a subformula, then ψ is almost unambiguous. By Lemma 2, $[\![\forall x.\psi]\!]$ and $[\![\neg\psi]\!]$ are regular. Hence $[\![\varphi]\!]$ is regular. In case of Theorem 4, we show for any $\varphi \in$ EMSO($\mathbb{K}, \Delta$) that $[\![\varphi]\!]$ is a regular step function; then by Lemma 1, $[\![\varphi]\!]$ is regular. For this, we apply Lemma 2 in combination with Theorem 1 for existential quantifications and universal first order quantifications.

As a consequence, we obtain that the lattice-valued implication problem is decidable.

Corollary 1. *Let $\mathcal{L}$ be a bounded lattice. Given two arbitrary MSO-sentences $\varphi, \psi \in$ MSO($\mathcal{L}$), it is decidable whether $[\![\varphi]\!] \leq [\![\psi]\!]$.*

Proof. All our proofs for Theorem 4, implication (2) $\to$ (1), are constructive. Hence we effectively obtain representations of $[\![\varphi]\!]$ and $[\![\psi]\!]$ as regular step functions, i.e. we obtain their finitely many values assumed and automata for the languages where they are assumed. Then it is easy to check whether $[\![\varphi]\!](nw) \leq [\![\psi]\!](nw)$ for each $nw \in NW(\Delta)$.

5   Conclusion

We have shown that weighted nested word automata and suitably restricted weighted MSO logic are expressively equivalent where the weights can be taken from any strong bimonoid. Moreover, this equivalence extends to the full weighted existential MSO logic if the operations of the weights satisfy local finiteness assumptions.

In the theory of formal power series, weights were always assumed to be taken from semirings, i.e., *distributive* strong bimonoids. This assumption ensures that the usual multiplication of matrices is associative which enables the development of deep algebraic treatments of weighted automata, cf. [12,25,15,5,9], also [7,8] for weighted automata and logics on words and [19] for the commutative semiring-weighted nested word setting. In [11], it was recently shown that the classical Büchi-Elgot-type result and the results of [7,8] also extend to weighted automata and logics over words where the weights are taken from a strong bimonoid. Here, we have developed this result for the nested word setting. Our arguments employed explicit automata constructions. Together with the results of [11], this raises the question which other results and algorithms ([21]) from the rich theory of semiring-weighted automata can be extended to more general quantitative settings.

References

1. Alur, R., Arenas, M., Barceló, P., Etessami, K., Immerman, N., Libkin, L.: First-order and temporal logics for nested words. Logical Methods in Computer Science 4(4:11), 1–44 (2008)
2. Alur, R., Madhusudan, P.: Adding nesting structure to words. Journal of the ACM 56(3), article 16, 1–43 (2009)
3. Birkhoff, B., von Neumann, J.: The logic of quantum mechanics. Annals of Math. 37, 823–843 (1936)
4. Bollig, B., Gastin, P.: Weighted versus Probabilistic Logics. In: Diekert, V., Nowotka, D. (eds.) DLT 2009. LNCS, vol. 5583, pp. 18–38. Springer, Heidelberg (2009)
5. Berstel, J., Reutenauer, C.: Rational Series and Their Languages. Monographs in Theoretical Computer Science, vol. 12. Springer (1988)
6. Ćirić, M., Droste, M., Ignjatović, J., Vogler, H.: Determinization of weighted finite automata over strong bimonoids. Inform. Sciences 180, 3497–3520 (2010)
7. Droste, M., Gastin, P.: Weighted automata and weighted logics. Theoretical Computer Science 380, 69–86 (2007); Special issue of ICALP 2005
8. Droste, M., Gastin, P.: Weighted automata and weighted logics. In: [9], ch. 5
9. Droste, M., Kuich, W., Vogler, H. (eds.): Handbook of Weighted Automata. EATCS Monographs in Theoretical Computer Science. Springer (2009)
10. Droste, M., Stüber, T., Vogler, H.: Weighted automata over strong bimonoids. Information Sciences 180, 156–166 (2010)
11. Droste, M., Vogler, H.: Weighted automata and multi-valued logics over arbitrary bounded lattices. Theoretical Computer Science 418, 14–36 (2012); Extended abstract in DLT 2010

12. Eilenberg, S.: Automata, Languages, and Machines, Volume A. Pure and Applied Mathematics, vol. 59. Academic Press (1974)
13. Gottwald, S.: A Treatise on Many-Valued Logics. Studies in Logic and Computation. Research Studies Press LTD, Hertfordshire (2001)
14. Grätzer, G.: General Lattice Theory. Birkhäuser Verlag, Basel (2003)
15. Kuich, W., Salomaa, A.: Semirings, Automata, Languages. Monographs in Theoretical Computer Science. An EATCS Series, vol. 6. Springer (1986)
16. Lengauer, T., Theune, D.: Unstructured Path Problems and the Making of Semirings. In: Dehne, F., Sack, J.-R., Santoro, N. (eds.) WADS 1991. LNCS, vol. 519, pp. 189–200. Springer, Heidelberg (1991)
17. Li, Y.M.: Finite automata based on quantum logic and monadic second-order quantum logic. Science China Information Sciences 53, 101–114 (2010)
18. Mallya, A.: Deductive Multi-valued Model Checking. In: Gabbrielli, M., Gupta, G. (eds.) ICLP 2005. LNCS, vol. 3668, pp. 297–310. Springer, Heidelberg (2005)
19. Mathissen, C.: Weighted logics for nested words and algebraic formal power series. Logical Methods Computer in Science 6, 1–34 (2010); Special issue of ICALP 2008
20. Mohri, M.: Minimization algorithms for sequential transducers. Theoretical Computer Science 234, 177–201 (2000)
21. Mohri, M.: Weighted automata algorithms. In: [9], ch. 6
22. Qiu, D.: Automata theory based on quantum logic: some characterizations. Information and Computation 190, 179–195 (2004)
23. Qiu, D.: Automata theory based on quantum logic: Reversibilities and pushdown automata. Theoretical Computer Science 386, 38–56 (2007)
24. Schützenberger, M.P.: On the definition of a family of automata. Information and Control 4, 245–270 (1961)
25. Salomaa, A., Soittola, M.: Automata-Theoretic Aspects of Formal Power Series. Texts and Monographs in Computer Science. Springer (1978)
26. Ying, M.: Automata theory based on quantum logic (I) and (II). Int. J. of Theoret. Physics 39, 985–995, 2545-2557 (2000)
27. Ying, M.: A theory of computation based on quantum logic (I). Theoretical Computer Science 344, 134–207 (2005)

A Fast Suffix Automata Based Algorithm for Exact Online String Matching

Simone Faro[1] and Thierry Lecroq[2]

[1] Università di Catania, Viale A.Doria n.6, 95125 Catania, Italy
[2] Université de Rouen, LITIS EA 4108, 76821 Mont-Saint-Aignan Cedex, France
faro@dmi.unict.it, thierry.lecroq@univ-rouen.fr

Abstract. Searching for all occurrences of a pattern in a text is a fundamental problem in computer science with applications in many other fields, like natural language processing, information retrieval and computational biology. Automata play a very important role in the design of efficient solutions for the exact string matching problem. In this paper we propose a new very simple solution which turns out to be very efficient in practical cases. It is based on a suitable factorization of the pattern and on a straightforward and light encoding of the suffix automaton. It turns out that on average the new technique leads to longer shift than that proposed by other known solutions which make use of suffix automata.

1 Introduction

The *string matching* problem consists in finding all the occurrences of a pattern P of length m in a text T of length n, both defined over an alphabet Σ of size σ. Automata play a very important role in the design of efficient string matching algorithms. For instance, the Knuth-Morris-Pratt algorithm [6] (KMP) was the first linear-time solution, whereas the Backward-DAWG-Matching algorithm [3] (BDM) reached the optimal $\mathcal{O}(n \log_\sigma(m)/m)$ lower bound time complexity on the average. Both the KMP and the BDM algorithms are based on finite automata; in particular, they respectively simulate a deterministic automaton for the language $\Sigma^* P$ and the deterministic suffix automaton of the reverse of P.

The efficiency of string matching algorithms depends on the underlying automaton used for recognizing the pattern P and on the encoding used for simulating it. The efficient simulation of nondeterministic automata can be performed by using the *bit parallelism* technique [1]. For instance the Shift-Or algorithm, presented in [1], simulates the nondeterministic version of the KMP automaton while a very fast BDM-like algorithm, (BNDM), based on the bit-parallel simulation of the nondeterministic suffix automaton, was presented in [8].

Specifically the bit-parallelism technique takes advantage of the intrinsic parallelism of the bitwise operations inside a computer word, allowing to cut down the number of operations that an algorithm performs by a factor up to w, where w is the number of bits in the computer word. However the correspondent encoding requires one bit per pattern symbol, for a total of $\lceil m/w \rceil$ computer words. Thus, as long as a pattern fits in a computer word, bit-parallel algorithms are

N. Moreira and R. Reis (Eds.): CIAA 2012, LNCS 7381, pp. 149–158, 2012.
© Springer-Verlag Berlin Heidelberg 2012

extremely fast, otherwise their performances degrades considerably as $\lceil m/\omega \rceil$ grows. Though there are a few techniques [9,2,4] to maintain good performance in the case of long patterns, such limitation is intrinsic.

In this paper we present a new algorithm based on the efficient simulation of a suffix automaton constructed on a substring of the pattern extracted after a suitable factorization. The new algorithm is based on a simple encoding of the underlying automaton and turns out to be very fast in most practical cases, as we show in our experimental results.

The paper is organized as follows. In Section 2 we briefly introduce the basic notions which we use along the paper. In Section 3 we review the previous results known in literature based on the simulation of the suffix automaton of the searched pattern. Then in Section 4 we present the new algorithm and some efficient variants of it. In Section 5 we compare the newly presented solutions with the suffix automata based algorithms known in literature. We draw our conclusions in Section 6.

2   Basic Notions and Definitions

Given a finite alphabet Σ, we denote by Σ^m, with $m \geq 0$, the set of strings of length m over Σ and put $\Sigma^* = \bigcup_{m \in \mathbb{N}} \Sigma^m$. We represent a string $P \in \Sigma^m$, also called an m-gram, as an array $P[0..m-1]$ of characters of Σ and write $|P| = m$ (in particular, for $m = 0$ we obtain the empty string ε). Thus, $P[i]$ is the $(i+1)$-st character of P, for $0 \leqslant i < m$, and $P[i..j]$ is the substring of P contained between its $(i+1)$-st and the $(j+1)$-st characters, for $0 \leqslant i \leqslant j < m$. For any two strings P and P', we say that P' is a suffix of P if $P' = P[i..m-1]$, for some $0 \leqslant i < m$, and write $Suff(P)$ for the set of all suffixes of P. Similarly, P' is a prefix of P if $P' = P[0..i]$, for some $0 \leqslant i < m$. In addition, we write $P \cdot P'$, or more simply PP', for the concatenation of P and P', and P^r for the reverse of the string P, i.e. $P^r = P[m-1]P[m-2]\cdots P[0]$.

For a string $P \in \Sigma^m$, the suffix automaton of P is an automaton which recognizes the language $Suff(P)$ of the suffixes of P.

Finally, we recall the notation of some bitwise infix operators on computer words, namely the bitwise and "&", the bitwise or "|", the left shift "≪" operator (which shifts to the left its first argument by a number of bits equal to its second argument), and the unary bitwise not operator "∼".

3   Previous Efficient Suffix Automaton Based Solutions

In this section we present the known solutions for the online string matching problem which make use of the suffix automaton for searching for all occurrences of the pattern. Most of them are filtering based solutions, thus they use the suffix automaton for finding candidate occurrences of the pattern and then perform an additional verification phase based on a naive algorithm.

The Backward DAWG Matching Algorithm

One of the first application of the suffix automaton to get optimal pattern matching algorithms on the average was presented in [3]. The algorithm which makes use of the suffix automaton of the reverse pattern is called Backward-DAWG-Matching algorithm (BDM). Such algorithm moves a window of size m on the text. For each new position of the window, the automaton of the reverse of P is used to search for a factor of P from the right to the left of the window. The basic idea of the BDM algorithm is that if the backward search failed on a letter c after the reading of a word u then cu is not a factor of p and moving the beginning of the window just after c is secure. If a suffix of length m is recognized then an occurrence of the pattern was found.

The Backward Nondeterministic DAWG Matching Algorithm

The BNDM algorithm [8] simulates the suffix automaton for P^r with the bit-parallelism technique, for a given string P of length m. The bit-parallel representation uses an array B of $|\Sigma|$ bit-vectors, each of size m, where the i-th bit of $B[c]$ is set iff $P[i] = c$, for $c \in \Sigma$, $0 \leqslant i < m$. Automaton configurations are then encoded as a bit-vector D of m bits, where each bit corresponds to a state of the suffix automaton (the initial state does not need to be represented, as it is always active). In this context the i-th bit of D is set iff the corresponding state is active. D is initialized to 1^m and the first transition on character c is implemented as $D \leftarrow (D \ \& \ B[c])$. Any subsequent transition on character c can be implemented as $D \leftarrow ((D \ll 1) \ \& \ B[c])$.

The BNDM algorithm works by shifting a window of length m over the text. Specifically, for each window alignment, it searches the pattern by scanning the current window backwards and updating the automaton configuration accordingly. Each time a suffix of P^r (i.e., a prefix of P) is found, namely when prior to the left shift the m-th bit of $D\&B[c]$ is set, the window position is recorded. A search ends when either D becomes zero (i.e., when no further prefixes of P can be found) or the algorithm has performed m iterations (i.e., when a match has been found). The window is then shifted to the start position of the longest recognized proper prefix.

When the pattern size m is larger than ω, the configuration bit-vector and all auxiliary bit-vectors need to be split over $\lceil m/\omega \rceil$ multiple words. For this reason the performance of the BNDM algorithm degrades considerably as $\lceil m/\omega \rceil$ grows. A common approach to overcome this problem consists in constructing an automaton for a substring of the pattern fitting in a single computer word, to filter possible candidate occurrences of the pattern. When an occurrence of the selected substring is found, a subsequent naive verification phase allows to establish whether this belongs to an occurrence of the whole pattern.

However, besides the costs of the additional verification phase, a drawback of this approach is that, in the case of the BNDM algorithm, the maximum possible shift length cannot exceed ω, which could be much smaller than m.

The Long BNDM Algorithm

Peltola and Tarhio presented in [9] an efficient approach for simulating the suffix automaton using bit-parallelism in the case of long patterns. Specifically the algorithm (called LBNDM) works by partitioning the pattern in $\lfloor m/k \rfloor$ consecutive substrings, each consisting in $k = \lfloor (m-1)/\omega \rfloor + 1$ characters. The $m - k\lfloor m/k \rfloor$ remaining characters are left to either end of the pattern. Then the algorithm constructs a superimposed pattern P' of length $\lfloor m/k \rfloor$, where $P'[i]$ is a class of characters including all characters in the i-th substring, for $0 \leq i < \lfloor m/k \rfloor$.

The idea is to search first the superimposed pattern in the text, so that only every k-th character of the text is examined. This filtration phase is done with the standard BNDM algorithm, where only the k-th characters of the text are inspected. When an occurrence of the superimposed pattern is found the occurrence of the original pattern must be verified. The time for its verification phase grows proportionally to m/ω, so there is a threshold after which the performance of the algorithm degrades significantly.

The BNDM Algorithm with Extended Shift

Durian *et al.* presented in [4] another efficient algorithm for simulating the suffix automaton in the case of long patterns. The algorithm is called BNDM with eXtended Shift (BXS). The idea is to cut the pattern into $\lceil m/\omega \rceil$ consecutive substrings of length w except for the rightmost piece which may be shorter. Then the substrings are superimposed getting a superimposed pattern of length ω. In each position of the superimposed pattern a character from any piece (in corresponding position) is accepted. Then a modified version of BNDM is used for searching consecutive occurrences of the superimposed pattern using bit vectors of length ω but still shifting the pattern by up to m positions. The main modification in the automaton simulation consists in moving the rightmost bit, when set, to the first position of the bit array, thus simulating a circular automaton. Like in the case of the LBNDM, algorithm the BXS algorithm works as a filter algorithm, thus an additional verification phase is needed when a candidate occurrence has been located.

The Factorized BNDM Algorithm

Cantone *et al.* presented in [2] an alternative technique, still suitable for bit-parallelism, to encode the nondeterministic suffix automaton of a given string in a more compact way. Their encoding is based on factorizations of strings in which no character occurs more than once in any factor. It turns out that the nondeterministic automaton can be encoded with k bits, where k is the size of the factorization. Though in the worst case $k = m$, on the average k is much smaller than m, making it possible to encode large automata in a single or few computer words. As a consequence, their bit-parallel variant of the BNDM, called Factorized BNDM algorithm (F-BNDM) based on such approach tends to be faster in the case of sufficiently long patterns.

4 A New Fast Suffix Automaton Based Algorithm

The efficiency of suffix automata based algorithms for the exact string matching problem resides in two main features of the underlying automaton: the efficiency of the adopted encoding and the size of the automaton itself.

Regarding the first point it turns out that automata admitting simpler encoding turns out to be more efficient in practice. This is the case, for instance, of the automata which admit a bit parallel encoding. Moreover longer automata lead to larger shifts during the searching phase when a backward scan of the window is performed.

In this section we present a new algorithm for the online exact string matching problem based on the simulation of a suffix automaton constructed on the pattern P. The basic idea behind the new algorithm is straightforward but efficient. It consists in constructing the suffix automaton of a substring of the pattern in which each character is repeated at most once. This leads to a simple encoding and, by convenient alphabet transformations, to quite long automata.

The resulting algorithm is named Backward-SNR-DAWG-Matching (BSDM), where SNR is the acronym of *substring with no repetitions*. In what follows we describe separately the preprocessing and the searching phase of the algorithm.

The Preprocessing Phase

Given a pattern P, of length m, over an alphabet Σ of size σ, we say that a substring $S = P[i .. j]$ of P is a *substring with no repetitions* (SNR) if any character $c \in \Sigma$ appears at most once in S. It turns out trivially that $|S| \leq \min\{\sigma, m\}$. Moreover an SNR admits a suffix automaton where do not exist two states having incoming transitions labeled with the same character.

The preprocessing phase of the BSDM algorithm consists in finding the maximal SNR of the pattern, i.e. an SNR with the maximal length. In particular it finds a pair of integers value (s, ℓ), where $0 \leq s < m$ is the starting position of the maximal SNR of P, and $1 \leq \ell \leq m - s$ is the length of such a substring.

For instance, given the pattern $P = \texttt{abcabdcbabd}$, we have that abc, abdc, cba are all SNR of P. The substring abdc, of length 4, is a maximal SNR of P.

In many practical cases the length of the maximal SNR is not large enough if compared with the size of the pattern. This happens especially for patterns over small alphabets, as in the case of genome sequences, or for patterns with characters occurring many times, as in the case of a natural language text.

In order to allow longer SNR it is convenient to use a condensed alphabet whose characters are obtained by combining groups of q characters, for a fixed value q. A hash function $hash : \Sigma^q \leftarrow \{0, \ldots, \text{MAX}-1\}$ can be used for combining the group of characters, for a fixed constant value MAX. Thus a new condensed pattern P_q of length $m - q + 1$, over the alphabet $\{0, \ldots, \text{MAX} - 1\}$, is obtained from P. Specifically we have $P_q[i .. j] = hash(P[i] \cdots P[i+q-1]) \cdots hash(P[j] \cdots P[j+q-1])$ for $0 \leq i, j \leq m - q$, where $P_q = P_q[0 .. m - q]$. The maximal SNR is then computed on P_q to get a longer suffix automaton.

Table 1. The average length of the maximal SNR in patterns randomly extracted from a genome sequence (on top), a protein sequence (in the middle) and a natural language text (on bottom). The SNR have been computed using condensed alphabets on q characters, where q ranges from 1 to 8.

q/m	2	4	8	16	32	64	128	256	512	1024	2048	4096
1	1.72	2.62	3.20	3.64	3.89	3.99	4.00	4.00	4.00	4.00	4.00	4.00
2	1.00	2.86	5.45	7.61	9.19	10.41	11.23	11.97	12.64	13.09	13.21	13.27
4	-	1.00	4.94	12.33	22.91	32.75	39.89	45.12	50.84	54.29	57.17	59.82
6	-	-	3.00	10.81	24.56	42.19	55.69	66.31	74.35	82.48	88.50	97.82
8	-	-	1.00	8.98	24.50	51.55	88.03	116.33	140.82	163.20	175.24	183.42

q/m	2	4	8	16	32	64	128	256	512	1024	2048	4096
1	1.91	3.46	5.43	6.98	8.20	9.27	10.08	10.95	11.70	12.27	12.91	13.69
2	1.00	2.96	6.53	12.03	17.28	21.04	24.24	27.32	30.02	32.30	34.84	36.61
4	-	1.00	4.99	12.84	27.29	49.85	71.44	88.44	99.13	111.73	125.03	132.34
6	-	-	2.99	10.85	25.03	45.34	62.62	73.88	82.87	90.78	99.09	106.52
8	-	-	1.00	8.99	24.62	53.25	92.54	126.52	152.86	172.15	195.92	217.41

q/m	2	4	8	16	32	64	128	256	512	1024	2048	4096
1	1.99	3.81	6.25	7.83	8.96	9.83	10.46	11.07	11.51	12.18	12.91	14.42
2	1.00	2.99	6.84	12.98	19.01	23.30	26.77	29.68	32.79	35.38	37.80	40.03
4	-	1.00	5.00	12.94	26.86	43.01	55.50	64.67	72.94	79.03	87.22	97.85
6	-	-	3.00	10.99	26.42	50.79	73.93	92.90	104.13	115.86	132.07	148.79
8	-	-	1.00	9.00	24.83	53.98	96.14	128.96	152.85	175.50	189.52	203.14

For instance if $q = 3$ the pattern $P = \texttt{abcabdcb}$ is condensed in a new pattern $P_3 = hash(\texttt{abc}) \cdot hash(\texttt{bca}) \cdot hash(\texttt{cab}) \cdot hash(\texttt{abd}) \cdot hash(\texttt{bdc}) \cdot hash(\texttt{dcb})$.

The size MAX of the new condensed alphabet depends on the available memory and on the size of the original alphabet Σ. An efficient method for computing a condensed alphabet was introduced by Wu and Manber [10], and then adopted also in [7]. It computes the shift value by using a **shift-and-addition** procedure and in particular $hash(c_1, c_2, \ldots, c_q) = \left(\sum_{i=1}^{q}(c_i \ll sh^{q-i})\right)$ mod MAX where $c_i \in \Sigma$ for $i = 1, \ldots, q$. The value of the shift sh depends on MAX and q.

Table 1 shows the average length of the maximal SNR in patterns randomly extracted from a genome sequence, a protein sequence and a natural language text, for different values of q and m, and with MAX $= 2^{16}$. When $1 \leq q \leq 4$ we use the value $sh = 2$ for computing the hash value, while we use $sh = 1$ when $q > 4$. It turns out that the length of the maximal SNR, though quite less than m in most cases, is quite larger than the size of a computer word (which typically is 32 or 64). This leads to larger shift in a suffix automata based algorithm.

The procedure which computes the maximal SNR of P using a condensed alphabet is shown in Fig. 1. The procedure iterates two indices, i and j along the pattern, starting from the leftmost character. At each iteration the value of i is incremented by one position, and the value of j is incremented in order to make the substring $P_q[j \mathinner{..} i]$ an SNR of P_q. At the end of each iteration, if the substring $P_q[j \mathinner{..} i]$ is longer than the temporary maximal SNR found in the previous iterations, then the values of s and ℓ are updated accordingly.

The time complexity of the resulting procedure is $\mathcal{O}(m)$ while the space required is $\mathcal{O}(\text{MAX})$.

```
HASH(P, i, q, b)                          POSITIONS(P, s, ℓ, q, b)
1. c ← P[i]                               1. for c ← 0 to MAX − 1 do pos(c) = −1
2. for j ← i + 1 to i + q − 1 do          2. for i ← 0 to ℓ − 1 do
3.      c ← (c ≪ b) + P[j]                3.      c ← HASH(P, s + i, q, b)
4. return c                               4.      pos(c) ← i
                                          5. return pos
MAXSNR(P, m, q, b)
1. for c ← 0 to MAX − 1 do δ(c) ← FALSE   BSDM(P, m, T, n, q, b)
2. s ← ℓ ← 0                              1. (s, l) ← MAXSNR(P, m, q, b)
3. j ← 0                                  2. pos ← POSITIONS(P, s, ℓ, q, b)
4. for i ← 0 to m − q do                  3. j ← ℓ − 1
5.      c ← HASH(P, i, q, b)              4. r ← s + ℓ
6.      if δ(c) then                      5. while j < n do
7.          d ← HASH(P, j, q, b)          6.      c ← HASH(T, j, q, b)
8.          while d ≠ c do                7.      i ← pos(c)
9.              δ(d) ← FALSE              8.      if i ≥ 0 then
10.             j ← j + 1                 9.          k ← 1
11.             d ← HASH(P, j, q, b)      10.         while k ≤ i and P[s + i − k] = T[j − k] do
12.         δ(d) ← FALSE                  11.             k ← k + 1
13.         j ← j + 1                     12.         if k > i then
14.      δ(c) ← TRUE                      13.             if k = ℓ then
15.      if ℓ < i − j + 1 then            14.                 if P = T[j − r + 1 .. j − r + m]
16.          ℓ ← i − j + 1               15.                 then output (j − s − ℓ + 1)
17.          s ← j                        16.             else j ← j − k
18. return (s, ℓ)                         17.      j ← j + ℓ
```

Fig. 1. The pseudocode of the algorithm BSDM and its auxiliary procedures. The input parameters q and b represent, respectively, the size of the group of characters used in the condensed alphabet and the value sh used for computing the hash function.

The Searching Phase

Let P be a pattern of length m over an alphabet Σ of size σ, and let P_q be the corresponding pattern, of length $m - q + 1$, obtained from P by using a condensed alphabet. Let s and ℓ be, respectively, the starting position and the length of the maximal SNR in the P_q. During the searching phase the BSDM algorithm works using a filtering method. Specifically it first searches for all occurrences of the substring $P_q[s .. s + \ell - 1]$ in the text. For this purpose the text is also scanned by using a condensed alphabet. When an occurrence is found, ending at position j of the text, the algorithm naively checks for the whole occurrence of the pattern, i.e. if $P = T[j - s - \ell + 1 .. j - s - \ell + m]$.

A function $pos : \{0, \ldots, \text{MAX} - 1\} \to \{0, \ldots, \ell - 1\}$ is defined for all characters in the condensed alphabet. In particular for each $0 \leqslant c < \text{MAX}$ the value of $pos(c)$ is defined as the relative position in $P_q[s .. s + \ell - 1]$ where the character c appears, if such position exists. Otherwise $pos(c)$ is set to -1. More formally $pos(c) = i$ if there exists $i < \ell$ such that $P_q[s + i] = c$ and -1 otherwise, for $0 \leqslant c < \text{MAX}$. Observe that if position i exists such that $i < \ell$ and $P_q[s + i] = c$, then it is unique, since the substring $P_q[s .. s + \ell - 1]$ has no repetitions of characters. The function pos is computed in $\mathcal{O}(m)$ time and $\mathcal{O}(\text{MAX})$ space by using the procedure POSITION shown in Fig. 1.

The pos function defined above is then used during the searching phase for simulating the suffix automaton of the maximal SNR of the pattern. Observe that, since there is no repetition of characters, at most a single state could be active at any time. Thus the configuration of the suffix automaton can be encoded by using a single integer value of $\lceil \log \ell \rceil$ bits, which simply indicates the active state of the automaton, if any. Otherwise it is set to -1.

The algorithm works by sliding a window on length $\ell + q - 1$ along the text. At each attempt a condensed character c is computed from the rightmost q characters of the window. If c is not present in the maximal SNR of the pattern, i.e. if $pos(c) = -1$, then the window is advanced ℓ positions to the right. Otherwise (if $pos(c) \geqslant 0$) the position i where character c appears in the maximal SNR is computed by setting $i = pos(c)$. Then the text and the pattern are compared, character by character, from positions $s + i$ and $j - s - \ell + i$, respectively, until a mismatch occurs or until position s in the pattern is passed.

If a mismatch occurs, no prefix of the substring has been recognized and the window is simply advanced ℓ positions to the right.

Otherwise, if position s in the pattern is passed, then a prefix of the substring has been recognized. If we read exactly ℓ characters in T then an occurrence of the substring has been found and a naive verification follows in order to check the occurrence of the whole pattern. If we read less than ℓ characters we recognized a prefix of the substring and the window is advanced in order to align he character of position s in the pattern with the starting position of the recognized prefix in the text. The searching phase of the algorithm is shown in Fig. 1. It has a $\mathcal{O}(nm)$ worst case time complexity and requires $O(\text{MAX})$ space.

5   Experimental Results

In this section we briefly present experimental evaluations in order to understand the performances of the newly presented algorithm and to compare it against the best on suffix (factor) automata based string matching algorithms. In particular we tested the following algorithms: the Backward-DAWG-Matching algorithm [3] (BDM); the Backward-Nondeterministic-DAWG-Matching algorithm [8] (BNDM); the Simplified version of the BNDM algorithm [9] (SBNDM); the BNDM for algorithm long patterns [9] (LBNDM); the Factorized BNDM algorithm [2] (F-BNDM); the BNDM algorithm with Extended Shift [4] (BXS); and the new Backward-SNR-DAWG-Matching algorithm using condensed alphabets with groups of q characters, with $1 \leq q \leq 8$ (BSDM$_q$)

All the algorithms listed above could be enhanced by using fast loops, q-grams and other efficient techniques. However this type of code tuning goes beyond the scope of this paper. Thus we tested only the original versions of the algorithms.

All algorithms have been implemented in the C programming language and have been tested using the SMART tool [5]. The experiments were executed locally on an MacBook Pro with 4 Cores, a 2 GHz Intel Core i7 processor, 4 GB RAM 1333 MHz DDR3, 256 KB of L2 Cache and 6 MB of Cache L3. Algorithms have been compared in terms of running times, including any preprocessing time.

For the evaluation we use a genome sequence, a protein sequence and a natural language text (English language), all sequences of 4MB. The sequences are provided by the SMART research tool. In all cases the patterns were randomly extracted from the text and the value m was made ranging from 2 to 4096. For each case we reported the mean over the running times of 500 runs.

Table 2. Experimental results on a genome sequence (on top), a protein sequence (in the middle) and natural language text (on bottom)

Experimental results on a genome sequence											
m	2	4	8	16	32	64	128	256	512	1024	2048	4096
BDM	21.04	14.66	9.90	7.40	5.94	5.15	4.79	4.68	4.90	5.42	7.22	10.59
BNDM	19.52	12.56	8.92	6.72	5.50	5.55	5.51	5.47	5.58	5.49	5.50	5.49
SBNDM	12.25	9.13	7.66	6.27	5.14	5.14	5.12	5.13	5.12	5.13	5.13	5.14
BXS	19.57	13.88	9.27	6.88	5.47	5.15	4.99	5.52	523.2	-	-	-
F-BNDM	15.49	10.74	8.71	7.09	5.78	5.10	5.03	5.03	5.02	5.03	5.05	5.05
LBNDM	27.62	15.24	9.79	7.28	5.80	5.38	5.36	8.45	26.93	25.28	22.50	20.67
BSDM	20.94	17.49	14.42	13.05	11.99	11.52	11.55	11.39	11.42	11.38	11.46	11.50
$BSDM_2$	**11.43**	9.26	8.66	8.31	7.82	7.44	7.15	6.89	6.60	6.51	6.37	6.24
$BSDM_3$	-	**7.44**	5.92	5.57	5.43	5.38	5.33	5.31	5.28	5.30	5.28	5.28
$BSDM_4$	-	9.67	**5.61**	**4.99**	4.79	4.73	4.66	4.64	4.63	4.63	4.66	4.66
$BSDM_5$	-	-	5.99	5.00	4.77	4.66	4.61	4.58	4.57	4.56	4.58	4.58
$BSDM_6$	-	-	6.86	5.09	**4.69**	4.58	4.53	4.49	4.50	4.47	4.50	4.50
$BSDM_7$	-	-	8.81	5.25	4.71	**4.55**	**4.51**	**4.47**	**4.45**	**4.47**	**4.47**	**4.49**
$BSDM_8$	-	-	14.88	5.57	4.80	4.56	**4.51**	4.50	4.48	4.48	4.49	4.50

Experimental results on a protein sequence											
m	2	4	8	16	32	64	128	256	512	1024	2048	4096
BDM	9.82	8.20	7.05	5.80	4.91	4.59	4.53	4.57	4.73	5.30	7.14	10.60
BNDM	9.27	7.67	6.74	5.61	4.81	4.83	4.80	4.80	4.81	4.80	4.82	4.83
SBNDM	9.25	**5.93**	**4.96**	**4.59**	**4.41**	4.57	4.57	4.57	4.58	4.57	4.62	4.58
BXS	8.41	7.19	6.41	5.45	4.69	**4.53**	**4.39**	**4.29**	**4.27**	**4.17**	**4.28**	105.3
F-BNDM	11.94	8.06	6.22	5.32	4.91	4.79	4.63	4.64	4.62	4.65	4.64	4.65
LBNDM	19.66	12.60	8.84	6.51	5.79	4.88	4.54	4.40	4.34	4.46	6.20	10.44
BSDM	8.37	7.58	7.15	6.89	6.63	6.37	6.14	5.92	5.71	5.56	5.47	5.37
$BSDM_2$	**8.29**	6.04	5.44	5.15	5.07	4.99	4.99	4.97	4.95	4.93	4.94	4.95
$BSDM_3$	-	6.58	5.25	4.85	4.71	4.64	4.62	4.59	4.59	4.58	4.60	4.60
$BSDM_4$	-	9.71	5.49	4.89	4.68	4.59	4.56	4.53	4.52	4.52	4.50	**4.53**
$BSDM_5$	-	-	6.04	5.07	4.79	4.68	4.65	4.61	4.61	4.61	4.62	4.64
$BSDM_6$	-	-	7.02	5.19	4.79	4.64	4.60	4.58	4.59	4.57	4.58	4.61
$BSDM_7$	-	-	9.02	5.38	4.82	4.64	4.62	4.58	4.58	4.60	4.58	4.59
$BSDM_8$	-	-	15.11	5.68	4.94	4.70	4.64	4.63	4.62	4.61	4.59	4.58

Experimental results on a natural language text											
m	2	4	8	16	32	64	128	256	512	1024	2048	4096
BDM	10.50	9.27	7.89	6.29	5.33	4.93	4.73	4.99	4.98	5.51	7.34	10.82
BNDM	10.02	8.74	7.50	6.06	5.20	5.25	5.23	5.25	5.23	5.25	5.25	5.26
SBNDM	9.68	6.39	5.47	5.01	4.76	4.99	4.99	4.99	4.99	4.97	4.98	4.98
BXS	9.12	8.25	7.20	5.91	5.06	4.76	**4.50**	**4.35**	**4.27**	**4.09**	**3.92**	**3.90**
F-BNDM	12.36	8.45	6.64	5.75	5.30	5.04	4.72	4.67	4.66	4.67	4.67	4.67
LBNDM	20.36	13.38	9.38	6.83	5.56	4.99	4.63	4.44	4.35	4.38	4.62	5.69
BSDM	8.90	8.35	7.72	7.15	6.71	6.43	6.16	6.01	5.85	5.79	5.69	5.61
$BSDM_2$	**8.41**	**6.24**	5.62	5.37	5.27	5.23	5.18	5.14	5.11	5.09	5.08	5.08
$BSDM_3$	-	6.76	**5.40**	5.00	4.85	4.79	4.74	4.71	4.69	4.67	4.70	4.71
$BSDM_4$	-	9.88	5.62	**4.95**	4.76	4.65	4.62	4.61	4.61	4.56	4.57	4.62
$BSDM_5$	-	-	6.02	5.00	**4.75**	4.65	4.62	4.59	4.59	4.55	4.60	4.63
$BSDM_6$	-	-	7.05	5.16	4.78	**4.64**	4.59	4.57	4.58	4.54	4.58	4.60
$BSDM_7$	-	-	9.26	5.41	4.84	4.66	4.60	4.56	4.59	4.56	4.57	4.58
$BSDM_8$	-	-	16.07	5.80	4.96	4.72	4.67	4.62	4.60	4.54	4.53	4.55

Table 2 shows experimental results on the three different sequences. Running times are expressed in thousands of seconds. Best times have been boldfaced and underlined. It turns out that the BSDM algorithm is really competitive against the most efficient algorithms which make use of suffix automata. The versions based on condensed characters obtain in many cases the best results, especially in the case of the genome sequence and of the natural language text. Otherwise SBNDM and BXS obtain the best times for short and long patterns, respectively.

6   Conclusions and Future Works

We presented a new simple and efficient algorithm, named Backward-SNR-DAWG-Matching, based on suffix automata. It uses a very simple encoding of the automaton, consisting in a single integer value, but obtains larger shift on average than that obtained by algorithms based on the bit parallel encoding.

In our future works we intend to tune the algorithm in order to make it competitive with the most efficient algorithms in practical cases. This includes the use of fast loops, q-grams and most efficient hash functions for implementing the condensed alphabets. We would also investigate the possibility of tuning the hash function in order to reflect only the size of the set of characters appearing in the pattern. Another possible line for enhancing the performances of the algorithm is to make it recognize factors instead of suffixes.

References

1. Baeza-Yates, R., Gonnet, G.H.: A new approach to text searching. Commun. ACM 35(10), 74–82 (1992)
2. Cantone, D., Faro, S., Giaquinta, E.: A Compact Representation of Nondeterministic (Suffix) Automata for the Bit-Parallel Approach. In: Amir, A., Parida, L. (eds.) CPM 2010. LNCS, vol. 6129, pp. 288–298. Springer, Heidelberg (2010)
3. Crochemore, M., Rytter, W.: Text algorithms, Oxford University Press (1994)
4. Ďurian, B., Peltola, H., Salmela, L., Tarhio, J.: Bit-Parallel Search Algorithms for Long Patterns. In: Festa, P. (ed.) SEA 2010. LNCS, vol. 6049, pp. 129–140. Springer, Heidelberg (2010)
5. Faro, S., Lecroq, T.: Smart: a string matching algorithm research tool. Univ. of Catania and Univ. of Rouen (2011), http://www.dmi.unict.it/~faro/smart/
6. Knuth, D.E., Morris Jr., J.H., Pratt, V.R.: Fast pattern matching in strings. SIAM J. Comput. 6(1), 323–350 (1977)
7. Lecroq, T.: Fast exact string matching algorithms. Inf. Process. Lett. 102(6), 229–235 (2007)
8. Navarro, G., Raffinot, M.: A Bit-Parallel approach to Suffix Automata: Fast Extended String Matching. In: Farach-Colton, M. (ed.) CPM 1998. LNCS, vol. 1448, pp. 14–33. Springer, Heidelberg (1998)
9. Peltola, H., Tarhio, J.: Alternative Algorithms for Bit-Parallel String Matching. In: Nascimento, M.A., de Moura, E.S., Oliveira, A.L. (eds.) SPIRE 2003. LNCS, vol. 2857, pp. 80–93. Springer, Heidelberg (2003)
10. Wu, S., Manber, U.: A fast algorithm for multi-pattern searching. Report TR-94-17, Depart. of Computer Science, University of Arizona, Tucson, AZ (1994)

P(l)aying for Synchronization*

Fedor Fominykh and Mikhail Volkov

Institute of Mathematics and Computer Science,
Ural Federal University, 620000 Ekaterinburg, Russia
FedorFo@yandex.ru, Mikhail.Volkov@usu.ru

Abstract. Two topics are presented: synchronization games and synchronization costs. In a synchronization game on a deterministic finite automaton, there are two players, Alice and Bob, whose moves alternate. Alice wants to synchronize the given automaton, while Bob aims to make her task as hard as possible. We answer a few natural questions related to such games. Speaking about synchronization costs, we consider deterministic automata in which each transition has a certain price. The problem is whether or not a given automaton can be synchronized within a given budget. We determine the complexity of this problem.

1 Introduction and Overview

A complete deterministic finite automaton (DFA) $\mathscr{A} = (Q, \Sigma)$ (here and below Q stands for the state set and Σ for the input alphabet) is called *synchronizing* if there exists a word $w \in \Sigma^*$ whose action brings $\mathscr{A}$ to one particular state no matter at which state w is applied: $q \cdot w = q' \cdot w$ for all $q, q' \in Q$. Any word w with this property is said to be a *reset* word for the automaton.

Synchronizing automata serve as transparent and natural models of error-resistant systems in many applications (coding theory, robotics, testing of reactive systems) and reveal interesting connections with symbolic dynamics, substitution systems and other parts of mathematics. The literature on synchronizing automata and their applications is rapidly growing so that even the most recent surveys [12,15] are becoming obsolete. A majority of research in the area focuses on the so-called Černý conjecture but the theory of synchronizing automata also offers many other interesting questions. In the present paper we introduce two new directions of the theory and obtain some initial results in these directions.

Section 2 concerns with synchronization games on DFAs. In such a game on a DFA $\mathscr{A}$, there are two players, Alice (Synchronizer) and Bob (Desynchronizer), whose moves alternate. Alice who plays first wants to synchronize $\mathscr{A}$, while Bob aims to prevent synchronization or, if synchronization is unavoidable, to delay it as long as possible. Provided that both players play optimally, the outcome of such a game depends only on the underlying automaton so studying synchronization games may be considered as a way to study synchronizing automata. The most natural questions here are the following. Given a DFA $\mathscr{A}$, how to

---

* Supported by the Russian Foundation for Basic Research, grant 10-01-00793.

decide who wins in the synchronization game on $\mathscr{A}$? If Alice wins, how many moves may she need in the worst case, in particular, is there a polynomial of n that bounds from above the number of moves in any game on a DFA with n states for which Alice has a winning strategy? It turns out that these questions can be answered by applying more or less standard techniques. This may be a bit disappointing but as a byproduct, we reveal a somewhat unexpected relation between synchronization games and a version of the Černý conjecture.

In Section 3 we consider weighted automata. A *deterministic weighted automaton* (DWA) is a DFA $\mathscr{A} = (Q, \Sigma)$ endowed with a function $\gamma : Q \times \Sigma \to \mathbb{Z}_+$ where $\mathbb{Z}_+$ stands for the set of all positive integers. In other words, each transition of a DWA has a certain price being a positive integer. Then every computation performed by $\mathscr{A}$ also gets a certain cost, namely, the sum of the costs of the transitions involved. If a DWA happens to be synchronizing and $w \in \Sigma^*$ is its reset word, then one can assign to w a cost measured, say, by the maximum among all costs of applying the word w at a state in Q. While in the non-weighted case one is usually interested in minimizing synchronization time, that is, the length of reset words, in the weighted case it is quite natural to minimize synchronization costs. A basic problem here is to determine, whether or not a given DWA can be synchronized within a given budget $B \in \mathbb{Z}_+$, in other words, whether or not $\mathscr{A}$ admits a reset word whose cost does not exceed B. We demonstrate that this problem is PSPACE-complete.

Besides initial questions discussed in this paper, each of the two outlined research directions leads to several intriguing open problems. We present and briefly discuss three such problems in Section 4.

2   Playing for Synchronization

The idea to consider synchronization as a game has independently arisen in [1] and [3]. In [1] a one-player game has been used to prove a lower bound on the minimum length of reset words for a certain series of 'slowly' synchronizing automata. In [3] a specific synchronization process arising in software testing has been analyzed in terms of a two-player game. The game that we consider here basically follows the model of [1] but is a two-player game as in [3].

Now we describe the rules of our synchronization game. It is played by two players, Alice and Bob say, on an arbitrary but fixed DFA $\mathscr{A} = (Q, \Sigma)$. In the initial position each state in Q holds a coin but, as the game progresses, some coins may be removed. The game is won by Alice when all but one coins are removed. Bob wins if he can keep at least two coins unremoved indefinitely long.

Alice moves first, then players alternate moves. The player whose turn it is to move proceeds by selecting a letter $a \in \Sigma$. Then, for each state $q \in Q$ that held a coin before the move, the coin advances to the state $q \cdot a$. (In the standard graphical representation of $\mathscr{A}$ as the labelled digraph with Q as the vertex set and the labelled edges of the form $q \xrightarrow{a} q \cdot a$, one can visualize the move as follows: all coins simultaneously slide along the edges labelled a.) If after this several coins happen to arrive at the same state, all of them but one are removed so that when the move is completed, each state holds at most one coin.

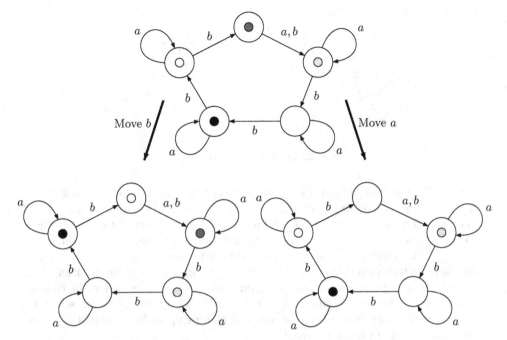

Fig. 1. Moves in a synchronization game

Fig. 1 illustrates the rules. Its upper part shows a typical position in a game on a 5-state automaton with 2 input letters a and b. The left lower part shows the effect of the move b while the right lower part demonstrates the result of the move a. Observe that in the latter case the dark-gray coin has been removed because it and the light-gray coin had arrived at the same state.

Let $a_1, a_2, \ldots, a_k$ with $a_i \in \Sigma$ be a sequence of moves in the synchronization game on $\mathscr{A} = (Q, \Sigma)$ and let $w = a_1 a_2 \cdots a_k$. It is easy to see that the set of states holding coins after this sequence of moves coincides with the image of Q under the action of the word w. Thus, sequences of moves that lead to Alice's win correspond precisely to reset words for $\mathscr{A}$. Therefore Bob wins on each DFA which is not synchronizing. Can he win on a synchronizing automaton? Yes, he can and, for instance, we show that Bob wins on synchronizing automata in the famous Černý series.

Černý [4] found for each $n > 1$ a synchronizing automaton $\mathscr{C}_n$ with n states and 2 input letters whose shortest reset word has length $(n-1)^2$. The states of $\mathscr{C}_n$ are the residues modulo n and the input letters a and b act as follows:

$$m \cdot a = \begin{cases} 1 & \text{for } m = 0, \\ m & \text{for } 1 \leq m < n; \end{cases} \qquad m \cdot b = m + 1 \pmod{n}.$$

The automaton is shown in Fig. 2.

Example 1. For each $n > 3$, Bob has a winning strategy in the synchronization game on $\mathscr{C}_n$.

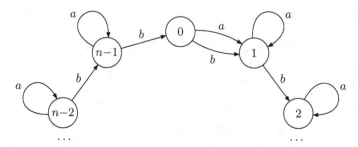

Fig. 2. The Černý automaton $\mathscr{C}_n$

Proof. Observe that in $\mathscr{C}_n$, the only state where two coins can meet is the state 1; moreover, this can happen only provided the move a has been played and before the move the states 0 and 1 both held a coin. We may assume for certainty that in this situation it is the coin arriving from 0 that is removed after the move.

Under this convention, the winning strategy for Bob is as follows. Bob only has to trace the coins that cover the states $n-1$ and 1 in the initial position. For his moves he must always select the letter a except two cases: when the chosen coins cover either $n-2$ and 0 or 0 and 2 in which cases Bob must select b. This way Bob can always keep the coins two steps apart from each other thus preventing them of being removed.

On the other hand, it is easy to find DFAs on which Alice has a winning strategy. For instance, a DFA $\mathscr{A}$ is called *definite* in [9] if there exists an $n > 0$ such that every input word of length at least n is a reset word for $\mathscr{A}$. Clearly, on each definite automaton, Alice always wins by selecting her moves at random.

The rules of our game readily guarantee that, given a DFA $\mathscr{A} = (Q, \Sigma)$, one of the players must have a winning strategy in the synchronization game on $\mathscr{A}$. If Alice has a winning strategy, consider a shortest winning sequence of her moves. Then it is clear that each move in this sequence creates a position that could not have appeared after an earlier move. However, the number of possible positions of the game does not exceed $2^{|Q|} - 1$ since each position is specified by the subset of states that currently hold coins. Therefore, if Alice has a winning strategy, she should be able to win after less than $2^{|Q|}$ moves. Thus, one can decide which player has a winning strategy in the game on $\mathscr{A}$ by an exhaustive search through all $|\Sigma|^{2^{|Q|+1}}$ words of length $2^{|Q|+1}$ over Σ in which letters in the odd positions are Alice's moves and ones in the even positions are Bob's replies. Of course, this brute force procedure is extremely inefficient and it is natural to ask whether an efficient—say, polynomial in the number of states—algorithm exists. A positive answer can be deduced from the next observation.

Lemma 1. *Alice has a winning strategy in the synchronization game on a DFA if and only if she has a winning strategy in every position in which only two states of the DFA hold coins.*

Proof. If Alice has no winning strategy for a position P with two coins, C and C' say, then Bob has a winning strategy for P. If Bob plays in the initial position according to this strategy, that is, selects his moves only on the basis of the location of the coins C and C', as if there were no other coins, the two coins persist forever so that Alice loses the game. (Here we assume that whenever one of the coins C and C' meets some third coin on some state in the course of the game, then it is this third coin that gets removed.)

Conversely, if Alice can win in every position in which only two states hold coins, she can use the following strategy. In the initial position she chooses a pair of coins, C and C' say, and plays as if there were no other coins, that is, she applies her winning strategy for the position in which C and C' cover the same states as they do in the initial position and all other coins are removed. This brings the game to a position in which either C or C' is removed. Then Alice chooses another pair of coins and again plays as if these were the only coins, and so on. Since at least one coin is removed in each round, Alice eventually wins.

Observe that Lemma 1 implies a cubic (in the number n of states of the underlying DFA) upper bound on the number of moves in any game that Alice wins. Indeed, suppose she uses the strategy just described and works with a pair of coins C and C'. Let q_i and q'_i be the states holding the coins C and C' after the ith move of Alice. Then if Alice plays optimally, we must have $\{q_i, q'_i\} \neq \{q_{i+j}, q'_{i+j}\}$ whenever $j > 0$. Indeed, the equality $\{q_i, q'_i\} = \{q_{i+j}, q'_{i+j}\}$ means that wherever Alice moves C and C' by her $(i+1)^{th}$, ..., $(i+j-1)^{th}$ moves, Bob can force Alice to return the coins by her $(i+j)^{th}$ move to the same states that the coins occupied after her i^{th} move. Then Bob can force Alice to return C and C' to the same states also by her $(i+2j)^{th}$, $(i+3j)^{th}$, ... moves, whence none of the two coins can ever be removed, a contradiction.

Hence the number of Alice's moves in any round in which she works with any fixed pair of coins does not exceed $\binom{n}{2}$. Moreover, in every synchronizing automaton there exist states q and q' such that $q \cdot a = q' \cdot a$ for some letter a. Therefore Alice can remove one coin by her first move. After that she needs at most $n-2$ rounds to remove $n-2$ of the remaining $n-1$ coins. We thus obtain:

Corollary 1. *If Alice has a winning strategy in the synchronization game on a DFA with n states, she can win in at most $\binom{n}{2}(n-2)+1$ moves.*

Now we return to the decidability question.

Theorem 1. *Let $\mathscr{A} = (Q, \Sigma)$ be a DFA with $|Q| = n$ and $|\Sigma| = k$. There exists an algorithm that in $O(n^2 k)$ time decides who has a winning strategy in the synchronization game on $\mathscr{A}$.*

Proof. We describe the algorithm rather informally. First we construct a new DFA $\mathscr{P} = (P \times \{0,1\} \cup \{s\}, \Sigma)$ where P is the set of all positions with two coins (each such position is specified by a couple of states holding coins) and s is an extra state. The action of the letters is defined as follows: all letters fix s and if $p \in P$ is the position in which two states $q, q' \in Q$ hold coins, $x \in \{0,1\}$, and $a \in \Sigma$, then

$$(p, x) \cdot a = \begin{cases} (p', 1-x) & \text{if } q \cdot a \neq q' \cdot a, \\ s & \text{otherwise,} \end{cases}$$

where p' is the position in which $q \cdot a$ and $q' \cdot a$ hold coins. Thus, the automaton $\mathscr{P}$ encodes 'transcripts' of all games starting in positions in P; the extra bit x controls whose turn it is to move: Alice moves if $x = 0$ and Bob moves if $x = 1$. Clearly, $\mathscr{P}$ has $n^2 - n + 1$ states and $k(n^2 - n + 1)$ edges (transitions).

We mark the state s and then recursively propagate the marking to $P \times \{0, 1\}$: a state of the form $(p, 0)$ is marked if and only if there is an $a \in \Sigma$ such that $(p, 0) \cdot a$ is marked and a state of the form $(p, 1)$ is marked if and only if for all $a \in \Sigma$ the states $(p, 1) \cdot a$ are marked. Clearly, the marking can be done by a breadth-first search in the underlying digraph of $\mathscr{P}$ with all edges reversed. The well known time estimate for breadth-first search in a graph with v vertices and e edges is $O(v + e)$, see, e.g., [5, Section 22.2], whence we conclude that the marking can be completed in $O(n^2 k)$ time. It follows from the construction of $\mathscr{P}$ and from the marking rules that Alice can win in the game starting at a position $p \in P$ if and only if the state $(p, 0)$ is marked. This and Lemma 1 readily imply that Alice has a winning strategy in the game on $\mathscr{A}$ if and only if all states of the form $(p, 0)$ get marked (or, equivalently, all states of $\mathscr{P}$ get marked).

Though Corollary 1 and Theorem 1 are worth being registered (as they answer the most natural questions related to synchronization games), the reader acquainted with the theory of synchronizing automata immediately realizes that these results closely follow some more or less standard patterns. Now we proceed with a more original contribution.

Suppose that Alice has a winning strategy in a synchronization game on an n-state DFA. Corollary 1 provides an cubic upper bound for the number of moves in the game. What about lower bounds? Our next result provides a transparent construction from which we can extract a quadratic lower bound.

Theorem 2. *Let $\mathscr{A} = (Q, \Sigma)$ be a synchronizing automaton with $|Q| = n$, $|\Sigma| \geq 2$ and let ℓ be the minimum length of reset words for $\mathscr{A}$. There exists a DFA $\mathscr{D}$ with $2n$ states such that Alice wins in the synchronization game on $\mathscr{D}$ but needs at least ℓ moves for this.*

Proof. We fix a letter $b \in \Sigma$ and a state $q_0 \in Q$. Now let $\mathscr{D} = (Q \times \{0, 1\}, \Sigma)$ where for each $q \in Q$ the action of an arbitrary letter $a \in \Sigma$ is defined as follows:

$$(q, 0) \cdot a = (q \cdot a, 1), \qquad (q, 1) \cdot a = \begin{cases} (q, 0) & \text{if } a = b, \\ (q_0, 1) & \text{otherwise.} \end{cases}$$

We call $\mathscr{D}$ the *duplication* of $\mathscr{A}$. Fig. 3 shows the duplication of the Černý automaton $\mathscr{C}_n$ from Fig. 2 (with the state 0 in the role of q_0).

Suppose that Alice opens the game by selecting a letter $a \neq b$. After that only states of the form $(q, 1)$ hold coins. Bob must reply with the move b since he loses immediately otherwise. After that coins cover the states $(q, 0)$ with $q \in Q \cdot a \cup \{q_0\}$. Now if Alice spells out a reset word for $\mathscr{A}$, she wins. Indeed, as soon as Bob selects a letter different from b, he loses immediately, and if he replies with b to all Alice's moves, each pair (Alice's move, Bob's move) has the same effect as applying the letter selected by Alice in the DFA $\mathscr{A}$.

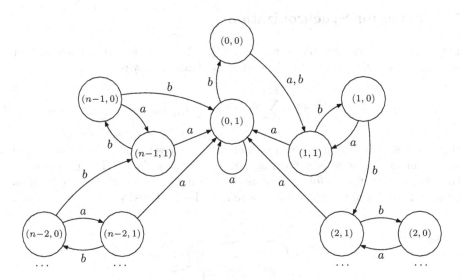

Fig. 3. The duplication of the automaton $\mathscr{C}_n$

On the other hand, Alice needs at least ℓ moves to win if Bob replies with b to each of her moves. Indeed, if Bob plays this way and a winning sequence of Alice's moves forms a word $w \in \Sigma^*$, then after the last move of the sequence every state $(q, 1)$ with $q \in Q \cdot w$ still holds a coin. Thus, for Alice to win, w must be a reset word for $\mathscr{A}$, whence the length of w is at least ℓ.

We denote by $\mathscr{D}_n$ the duplication of the Černý automaton $\mathscr{C}_n$. Combining Theorem 2 and the fact that the minimum length of reset words for $\mathscr{C}_n$ is $(n-1)^2$, we obtain that Alice needs at least $(n-1)^2$ moves to win on $\mathscr{D}_n$. (In fact, the exact number of moves needed is easily seen to be $(n-1)^2 + 1$.) Thus, we have found a series of k-state DFAs ($k = 2n$ is even) on which Alice's win requires a quadratic in k number of moves. A similar series can be constructed for odd k: we can just add an extra state to $\mathscr{D}_n$ and let both a and b send this added state to the state $(q_0, 1)$.

We notice that the duplication of an arbitrary DFA belongs to a very special class of synchronizing automata as it can be reset by a word of length 2. A somewhat unexpected though immediate consequence of Theorem 2 is that a progress in understanding synchronization games within this specific class may lead to a solution of a major problem in the theory of synchronizing automata.

Corollary 2. *If for every n-state synchronizing automaton with a reset word of length 2 on which Alice can win, she has a winning strategy with $O(n^2)$ moves, then every n-state synchronizing automaton has a reset word of length $O(n^2)$.*

Recall that all known results on synchronization of n-state DFAs (see [10] and [14] for the best bounds) guarantee only the existence of reset words of length $\Omega(n^3)$.

3   Paying for Synchronization

Let $\mathscr{A} = (Q, \Sigma, \gamma)$ be a DWA, where $\gamma : Q \times \Sigma \to \mathbb{Z}_+$ is a cost function. For $w = a_1 \cdots a_k \in \Sigma^*$ and $q \in Q$, the cost of applying w at q is

$$\gamma(q, w) = \sum_{i=0}^{k-1} \gamma\big(q \cdot (a_1 \cdots a_i), a_{i+1}\big).$$

If $\mathscr{A}$ is a synchronizing automaton and w is its reset word, then the cost of synchronizing $\mathscr{A}$ by w is defined as $\gamma(w) = \max_{q \in Q} \gamma(q, w)$. The intuition for this choice of $\gamma(w)$ is as follows: we use w to bring $\mathscr{A}$ to a certain state from an unknown state, and therefore, we have to take the most costly case into account[1].

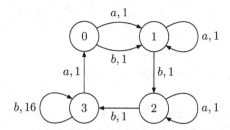

Fig. 4. A deterministic weighted automaton

Fig. 4 shows a DWA (transition costs are included in the labels) and illustrates the difference between two optimization problems: minimizing synchronization cost and minimizing the length of reset words. The shortest reset word for the DWA is b^3 but the cost of synchronizing by b^3 is 48. On the other hand, the longer word $a^2 baba^2$ manages to avoid the 'expensive' loop at the state 3 whence the cost of synchronizing by $a^2 baba^2$ is only 7.

We study in the computational complexity of the following decision problem:

SYNCHRONIZING ON BUDGET: *Given a DWA $\mathscr{A} = (Q, \Sigma, \gamma)$ and a positive integer B, is it true that $\mathscr{A}$ has a reset word w with $\gamma(w) \le B$?*

Here we assume that the values of γ and the number B are given in binary. (The unary version of SYNCHRONIZING ON BUDGET can be easily shown to be NP-complete on the basis of the NP-completeness of the problem SHORT RESET WORD [11,6]: given a DFA $\mathscr{A}$ and a positive integer ℓ, is it true that $\mathscr{A}$ has a reset word of length ℓ?)

---

[1] Of course, in some situations other definitions of the cost of synchronization may make sense. For instance, if we treat synchronization in the flavor of Section 2, that is, as the process of moving coins initially placed on all states in Q to a certain state, it is natural to define the cost of the process as $\sum_{q \in Q} \gamma(q, w)$. The results that follow can be adapted to this setting mutatis mutandis.

Theorem 3. SYNCHRONIZING ON BUDGET *is PSPACE-complete.*

Proof. By Savitch's theorem [8, Section 4.3], in order to show that SYNCHRO-NIZING ON BUDGET lies in the class PSPACE, it suffices to solve this problem in polynomial space by a non-deterministic algorithm. A small difficulty is that for some instances $(Q, \Sigma, \gamma; B)$ of SYNCHRONIZING ON BUDGET, every reset word w satisfying $\gamma(w) \leq B$ may be exponentially long in $|Q|$ and so even if our algorithm correctly guesses such a w, it would not have enough space to store its guess. To bypass the difficulty, the algorithm should guess w letter by letter. It guesses the first letter of w (say, a), applies a at every state $q \in Q$ and saves two arrays: $\{q \cdot a\}$ and $\{\gamma(q, a)\}$. Each of the arrays clearly requires only poly-nomial space. Then the algorithm guesses the second letter of w and updates both arrays, etc. At the end of the guessing steps the algorithm check whether all entries of the first array are equal (if so, then w is indeed a reset word for (Q, Σ)) and whether the maximum number in the second array is less than or equal to B (if so, then synchronization is indeed achieved within the budget B).

To show that SYNCHRONIZING ON BUDGET is PSPACE-complete, we use a re-duction from a problem concerning partial automata. A *partial* finite automaton (PFA) is a pair $\mathscr{A} = (Q, \Sigma)$, where Q is the state set and Σ is the input alphabet whose letters act on Q as partial transformations. Such a PFA is said to be *carefully synchronizing* if there exists $w = a_1 \cdots a_\ell$ with $a_1, \ldots, a_\ell \in \Sigma$ such that $q \cdot a_i$ with $1 \leq i \leq \ell$ is defined for all $q \in Q \cdot (a_1 \cdots a_{i-1})$ and $|Q \cdot w| = 1$. Every word w with these properties is called a *careful reset word* for $\mathscr{P}$. Informally, a careful reset word synchronizes $\mathscr{A}$ and manages to avoid any undefined transition.

Martyugin [7] has recently proved that the next problem is PSPACE-complete:

CAREFUL SYNCHRONIZATION: *Is a given PFA carefully synchronizing?*

It is the problem that we reduce to SYNCHRONIZING ON BUDGET. Our reduction relies on a known fact whose proof is included for the reader's convenience.

Lemma 2. *The minimum length of careful reset words for carefully synchroniz-ing PFAs with n states does not exceed $2^n - n - 1$.*

Proof. Given a PFA $\mathscr{A} = (Q, \Sigma)$ with $|Q| = n$, consider the set of the non-empty subsets of Q and let each $a \in \Sigma$ act on $P \subseteq Q$ as follows:

$$P \cdot a = \begin{cases} \{q \cdot a \mid a \in \Sigma\} & \text{provided } q \cdot a \text{ is defined for all } q \in P, \\ \text{undefined} & \text{otherwise.} \end{cases}$$

We obtain a new PFA $\mathscr{P}$, and it is clear that $w \in \Sigma^*$ is a careful reset word for $\mathscr{A}$ if and only if w labels a path in $\mathscr{P}$ starting at Q and ending at a singleton. A path of minimum length does not visit any state of $\mathscr{P}$ twice and stops as soon as it reaches a singleton. Hence the length of the path does not exceed the number of non-empty and non-singleton subsets of Q, that is, $2^n - n - 1$.

Now take an arbitrary instance of CAREFUL SYNCHRONIZATION, that is, a PFA $\mathscr{A} = (Q, \Sigma)$. We assign to $\mathscr{A}$ an instance of SYNCHRONIZING ON BUDGET as follows. First, extend the action of each letter $a \in \Sigma$ to the whole set Q letting

$$q \odot a = \begin{cases} q \cdot a & \text{if } q \cdot a \text{ is defined in } \mathscr{A}, \\ q & \text{otherwise.} \end{cases}$$

These extended actions give rise to a DFA $\mathscr{A}'$ with the same state set Q and input alphabet Σ. Further, let $|Q| = n$, and define $\gamma : Q \times \Sigma \to \mathbb{Z}_+$ by the rule:

$$\gamma(q, a) = \begin{cases} 1 & \text{if } q \cdot a \text{ is defined in } \mathscr{A}, \\ 2^n & \text{otherwise.} \end{cases}$$

This makes $\mathscr{A}'$ a DWA. The construction is illustrated by Fig. 5. Finally, let $B = 2^n - 1$. Observe that the binary presentations of B and of the values of γ are of a linear in n size so that the construction requires only polynomial time in the size of the PFA $\mathscr{A}$.

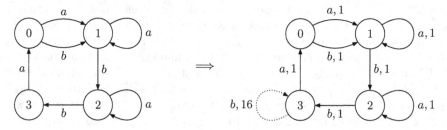

Fig. 5. Transforming a partial automaton into a weighted automaton

We aim to show that the PFA $\mathscr{A}$ is carefully synchronizing if and only if the DWA $\mathscr{A}'$ can be synchronized within the budget B. Indeed, if w is a careful reset word for $\mathscr{A}$, then w can be applied to every state in $\mathscr{A}$. This implies that w labels the same paths in $\mathscr{A}'$ as it does in $\mathscr{A}$ whence w synchronizes $\mathscr{A}'$ and involves only transitions with cost 1. Therefore $\gamma(q, w)$ is equal to the length of w for each $q \in Q$ and so is $\gamma(w) = \max_{q \in Q} \gamma(q, w)$. By Lemma 2 w can be chosen to be of length at most $2^n - n - 1$, whence $\gamma(w) \leq 2^n - n - 1 < 2^n - 1 = B$. Conversely, if w is a reset word for $\mathscr{A}'$ with $\gamma(w) \leq B$, then $\gamma(q, w) \leq 2^n - 1$ for each $q \in Q$, whence no path labelled w and starting at q involves any transition with cost 2^n. This means every transition in such a path is induced by a transition with the same effect defined in $\mathscr{A}$. Therefore w can be applied to every state in $\mathscr{A}$. Since all paths labelled w are coterminal in $\mathscr{A}'$, they have the same property in $\mathscr{A}$ and w is a careful reset word for $\mathscr{A}$.

4   Open Problems

Complexity of synchronization games. In the context of synchronization games, many natural complexity-theoretical questions arise. As an example, consider the following decision problem in the flavor of the problem SHORT RESET WORD mentioned in Section 3:

SHORT SYNCHROGAME: *given a DFA $\mathscr{A}$ and a positive integer ℓ, is it true that Alice can win the synchronization game on $\mathscr{A}$ in at most ℓ moves?*

Problem 1. Find the computational complexity of SHORT SYNCHROGAME.

One can use the duplication construction from the proof of Theorem 2 to reduce SHORT RESET WORD to SHORT SYNCHROGAME. This implies that the latter problem is NP-hard. However, it is not likely that SHORT SYNCHROGAME lies in NP, and moreover, we suspect that the problem is PSPACE-complete.

Road Coloring Games. A digraph G in which each vertex has the same out-degree k is called a *digraph of out-degree k*. If we take an alphabet Σ of size k, then we can label the edges of such G by letters of Σ such that the resulting automaton will be complete and deterministic. Any DFA obtained this way is referred to as a *coloring* of G.

The famous Road Coloring Problem asked for necessary and sufficient conditions on a digraph G to admit a synchronizing coloring. The problem has been recently solved by Trahtman [13] and the solution implies that if G has a synchronizing coloring, then such a coloring can be found in $O(n^2k)$ time where n is the number of vertices and k is the out-degree of G, see [2].

Now consider the following *Rood Coloring game*. Alice and Bob alternately label the edges of a given digraph G of out-degree k by letters from an alphabet Σ of size k (observing the rule that no edges leaving the same vertex may get the same label) until G becomes a DFA. Alice who plays first wins if the resulting DFA is synchronizing, and Bob wins otherwise.

Problem 2. Is there an algorithm that, given a digraph G of constant out-degree, decides in polynomial in the size of G time which player has a winning strategy in the Road Coloring game on G?

Observe that there are digraphs on which Alice wins by making random moves (for instance, the underlying digraphs of the automata in the Černý series can be shown to have this property); on the other hand, Bob can win on some digraphs admitting synchronizing colorings.

Synchronization Games on Weighted Automata. As a synthesis of the two topics of this paper, one can consider synchronization games on DWAs where the aim of Alice is to minimize synchronization costs while Bob aims to prevent synchronization or at least to maximize synchronization costs[2]. In particular, we suggest to investigate the following problem that can be viewed as a common generalization of SHORT SYNCHROGAME and SYNCHRONIZING ON BUDGET.

SYNCHROGAME ON BUDGET: *Given a DWA $\mathscr{A} = (Q, \Sigma, \gamma)$ and a positive integer B, is it true that Alice can win the synchronization game on $\mathscr{A}$ with a sequence w of moves satisfying $\gamma(w) \leq B$?*

Problem 3. Find the computational complexity of SYNCHROGAME ON BUDGET.

---

[2] Such games may resemble the current economic games within the European Union.

Acknowledgement. The authors are grateful to the anonymous referees for their remarks and a number of useful suggestions that have been incorporated in the present version of the paper.

References

1. Ananichev, D.S., Volkov, M.V., Zaks, Y.I.: Synchronizing automata with a letter of deficiency 2. Theor. Comput. Sci. 376, 30–41 (2007)
2. Béal, M.-P., Perrin, D.: A quadratic algorithm for road coloring. Technical report, Université Paris-Est (2008), http://arxiv.org/abs/0803.0726
3. Blass, A., Gurevich, Y., Nachmanson, L., Veanes, M.: Play to Test. In: Grieskamp, W., Weise, C. (eds.) FATES 2005. LNCS, vol. 3997, pp. 32–46. Springer, Heidelberg (2006)
4. Černý, J.: Poznámka k homogénnym eksperimentom s konečnými automatami. Matematicko-fyzikalny Časopis Slovensk. Akad. Vied 14(3), 208–216 (1964) (in Slovak)
5. Cormen, T.H., Leiserson, C.E., Rivest, R.L., Stein, C.: Introduction to Algorithms, 3rd edn. MIT Press, McGraw-Hill, Cambridge (2009)
6. Eppstein, D.: Reset sequences for monotonic automata. SIAM J. Comput. 19, 500–510 (1990)
7. Martyugin, P.V.: Complexity of Problems Concerning Carefully Synchronizing Words for PFA and Directing Words for NFA. In: Ablayev, F., Mayr, E.W. (eds.) CSR 2010. LNCS, vol. 6072, pp. 288–302. Springer, Heidelberg (2010)
8. Papadimitriou, C.H.: Computational Complexity. Addison-Wesley (1994)
9. Perles, M., Rabin, M.O., Shamir, E.: The theory of definite automata. IEEE Trans. Electronic Comput. 12, 233–243 (1963)
10. Pin, J.-E.: On two combinatorial problems arising from automata theory. Ann. Discrete Math. 17, 535–548 (1983)
11. Rystsov, I.K.: On minimizing length of synchronizing words for finite automata. In: Theory of Designing of Computing Systems, pp. 75–82. Institute of Cybernetics of Ukrainian Acad. Sci. (1980) (in Russian)
12. Sandberg, S.: Homing and Synchronizing Sequences. In: Broy, M., Jonsson, B., Katoen, J.-P., Leucker, M., Pretschner, A. (eds.) Model-Based Testing of Reactive Systems. LNCS, vol. 3472, pp. 5–33. Springer, Heidelberg (2005)
13. Trahtman, A.: The Road Coloring Problem. Israel J. Math. 172(1), 51–60 (2009)
14. Trahtman, A.N.: Modifying the Upper Bound on the Length of Minimal Synchronizing Word. In: Owe, O., Steffen, M., Telle, J.A. (eds.) FCT 2011. LNCS, vol. 6914, pp. 173–180. Springer, Heidelberg (2011)
15. Volkov, M.V.: Synchronizing Automata and the Černý Conjecture. In: Martín-Vide, C., Otto, F., Fernau, H. (eds.) LATA 2008. LNCS, vol. 5196, pp. 11–27. Springer, Heidelberg (2008)

Synchronizing Automata of Bounded Rank[*]

Vladimir V. Gusev

Institute of Mathematics and Computer Science,
Ural Federal University, 620083 Ekaterinburg, Russia
vl.gusev@gmail.com

Abstract. We reduce the problem of synchronization of an n-state automaton with letters of rank at most $r < n$ to the problem of synchronization of an r-state automaton with constraints given by a regular language. Using this technique we construct a series of synchronizing n-state automata in which every letter has rank $r < n$ and whose reset threshold is at least $r^2 - r - 1$ Moreover, if $r > \frac{n}{2}$, such automata are strongly connected.

1   Introduction

A complete deterministic finite automaton $\mathscr{A}$ is called *synchronizing* if the action of some word w resets $\mathscr{A}$, that is, leaves the automaton in one particular state no matter at which state w is applied. Any such word w is said to be a *reset word* for the automaton. The minimum length of reset words for $\mathscr{A}$ is called the *reset threshold* of $\mathscr{A}$ and is denoted by rt($\mathscr{A}$). Synchronizing automata often serve as natural models of error-resistant systems. For a brief introduction to the theory of synchronizing automata we refer the reader to the recent surveys [12,15]. The interest to the field is heated by the famous Černý conjecture.

In 1964 Jan Černý [5] constructed for each $n > 1$ a synchronizing automaton $\mathscr{C}_n$ with n states whose reset threshold is equal to $(n-1)^2$. Soon after that he conjectured that these automata represent the worst possible case. In other words, every synchronizing automaton with n states can be synchronized by a word of length $(n-1)^2$. Despite intensive research, the best upper bound on the reset threshold of synchronizing automata with n states achieved so far is cubic. Classical bound is $\frac{n^3-n}{6}$, see [10]. A slightly better bound $\frac{n(7n^2+6n-16)}{48}$ has been claimed in [14]. Though the Černý conjecture is open in general, it has been confirmed for some restricted classes of synchronizing automata, see [1,6,8,13,16]. For some classes quadratic upper bound is established, see [4,11].

In the present paper we approach the following question: how does the reset threshold of an automaton $\mathscr{A}$ depend on the ranks of letters in $\mathscr{A}$? More formally, consider a synchronizing n-state automaton $\mathscr{A}$ with k letters and let $r_1, r_2, \ldots, r_k$ be their ranks. What is the upper bound on the reset threshold of $\mathscr{A}$ in terms of

---

[*] Supported by the Russian Foundation for Basic Research, grant 10-01-00793, and by the Presidential Program for young researchers, grant MK-266.2012.1. Author is also grateful to Erasmus Mundus Action 2 Partnerships — Triple I.

N. Moreira and R. Reis (Eds.): CIAA 2012, LNCS 7381, pp. 171–179, 2012.
© Springer-Verlag Berlin Heidelberg 2012

n and $r_1, r_2, \ldots, r_k$? In full generality, this question appears to be very hard. In this paper we focus on examples with "large" reset threshold and bounded rank. There have already been some related results in the literature. Synchronizing automata with a letter of deficiency 2 were considered in [3] while in [2] a series of slowly synchronizing automata in which all letters deficiency 2 was reported.

We say that an automaton $\mathscr{A}$ is *of bounded rank* r if every letter of $\mathscr{A}$ has rank at most r. We present a uniform approach to deal with such automata. We illustrate this approach by constructing a series of automata of bounded rank r with reset threshold at least $r^2 - r - 1$.

2   Preliminaries

Let $\mathscr{A} = (Q, \Sigma)$ be a complete deterministic finite automaton. Here Σ is an *input alphabet* and Q is a *set of states*. For an arbitrary state $q \in Q$ and a letter $a_\ell \in \Sigma$ we denote by $q \cdot a_\ell$ the image of the state q under the action of the letter a_ℓ. Let $S \cdot a_\ell = \bigcup_{q \in S} q \cdot a_\ell$ for any $S \subseteq Q$. The *rank* of a letter a_ℓ denoted by $\mathrm{rk}(a_\ell)$ is the cardinality of the set $Q \cdot a_\ell$. The *defect* of a letter a_ℓ is equal to $|Q| - \mathrm{rk}(a_\ell)$, where $|Q|$ is the cardinality of Q.

Let us fix some orderings of $Q = \{q_1, q_2, \ldots q_n\}$ and $\Sigma = \{a_1, \ldots, a_k\}$. We associate with an arbitrary letter a_ℓ a square $(0, 1)$-matrix $M(a_\ell)$ of order n by the following rule: $M(a_\ell)[i, j] = 1$ if $q_i \cdot a_\ell = q_j$, otherwise $M(a_\ell)[i, j] = 0$. We call the set $\langle M(a_1), \ldots, M(a_k) \rangle$ the *matrix representation* of the automaton $\mathscr{A}$ and write $\mathscr{A} = \langle M(a_1), \ldots, M(a_k) \rangle$. We can uniquely extend the domain of mapping the $M(\cdot)$ from Σ to Σ^* in accordance with the equation $M(uv) = M(u)M(v)$, where $u, v \in \Sigma^*$. It is not hard to see the following important property of the matrix representation: for every word w we have $M(w)[i, j] = 1$ if $q_i \cdot w = q_j$, otherwise $M(w)[i, j] = 0$. This immediately implies that a word w is a reset word of $\mathscr{A}$ if and only if $M(w)$ has a column of 1's.

We will denote the rank of a matrix M by $\mathrm{rk}(M)$. Note that the rank of a letter a_ℓ and the rank of the corresponding matrix $M(a_\ell)$ are equal, i.e. $\mathrm{rk}(a_\ell) = \mathrm{rk}(M(a_\ell))$. Throughout the paper we also make use of the following notation. Let $\mathscr{M}$ be a set of square matrices. Then define $\mathscr{M}^k = \{ M_1 \cdot M_2 \cdot \ldots \cdot M_k \mid M_i \in \mathscr{M} \text{ for } 1 \le i \le k \}$ and $\mathscr{M}^* = \bigcup_{k=1}^{\infty} \mathscr{M}^k$.

Let us introduce the matrix representation of the Černý automaton $\mathscr{C}_n = \langle A_n, B_n \rangle$ of order n. We denote the i-th row of the matrix A_n by $A_n[i, .]$. Let e_i denote the row vector whose only non-zero entry is equal to 1 and is located in position i.

$$A_n[i, .] = \begin{cases} e_2, & \text{if } i = 1 \\ e_i, & \text{if } i \ne 1 \end{cases}, \quad B_n[i, .] = \begin{cases} e_{i+1}, & \text{if } i \ne n \\ e_1, & \text{if } i = n \end{cases}.$$

The matrix A_n corresponds to the action of the letter a, the matrix B_n corresponds to the action of the letter b. Later on we will need the following statement concerning reset words of $\mathscr{C}_n$ (Proposition 3 in [7]).

Proposition 1. *Every reset word of the automaton $\mathscr{C}_n$ contains at least $n^2 - 3n + 2$ occurrences of the letter b and at least $n - 1$ occurrences of the letter a.*

3   Main Results

For the sake of simplicity throughout this section we consider automata only over two letters. All definitions and propositions can be easily generalized. Consider an n-state automaton $\mathscr{A}$ given by a pair of matrices A and B. We say that the set of matrix pairs $\sigma = \langle (X, Y), (\Gamma, \Delta) \rangle$ is a *decomposition* of the automaton $\mathscr{A}$ if:

  (i) X, Γ are $(0, 1)$-matrices of size n by r
  (ii) Y, Δ are $(0, 1)$-matrices of size r by n
  (iii) $A = XY$, $B = \Gamma \Delta$
  (iv) every row of X, Y, Γ, Δ has only one occurrence of 1.

We say that decomposition is *unavoidable* if every column of X and Γ also contains at least one occurrence of 1.

The introduced definition implies that for any decomposition we have $r \geq max\{\text{rk}(A), \text{rk}(B)\}$. Moreover, for any such r we can easily construct some decomposition. For example, matrix Y may consist of different rows of A in lexicographic order (or any other order). The number of different rows is exactly $\text{rk}(A)$. Thus, if $r > \text{rk}(A)$, then the lexicographically largest row appears in Y several times. Let j be the position of the i-th row of A in lexicographic order. Then the i-th row of X is equal to e_j. It is not hard to see that defining in the same way matrices Γ and Δ gives us a decomposition of $\mathscr{A}$. Moreover, if $r = \text{rk}(A) = \text{rk}(B)$, then this decomposition is unavoidable. We notice that even in case $r = \text{rk}(A) = \text{rk}(B)$ there could exist several different decompositions.

Suppose now that we are given a decomposition $\sigma = \langle (X, Y), (\Gamma, \Delta) \rangle$ of an automaton $\mathscr{A}$, where X has size n by r. We are to define the *reduced automaton* $\mathscr{A}_\sigma$ that is going to be a key object of this paper. The automaton $\mathscr{A}_\sigma$ has $\{1, \ldots, r\}$ as the set of states. The action of the input alphabet $\Sigma' = \{y\gamma, yx, \delta x, \delta\gamma\}$ (these are letters, but expressed as two symbols for convenience) is given by the corresponding $(0, 1)$-matrices $\mathcal{M} = \{Y\Gamma, YX, \Delta X, \Delta\Gamma\}$ of size r by r. We notice that every row of the matrices X, Y, Γ, Δ contains only one occurrence of 1, so their products have the same property. Thus automaton $\mathscr{A}_\sigma$ is deterministic and complete.

Example 1. Let $\mathscr{D}$ be the automaton shown in Fig. 1 on the left (this automaton first appeared in [2] as $\mathscr{D}_5''$). We can define two decompositions of $\mathscr{D}$ as follows:

$$\sigma_1: \quad A = \begin{pmatrix} 1&0&0&0 \\ 0&1&0&0 \\ 0&0&1&0 \\ 0&0&0&1 \\ 1&0&0&0 \end{pmatrix} \begin{pmatrix} 0&1&0&0&0 \\ 0&0&1&0&0 \\ 0&0&0&1&0 \\ 0&0&0&0&1 \end{pmatrix}, \quad B = \begin{pmatrix} 1&0&0&0 \\ 1&0&0&0 \\ 0&1&0&0 \\ 0&0&1&0 \\ 0&0&0&1 \end{pmatrix} \begin{pmatrix} 0&0&1&0&0 \\ 0&0&0&1&0 \\ 0&0&0&0&1 \\ 1&0&0&0&0 \end{pmatrix};$$

$$\sigma_2: \quad A = \begin{pmatrix} 1&0&0&0 \\ 0&1&0&0 \\ 0&0&1&0 \\ 0&0&0&1 \\ 1&0&0&0 \end{pmatrix} \begin{pmatrix} 0&1&0&0&0 \\ 0&0&1&0&0 \\ 0&0&0&1&0 \\ 0&0&0&0&1 \end{pmatrix}, \quad B = \begin{pmatrix} 0&1&0&0 \\ 0&1&0&0 \\ 0&0&1&0 \\ 0&0&0&1 \\ 1&0&0&0 \end{pmatrix} \begin{pmatrix} 1&0&0&0&0 \\ 0&0&1&0&0 \\ 0&0&0&1&0 \\ 0&0&0&0&1 \end{pmatrix}.$$

The reduced automaton $\mathscr{D}_{\sigma_1}$ is shown in the middle of Fig. 1. The letter $y\gamma$ is omitted, since it acts as identity mapping. The action of the letter $\delta\gamma$ coincides

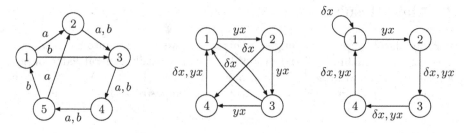

Fig. 1. Automata $\mathcal{D}$, $\mathcal{D}_{\sigma_1}$ and $\mathcal{D}_{\sigma_2}$

with the action of yx. The reduced automaton $\mathcal{D}_{\sigma_2}$ is shown in Fig. 1(right). Since the actions of letters $yx, y\gamma, \delta\gamma$ coincide, we keep only yx on the picture.

This example shows that reduced automata may be significantly different.

Recall that a *permutation matrix* is a square $(0,1)$-matrix with exactly one 1 in each row and exactly one 1 in each column. It is not hard to see that every permutation matrix P satisfies $PP^T = I$, where P^T is a transpose and I is the identity matrix. The following simple proposition gives a way to obtain several decompositions of a given automaton. The proof of this fact is straightforward.

Proposition 2. *Let $\mathscr{A}$ be an automaton with a decomposition $\langle (X, Y), (\Gamma, \Delta) \rangle$. Then for any pair of permutation matrices P and Q of appropriate size,*

$$\langle (XP, P^T Y), (\Gamma Q, Q^T \Delta) \rangle$$

is also a decomposition of $\mathscr{A}$.

The next propositions show the reason of our interest in the automaton $\mathscr{A}_\sigma$. It turns out that the reduced automaton inherits important properties of the original automaton $\mathscr{A}$.

Proposition 3. *Let $\mathscr{A}$ be a synchronizing automaton given by a pair of matrices A and B. Then for every decomposition $\sigma = \langle (X, Y), (\Gamma, \Delta) \rangle$ the reduced automaton $\mathscr{A}_\sigma$ is synchronizing and $\mathrm{rt}(\mathscr{A}_\sigma) \leq \mathrm{rt}(\mathscr{A}) + 1$. Moreover, if the decomposition σ is unavoidable, then $\mathrm{rt}(\mathscr{A}_\sigma) \leq \mathrm{rt}(\mathscr{A})$.*

Proof. Since $\mathscr{A}$ is synchronizing, there is a matrix $W \in \{A, B\}^{\mathrm{rt}(\mathscr{A})}$ such that it contains a column of 1's. We can represent W as a product of matrices X, Y, Γ, Δ in accordance with σ. Thus, $W = SW'T$, where $S \in \{X, \Gamma\}$, $T \in \{Y, \Delta\}$ and $W' \in \mathcal{M}^{\mathrm{rt}(\mathscr{A})-1}$. Observe that the matrix YWX is of order r and also contains a column of 1's. Moreover, it can be represented as a product of matrices in $\mathcal{M}$ of length $\mathrm{rt}(\mathscr{A}) + 1$. Hence, the automaton $\mathscr{A}_\sigma$ is synchronizing and its reset threshold is at most $\mathrm{rt}(\mathscr{A}) + 1$. In case of an unavoidable decomposition the matrix $W'T$ has to contain a column of 1's. Thus $W'TX \in \mathcal{M}^{\mathrm{rt}(\mathscr{A})}$ can play a role of reset word. So, the inequality $\mathrm{rt}(\mathscr{A}_\sigma) \leq \mathrm{rt}(\mathscr{A})$ holds true.

Proposition 4. *Let $\mathscr{A}$ be a strongly connected automaton given by a pair of matrices A and B. Then for every unavoidable decomposition $\sigma = \langle (X, Y), (\Gamma, \Delta) \rangle$ the reduced automaton $\mathscr{A}_\sigma$ is also strongly connected.*

Proof. Let us fix arbitrary $i, j \in \overline{1, r}$. Now we are to construct a path from i to j in automaton $\mathscr{A}_\sigma$. Since every row of Y contains 1, there exists i' such that $Y[i, i'] = 1$. The decomposition σ is unavoidable, thus there exists j' such that $X[j', j] = 1$. Since $\mathscr{A}$ is strongly connected, we have a matrix $W \in \{A, B\}^*$ such that $W[i', j'] = 1$. It is clear that $YWX[i, j] = 1$. Moreover, $YWX \in \mathcal{M}^*$. Therefore, $\mathscr{A}_\sigma$ is strongly connected.

In order to obtain an upper bound on $\mathrm{rt}(\mathscr{A})$ using $\mathrm{rt}(\mathscr{A}_\sigma)$, we introduce a new notion. Let $\langle \mathscr{A}, \mathscr{F} \rangle$ be a pair of automata, where $\mathscr{F}$ has initial and final states. We say that $\mathscr{A}$ is *synchronizing with constraint* $\mathscr{F}$ if there is a word w such that w resets $\mathscr{A}$ and is also accepted by $\mathscr{F}$. The minimum length of such a word w is called the *constrained reset threshold with respect to* $\mathscr{F}$ and is denoted by $\mathrm{rt}_c(\mathscr{A}, \mathscr{F})$. We will often omit the explicit reference to the automaton $\mathscr{F}$, since it will be known from the context.

We would like to note that the constrained reset threshold can grow exponentially with the number of states in the automaton $\mathscr{A}$ even if the automaton $\mathscr{F}$ is fixed. Such an example can be obtained by a slight modification of the 3-letter automaton $\mathscr{A}^3_{pfa}(n)$ from [9]. Let us define automaton $\mathscr{A}(k) = (Q_k, \Sigma)$. Alphabet Σ is equal to $\{a, b, c\}$. Let p_i be i-th prime number ($p_1 = 2$). Then $Q_k = \{(i, j) \mid 1 \le i \le k, \; 0 \le j \le p_i - 1\}$. Actions of letters are defined as follows:

$$(i, j) \cdot a = \begin{cases} (i, j+1), & j < p_i - 1, \\ (i, 0), & j = p_i - 1; \end{cases} \qquad (i, j) \cdot b = (i, 0);$$

$$(i, j) \cdot c = \begin{cases} (i, j), & j < p_i - 1, \\ (1, 1), & j = p_i - 1. \end{cases}$$

Let $\mathscr{F}$ be the minimal automaton of the language ba^*c. We claim that constrained reset threshold of $\mathscr{A}(k)$ with respect to $\mathscr{F}$ grow exponentially with the number of states in the automaton $\mathscr{A}(k)$.

Let $ba^\ell c$ be the shortest constrained reset word. Note that $Q_k \cdot b = \{(i, 0) \mid 1 \le i \le k\}$. The action of letter a preserves the first component of any state. Therefore, $Q_k \cdot ba^\ell = \{(i, x_i) \mid 1 \le i \le k\}$, where x_i is uniquely defined. Since $|Q_k \cdot ba^\ell c| = 1$ we have $x_i = p_i - 1$. It is not hard to see that $(i, 0) \cdot ba^\ell = (i, p_i - 1)$ if and only if $\ell \equiv p_i - 1 \pmod{p_i}$. Thus, ℓ is the smallest non-negative solution of the following system of simultaneous congruences: $\ell \equiv p_i - 1 \pmod{p_i}$ for all $1 \le i \le k$. It follows from Chinese remainder theorem that $\ell = p_1 p_2 \ldots p_k - 1$. Since $\mathscr{A}(k)$ has only $p_1 + p_2 + \ldots p_k$ states our claim follows. Growth rate is estimated in [9].

Let $\mathscr{R}$ be the automaton shown in Fig. 2. It has $\{S, YX, Y\Gamma, \Delta X, \Delta\Gamma, Z\}$ as the set of states. For the sake of readability, zero state Z is not presented in the picture. For all letters $\ell \in \Sigma'$ we have $Z \cdot \ell = Z$. If the action of a letter ℓ on a state q is not shown in the picture then $q \cdot \ell = Z$. The state S is the initial state of the automaton $\mathscr{R}$ and the states $YX, Y\Gamma, \Delta X, \Delta\Gamma$ are final. From now on the automaton $\mathscr{R}$ will be the only constraint automaton we consider.

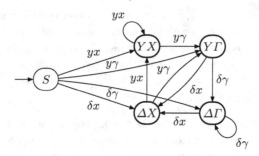

Fig. 2. Automaton $\mathscr{R}$

Proposition 5. *Let $\mathscr{A}$ be an automaton given by a pair of matrices A and B. Then for every decomposition $\sigma = \langle (X, Y), (\Gamma, \Delta) \rangle$ automaton $\mathscr{A}_\sigma$ is synchronizing with constraint $\mathscr{R}$ if and only if $\mathscr{A}$ is synchronizing. Moreover, $\mathrm{rt}_c(\mathscr{A}_\sigma) - 1 \leq \mathrm{rt}(\mathscr{A}) \leq \mathrm{rt}_c(\mathscr{A}) + 1$. If the decomposition σ is unavoidable then $\mathrm{rt}_c(\mathscr{A}_\sigma) \leq \mathrm{rt}(\mathscr{A})$.*

Proof. If $\mathscr{A}$ is synchronizing, then synchronizability of $\mathscr{A}_\sigma$ with constraint $\mathscr{R}$ and lower bounds on the reset threshold of $\mathscr{A}$ immediately follow from the proof of Proposition 3. Representations of the constructed matrices YWX and $W'TX$ as a product of matrices in $\mathcal{M}$ satisfy the constraint $\mathscr{R}$.

Now let the matrix $W = SW'P$ be a matrix corresponding to some constrained reset word w of $\mathscr{A}_\sigma$, where $S \in \{Y, \Delta\}$ and $P \in \{X, \Gamma\}$. We notice that W can be represented as a product of matrices in $\mathcal{M}$ that the automaton $\mathscr{R}$ accepts. Then it is not hard to see that the matrix $S'WP'$ has a column of 1's and can be represented as a product of matrices A and B, where S', P' are defined as follows:

$$S' = \begin{cases} X, & \text{if } S = Y, \\ \Gamma, & \text{if } S = \Delta; \end{cases} \qquad P' = \begin{cases} Y, & \text{if } P = X, \\ \Delta, & \text{if } P = \Gamma. \end{cases}$$

Thus, $S'WP'$ can play the role of a reset word for $\mathscr{A}$ and we get $\mathrm{rt}(\mathscr{A}) \leq \mathrm{rt}_c(\mathscr{A}) + 1$.

Proposition 5 immediately implies that there is a function $f(r)$ such that $\mathrm{rt}(\mathscr{A}) \leq f(r)$ for any 2-letter automaton $\mathscr{A}$ of bounded rank r. Note that automaton $\mathscr{A}$ can have arbitrary many states. Proposition 5 can also be used in order to give a tight estimate for the reset threshold of an automaton of bounded rank. Let $\mathscr{E}_n$ be defined as follows:

$$i \cdot a = \begin{cases} 2, & \text{if } i = 1, \\ 3, & \text{if } i = 2, \\ i, & \text{if } i > 2; \end{cases} \qquad i \cdot b = \begin{cases} 3, & \text{if } i = 1, \\ i + 1, & \text{if } 1 < i < n, \\ 1, & \text{if } i = n. \end{cases}$$

Automaton $\mathscr{E}_5$ is shown in Fig. 3.

Let $\langle A, B \rangle$ be the matrix representation of $\mathscr{E}_n$ (in accordance with given numbering). We define the decomposition σ_n as follows: the matrix Y is obtained from A by deleting the second row and the matrix Δ is obtained from B by deleting the first row. The matrices X, Γ are uniquely determined by Y and Δ. For example for $n = 5$ we get:

$$\sigma_5: \quad A = \begin{pmatrix} 1&0&0&0 \\ 0&1&0&0 \\ 0&1&0&0 \\ 0&0&1&0 \\ 0&0&0&1 \end{pmatrix} \begin{pmatrix} 0&1&0&0&0 \\ 0&0&1&0&0 \\ 0&0&0&1&0 \\ 0&0&0&0&1 \end{pmatrix}, \quad B = \begin{pmatrix} 1&0&0&0 \\ 1&0&0&0 \\ 0&1&0&0 \\ 0&0&1&0 \\ 0&0&0&1 \end{pmatrix} \begin{pmatrix} 0&0&1&0&0 \\ 0&0&0&1&0 \\ 0&0&0&0&1 \\ 1&0&0&0&0 \end{pmatrix}.$$

By a straightforward computation we get $YX = A_{n-1}$, $Y\Gamma = I_{n-1}$, $\Delta X = \Delta\Gamma = B_{n-1}$, where I_{n-1} is the identity matrix of order $n - 1$. The reduced automaton for $\mathscr{E}_5$ is shown on the right in Fig. 3. We have omitted the actions of $y\gamma, \delta\gamma$ to improve readability. Notice that $\langle YX, \Delta\Gamma \rangle$ is a matrix representation of the automaton $\mathscr{C}_{n-1}$. We can use this fact and Proposition 1 to estimate the constrained reset threshold. So, let $w \in \Sigma'$ be a constrained reset word of minimal length for $\mathscr{E}_{\sigma_n}$. By Proposition 1 w has at least $(n-1)^2 - 3(n-1) + 2 = n^2 - 5n + 6$ occurrences of $\delta x, \delta\gamma$ and $n-2$ occurrences of yx. Notice that $(yx)^2$ is not a factor of w, otherwise we could obtain a shorter constrained reset word by reducing $(yx)^2$ to just yx. Therefore, after every occurrence of yx, except maybe the last one, there is an occurrence of δx, $\delta\gamma$ or $y\gamma$. Since w is accepted by $\mathscr{R}$ we conclude that yx is followed by $y\gamma$. Hence, we get $|w| \geq n^2 - 5n + 6 + n - 2 + n - 3 = n^2 - 3n + 1$. Proposition 5 implies $n^2 - 3n + 1 \leq \mathrm{rt}(\mathscr{E}_n) \leq n^2 - 3n + 2$.

We notice that there is also another way to determine the reset threshold of $\mathscr{E}_n$. It is presented in an extended version of [2] (submitted).

Now we use the series $\mathscr{E}_n$ as a base for constructing 2-letter automata of bounded rank r with "large" reset threshold. The automaton $\mathscr{E}_n$ has been chosen for the following reason. Computational experiments show that for small number of states n, upper bound $n^2 - 3n + 2$ on reset threshold is optimal for 2-letter automata of bounded rank $n-1$. Thus, we hope that generalizations of the series $\mathscr{E}_n$ will have reset thresholds close to be optimal.

Proposition 6. *For every $r < n$ there is a synchronizing 2-letter n-state automaton of bounded rank r whose reset threshold is at least $r^2 - r - 1$. If $r > \frac{n}{2}$, then such an automaton is strongly connected.*

Let us fix the number of states n, rank r and defect $d = n - r$. The automaton $\mathscr{E}_n$ essentially was reduced to $\mathscr{C}_{n-1}$. Now we are going to reverse this procedure.

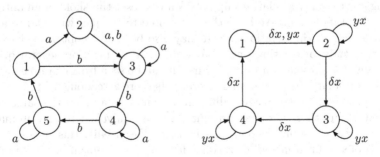

Fig. 3. The automaton $\mathscr{E}_5$ and its reduced automaton

We look for an automaton that can be reduced to $\mathscr{C}_r$. More precisely, consider a solution X, Y, Γ, Δ of the following system of matrix equations: $YX = A_r$, $Y\Gamma = I_r$, $\Delta X = \Delta\Gamma = B_r$, such that:

(i) X, Γ are $(0,1)$-matrices of size n by r
(ii) Y, Δ are $(0,1)$-matrices of size r by n
(iii) every row of X, Y, Γ, Δ has only one occurrence of 1
(iv) every column of X and Γ contains at least one occurrence of 1.

Arguing as before, we get $\mathrm{rt}_c(YX, Y\Gamma, \Delta X, \Delta\Gamma) \geq r^2 - r - 1$. Consider the automaton $\mathscr{A} = \langle XY, \Gamma\Delta \rangle$. It has n states and the rank of each letter is bounded by r. Since $\langle (X,Y), (\Gamma, \Delta) \rangle$ is an unavoidable decomposition of $\mathscr{A}$, then by Proposition 5 we conclude that $\mathscr{A}$ is synchronizing. Moreover, its reset threshold is at least $r^2 - r - 1$. A simple solution with required properties can be found in the following form:

$$X = \begin{pmatrix} X' \\ e_1 \\ A_r \end{pmatrix}, \quad Y = \begin{pmatrix} 0_{r,d} \mid I_r \end{pmatrix}, \quad \Gamma = \begin{pmatrix} \Gamma' \\ e_1 \\ I_r \end{pmatrix}, \quad \Delta = \begin{pmatrix} 0_{r-1,d+1} \mid I_{r-1} \\ \hline e_d \end{pmatrix},$$

where X', Γ' are arbitrary $(0,1)$-matrices of size $d-1$ by r with unique occurrence of 1 in each row. Here $0_{r,d}$ denotes zero matrix of size r by d. This completes first part of the proof.

We need a more complicated solution in order to get a strongly connected automaton. If $r = n - 1$, then the automaton $\mathscr{E}_n$ satisfies the conditions of our proposition. Suppose now that $r \neq n - 1$. Then the desired solution is the following:

$$X = \begin{pmatrix} M \\ e_1 \\ A_r \end{pmatrix}, Y = \begin{pmatrix} 0_{r,d} \mid I_r \end{pmatrix}, \Gamma = \begin{pmatrix} M \\ e_1 \\ I_r \end{pmatrix}, \Delta = \begin{pmatrix} I_{d-1} \mid 0_{d-1,r+1} \\ \hline 0_{r-d,2d} \mid I_{r-d} \\ \hline e_d \end{pmatrix},$$

where $M = \begin{pmatrix} e_2 \\ e_3 \\ \dots \\ e_d \end{pmatrix}$ is matrix of size $d - 1$ by r. The proof that the automaton $\langle XY, \Gamma\Delta \rangle$ is strongly connected is straightforward and technical. We will omit it.

4   Conclusion and Discussion

We suggest to study a relationship between the reset threshold of an automaton and ranks of its letters. We hope that presented techniques will lead to a better understanding of this relationship. It may also be used to obtain new examples or in the study of existing ones. We would like to mention several questions concerning the presented approach. First, it can be seen from example 1 that the reduced automaton is not uniquely defined. Is there a "canonical" reduction or a most "convenient" one? Second, different automata may have the same reduced automaton. In other words, an automaton $\mathscr{A}$ is equivalent to an automaton $\mathscr{B}$ if $\mathscr{A}_\sigma = \mathscr{B}_{\sigma'}$ for some decompositions σ, σ'. Essentially, this means that the problem of synchronization of $\mathscr{A}$ coincides with the problem of synchronization of $\mathscr{B}$. Is there a transparent characterization of such equivalence classes?

References

1. Almeida, J., Steinberg, B.: Matrix Mortality and the Černý-Pin Conjecture. In: Diekert, V., Nowotka, D. (eds.) DLT 2009. LNCS, vol. 5583, pp. 67–80. Springer, Heidelberg (2009)
2. Ananichev, D., Gusev, V., Volkov, M.: Slowly Synchronizing Automata and Digraphs. In: Hliněný, P., Kučera, A. (eds.) MFCS 2010. LNCS, vol. 6281, pp. 55–65. Springer, Heidelberg (2010)
3. Ananichev, D.S., Volkov, M.V., Zaks, Y.I.: Synchronizing automata with a letter of deficiency 2. Theor. Comput. Sci. 376, 30–41 (2007)
4. Béal, M.-P., Berlinkov, M.V., Perrin, D.: A quadratic upper bound on the size of a synchronizing word in one-cluster automata. Int. J. Found. Comput. Sci. 22(2), 277–288 (2011)
5. Černý, J.: Poznámka k homogénnym eksperimentom s konečnými automatami. Matematicko-fyzikalny Časopis Slovensk. Akad. Vied 14(3), 208–216 (1964) (in Slovak)
6. Dubuc, L.: Sur les automates circulaires et la conjecture de Černý. RAIRO Inform. Théor. Appl. 32, 21–34 (1998) (in French)
7. Gusev, V.V.: Lower Bounds for the Length of Reset Words in Eulerian Automata. In: Delzanno, G., Potapov, I. (eds.) RP 2011. LNCS, vol. 6945, pp. 180–190. Springer, Heidelberg (2011)
8. Kari, J.: Synchronizing finite automata on Eulerian digraphs. Theoret. Comput. Sci. 295, 223–232 (2003)
9. Martyugin, P.V.: Lower bounds for the length of the shortest carefully synchronizing words for two- and three-letter partial automata. Diskretn. Anal. Issled. Oper. 15(4), 44–56 (2008) (in Russian)
10. Pin, J.-E.: On two combinatorial problems arising from automata theory. Ann. Discrete Math. 17, 535–548 (1983)
11. Rystsov, I.K.: Estimation of the length of reset words for automata with simple idempotents. Cybernetics and Systems Analysis 36(3), 339–344 (2000)
12. Sandberg, S.: 1 Homing and Synchronizing Sequences. In: Broy, M., Jonsson, B., Katoen, J.-P., Leucker, M., Pretschner, A. (eds.) Model-Based Testing of Reactive Systems. LNCS, vol. 3472, pp. 5–33. Springer, Heidelberg (2005)
13. Trahtman, A.N.: The Černý conjecture for aperiodic automata. Discrete Math. Theor. Comput. Sci. 9(2), 3–10 (2007)
14. Trahtman, A.N.: Modifying the Upper Bound on the Length of Minimal Synchronizing Word. In: Owe, O., Steffen, M., Telle, J.A. (eds.) FCT 2011. LNCS, vol. 6914, pp. 173–180. Springer, Heidelberg (2011)
15. Volkov, M.V.: Synchronizing Automata and the Černý Conjecture. In: Martín-Vide, C., Otto, F., Fernau, H. (eds.) LATA 2008. LNCS, vol. 5196, pp. 11–27. Springer, Heidelberg (2008)
16. Volkov, M.V.: Synchronizing automata preserving a chain of partial orders. Theoret. Comput. Sci. 410, 2992–2998 (2009)

Automatic Theorem-Proving
in Combinatorics on Words

Daniel Goč, Dane Henshall, and Jeffrey Shallit

School of Computer Science, University of Waterloo, Waterloo, ON N2L 3G1 Canada
{dhenshall,dgoc,shallit}@uwaterloo.ca

Abstract. We describe a technique for mechanically proving certain
kinds of theorems in combinatorics on words, using finite automata and a
package for manipulating them. We illustrate our technique by applying
it to (a) solve an open problem of Currie and Saari on the lengths of
unbordered factors in the Thue-Morse sequence; (b) verify an old result
of Prodinger and Urbanek on the paperfolding sequence and (c) find
an explicit expression for the recurrence function for the Rudin-Shapiro
sequence. All results were obtained by machine computations.

Dedicated to the memory of Sheng Yu (1950–2012): friend and colleague

1   Introduction

The title of this paper is a bit of a pun. On the one hand, we are concerned
with certain natural questions about *automatic sequences*: sequences over a finite
alphabet where the n'th term of the sequence is expressible as a finite-state
function of the base-k representation of n. On the other hand, we are interested
in answering these questions purely mechanically, in an *automated* fashion.

Let $\mathbf{x} = (a(n))_{n \geq 0}$ be an infinite sequence over a finite alphabet Δ. Then $\mathbf{x}$ is said
to be k-*automatic* if there is a deterministic finite automaton M taking as input the
base-k representation of n, and having $a(n)$ as the output associated with the last
state encountered [2]. In this case, we say that M *generates* the sequence $\mathbf{x}$.

We write $\mathbf{x}[i] = a(i)$, and we let $\mathbf{x}[i..i+n-1]$ denote the *factor* of length n
beginning at position i in $\mathbf{x}$. A sequence is said to be *squarefree* if it contains no
factor of the form xx, where x is a nonempty word, and is said to be *overlapfree*
if it contains no factor of the form $ayaya$, where a is a single letter and y is a
possibly empty word.

In Figure 1, we give, as an example, an automaton generating the well-known
Thue-Morse sequence $\mathbf{t} = t(0)t(1)t(2)\cdots = 011010011001\cdots$ [3]. The input is
n, expressed in base 2, and the output is the number contained in the state last
reached. Thus $t(n)$ is the sum, modulo 2, of the binary digits of n. In a celebrated
result, Thue proved [27,28,4] that the sequence $\mathbf{t}$ is overlapfree.

For at least 25 years, researchers have been interested in the algorithmic decidability of assertions about automatic sequences. For example, in one of the
earliest results, Honkala [19] showed that, given an automaton M, it is decidable
if the sequence generated by M is ultimately periodic.

N. Moreira and R. Reis (Eds.): CIAA 2012, LNCS 7381, pp. 180–191, 2012.

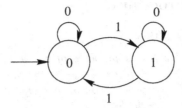

Fig. 1. A finite automaton generating the Thue-Morse sequence

Recently, Allouche et al. [1] found a different proof of Honkala's result using a more general technique. Using this technique, they were able to give algorithmic solutions to many classical problems from combinatorics on words such as

Given an automaton, is the generated sequence squarefree? Or overlapfree?

The technique of Allouche et al. is at its core, very similar to work of Büchi, Bruyère, Michaux, Villemaire, and others, involving formal logic; see, e.g., [5]. The basic idea is as follows: given the automaton M, and some predicate $P(n)$ we want to check, we alter M by a series of transformations to a new automaton M' that accepts the base-k representations of those integers n for which $P(n)$ is true. Then we can check the assertion "$\exists n\ P(n)$" simply by checking if M' accepts anything (which can be done by a standard depth-first search on the underlying directed graph of the automaton). We can check the assertion "$\forall n\ P(n)$" by checking if M' accepts everything. And we can check assertions like "$P(n)$ holds for infinitely many n" by checking if M' has a reachable cycle from which a final state is reachable.

Using this idea, Allouche et al. were able to show to reprove, purely mechanically using a computer program, Thue's classic result on the overlapfreeness of the the Thue-Morse sequence.

Later, the technique was applied to give decision procedures for other properties of automatic sequences. For example, Charlier et al. [6] showed that it can be used to decide if a given k-automatic sequence

- contains powers of arbitrarily large exponent;
- is recurrent;
- is uniformly recurrent.

A sequence is said to be *recurrent* if every factor that occurs, occurs infinitely often. A sequence **x** is said to be *uniformly recurrent* if it is recurrent and furthermore for each finite factor w occurring in **x**, there is a constant $c(w)$ such that two consecutive occurrences of w are separated by at most $c(w)$ positions.

More recently, variations of the technique have been used to

- compute the critical exponent;
- compute the initial critical exponent;
- decide if a sequence is linearly recurrent;
- compute the Diophantine exponent.

(For definitions of these terms see [25].)

2   The Decision Procedure

In [6] we have the following theorem:

Theorem 1. *If we can express a property of a k-automatic sequence* **x** *using quantifiers, logical operations, integer variables, the operations of addition, subtraction, indexing into* **x**, *and comparison of integers or elements of* **x**, *then this property is algorithmically decidable.*

Let us outline how the decision procedure works. First, the input to the decision procedure: an automaton $M = (Q, \Sigma_k, \Delta, \delta, q_0, \tau)$ generating the k-automatic sequence **x**. Here

- Q is a nonempty set of states;
- $\Sigma_k := \{0, 1, \ldots, k-1\}$;
- Δ is the output alphabet;
- $\delta : Q \times \Sigma \to Q$ is the transition function;
- q_0 is the initial state; and
- $\tau : Q \to \Delta$ is the output mapping.

In this paper, we assume that the automaton takes as input the representation of n in base k, *starting with the least significant digit*; we call this the *reversed representation* of n and write it as $(n)_k$. We allow leading zeroes in the representation (which, because of our convention, are actually trailing zeroes). Thus, for example, 011 and 01100 are both acceptable representations for 6 in base 2.

We might also need to encode pairs, triples, or r-tuples of integers. We handle these by first padding the reversed representation of the smaller integer with trailing zeroes, and then coding the r-tuple as a word over Σ_k^r. For example, the pair $(20, 13)$ could be represented in base-2 as

$$[0, 1][0, 0][1, 1][0, 1][1, 0],$$

where the first components spell out 00101 and the second components spell out 10110. Of course, there are other possible representations, such as

$$[0, 1][0, 0][1, 1][0, 1][1, 0][0, 0],$$

which correspond to non-canonical representations having trailing zeroes; these are also permitted.

Rather than present a detailed proof, we illustrate the idea of the decision procedure in the proof of the following new result:

Theorem 2. *The following problem is algorithmically decidable: given two k-automatic sequences* **x** *and* **y**, *generated by automata M_1 and M_2, respectively, decide if* **x** *is a shift of* **y** *(that is, decide if there exists a constant c such that* $\mathbf{x}[n] = \mathbf{y}[n+c]$ *for all $n \geq 0$).*

Proof. We first create an NFA M that accepts the language

$$\{(c)_k \;:\; \exists\, n \text{ such that } \mathbf{x}[n] \neq \mathbf{y}[n+c]\}.$$

To do so, on input $(c)_k$, M

- guesses $w_1 = (n)_k$ nondeterministically (perhaps with trailing zeroes appended),
- simulates M_1 on w_1,
- adds n to c and computes the base-k representation of $w_2 = (n + c)_k$ digit-by-digit "on the fly", keeping track of carries, as necessary, and simulates M_2 on w_2, and
- accepts if the outputs of both machine differ.

We now convert M to a DFA M', and change final states to non-final (and vice versa). Then M' accepts the language

$$\{(c)_k \ : \ \mathbf{x}[n] = \mathbf{y}[n + c] \text{ for all } n \geq 0\}.$$

Thus, $\mathbf{x}$ is a shift of $\mathbf{y}$ if and only if M' accepts any word, which is easily checked through depth-first search.   □

Remark 1. As we can see, the size of the automata involved depends, in an unpleasant way, on the number of quantifiers needed to state the logical expression characterizing the property being checked, because existential quantifiers are implemented through nondeterminism, and universal quantifiers are implemented through nondeterminism and complementation (which is implemented in a DFA by exchange of the role final and non-final states). Thus each new quantifier could increase the current number of states, say n, to 2^n using the subset construction. If the original automata have at most N states, it follows that the running time is bounded by an expression of the form

$$2^{2^{\cdot^{\cdot^{\cdot^{2^{p(N)}}}}}}$$

where p is a polynomial and the number of exponents in the tower is one less than the number of quantifiers in the logical formula characterizing the property being checked.

This extraordinary computational complexity raises the natural question of whether the decision procedure could actually be implemented for anything but toy examples. Luckily the answer seems to be yes — at least in some cases — as we will see below.

The algorithms we discuss were implemented by the first two authors, independently, using two different programs. The results in Sections 3 and 4 have been double-checked with these separate implementations, which should give some confidence about the results.

Remark 2. Prior art: as a referee points out, very similar ideas are contained in the work of Glenn and Gasarch [14,15] on implementing a decision procedure for WS1S, the weak second-order theory of one successor. The main differences between their work and ours are (a) we work with base-k encodings of integers, instead of unary encodings, and (b) we apply our ideas to solve some interesting open problems about automatic sequences, instead of checking randomly-generated sentences.

3   Borders

A word w is *bordered* if it begins and ends with the same word x with $0 < |x| \le |w|/2$; Otherwise it is *unbordered*. An example in English of a bordered word is **entanglement**. A bordered word is also called *bifix* in the literature, and unbordered words are also called *bifix-free* or *primary*.

Bordered and unbordered words have been actively studied in the literature, particularly with regard to the Ehrenfeucht-Silberger problem; see, for example, [13,20,10,11,16,17,7,18,22,12], just to name a few.

Currie and Saari [8] studied the unbordered factors of the Thue-Morse sequence **t**. They proved that if $n \not\equiv 1 \pmod 6$, then **t** has an unbordered factor of length n. (Also see [24, Lemma 4.10 and Problem 4.1].) However, this is not a necessary condition, as

$$\mathbf{t}[39..69] = 0011010010110100110010110100101,$$

which is an unbordered factor of length 31. Currie and Saari left it as an open problem to give a complete characterization of the integers n for which **t** has an unbordered factor of length n.

The following theorem and proof, quoted practically verbatim from [6], shows that, more generally, the characteristic sequence of n for which a given k-automatic sequence has an unbordered factor of length n, is itself k-automatic:

Theorem 3. *Let* $\mathbf{x} = a(0)a(1)a(2) \cdots$ *be a k-automatic sequence. Then the associated infinite sequence* $\mathbf{b} = b(0)b(1)b(2) \cdots$ *defined by*

$$b(n) = \begin{cases} 1, & \text{if } \mathbf{x} \text{ has an unbordered factor of length } n; \\ 0, & \text{otherwise}; \end{cases}$$

is k-automatic.

Proof. The sequence $\mathbf{x}$ has an unbordered factor of length n

 iff

$\exists\, j \ge 0$ such that the factor of length n beginning at position j of $\mathbf{x}$ is unbordered

 iff

there exists an integer $j \ge 0$ such that for all possible lengths l with $1 \le l \le n/2$, there is an integer i with $0 \le i < l$ such that the supposed border of length l beginning and ending the factor of length n beginning at position j of $\mathbf{x}$ actually differs in the i'th position

 iff

there exists an integer $j \ge 0$ such that for all integers l with $1 \le l \le n/2$ there exists an integer i with $0 \le i < l$ such that $\mathbf{x}[j+i] \ne \mathbf{x}[j+n-l+i]$.

Now assume $\mathbf{x}$ is a k-automatic sequence, generated by some finite automaton. We show how to implement the characterization given above with an automaton.

We first create an NFA that given the $(j, l, n)_k$ guesses the base-k representation of i, digit-by-digit, checks that $i < l$, computes $j + i$ and $j + n - l + i$ on the fly, and checks that $\mathbf{x}[j+i] \neq \mathbf{x}[j+n-l+i]$. If such an i is found, it accepts. We then convert this to a DFA, and interchange accepting and nonaccepting states. This DFA M_1 accepts $(j, l, n)_k$ such that there is no i, $0 \leq i < l$ such that $\mathbf{x}[j+i] = \mathbf{x}[j+n-l+i]$. We then use M_1 as a subroutine to build an NFA M_2 that on input $(j, n)_k$ guesses l, checks that $1 \leq l \leq n/2$, and calls M_1 on the result. We convert this to a DFA and interchange accepting and nonaccepting states to get M_3. Finally, this M_3 is used as a subroutine to build an NFA M_4 that on input n guesses j and calls M_3.

The characteristic sequence of these integers n is therefore k-automatic.   □

Since the proof is constructive, one can, in principle, carry out the construction to get an explicit description of the lengths for which the Thue-Morse sequence has an unbordered factor.

Doing so results in the following theorem:

Theorem 4. *There is an unbordered factor of length n in $\mathbf{t}$ if and only if the base-2 representation of n (starting with the most significant digit) is not of the form $1(01^*0)^*10^*1$.*

Proof. The proof of this theorem is purely mechanical, and it involves performing a sequence of operations on finite automata. The second author wrote a program in C++, using his own automata package, to perform these operations. There are four stages to the computation, which are described in detail below.

Stage 1

Let T be the automaton of Figure 1 generating the Thue-Morse sequence $\mathbf{t}$. Stage 1 takes T as input and outputs an automaton M_1, where M_1 accepts $w \in (\{0, 1\}^4)^*$ if and only if w is the base-2 representation of some $(n, j, l, i) \in S_1$, where

$$S_1 = \{(n, j, l, i) \; : \; 0 < l \leq n/2 \text{ and } i < j \text{ and } \mathbf{t}[j+i] \neq \mathbf{t}[n+j-l+i]\}. \quad (1)$$

The size of M_1 was only 102 states. However, since the input alphabet for M_1 is of size $2^4 = 16$, a considerable amount of complexity is being stored in the transition matrix. Stage 1 passed all 1.3 million tests meant to ensure that M_1 corresponds to S_1.

Stage 2

The purpose of Stage 2 is to remove the variable i by simulating it. The resulting machine, after being negated, accepts (n, j, l) iff the length n factor of t starting at index j has a border of length l. So Stage 2 produces the automaton M_2, which is the negation of the result of simulating i. More formally, M_2 accepts a word $w \in (\{0, 1\}^3)^*$ if and only if w is the base-2 representation of some $(n, j, l) \in S_2$, where

$$S_2 = \{(n, j, l) : \nexists \, i \text{ for which } (n, j, l, i) \in S_1\} \quad (2)$$

The size of M_2 after subset construction was 8689 states, and it minimized down to 127 states. The output of Stage 2 passed all 1.6 million tests meant to ensure that M_2 corresponds to S_2.

Stage 3

The purpose of Stage 3 is to remove l by simulating it. By the end of Stage 3, most of the work has already been done. The output of Stage 3, M_3, accepts an input word $w \in (\{0,1\}^2)^*$ if and only if w is the base-2 representation of some $(n, j) \in S_3$, where

$$S_3 = \{(n, j) : \nexists\ l \text{ such that } (n, j, l) \in S_2\} \tag{3}$$

or, in other words

$$S_3 = \{(n, j) : \mathbf{t} \text{ has an unbordered factor of length } n \text{ at index } j\}. \tag{4}$$

The size of M_3 after subset construction was 1987 states, and it minimized down to 263 states. The output of Stage 3 passed all 1.9 million tests meant to ensure that M_3 corresponds to S_3.

Stage 4

Finally, Stage 4 simulates j on M_3 and negates the result. So the output of Stage 3 is an automaton that accepts the binary representation of a positive integer $n > 1$ if and only if the Thue-Morse word has no unbordered factor of length n. Formally put, the automaton M_4 produced by Stage 4 accepts a word $w \in \{0,1\}^*$ if and only if w is the base-2 representation of some $n \in S_4$, where

$$S_4 = \{n \in \mathbb{N} : n > 1, \nexists\ j \text{ for which } (n, j) \in S_3\}. \tag{5}$$

The size of M_4 after subset construction is 2734 states, and it minimized to 7 states. M_4 accepts the reverse of $1(01^*0)^*10^*1$. Therefore the Thue-Morse word has an unbordered factor of length n if and only if the base-2 representation of n (starting with the most significant digit) is not of the form $1(01^*0)^*10^*1$.

The total computation took 9 seconds of CPU time on a 2.9GHz Dell XPS laptop.   □

Remark 3. Here are some additional implementation details.

In order to implement the needed operations on automata, we must decide on an encoding of elements of $(\Sigma_k^n)^*$. We could do this by performing a perfect shuffle of each individual word over Σ_k^*, or by letting the alphabet itself be represented by k-tuples. The decision represents a tradeoff between state size and alphabet size. We used the latter representation, since (a) it makes the algorithms considerably easier to implement and understand and (b) decreases the number of states needed.

It was mentioned earlier how many tests were passed in each stage. In order to make sure that the final automaton is what we expect, a number of tests are run after each stage on the output of that stage.

For example, let $\mathbf{x}$ be an automatic sequence. The testing framework requires a C++ function which given n computes $\mathbf{x}[n]$. Before any operations are done, the automaton given for $\mathbf{x}$ is tested against the C++ function to make sure that they match for the first 10,000 elements. Then, at each stage before Stage 4 the resulting automaton is tested to give confidence that the operations on the automata are giving the desired results.

For example, after Stage 2 of computing the set of lengths for which there exists an unbordered factor of an automatic sequence $\mathbf{x}$, we expect the machine M_2 to accept the language S_2, where

$$S_2 = \{(n, j, l) : \not\exists \ i \text{ for which } \mathbf{x}[j + i] = \mathbf{x}[n + j - l + i]\} \tag{6}$$

This is then tested by making sure M_2 accepts $(n, j, l)_k$ if and only if $(n, j, l) \in S_2$ for all $n, j, l \leq 1400$. These tests were invaluable to debugging, and provide confidence in the final result of the computation.

Finally, we have to address the issue of multiple representations. It is easy to forget that automata accept words in $\Sigma_k{}^*$, and not integers. For some operations, such as complement and intersection, it is crucial that if one binary representation is accepted by the automaton, then all binary representations must be accepted.

4   Additional Results on Unbordered Words

We also applied our decision procedure above to two other famous sequences: the Rudin-Shapiro sequence [23,26] and the paperfolding sequence [9].

For a word $w \in 1(0 + 1)^*$, we define $a_w(n)$ to be the number of (possibly overlapping) occurrences of w in the (ordinary, unreversed) base-2 representation of n. Thus, for example, $a_{11}(7) = 2$.

The *Rudin-Shapiro* sequence $\mathbf{r} = r(0)r(1)r(2) \cdots$ is then defined to be $r(n) = (-1)^{a_{11}(n)}$. It is a 2-automatic sequence generated by an automaton of four states.

The *paperfolding sequence* $\mathbf{p} = p(0)p(1)p(2) \cdots$ is defined as follows: writing $(n)_2 00$ as $1^i 0aw$ for some $i \geq 0$ some $a \in \{0, 1\}$, and some $w \in \{0, 1\}^*$, we have $p(n) = (-1)^a$. It is a 2-automatic sequence generated by an automaton of four states.

Theorem 5. *The Rudin-Shapiro sequence has an unbordered factor of every length.*

Proof. We applied the same technique discussed previously for the Thue-Morse sequence.

Here is a summary of the computation:

Stage 1: 269 states
Stage 2: 85313 states minimized to 1974
Stage 3: 48488 states minimized to 6465
Stage 4: 6234 states.

The Stage 4 NFA has 6234 states. We were unable to determinize this automaton directly (using two different programs) due to an explosion in the number of states created. Instead, we reversed the NFA (creating an NFA for L^R) and determinized this instead. The resulting DFA has 30 states, and upon minimization, gives a 1-state automaton accepting all strings.   □

Theorem 6. *The paperfolding sequence has an unbordered factor of length n if and only if the reversed representation $(n)_2$ is rejected by the automaton given in Figure 4.*

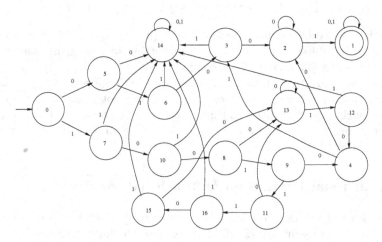

Fig. 2. A finite automaton for unbordered factors in the paperfolding word

Proof. We applied the same technique discussed previously for the Thue-Morse sequence.

Here is a summary of the computation: 6 seconds cpu time on a 2.9GHz Dell XPS laptop.

Stage 1, 159 states
Stage 2, 1751 minimized down to 89 states
Stage 3, 178 minimized down to 75 states
Stage 4, 132 minimize down to 17 states .   □

5   Other Problems

We applied our technique to some other problems. First, we considered the squares in the paperfolding sequence. In 1979, Prodinger and Urbanek [21] characterized the squares in the paperfolding sequence, using a case analysis. We verified this by creating an automaton to accept the language

$$\{(n)_2 \ : \ \exists \, i \ \mathbf{p}[i..i+n-1] = \mathbf{p}[i+n..i+2n-1]\}.$$

The resulting automaton (most significant-digit first) is depicted below, from which we recover the Prodinger-Urbanek result that the only squares xx in $\mathbf{x}$ have lengths $|x| = 1, 3,$ or 5.

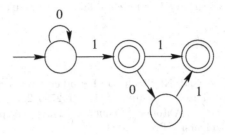

Fig. 3. Lengths of squares in the paperfolding sequence

Next, we computed a new explicit expression for the recurrence function $R_{\mathbf{r}}(n)$ and recurrence quotient for the Rudin-Shapiro sequence $\mathbf{r}$. Here $R_{\mathbf{r}}(n)$ is the smallest integer m such that every factor of $\mathbf{r}$ of length m contains as a factor all the factors of length n. Allouche and Bousquet-Mélou gave the estimate $R_{\mathbf{r}}(n+1) < 172n$ for $n \geq 1$. (Actually, their result was more general, as it applies to any "generalized" Rudin-Shapiro sequence.) We used our method to prove the following result:

Theorem 7. *Let* $\mathbf{r} = (r(n))_{n \geq 0}$ *be the Rudin-Shapiro sequence. Then*

$$R_{\mathbf{r}}(n) = \begin{cases} 5, & \text{if } n = 1; \\ 19, & \text{if } n = 2; \\ 25, & \text{if } n = 3; \\ 20 \cdot 2^t + n - 1, & \text{if } n \geq 4 \text{ and } t = \lceil \log_2(n-1) \rceil. \end{cases}$$

Furthermore, the recurrence quotient

$$\sup_{n \geq 1} \frac{R_{\mathbf{r}}(n)}{n}$$

is equal to 41; it is not attained.

Proof. We created a DFA to accept

$$\{(m,n)_2 \ : \ (m - 20 \cdot 2^t - n + 1, n) \ : \ n \geq 4 \text{ and } m = R(n) \text{ and } t = \lceil \log_2(n-1) \rceil\}.$$

We then verified that the resulting DFA accepted only pairs of the form $(0, n)_2$ for $n \geq 4$.

For the recurrence quotient, the local maximum is evidently achieved when $n = 2^r + 2$ for some $r \geq 1$; here it is equal to $(41 \cdot 2^r + 2)/(2^r + 2)$. As $r \to \infty$, this clearly approaches 41 from below.   □

6 Open Problems

Which of the problems mentioned in § 1 are algorithmically decidable for the more general class of morphic sequences?

Can the techniques be applied to detect abelian powers in automatic sequences?

References

1. Allouche, J.P., Rampersad, N., Shallit, J.: Periodicity, repetitions, and orbits of an automatic sequence. Theoret. Comput. Sci. 410, 2795–2803 (2009)
2. Allouche, J.P., Shallit, J.: Automatic Sequences: Theory, Applications, Generalizations. Cambridge University Press (2003)
3. Allouche, J.P., Shallit, J.O.: The ubiquitous Prouhet-Thue-Morse sequence. In: Ding, C., Helleseth, T., Niederreiter, H. (eds.) Sequences and Their Applications, Proceedings of SETA 1998, pp. 1–16. Springer, Heidelberg (1999)
4. Berstel, J.: Axel Thue's Papers on Repetitions in Words: a Translation, vol. 20. Publications du Laboratoire de Combinatoire et d'Informatique Mathématique, Université du Québec à Montréal (February 1995)
5. Bruyère, V., Hansel, G., Michaux, C., Villemaire, R.: Logic and p-recognizable sets of integers. Bull. Belgian Math. Soc. 1, 191–238 (1994); Corrigendum, Bull. Belg. Math. Soc. 1, 577 (1994)
6. Charlier, É., Rampersad, N., Shallit, J.: Enumeration and Decidable Properties of Automatic Sequences. In: Mauri, G., Leporati, A. (eds.) DLT 2011. LNCS, vol. 6795, pp. 165–179. Springer, Heidelberg (2011)
7. Costa, J.C.: Biinfinite words with maximal recurrent unbordered factors. Theoret. Comput. Sci. 290, 2053–2061 (2003)
8. Currie, J.D., Saari, K.: Least periods of factors of infinite words. RAIRO Inform. Théor. App. 43, 165–178 (2009)
9. Dekking, F.M., Mendès France, M., van der Poorten, A.J.: Folds! Math. Intelligencer 4, 130–138, 173–181, 190–195 (1982), erratum 5 (1983)
10. Duval, J.P.: Une caractérisation de la période d'un mot fini par la longueur de ses facteurs primaires. C. R. Acad. Sci. Paris 290, A359–A361 (1980)
11. Duval, J.P.: Relationship between the period of a finite word and the length of its unbordered segments. Discrete Math. 40, 31–44 (1982)
12. Duval, J.P., Harju, T., Nowotka, D.: Unbordered factors and Lyndon words. Discrete Math. 308, 2261–2264 (2008)
13. Ehrenfeucht, A., Silberger, D.M.: Periodicity and unbordered segments of words. Discrete Math. 26, 101–109 (1979)
14. Glenn, J., Gasarch, W.I.: Implementing WS1S Via Finite Automata. In: Raymond, D.R., Yu, S., Wood, D. (eds.) WIA 1996. LNCS, vol. 1260, pp. 50–63. Springer, Heidelberg (1997)
15. Glenn, J., Gasarch, W.I.: Implementing WS1S Via Finite Automata: Performance Issues. In: Wood, D., Yu, S. (eds.) WIA 1997. LNCS, vol. 1436, pp. 75–86. Springer, Heidelberg (1998)
16. Harju, T., Nowotka, D.: Periodicity and unbordered words: a proof of the extended duval conjecture. J. Assoc. Comput. Mach. 54, 1–20 (2007)
17. Holub, S.: A proof of the extended Duval's conjecture. Theoret. Comput. Sci. 339, 61–67 (2005)

18. Holub, S., Nowotka, D.: On the relation between periodicity and unbordered factors of finite words. Internat. J. Found. Comp. Sci. 21, 633–645 (2010)

19. Honkala, J.: A decision method for the recognizability of sets defined by number systems. RAIRO Inform. Théor. App. 20, 395–403 (1986)

20. Nielsen, P.T.: A note on bifix-free sequences. IEEE Trans. Inform. Theory IT-19, 704–706 (1973)

21. Prodinger, H., Urbanek, F.J.: Infinite 0–1-sequences without long adjacent identical blocks. Discrete Math. 28, 277–289 (1979)

22. Rampersad, N., Shallit, J., Wang, M.W.: Inverse star, borders, and palstars. Inform. Process. Lett. 111, 420–422 (2011)

23. Rudin, W.: Some theorems on Fourier coefficients. Proc. Amer. Math. Soc. 10, 855–859 (1959)

24. Saari, K.: On the Frequency and Periodicity of Infinite Words. Ph.D. thesis, University of Turku, Finland (2008)

25. Shallit, J.: The critical exponent is computable for automatic sequences. In: Ambrož, P., Holub, S., Masáková, Z. (eds.) WORDS 2011: 8th International Conference. Elect. Proc. Theor. Comput. Sci., pp. 231–239 (2011), revised version, with L. Schaeffer, http://arxiv.org/abs/1104.2303v2

26. Shapiro, H.S.: Extremal problems for polynomials and power series. Master's thesis, MIT (1952)

27. Thue, A.: Über unendliche Zeichenreihen. Norske vid. Selsk. Skr. Mat. Nat. Kl. 7, 1–22 (1906); reprinted in Nagell, T. (ed.) Selected Mathematical Papers of Axel Thue, pp. 139–158. Universitetsforlaget, Oslo (1977)

28. Thue, A.: Über die gegenseitige Lage gleicher Teile gewisser Zeichenreihen. Norske vid. Selsk. Skr. Mat. Nat. Kl. 1, 1–67 (1912); reprinted in Nagell, T. (ed.) Selected Mathematical Papers of Axel Thue, pp. 413–478. Universitetsforlaget, Oslo (1977)

How to Synchronize the Heads
of a Multitape Automaton

Oscar H. Ibarra[1,*] and Nicholas Q. Tran[2]

[1] Department of Computer Science,
University of California, Santa Barbara, CA 93106, USA
ibarra@cs.ucsb.edu
[2] Department of Mathematics & Computer Science,
Santa Clara University, Santa Clara, CA 95053, USA
ntran@math.scu.edu

Abstract. Given an n-tape automaton M with a one-way read-only head per tape and a right end marker $ on each tape, we say that M is aligned or 0-synchronized (or simply, synchronized) if for every n-tuple $x = (x_1, \ldots, x_n)$ that is accepted, there is a computation on x such that at any time during the computation, all heads, except those that have reached the end marker, are on the same position. When a head reaches the marker, it can no longer move. As usual, an n-tuple $x = (x_1, \ldots, x_n)$ is accepted if M eventually reaches the configuration where all n heads are on $ in an accepting state. In two recent papers, we looked at the problem of deciding, given an n-tape automaton of a given type, whether there exists an equivalent synchronized n-tape automaton of the same type. In this paper, we exhibit various classes of multitape automata which can(not) be converted to equivalent synchronized multitape automata.

Keywords: multitape automata, aligned, synchronized, semilinear, decidable, undecidable, 1-reversal counters, reversal-bounded counters.

1 Introduction

Motivated by reachability analyses of string programs using multitrack finite-state automata [9], we recently introduced and studied multitape finite automata whose heads are k-synchronized, i.e, never more than k cells apart for some fixed k. Synchronized multitape machines are easier to work with and yet they can be converted to multitrack machines and hence also have decidable decision problems. The following decision question, namely synchronizability, naturally arises regarding multitape machines:

- given a multitape machine M, is there a k-synchronized machine M' of the same type, for some k, such that $L(M) = L(M')$?

---

* Supported in part by NSF Grants CCF-1143892 and CCF-1117708.

N. Moreira and R. Reis (Eds.): CIAA 2012, LNCS 7381, pp. 192–204, 2012.

Formally, a (one-way) n-tape deterministic finite automaton (DFA) M is a finite automaton with n tapes where each tape contains a string over input alphabet Σ. Each tape is read-only and has an associated one-way input head. We assume that each tape has a right end marker $\$$ (not in Σ). On a given n-tuple input $x = (x_1, \ldots, x_n)$, M starts in initial state q_0 with all the heads on the first symbols of their respective tapes. The transition function of M consists of rules of the form $\delta(q, a_i, \ldots, a_n) = (p, d_1, \ldots, d_n)$ (resp., $= \varnothing$). Such a rule means that if M is in state q, with head H_i on symbol a_i, then the machine moves H_i one cell to the right on its input tape if $d_i = 1$ (resp., does not move H_i if $d_i = 0$), and enters state p (resp., halts). When a head reaches the end marker $\$$, that head has to remain on the end marker. The input x is accepted if M eventually reaches the configuration where all n heads are on $\$$ in an accepting state. Let M be an n-tape DFA and $k \geq 0$. M is k-synchronized if for every n-tuple $x = (x_1, \ldots, x_n)$ that is accepted, the (unique) computation on x is such that at any time, no pair of input heads, neither of which is on $\$$, are more than k cells apart. Notice that, since the condition in the definition concerns pairs of heads that are both on symbols in Σ, if one of these two heads is on $\$$, then we can stipulate that the condition is automatically satisfied, irrespective of the distance between the heads. In particular, if $k = 0$, then all heads move to the right synchronously at the same time (except for heads that reach the right end marker early).

The above definitions generalize to n-tape nondeterministic finite automata (NFAs). Now, k-synchronized requires that every n-tuple $x = (x_1, \ldots, x_n)$ that is accepted has an accepting computation on x such that at any time during the computation, no pair of input heads, neither of which is on $\$$, are more than k cells apart. The definitions can also be generalized to n-tape deterministic pushdown automata (DPDAs) and n-tape nondeterministic pushdown automata (NPDAs), which may even be augmented with a finite number of reversal-bounded counters. At each step, each counter (which is initially set to zero) can be incremented by 1, decremented by 1, or left unchanged and can be tested for zero. The counters are reversal-bounded in the sense that there is a specified r such that during any computation, no counter can change mode from increasing to decreasing and vice-versa more than r of times. A counter is 1-reversal if once it decrements, it can no longer increment. It is easy to show that an r-reversal counter can be simulated by $\lceil (r + 1)/2 \rceil$ 1-reversal counters (see, e.g., [4]).

A nondeterministic counter machine (NCM) is an NFA augmented with a counter. Note that an NCM is a special case of an NPDA where the stack alphabet consists of only one symbol, in addition to a distinguished bottom-of-the-stack symbol B which is never modified. Hence the stack can be thought of as a counter since it can only push or pop the same symbol, which would correspond to incrementing or decrementing the stack height by 1. The count is zero when the stack contains only the bottom symbol B. DCM is the deterministic version of NCM.

A machine is *k-ambiguous* if there are at most k accepting computations for any input. Note that unambiguous is the same as 1-ambiguous, and deterministic is a special case of unambiguous.

In two recent papers [2,6], we showed the following theorems (where "synchronized" was called "weakly synchronized"):

Theorem 1 ([2]). *The following problems are undecidable, given a 2-ambiguous 2-tape (and hence multitape) NFA M:*

1. *Is M k-synchronized for a given k?*
2. *Is M k-synchronized for some k?*
3. *Is there a 2-tape (multitape) NFA M' that is 0-synchronized (or k-synchronized for a given k, or k-synchronized for some k) such that $L(M') = L(M)$?*

The stack (resp., counter) of an NPDA (resp., NCM) is *r-reversal* if the number of times it changes mode from pushing to popping (resp., incrementing to decrementing) and vice-versa during any computation is at most r times.

Theorem 2 ([6]). *The following problems are undecidable, given a 2-ambiguous 2-tape 3-reversal NPDA M over $\Sigma^* \times a^*$, where Σ is an alphabet with at least 2 symbols and a is a symbol.*

1. *Is M k-synchronized for a given k?*
2. *Is M k-synchronized for some k?*
3. *Is there a 2-tape NPDA M' that is 0-synchronized (or k-synchronized for a given k, or k-synchronized for some k) such that $L(M') = L(M)$?*

Theorem 3 ([6]). *The following problems are undecidable, given a 3-ambiguous 2-tape NCM (resp., a 2-tape 1-reversal NCM) M over $\Sigma^* \times a^*$, where Σ is an alphabet with at least 2 symbols and a is a symbol.*

1. *Is M k-synchronized for a given k?*
2. *Is M k-synchronized for some k?*
3. *Is there a 2-tape NCM or 2-tape NPDA M' that is 0-synchronized (or k-synchronized for a given k, or k-synchronized for some k) such that $L(M') = L(M)$?*

We note that parts (1) and (2) of the above theorems have been shown to be decidable when the machine is unambiguous (i.e., 1-ambiguous) [2,6].

In this paper, we exhibit various classes of multitape automata which can or cannot be converted to equivalent synchronized multitape automata. In particular, we show that if M is a 2-tape NPDA (even when augmented with several 1-reversal counters) over $x_1^* \times x_2^*$ for some strings x_1, x_2, then M can always be converted to a 0-synchronized 2-tape 1-reversal NCM (i.e., once the counter decrements, it can no longer increment). This result is tight in the sense that there are 2-tape DFAs over $a^* \times a^*$, 3-tape DFAs over $a^* \times a^* \times a^*$, and 2-tape DPDAs over $a^* b^* \times a^*$, (where a, b are distinct symbols) that cannot be accepted by 0-synchronized 2-tape NFAs, 3-tape NPDAs, and 2-tape NPDAs, respectively. We also show, rather unexpectedly, that an n-tape NPDA augmented with 1-reversal counters over $B_1 \times \cdots \times B_n$, where each B_i is of the form $x_1^* \cdots x_k^*$ (for some nonnull strings $x_1, \ldots, x_k$), can always be converted to a synchronized n-tape DFA with 1-reversal counters (i.e., deterministic and no stack).

2   Preliminaries

A language L is *letter-bounded* if it is a subset of $a_1^* \cdots a_n^*$ for some distinct letters (symbols) $a_1, \ldots, a_n$. L is *bounded* if it is a subset of $w_1^* \cdots w_n^*$ for some (not necessarily distinct) nonnull strings $w_1, \ldots, w_n$. When there is no confusion from the context, we shall refer to letter-bounded also as bounded.

Given an n-tuple $(x_1, \ldots, x_n)$, denote by $AL(x_1, \ldots, x_n)$ an n-track string where the symbols of x_i's are left-justified (i.e., the symbols are aligned) and the shorter strings are right-filled with blanks (λ) to make all tracks the same length. For example, $AL(01, 1111, 101)$ has $01\lambda\lambda$ on the upper track, 1111 on the middle track, and 101λ on the lower track. Given a set L of n-tuples, define $AL(L) = \{AL(x) \mid x \in L\}$.

The following lemma can be easily verified.

Lemma 1. *Let L a set of n-tuples.*

1. *L is accepted by a 0-synchronized n-tape NFA if and only if $AL(L)$ is regular.*
2. *L is accepted by a 0-synchronized n-tape NPDA if and only if $AL(L)$ is context-free.*

The next lemma says that a k-synchronized automaton can always be converted to an equivalent 0-synchronized automaton. The conversion uses a "finite buffer" in the states.

Lemma 2. *If L is accepted by a k-synchronized n-tape automaton (e.g., an n-tape NFA, n-tape NPDA, n-tape NCM, etc.) M for some k, then M can be converted to an equivalent 0-synchronized n-tape automaton of the same type.*

Let $\mathbb{N}$ be the set of nonnegative integers and k be a positive integer. A subset Q of $\mathbb{N}^k$ is a *linear set* if there exist vectors $v_0, v_1, \ldots, v_n$ in $\mathbb{N}^k$ such that $Q = \{v_0 + t_1 v_1 + \cdots + t_n v_n \mid t_1, \ldots, t_n \in \mathbb{N}\}$. The vectors v_0 (referred to as the *constant vector*) and $v_1, \ldots, v_n$ (referred to as the *periods*) are called the *generators* of the linear set Q. The set $Q \subseteq N^k$ is *semilinear* if it is a finite union of linear sets. The empty set is a trivial (semi)linear set, where the set of generators is empty. Every finite subset of $\mathbb{N}^k$ is semilinear – it is a finite union of linear sets whose generators are constant vectors. Semilinear sets are closed under (finite) union, complementation and intersection. It is known that the disjointness, containment, and equivalence problems for semilinear sets are decidable [3].

Let $\Sigma = \{a_1, \ldots, a_k\}$. For $w \in \Sigma^*$, let $|w|$ is the number of letters in w, and $|w|_{a_i}$ denote the number of occurrences of a_i in w. The *Parikh image* $P(w)$ of w is the vector $(|w|_{a_1}, \ldots, |w|_{a_k})$; similarly, the Parikh image of a language L is defined as $P(L) = \{P(w) \mid w \in L\}$.

It is known that the Parikh image of a language L accepted by an NPDA (i.e., L is context-free) is an effectively computable semilinear set [8]. This was generalized in [4]:

Theorem 4

1. If $L \subseteq \Sigma^*$ is accepted by an NPDA with reversal-bounded counters, then $P(L)$ is an effectively computable semilinear set.
2. If $L \subseteq w_1^* \cdots w_n^*$ is accepted by an NPDA with reversal-bounded counters (where $w_1, \ldots, w_n$ are nonnull strings), then $Q_L = \{(i_1, \ldots, i_n) \mid w_1^{i_1} \cdots w_n^{i_n} \in L\}$ is an effectively computable semilinear set.

We will need the following result from [4]:

Theorem 5. *The emptiness (Is $L(M) = \varnothing$?) and infiniteness (Is $L(M)$ infinite ?) problems for 1-tape NPDAs with reversal-bounded counters are decidable.*

3   Synchronizabilty of Multitape NPDAs

Our first result shows that a 2-tape NPDA (possibly augmented with reversal-bounded counters) on unary inputs can *always* be converted to an equivalent 0-synchronized 2-tape 1-reversal NCM.

Theorem 6. *Let $r, s \geq 1$. Every 2-tape NPDA M with reversal-bounded counters over $(a^r)^* \times (b^s)^*$ can be converted to an equivalent (i.e., they accept the same language) 0-synchronized 2-tape 1-reversal NCM M'.*

Proof. Let A be the 2-track symbol $AL(a, b)$ and B be the 2-track symbol $AL(a, \lambda)$. We construct from M a 1-tape NPDA M_1 with two additional counters, C_1 and C_2. M_1 accepts the bounded language (i.e., a subset of $A^* B^*$) $L_1 = \{AL((a^r)^m, (b^s)^n) \mid rm \geq sn, ((a^r)^m, (b^s)^n) \in L(M)\}$ as follows, when given input $A^i B^j$:

1. M_1 reads the input and checks that $i = sn$ for some n and $i + j = rm$ for some m. It also stores the value sn in counter C_1 and the value rm in counter C_2.
2. M_1 then simulates the computation of M on $((a^r)^m, (b^s)^n)$ by decrementing (i.e., reversing) the counters C_1 and C_2. When M accepts, M_1 accepts.

Now M_1 is an NPDA with reversal-bounded counters. From Theorem 4, the Parikh image of any language accepted by an NPDA with reversal-bounded counters over a bounded language (in this case over the language $A^* B^*$) is a semilinear set. Hence, the set $\{(i, j) \mid A^i B^j \in L(M_1)\}$ is semilinear, i.e., a finite union of linear sets. Let Q be one of these linear sets. Then it has the form $Q = \{(v_{01}, v_{02}) + t_1(v_{11}, v_{12}) + \cdots + t_n(v_{n1}, v_{n2}) \mid t_1, \ldots, t_n \in \mathbb{N}\}$. We now describe a 1-tape 1-reversal NCM accepting the language $\{A^{v_{01} + t_1 v_{11} + \cdots + t_n v_{n1}} B^{v_{02} + t_1 v_{12} + \cdots + t_n v_{n2}} \mid t_1, \ldots, t_n \in \mathbb{N}\}$. The v_{ij}'s can be hard-coded in the finite-state control. On input $A^i B^j$, the NCM first moves the input head v_{01} cells to the right and increments the counter by v_{02}. Then for $1 \leq i \leq n$, the machine executes the following process $t_i \geq 0$ times (where t_i is nondeterministically chosen): Moves the head v_{i1} cells to the right and increments the counter by v_{i2}. When the head reaches the end of the input segment A^i, the NCM checks that j (of B^j) is equal to the value stored in the counter. It follows (since 1-tape 1-reversal NCM languages are obviously closed under union) that $L(M_1)$ can be accepted by a 1-tape 1-reversal NCM M_2. From M_2, we can then directly construct a 0-synchronized 2-tape 1-reversal NCM M' accepting $\{((a^r)^m, (b^s)^n) \mid rm \geq sn, ((a^r)^m, (b^s)^n) \in L(M)\}$.

Similarly, we can construct a 0-synchronized 2-tape 1-reversal NCM M'' accepting $\{((a^r)^m, (b^s)^n)) \mid rm < sn, ((a^r)^m, (b^s)^n) \in L(M)\}$. Finally, we construct from M' and M'' a 0-synchronized 2-tape 1-reversal NCM M''' accepting $L(M') \cup L(M'') = L(M)$. $\qquad\square$

Corollary 1. *We can construct, given a 2-tape NPDA M with reversal-bounded counters over $w_1^* \times w_2^*$ for some nonnull strings w_1 and w_2, an equivalent 0-synchronized 2-tape 1-reversal NCM M'.*

Proof. Let $w_1 = a_1 \cdots a_r$ and $w_2 = b_1 \cdots b_s$ for some $r, s \geq 1$ and symbols $a_1, \ldots, a_r, b_1, \ldots, b_s$. We construct from M a 2-tape NPDA M_1 with reversal-bounded counters that accepts $\{((a^r)^m, (b^s)^n) \mid (w_1^m, w_2^n) \in L(M)\}$. Then from Theorem 6 we can construct a 0-synchronized 2-tape 1-reversal NCM M_2 equivalent to M_1. Finally, from M_2 we can construct a 0-synchronized 2-tape 1-reversal NCM M' equivalent to M. $\qquad\square$

The ideas in the proofs of Theorem 6 and Corollary 1 can be used to also show:

Corollary 2. *Every language over $w_1^* w_2^*$ accepted by a 1-tape NPDA with reversal-bounded counters can be accepted by a 1-tape 1-reversal NCM.*

The next result shows that Theorem 6 is tight in that it cannot be extended to the following cases: (1) one tape is binary, even under the restriction that the input on that tape is bounded, and (2) when there are three unary tapes.

Theorem 7

1. $L = \{(a^{3i}b^{2i}, c^i) \mid i > 0\}$ can be accepted by 2-tape 1-reversal DCM but cannot be accepted by a 0-synchronized 2-tape NPDA.
2. $L = \{(a^{3i}, b^{2i}, c^i) \mid i > 0\}$ can be accepted by 3-tape DFA but cannot be accepted by a 0-synchronized 3-tape NPDA.

Proof. For part 1, clearly L can be accepted by a 2-tape 1-reversal DCM. Suppose L can be accepted by a 0-synchronized (k-synchronized for a given k, or k-synchronized for some k) 2-tape NPDA M. Then $AL(L)$ is context-free by Lemma 1. Let

$$t = (a^{3m}b^{2m}, c^m)$$

be the tuple in L where m is the Ogden's pumping lemma constant for L.

We mark the m symbols $AL(a, c)$ in $AL(t)$. According to Ogden's lemma, $AL(t)$ can be written as $UVXYZ$ where VY has at least one marked position; VXY has at most m marked positions; and UV^kXY^kZ is in L for every $k \geq 0$.

We have the following cases:

1. Y has a marked position: this means that $|Y| \geq 1$, Y consists solely of $AL(a, c)$, and V is either empty or consists solely of $AL(a, c)$. The word $UVVXYYZ$ does not have the correct number of b's on its upper track, and cannot be in L, a contradiction.
2. V has a marked position: this means that $|V| \geq 1$, V consists solely of $AL(a, c)$, and Y is either empty or consists solely of $AL(a, c)$, $AL(a, \lambda)$, or $AL(b, \lambda)$.
   (a) Y is empty: same as case 1.
   (b) Y consists solely of $AL(a, c)$: same as case 1.
   (c) Y consists solely of $AL(a, \lambda)$: same as case 1.
   (d) Y consists solely of $AL(b, \lambda)$: the strings $UV^{k+1}XY^{k+1}Z$ are of the form

$$AL(xa^{3m+k|V|}b^{2m+k|Y|}, xc^{m+k|V|}),$$

where $3m + k|V| = 3(m + k|V|)$, which implies that $|V| = 0$, a contradiction.

For part (2), L can be accepted by a 3-tape DFA, but we can show (as in part 1) that L cannot be accepted by a 0-synchronized 3-tape NPDA.   □

Even though Theorem 7 is a negative result, we now show that synchronization of n-tape NPDAs (even with reversal-bounded counters) over $B_1 \times \cdots \times B_n$, where each B_i is of the form $x_{i_1}^* \cdots x_{i_{k_i}}^*$ (for some nonnull strings $x_{i_1}, \ldots, x_{i_{k_i}}$), is decidable.

Theorem 8. *It is decidable to determine, given an integer $k \geq 0$ and an n-tape NPDA M with reversal-bounded counters over $B_1 \times \cdots \times B_n$, whether M is k-synchronized.*

Proof. We prove the theorem for the case of 2 tapes. The same technique works for any number of tapes. Let M be a 2-tape NPDA, where the string on tape 1 is from $w_1^* \cdots w_r^*$ and the string on tape 2 is from $z_1^* \cdots z_s^*$, for some $r, s \geq 1$ and nonnull strings $w_1, \ldots, w_r, z_1, \ldots, z_s$. Construct a (1-tape) NPDA M_1 with $r + s$ 1-reversal counters $C_1, \ldots, C_r, D_1, \ldots, D_s$ that accepts the language $L(M_1) = \{w_1^{i_1} \cdots w_r^{i_r} z_1^{j_1} \cdots z_s^{j_s} \mid (w_1^{i_1} \cdots w_r^{i_r}, z_1^{j_1} \cdots z_s^{j_s}) \in L(M)\}$. M_1 operates as follows when given the input $w_1^{i_1} \cdots w_r^{i_r} z_1^{j_1} \cdots z_s^{j_s}$: It reads the input and guesses its decomposition and stores $i_1, \ldots, i_r, j_1, \ldots, j_s$ in counters $C_1, \ldots, C_r, D_1, \ldots, D_s$. Then M_1 simulates the computation of M on $(w_1^{i_1} \cdots w_r^{i_r}, z_1^{j_1} \cdots z_s^{j_s})$, using the integers stored in the counters. M_1 accepts if and only if M accepts. From part 2 of Theorem 4,

$$Q_{L(M_1)} = \{(i_1, \ldots, i_r, j_1, \ldots, j_s) \mid w_1^{i_1} \cdots w_r^{i_r} z_1^{j_1} \cdots z_s^{j_s} \in L(M_1)\}$$

is an effectively computable semilinear set.

Next, we construct from M a 2-tape NPDA M' which simulates M faithfully except that it halts and rejects if, during the computation, the heads of M

are more than k cells apart (when neither head is on \$). Note that the finite-state control can check that the distance between the heads is no more than k during the computation, since k is given. Clearly $L(M') \subseteq L(M)$, and M is k-synchronized iff $L(M) = L(M')$. To decide this condition, we construct from M' (as above), an NPDA with 1-reversal counters M'_1 such that $L(M'_1) = \{w_1^{i_1} \cdots w_r^{i_r} z_1^{j_1} \cdots z_s^{j_s} \mid (w_1^{i_1} \cdots w_r^{i_r}, z_1^{j_1} \cdots z_s^{j_s}) \in L(M')\}$. Then $Q_{(L(M'_1))}$ is also semilinear. Clearly, $L(M) = L(M')$ iff $L(M_1) = L(M'_1)$. The result follows, since the equivalence of semilinear sets is decidable [3]. □

Note that Theorem 8 is the best possible in the sense that if the string on one of the tapes is allowed to come from Σ^* (where Σ contains at least two letters), the problem becomes undecidable for 2-tape NPDA (even without counters), by Theorem 2, part 2. However, for n-tape NFAs over $\Sigma^* \times B_1 \times \cdots \times B_{n-1}$ (i.e., no stack and no counters), the problem is decidable [2].

The next proposition shows that the 2-tape NCM in Theorem 6 cannot be replaced with a 2-tape NFA.

Proposition 1. *The unary language $L = \{(a^m, b^n) \mid m > 0, n = 2m\}$ can be accepted by a 2-tape DFA M but cannot be accepted by any 0-synchronized 2-tape NFA.*

Proof. It is easy to see that L can be accepted by a 2-tape DFA. However, it cannot be accepted by any 0-synchronized 2-tape NFA M. For suppose M has p states. Consider the tuple (a^p, b^{2p}). Then for some $1 \leq k \leq p$, (a^{p+k}, b^{p+k+p}) will also be accepted, a contradiction. In fact L cannot be accepted by k-synchronized 2-tape NFA for any k, since we know that such a machine can be converted to be 0-synchronized. □

Note that by Theorem 6, the language L above can be accepted by a 0-synchronized 2-tape 1-reversal NCM M. Such a machine can be constructed as follows: Given (a^m, b^n), M moves both heads to the right simultaneously in (0-sync) while incrementing the counter for every a. When the first head reaches \$, the second head uses the counter (which would now have m) and checks that length of the remaining b's is m (i.e., $n = 2m$). Note that the M is actually a 2-tape 1-reversal counter DCM.

It was shown in [2] that it is decidable to determine, given a unary 2-tape NFA (i.e., the inputs are over $a^* \times b^*$), whether it is k-synchronized for some k (i.e., k is not specified). It is open whether this problem is decidable for unary 2-tape NPDAs and unary 2-tape NCMs. We will show that for some extensions of these models, the problem is undecidable.

An n-tape 1-reversal 2-pushdown automaton is an n-tape NFA augmented with two pushdown stacks, each of which makes at most one reversal. Similarly, a deterministic n-tape 2-counter machine is an n-tape DFA augmented with two counters.

The emptiness problem for 1-tape 1-reversal 2-pushdown automata is undecidable [1], and the halting problem for deterministic 2-counter machines without an input tape (where the counters are initially zero) is undecidable [7].

Proposition 2. *The following problems are undecidable, given a 2-tape 1-reversal 2-pushdown automaton (resp., a deterministic 2-tape 2-counter machine) M over unary inputs from $a^* \times b^*$:*

1. *Is M k-synchronized for a given k ?*
2. *Is M k-synchronized for some k ?*

Proof. Let T be a one-tape 1-reversal 2-pushdown automaton. We construct a 2-tape 1-reversal 2-pushdown automaton M which, when given input (a^i, b), operates as follows: Without moving the heads, M simulates the computation of T on some input y (by guessing the symbols comprising y) and if T accepts, M moves its first head to the right until it reaches the end marker and then moves its second head to the right end marker and accepts. Then either $L(T) = \varnothing$ and M is 0-synchronized, or $L(T) \neq \varnothing$ and M is not k-synchronized for any k. It follows that M is k-synchronized for some k (or a fixed k) iff the language accepted by T is empty, which is undecidable.

Similarly, if T is a deterministic 2-counter machine (without an input tape), we construct a deterministic 2-tape 2-counter machine M which, when given (a^i, b), first simulates T and if T halts, then M proceeds as above.   □

It follows from Theorem 2 part 1 (resp., part 2) that it is undecidable to determine, given a 2-ambiguous 2-tape 3-reversal NPDA M with 1-reversal counters over $\Sigma^* \times a^*$, whether M is k-synchronized for a given k (resp., for some k). However, part 3 of Theorem 2 is decidable for this machine, since the answer is always yes, as the following result indicates:

Theorem 9. *Every n-tape NPDA M with 1-reversal counters over $\Sigma^* \times B_1 \times \cdots \times B_{n-1}$, can be converted to an equivalent 0-synchronized n-tape NPDA M' with 1-reversal counters.*

Proof. (Sketch) We illustrate the construction for $n = 2$. Let M be a 2-tape NPDA over $\Sigma^* \times x_1^* \cdots x_r^*$. We construct a 2-tape NPDA M' which will accept $L(M)$ as follows. M' will need $2r$ additional 1-reversal counters, $C_1, \ldots, C_r$, $D_1, \ldots, D_r$. Given an input $(w, x_1^{i_1} \cdots x_r^{i_r})$, M' starts by nondeterministically guessing and storing i_1 in C_1 and D_1, i_2 in C_2 and D_2, $\ldots$, i_r in C_r and D_r. Then M' simulates the computation of M on $(w, x_1^{i_1} \cdots x_r^{i_r})$ moving the two heads in 0-sync using counters $C_1, \ldots, C_r$; whenever the second head of M has consumed an x_i segment, C_i is decremented. The second tape of M' and the counters $D_1, \ldots, D_r$ are used to verify the guessed values $i_1, \ldots, i_r$; the second head of M' moves in sync with the first head and decrements D_i whenever it moves right of an x_i segment. We omit the details.   □

If in Theorem 9, all input tapes are bounded, we can prove a stronger result. We will need the following lemma:

Lemma 3. *Let Σ be an alphabet and $L \subseteq x_1^* \cdots x_m^* \times y_1^* \cdots y_n^*$. Define the 2-track alphabet $\Delta = (\Sigma \cup \{\lambda\}) \times (\Sigma \cup \{\lambda\})$. Then for some k and nonnull strings $u_1, \ldots, u_k$ in Δ^*, $AL(L) \subseteq u_1^* \cdots u_k^*$.*

Proof. By induction. Recall that bounded languages (over subsets of $x_1^* \cdots x_m^*$, for some nonnull strings $x_1, \ldots, x_m$) are closed under union.

1. $m = 0$: $L \subseteq AL(\lambda, y_1)^* \ldots AL(\lambda, y_n)^*$.
2. $n = 0$: $L \subseteq AL(x_1, \lambda)^* \ldots AL(x_m, \lambda)^*$.
3. $m = n = 1$: Given a string $z = z_1 z_2 \ldots z_r$ and $0 \le k \le r$, define $pre(z, k)$ to be the prefix of x consisting of the first k symbols of z. In the following, let x and y be nonnull strings. Define

$$e_w = \frac{lcm(|x|, |y|)}{|w|}, \quad w \in \{x, y\};$$

$$v_{i,j,k,l} = AL(x^i pre(x, k), y^j pre(y, l))$$

$$\text{for } 0 \le i \le e_x, 0 \le j \le e_y, 0 \le k \le |x|, 0 \le l \le |y|;$$

$$B(x, y) = \bigcup_{i,j,k,l} v_{e_x, e_y, 0, 0}^* v_{i,j,k,l} AL(x, \lambda)^* AL(\lambda, y)^*.$$

Then $AL(L) \subseteq B(x_1, y_1)$, since for any $i, j \ge 0$

$$AL(x^i, y^j) = AL(x^{ke_x}, x^{ke_y}) \cdot AL(x^r, y^s)$$

for some $k, r, s \ge 0$, and either $r < e_x$ or $s < e_y$ ($\cdot$ denotes concatenation). The string on the right is a member of $v_{e_x, e_y, 0, 0}^* v_{r,s,0,0} AL(x, \lambda)^* \cdot AL(\lambda, y)^*$. Note that $B(x, y)$ contains more strings than $AL(x^i, y^j)$. The parameters l and k allow using a prefix of x or y instead of λ's as padding symbols in the alignment of x^i and y^j. These additional words will be used in the next case.

4. $mn > 1$: In the following, let $clr(z, k)$ be the circular left shift of a nonnull string z by k positions, and let $post(z, k)$ be the postfix consisting of the last k symbols of the string z. Define

$$C_k = B(x_1, y_1)B(\{clr(x_1, k), x_2, \ldots, x_m\}, \{y_2, \ldots, y_n\}), 0 \le k \le |x_1|$$
$$D_k = B(x_1, y_1)B(\{post(x_1, k), x_2, \ldots, x_m\}, \{y_2, \ldots, y_n\}), 0 \le k \le |x_1|$$
$$E_l = B(x_1, y_1)B(\{x_2, \ldots, x_m\}, \{clr(y_2, l), y_2, \ldots, y_n\}), 0 \le l \le |y_1|$$
$$F_l = B(x_1, y_1)B(\{x_2, \ldots, x_m\}, \{post(y_2, l), y_2, \ldots, y_n\}), 0 \le l \le |y_1|$$

and $B(\{x_1, \ldots, x_m\}, \{y_1, \ldots, y_n\}) = \bigcup_{k,l} (C_k \cup D_k \cup E_l \cup F_l)$. Then $AL(L) \subseteq B(\{x_1, \ldots, x_m\}, \{y_1, \ldots, y_n\})$. To see this, consider $AL(x_1^{i_1} \ldots x_m^{i_m}, y_1^{j_1} \ldots y_m^{j_m})$ where the exponents are nonnegative. There are two cases:

(a) $|x_1^{i_1}| \ge |y_1^{j_1}|$: there exist i and k such that $|x_1^i pre(x_1, k)| = |y_1^{j_1}|$ and

$$AL(x_1^{i_1} \ldots x_m^{i_m}, y_1^{j_1} \ldots y_m^{j_m}) = AL(x_1^i pre(x_1, k), y_1^{j_1}) \cdot AL(post(x_1,$$
$$|x_1| - k)x_1^{i_1 - i - 1} x_2^{i_2} \ldots x_m^{i_m}, y_2^{j_2} \ldots y_n^{j_n}).$$

Note that if $i_1 - i - 1 > 0$,

$$post(x_1, |x_1| - k)x_1^{i_1 - i - 1} = clr(x_1, k)^{i_1 - i - 1} post(clr(x_1, k), |x_1| - k),$$

and hence $AL(x_1^{i_1} \ldots x_m^{i_m}, y_1^{j_1} \ldots y_m^{j_m})$ is either in C_k (if $i_1 - i - 1 > 0$) or D_k (otherwise).

(b) $|x_1^{i_1}| < |y_1^{j_1}|$: same as the above case with the roles of x_1 and y_1 reversed.

$\square$

Theorem 10. *Every n-tape NPDA M with 1-reversal counters over $x_{11}^* \cdots x_{1m_1}^*$ $\times \cdots \times x_{n1}^* \cdots x_{nm_n}^*$ can be converted to an equivalent 0-synchronized DFA M' with 1-reversal counters. (Note that M' is deterministic and has no pushdown stack.)*

Proof. We prove the result when M has only two input tapes. The proof easily generalizes to multitapes.

Let $L = L(M) \subseteq x_1^* \cdots x_r^* \times y_1^* \cdots y_s^*$, where the x_i's and the y_i's are nonnull strings over some alphabet Σ.

1. From Theorem 9, L can be accepted by a 0-synchronized 2-tape NPDA M_1 with 1-reversal counters.
2. Define the 2-track alphabet $\Delta = (\Sigma \cup \{\lambda\}) \times (\Sigma \cup \{\lambda\})$.
3. From M_1, we can easily construct a 1-tape NPDA M_2 with 1-reversal counters such that $L(M_2) = AL((L(M_1)) = AL(L) \subseteq \Delta^*$.
4. From Lemma 3, $L(M_2) \subseteq u_1^* \cdots u_k^*$ for some k and nonnull strings $u_1, \ldots, u_k$ in Δ^*.
5. In a recent paper [5], it was shown that if a 1-tape NPDA with 1-reversal counters accepts a language that is a subset of $v_1^* \cdots v_t^*$, where $v_1, \ldots, v_t$ are (not necessarily distinct) nonnull strings over some alphabet, then the language is also be accepted by a DFA with 1-reversal counters (i.e., the machine is deterministic and no stack).
6. It then follows from (3), (4) and (5) that $L(M_2) = AL(L)$ can be accepted by a DFA M_3 with 1-reversal counters.
7. From M_3, we can then directly construct a 0-synchronized DFA M' with 1-reversal counters accepting L.   □

4   Synchronizability of 2-Tape NFAs with 1-Reversal Counters

The next result concerns 2-tape NFAs with 1-reversal counters.

Theorem 11. *It is undecidable to determine, given a 2-ambiguous 2-tape NFA M, whether there exists a 2-tape NFA M' with 1-reversal counters that is 0-synchronized (or k-synchronized for a given k, or k-synchronized for some k) such that $L(M') = L(M)$.*

Proof. Let $\Sigma = \{0, 1, \#\}$ and I be an instance of PCP over the alphabet $\{0, 1\}$. Define the language

$$L = \{(x\#z, y\#^i w) \mid x, y, z, w \in \{0,1\}^*, i > 1, x \neq y\} \cup$$
$$\{(x\#w, x\#^i w) \mid x, w \in \{0,1\}^*, i > 1, x \text{ is a solution to PCP instance } I\}.$$

Clearly, L can be accepted by a 2-tape NFA. Note that first part of L can be accepted deterministically in 0-sync mode. It is easily verified that if the PCP instance has no solution, then $L(M)$ can be accepted by a 0-synchronized 2-tape NFA (even without 1-reversal counters).

Now suppose the PCP instance I has a solution. Assume that L can be accepted by some 0-synchronized 2-tape NFA M' with 1-reversal counters. Then $AL(L)$ can be accepted a 1-tape 2-track NFA M'' with 1-reversal counters.

Let x be a solution to I. Fix this x and consider all the tuples of the form $(x\#w, x\#^{|w|+1}w)$, where $w \in \{0,1\}^+$. Then the 2-track strings of the form $AL((x\#w, x\#^{|w|+1}w))$ are accepted by M''. Now it is known [1] that any 1-tape NFA with 1-reversal counters can be converted to an equivalent machine that runs in linear time; so we may assume that M'' runs in linear time. Suppose M'' has s counters. A configuration of M'' is an $(s+1)$-tuple $(q, v_1, \ldots, v_s)$, where q is a state and $(v_1, \ldots, v_s)$ are the values of the counters.

Consider a 2-track input of the form $AL((x\#w, x\#^{|w|+1}w))$ where $|w| = n$ This input is accepted by M''. Clearly, the number of possible configurations when the input head of M'' reaches $AL((x\#w, x\#^{|w|+1}))$ is $O(n^s)$.

Now consider another input $AL((x\#u, x\#^{|u|+1}u))$ where $|u| = n$ and $u \neq w$. Since there are 2^n binary strings of length n, $AL((x\#u, x\#^{|u|+1}w))$ will also be accepted by M'' for n large enough. It follows that $(x\#u, x\#^{|w|+1}w)$ will be accepted by M'. This is a contradiction.

The 2-tape NFA M can be made 2-ambiguous. We omit the construction. □

5 Conclusions

We studied the decision problem of whether a given multitape machine M can be converted into an equivalent k-synchronized machine M' for various classes of NPDA and obtained answers ranging from 'always possible' to 'decidable' to 'undecidable'.

References

1. Baker, B.S., Book, R.V.: Reversal-bounded multipushdown machines. J. Computer and System Sciences 8, 315–332 (1974)
2. Eğecioğlu, Ö., Ibarra, O.H., Tran, N.Q.: Multitape NFA: Weak Synchronization of the Input Heads. In: Bieliková, M., Friedrich, G., Gottlob, G., Katzenbeisser, S., Turán, G. (eds.) SOFSEM 2012. LNCS, vol. 7147, pp. 238–250. Springer, Heidelberg (2012)
3. Ginsburg, G., Spanier, E.: Bounded Algol-like languages. Trans. of the Amer. Math. Society 113, 333–368 (1964)
4. Ibarra, O.H.: Reversal-bounded multicounter machines and their decision problems. J. Assoc. Comput. Math. 25, 116–133 (1978)
5. Ibarra, O.H., Seki, S.: Characterizations of bounded semilinear languages by one-way and two-way deterministic machines. In: Proc. 13th Int. Conf. on Automata and Formal Languages, AFL 2011 (2011)
6. Ibarra, O.H., Tran, N.Q.: Weak Synchronization and Synchronizability of Multitape Pushdown Automata and Turing Machines. In: Dediu, A.-H., Martín-Vide, C. (eds.) LATA 2012. LNCS, vol. 7183, pp. 337–350. Springer, Heidelberg (2012)

7. Minsky, M.: Recursive unsolvability of Post's problem of Tag and other topics in the theory of Turing machines. Ann. of Math. (74), 437–455 (1961)
8. Parikh, R.J.: On context-free languages. J. Assoc. Comput. Mach. 13, 570–581 (1966)
9. Yu, F., Bultan, T., Ibarra, O.H.: Relational String Verification Using Multi-track Automata. In: Domaratzki, M., Salomaa, K. (eds.) CIAA 2010. LNCS, vol. 6482, pp. 290–299. Springer, Heidelberg (2011); Extended version in International J. Found. of Comput. Sci. 22, 1909–1924 (2011)

Regular Ideal Languages
and Their Boolean Combinations

Franz Jahn, Manfred Kufleitner*, and Alexander Lauser*

FMI, University of Stuttgart, Germany
jahnfz@studi.informatik.uni-stuttgart.de,
{kufleitner, lauser}@fmi.uni-stuttgart.de

Abstract. We consider ideals and Boolean combinations of ideals. For the regular languages within these classes we give expressively complete automaton models. In addition, we consider general properties of regular ideals and their Boolean combinations. These properties include effective algebraic characterizations and lattice identities.

In the main part of this paper we consider the following deterministic one-way automaton models: unions of flip automata, weak automata, and Staiger-Wagner automata. We show that each of these models is expressively complete for regular Boolean combination of right ideals. Right ideals over finite words resemble the open sets in the Cantor topology over infinite words. An omega-regular language is a Boolean combination of open sets if and only if it is recognizable by a deterministic Staiger-Wagner automaton; and our result can be seen as a finitary version of this classical theorem. In addition, we also consider the canonical automaton models for right ideals, prefix-closed languages, and factorial languages.

In the last section, we consider a two-way automaton model which is known to be expressively complete for two-variable first-order logic. We show that the above concepts can be adapted to these two-way automata such that the resulting languages are the right ideals (resp. prefix-closed languages, resp. Boolean combinations of right ideals) definable in two-variable first-order logic.

1 Introduction

The Cantor topology over infinite words is an important concept for classifying languages over infinite words. For example, an ω-regular language is deterministic if and only if it is a countable intersection of open sets, *cf.* [18, Remark 5.1]. There are many other properties of ω-languages which can be described using the Cantor topology, see *e.g.* [12,15]. Ideals are the finitary version of open sets in the Cantor topology. A subset P of a monoid M is a right (resp. left, two-sided) *ideal* if $PM \subseteq P$ (resp. $MP \subseteq P$, $MPM \subseteq P$). In particular, a language $L \subseteq A^*$ is a right ideal if $LA^* \subseteq L$. A *filter* is the complement of an ideal. Thus over finite words, a language $L \subseteq A^*$ is a right filter if and only if it is *prefix-closed*,

---

* The last two authors were supported by the German Research Foundation (DFG) under grant DI 435/5-1.

N. Moreira and R. Reis (Eds.): CIAA 2012, LNCS 7381, pp. 205–216, 2012.

i.e., if $uv \in L$ implies $u \in L$. Prefix-closed languages correspond to closed sets in the Cantor topology. A language $L \subseteq A^*$ is a two-sided filter if and only if it is *factorial* (also known as *factor-closed* or *infix-closed*), *i.e.*, if $uvw \in L$ implies $v \in L$. Our first series of results gives effective algebraic characterizations of right (resp. left, two-sided) ideal languages and of Boolean combinations of such languages. In addition, we give lattice identities for each of the resulting language classes. As a byproduct, we show that a language is both regular and a Boolean combination of right (resp. left, two-sided) ideals if and only if it is a Boolean combination of regular right (resp. left, two-sided) ideals, *i.e.*, if $\mathcal{I}$ is the class of right (resp. left, two-sided) ideals and REG is the class of regular languages, then $\mathrm{REG} \cap \mathbb{B}\mathcal{I} = \mathbb{B}(\mathrm{REG} \cap \mathcal{I})$. Here, $\mathbb{B}$ denotes the Boolean closure.

The second contribution of this paper consists of expressively complete (one-way) automaton models for right ideals, for prefix-closed languages, for factorial languages, and for Boolean combinations of right ideals. The results concerning ideals and closed languages are straightforward and stated here only to draw a more complete picture. Our main original contribution are automaton models for regular Boolean combinations of right ideals. We always assume that every state in an automaton is reachable from some initial state, *i.e.*, all automata in this paper are accessible.

- A *flip automaton* is an automaton with no transitions from final states to non-final states, *i.e.*, it "flips" at most once from a non-final to a final state. Consequently, every minimal complete flip automaton has at most one final state which has a self-loop for each letter of the alphabet. Paz and Peleg have shown that if a language L is recognized by a complete deterministic automaton $\mathcal{A}$, then L is a right ideal if and only if $\mathcal{A}$ is a flip automaton [11]. A language is a regular Boolean combination of right ideals if and only if it is recognized by a union of flip automata (which do not have to be complete).
- An automaton is *fully accepting* if all states are final. A word u is rejected in a fully accepting automaton $\mathcal{A}$ if and only if there is no u-labeled path in $\mathcal{A}$ which starts in an initial state. Nondeterministic fully accepting automata are expressively complete for prefix-closed languages. Moreover, if a language L is recognized by a deterministic trim automaton $\mathcal{A}$, then L is prefix-closed if and only if $\mathcal{A}$ is fully accepting.
- A *path automaton* is an automaton $\mathcal{A}$ such that all states are both initial and final, *i.e.*, a word u is accepted by $\mathcal{A}$ if there exists a u-labeled path in $\mathcal{A}$. Both deterministic and nondeterministic path automata recognize exactly the class of regular factorial languages. This characterization can be implicitly found in the work of Avgustinovich and Frid [1].
- An automaton is *weak* if in each strongly connected component either all states are final or all states are non-final. Any run of a weak automaton flips only a bounded number of times between final and non-final states. Nondeterministic weak automata can recognize all regular languages. On the other hand, if a language L is recognized by a deterministic automaton $\mathcal{A}$, then L is a Boolean combination of right ideals if and only if $\mathcal{A}$ is weak. Weak automata have been introduced by Muller, Saoudi, and Schupp [10].

– *Deterministic Staiger-Wagner automata* over infinite words have been used for characterizing ω-languages $L \subseteq A^\omega$ such that both L and $A^\omega \setminus L$ are deterministic [16]. Acceptance of a run in a Staiger-Wagner automaton only depends on the set of states visited by the run (but not on their order or their number of occurrences). We show that, over finite words, deterministic Staiger-Wagner automata are expressively complete for Boolean combinations of right ideals. In particular, deterministic Staiger-Wagner automata and deterministic weak automata accept the same class of languages.

We note that flip automata, fully accepting automata, and weak automata yield effective characterizations of the respective language classes. For example, in order to check whether a deterministic automaton $\mathcal{A}$ recognizes a Boolean combination of right ideals, it suffices to test if $\mathcal{A}$ is weak. Moreover, the above automaton models can easily be applied to subclasses of automata such as counter-free automata [9]. This immediately yields results of the following kind: A regular language L is both star-free and a Boolean combination of right ideals if and only if its minimal automaton is weak and counter-free.

For some classes of languages it is more adequate to use two-way automata. The relation between two-way automata and ideals (resp. closed languages, Boolean combinations of ideals) is more complex than for one-way automata. In the last section, we consider deterministic partially ordered two-way automata (po2dfa). Partially ordered automata are also known as *very weak*, *1-weak*, or *linear* automata. We give restrictions of po2dfa's which define the right ideals (resp. prefix-closed languages, Boolean combinations of right ideals) inside the po2dfa-recognizable languages. The class of languages recognized by po2dfa has a huge number of equivalent characterizations; these include the variety **DA** of finite monoids, two-variable first-order logic, unary temporal logic, unambiguous polynomials, and rankers; see *e.g.* [17,4]. Some of these characterizations admit natural restrictions which are expressively complete for their ideal (resp. prefix-closed, Boolean combination of ideals) counterparts. We introduce one-pass flip po2dfa (resp. one-pass fully accepting po2dfa, one-pass po2dfa) as expressively complete automaton models for right ideals (resp. prefix-closed languages, Boolean combinations of right ideals) inside the class of po2dfa-recognizable languages. For definitions of these automaton models, we refer the reader to Section 5. The main challenge for each of the above automaton models is showing closure under union and intersection since standard techniques, such as sequentially executing one automaton after the other, cannot be applied. As a complementary result we see that weak one-pass two-way dfa's have the same expressive power as their one-way counterparts, *i.e.*, recognize regular Boolean combinations of right ideals.

2 Preliminaries

Throughout this paper, A is a finite alphabet. The set of finite words over the alphabet A is denoted by A^*; it is the free monoid over A. The neutral element is the empty word ε. The set of nonempty words is $A^+ = A^* \setminus \{\varepsilon\}$. If a language $L \subseteq A^*$ satisfies $LA^* \subseteq L$ (resp. $A^*L \subseteq L$, $A^*LA^* \subseteq L$), then L is a *right ideal* (resp. *left ideal, two-sided ideal*). If $L = A^* \setminus K$ for some right (resp. left, two-sided) ideal K, then L is *prefix-closed* (resp. *suffix-closed, factorial*). Factorial languages are also known as *factor-closed* or *infix-closed*. Boolean combinations consist of complementation, *finite* unions, and *finite* intersections.

   Green's relations on a monoid M are defined as follows. For $x, y \in M$ let $x \leq_{\mathcal{R}} y$ (resp. $x \leq_{\mathcal{L}} y$, $x \leq_{\mathcal{J}} y$) if there exist $s, t \in M$ such that $x = ys$ (resp. $x = ty$, $x = tys$). We set $x \mathrel{\mathcal{R}} y$ if both $x \leq_{\mathcal{R}} y$ and $y \leq_{\mathcal{R}} x$. The relations $\mathcal{L}$ and $\mathcal{J}$ are defined similarly involving $\leq_{\mathcal{L}}$ and $\leq_{\mathcal{J}}$, respectively. An element $x \in M$ is *idempotent* if $x = x^2$. In every finite monoid M there exists a number $\omega \geq 1$ such that x^ω is idempotent for all $x \in M$. A homomorphism $h : A^* \to M$ *recognizes* a language $L \subseteq A^*$ if $L = h^{-1}(P)$ for some $P \subseteq M$, i.e., $u \in L$ if and only if $h(u) \in P$. A monoid M recognizes L if there exists a homomorphism $h : A^* \to M$ recognizing L. For every regular language L there exists a unique minimal finite monoid $\mathrm{Synt}(L)$ which recognizes L (and which is effectively computable as the transition monoid of the minimal automaton). It is the *syntactic monoid* of L, and it is naturally equipped with a recognizing homomorphism $h_L : A^* \to \mathrm{Synt}(L)$, called the *syntactic homomorphism*. A language is regular if and only if its syntactic monoid is finite, see *e.g.* [12].

   Lattice identities are a tool for describing classes of languages (these language classes form so-called lattices). Lattice identities can be defined in the general setting of free profinite monoids [6]. In this paper, we only introduce the ω-notation. We inductively define ω-*terms* over a set of variables Σ: Every $x \in \Sigma$ is an ω-term; and if x and y are ω-terms, then so are xy and $(x)^\omega$. For a number $n \in \mathbb{N}$ and an ω-term x, we define $x(n)$ inductively by $(xy)(n) = x(n)y(n)$, $(x^\omega)(n) = x(n)^{n!}$, and $x(n) = x$ for $x \in \Sigma$, i.e., $x(n)$ is the word obtained by replacing all exponents ω in x by $n!$. Intuitively, x^ω is the idempotent element generated by x with respect to *all* regular languages. A regular language L satisfies the lattice identity $x \to y$ for ω-terms x and y if there exists $n_0 \in \mathbb{N}$ such that for all $n \geq n_0$ and for all homomorphisms $h : \Sigma^* \to A^*$ the implication $h(x(n)) \in L \Rightarrow h(y(n)) \in L$ holds. It satisfies $x \leftrightarrow y$ if $x \to y$ and $y \to x$.

3   Ideals and Their Boolean Combinations

Many interesting properties over finite words can be stated as follows: There exists a prefix (resp. suffix, factor) which has some desirable property $L \subseteq A^*$ and we do not care about subsequent actions. This immediately leads to the right ideal LA^* (resp. left ideal A^*L, two-sided ideal A^*LA^*). Such languages and their Boolean combinations arise naturally, see *e.g.* [2,7]. We give effective algebraic characterizations and lattice identities for the regular ideal languages (Proposition 1) and the regular Boolean combinations of ideals (Theorem 1). In the case of ideals, the proof is straightforward and relies on the following simple fact. If $h : M \to N$ is a surjective homomorphism between monoids and $I \subseteq M$ as well as $J \subseteq N$ are right ideals (resp. left ideals, two-sided ideals), then $h(I)$ and $h^{-1}(J)$ are also right ideals (resp. left ideals, two-sided ideals), *i.e.*, ideals are closed under homomorphic and inverse homomorphic images.

Proposition 1. *Let $L \subseteq A^*$ be a regular language recognized by a surjective homomorphism $h : A^* \to M$ onto a monoid M. The following are equivalent:*

1. *L is a right ideal (resp. left ideal, two-sided ideal).*
2. *$h(L)$ is a right ideal (resp. left ideal, two-sided ideal).*
3. *L satisfies the lattice identity $y \to yz$ (resp. $y \to xy$, $y \to xyz$).*

In particular, property (2) yields decidability of whether a given regular language is a (right, left, or two-sided) ideal of A^* because the syntactic homomorphism $h_L : A^* \to \mathrm{Synt}(L)$ and the set $h_L(L)$ are effectively computable. Moreover, regular (right, left, and two-sided) ideals are closed under union, intersection, and inverse homomorphisms. They do not form so-called *positive varieties* because they are not closed under residuals (even though right ideals are closed under left residuals, and left ideals are closed under right residuals), *cf.* [12]. An easy example is $L = abA^*$ over the alphabet $A = \{a, b\}$; we have $a \in Lb^{-1} = L \cup \{a\}$ and $aa \notin Lb^{-1}$, showing that Lb^{-1} is not a right ideal.

In the next theorem, we consider Boolean combinations of ideals. Note that if $h : M \to N$ is a surjective homomorphism and I, J are ideals of M, then in general, we have $h(I \setminus J) \neq h(I) \setminus h(J)$. Another obstacle for Boolean combinations of ideals is the following: If L is regular and a Boolean combination of ideals K_i, then the K_i need not be regular. As a byproduct of our characterization in Theorem 1, we see that in the above situation, one can find regular ideals K_i' such that L is a Boolean combination of the languages K_i'.

Theorem 1. *Let $L \subseteq A^*$ be a language recognized by a surjective homomorphism $h : A^* \to M$ onto a finite monoid M. Then the following are equivalent:*

1. *L is a Boolean combination of right (resp. left, two-sided) ideals.*
2. *$h(L)$ is a union of $\mathcal{R}$-classes (resp. $\mathcal{L}$-classes, $\mathcal{J}$-classes).*
3. *L satisfies the lattice identity $z(xy)^\omega x \leftrightarrow z(xy)^\omega$ (resp. $s(ts)^\omega z \leftrightarrow (ts)^\omega z$, $s(ts)^\omega z(xy)^\omega x \leftrightarrow (ts)^\omega z(xy)^\omega$).*

Since Theorem 1 (2) can be verified effectively for the syntactic homomorphism, it is decidable whether a given regular language is a Boolean combination of right ideals (resp. left ideals, two-sided ideals).

Every $\mathcal{R}$-class is the set difference between two right ideals. Thus if L is a Boolean combination of (arbitrary) right ideals and if L is recognized by $h : A^* \to M$, then by Theorem 1, the language L can also be written as a Boolean combination of right ideals K_i such that each K_i is recognized by h. The situation for Boolean combinations of left ideals (resp. two-sided ideals) is similar.

For finite monoids, $\mathcal{J}$ is the smallest equivalence relation such that $\mathcal{R} \subseteq \mathcal{J}$ and $\mathcal{L} \subseteq \mathcal{J}$, see e.g. [12, Proposition A.2.5 (2)]. Hence, it follows from Theorem 1 that a regular language L is a Boolean combination of two-sided ideals if and only if L is both a Boolean combination of right ideals and a Boolean combination of left ideals.

In Boolean combinations of right ideals, intuitively speaking, what happens is that the end of words is "concealed." Appending a new symbol as an end-marker to a language yields a Boolean combination of right ideals. Specifically, if L is language over $A \setminus \{a\}$, then La is a Boolean combination of right ideals of A^* because $La = LaA^* \setminus LaA^+$. In Section 5, we will avoid this "revealing" of the end of the word by the right end marker by considering one-pass automata.

4   One-Way Automaton Models

As usual, an *automaton* $\mathcal{A} = (Q, A, \delta, Q_0, F)$ is given by a finite set of states Q, an input alphabet A, a transition relation $\delta \subseteq Q \times A \times Q$, a set of initial states $Q_0 \subseteq Q$, and a set of final states $F \subseteq Q$. For transitions $(p, a, q) \in \delta$ we write $p \xrightarrow{a} q$ and we inductively extend the transition relation to words: $q \xrightarrow{\varepsilon} q$ for all $q \in Q$; and $p \xrightarrow{au} q$ if there exists some $r \in Q$ such that $p \xrightarrow{a} r \xrightarrow{u} q$. A *run* on a word $a_1 \cdots a_n$ with $a_i \in A$ is a sequence of states $q_0 q_1 \cdots q_n$ such that $q_0 \in Q_0$ and $q_{i-1} \xrightarrow{a_i} q_i$ for all i. We always assume that all states are accessible, *i.e.*, for every $q \in Q$ there exist $q_0 \in Q_0$ and $u \in A^*$ such that $q_0 \xrightarrow{u} q$. A word $u \in A^*$ is *accepted* by $\mathcal{A}$ if there exist $p \in Q_0$ and $q \in F$ such that $p \xrightarrow{u} q$. The language *recognized* by $\mathcal{A}$ is $L(\mathcal{A}) = \{u \in A^* \mid u \text{ is accepted by } \mathcal{A}\}$. The automaton $\mathcal{A}$ is *complete* if for every $p \in Q$ and for every $a \in A$ there exists at least one state $q \in Q$ such that $p \xrightarrow{a} q$; it is *trim* if for every $q \in Q$ there exists $u \in A^*$ and $p \in F$ such that $q \xrightarrow{u} p$; and it is *deterministic* if $|Q_0| = 1$ and for all $p \in Q$ and all $a \in A$ there is at most one state $q \in Q$ with $p \xrightarrow{a} q$.

In the remainder of the section, we give automaton models for regular right ideals, prefix-closed languages, factorial languages, and Boolean combinations of right ideals. The results concerning ideals and closed languages are straightforward and presented here only for the sake of completeness. Our main original contribution is Theorem 2, where we give three automaton descriptions of Boolean combinations of ideals: deterministic weak automata, deterministic Staiger-Wagner automata, and unions of deterministic flip automata.

A *flip automaton* is an automaton such that $p \in F$ and $p \xrightarrow{a} q$ implies $q \in F$. The idea is that, in every run, flip automata can "flip" at most once from non-accepting to accepting. Note that the language of a complete flip automata remains unchanged if we add a self-loop $q \xrightarrow{a} q$ for every state $q \in F$ and every letter $a \in A$.

Proposition 2. *Let $L \subseteq A^*$ be recognized by a complete deterministic automaton $\mathcal{A}$. Then the following are equivalent:*

1. *L is a right ideal.*
2. *$\mathcal{A}$ is a flip automaton.*
3. *L is recognized by some complete (nondeterministic) flip automaton.*

The equivalence of (1) and (2) in Proposition 2 is due to Paz and Peleg [11]. Of course, not every complete nondeterministic automaton which recognizes a right ideal has to be a flip automaton. Note that arbitrary (*i.e.*, non-complete and nondeterministic) flip automata can recognize all regular languages.

A *fully accepting automaton* is an automaton in which all states are final, *i.e.*, $F = Q$. The only possibility to reject a word is a missing outgoing transition at some point of the computation. Complementing Proposition 2 leads to the following characterization of fully accepting automata.

Corollary 1. *Let $L \subseteq A^*$ be recognized by a deterministic trim automaton $\mathcal{A}$. Then the following are equivalent:*

1. *L is prefix-closed.*
2. *$\mathcal{A}$ is fully accepting.*
3. *L is recognized by some (nondeterministic) fully accepting automaton.*

A *path automaton* is an automaton such that every state is both initial and final, i.e., $Q_0 = F = Q$. In particular, a path automaton accepts a word $u \in A^*$ if and only if there exists a path $p \xrightarrow{u} q$ for some $p, q \in Q$.

Corollary 2. *Let $L \subseteq A^*$ be a regular language. Then L is factorial if and only if L is recognized by a path automaton.*

For deterministic transition relations, the statement of Corollary 2 can be found implicitly in the work of Avgustinovich and Frid [1].

An automaton is *weak* if for every strongly connected component $C \subseteq Q$, we either have $C \subseteq F$ or $C \cap F = \emptyset$. The concept of weak automata has been introduced by Muller, Saoudi, and Schupp [10] for alternating tree automata. A *Staiger-Wagner automaton* is given by $\mathcal{B} = (Q, A, \delta, q_0, \mathcal{T})$ where $\mathcal{T} \subseteq 2^Q$. Acceptance of a run by a Staiger-Wagner automaton only depends on the set of states visited by the run. A run $q_0 q_1 \cdots q_n$ is *accepting* if $\{q_0, q_1, \ldots, q_n\} \in \mathcal{T}$; and a word is accepted if it has an accepting run.

Lemma 1. *Let $\mathcal{A} = (Q, A, \delta, Q_0, F)$ be a weak automaton. Then there exists $\mathcal{T}$ such that the Staiger-Wagner automaton $\mathcal{B} = (Q, A, \delta, Q_0, \mathcal{T})$ recognizes $L(\mathcal{A})$.*

Our next result shows that both deterministic weak automata and deterministic Staiger-Wagner automata are expressively complete for Boolean combinations of right ideals. Moreover, if a deterministic automaton $\mathcal{A}$ recognizes a Boolean combination of right ideals, then, by Lemma 1, the automaton $\mathcal{A}$ itself can be equipped with a Staiger-Wagner acceptance condition. A third automaton model for Boolean combinations of right ideals is given by unions of (not necessarily complete) deterministic flip automata. This last property follows form Theorem 1 since the inverse homomorphic image of every $\mathcal{R}$-class of a finite monoid is recognizable by a flip automaton.

Theorem 2. *Let $L \subseteq A^*$ be recognized by a deterministic automaton $\mathcal{A}$. Then the following are equivalent:*

1. *L is a Boolean combination of right ideals.*
2. *$\mathcal{A}$ is weak.*
3. *L is recognized by some deterministic Staiger-Wagner automaton.*
4. *L is a finite disjoint union of languages $L(\mathcal{B}_i)$ such that each $\mathcal{B}_i$ is a deterministic flip automaton.*

We note that both nondeterministic weak automata and nondeterministic Staiger-Wagner automata are expressively complete for the class of all regular languages.

Remark 1. Proposition 2 (resp. Corollary 1, Theorem 2) yields another decision procedure for the class of regular right ideals (resp. prefix-closed languages, Boolean combinations of right ideals). In the case of Proposition 2, this was first observed by Paz and Peleg [11]. Moreover, the above decidability results can often be combined with other automaton models. For example, a well-known result of McNaughton and Papert says that a language is definable in first-order logic if and only if its minimal automaton is counter-free [9]. Together with Theorem 2, we see that a language L is a first-order definable Boolean combination of right ideals if and only if the minimal automaton of L is weak and counter-free.   ◇

5   Two-Way Automaton Models and Languages in $\mathcal{DA}$

The results in the previous section can easily be translated into characterizations of regular left ideals (resp. suffix-closed languages, Boolean combinations of left ideals) by considering automata which read the input from right to left. Varying the direction of the head movement naturally leads to two-way automata. The situation for arbitrary two-way automata is more involved than for one-way automata; the main reason is that two-way automata are usually defined using left and right end markers. On the other hand, if $L \subseteq (A \setminus \{a\})^*$, then $La = LaA^* \setminus LaA^+$. This shows that by adding an explicit end marker, every language becomes a Boolean combination of right ideals. To overcome this, we introduce the notion of *one-pass* two-way automata; these automata stop processing the input as soon as they read the right end marker. Now, the problem with classes of one-pass two-way automata is that, in general, they are not closed under union and intersection (since standard techniques, such as executing one automaton after the other, cannot be applied). We have no satisfactory solution for arbitrary two-way automata, but we show that the concepts of Section 4 can be adapted to a well-known subclass of two-way automata, namely deterministic partially ordered two-way automata (po2dfa). The class of languages recognized by po2dfa is a natural subclass of the star-free languages which has a huge number of different characterizations, see *e.g.* [17,4]. The most prominent of these characterizations is definability in two-variable first-order logic. By a description of algebraic means, it is the language variety $\mathcal{DA}$, *i.e.*, the class of regular languages satisfying the lattice identity $p(xy)^\omega q \leftrightarrow p(xy)^\omega x(xy)^\omega q$. As a byproduct, we show that some of the other characterizations of po2dfa recognizable languages also admit natural counterparts for right ideals and their Boolean combinations.

A *two-way automaton* is a tuple $\mathcal{A} = (Z, A, \delta, X_0, F)$. The finite set of states $Z = X \dot\cup Y$ is partitioned into *right-moving* states X (for neXt) and *left-moving* states Y (for Yesterday). The states in $X_0 \subseteq X$ are initial and states in $F \subseteq Z$ are final. On input $u \in A^*$, the tape content is $\triangleright u \triangleleft$ where $\triangleright$ and $\triangleleft$ are new symbols marking the left and right end of the tape, respectively. Initially, the head is at the first letter of u. The direction in which the input is processed can be controlled by $\mathcal{A}$. The idea is that *before* a transition is made, the head movement is performed, and the direction of the movement depends only on the *destination state* of the transition. The left end marker $\triangleright$ must not be overrun. More formally, the transition relation satisfies $\delta \subseteq (Z \times A \times Z) \cup (Y \times \{\triangleright\} \times X) \cup (X \times \{\triangleleft\} \times Z)$. As for one-way automata, we write $z \xrightarrow{a} z'$ instead of $(z, a, z') \in \delta$. More formally, a *configuration* is a pair $(z, i) \in Z \times \mathbb{N}$ where z is the current state and i is the current position on the tape. Suppose position i is labeled by $a \in A \cup \{\triangleright, \triangleleft\}$. Then a transition $(z, i) \vdash_{\mathcal{A}} (z', j)$ between configurations exists if $z \xrightarrow{a} z'$ and $j = i + 1$ (for $z' \in X$) or $j = i - 1$ (for $z' \in Y$). A *computation* of $\mathcal{A}$ on input u is a sequence

$$(z_0, i_0) \vdash_{\mathcal{A}} \cdots \vdash_{\mathcal{A}} (z_t, i_t)$$

of configurations such that $z_0 \in X_0$, $i_0 = 1$, $i_k \in \{0, \ldots, |u| + 1\}$ for $1 \leq k < t$, and $i_t = |u| + 2$. Note that position 0 is labeled with the left end marker $\triangleright$ and the position $|u| + 1$ is labeled with the right end marker $\triangleleft$. The computation is *accepting* if $z_t \in F$ is final and the input u is accepted if there exists an accepting computation for it. Note that by the signature of the transition relation, the left end marker $\triangleright$ cannot be trespassed. One-way automata may be seen as special cases with $Y = \emptyset$. The language $L(\mathcal{A})$ *recognized* by $\mathcal{A}$ is $L(\mathcal{A}) = \{u \in A^* \mid \mathcal{A} \text{ accepts } u\}$.

A two-way automaton is *deterministic* if $|X_0| = 1$ and for all $z \in Z$ and all $a \in A \cup \{\triangleright, \triangleleft\}$ there exists at most one $z' \in Z$ with $z \xrightarrow{a} z'$. For technical reasons, we also consider the empty automaton ($Z = \delta = X_0 = F = \emptyset$) as deterministic. It is *complete* if for all $z \in Z$ and all a there exists $z' \in Z$ with $z \xrightarrow{a} z'$ (more precisely, we require the existence of z' if either $z \in Y$ and $a \in A \cup \{\triangleright\}$ or if $z \in X$ and $a \in A \cup \{\triangleleft\}$). A two-way automaton is *one-pass* if $z \xrightarrow{\triangleleft} z'$ implies $z = z'$. The idea is that a two-way automaton has finished "one pass" when it encounters the right end marker $\triangleleft$ for the first time; hence for a one-pass automaton, the acceptance of a word is determined by the state when scanning $\triangleleft$ for the first time. The automaton is *partially ordered* if there exists a partial ordering $\sqsubseteq$ of the states such that transitions are non-descending, *i.e.*, if $z \xrightarrow{a} z'$, then $z \sqsubseteq z'$. In other words, once a state is left in a partially ordered automaton, it is never re-entered. We abbreviate "deterministic partially ordered two-way automaton" by *po2dfa*.

Schwentick, Thérien, and Vollmer [14] have shown that po2dfa are expressively complete for $\mathcal{DA}$. The main result of this section is a characterization of right ideals (resp. prefix-closed languages, Boolean combinations of right ideals) in $\mathcal{DA}$ in terms of subclasses of one-pass po2dfa.

As for one-way automata in Section 4, we get right ideals in $\mathcal{DA}$ if the recognizing automaton is a flip automaton. For a *flip automaton*, a transition $z \xrightarrow{a} z'$ with final state z implies that z' is final. As an intermediate step, we get a characterization in terms of unambiguous monomials. A *monomial* is a language $P = A_1^* a_1 \cdots A_k^* a_k A_{k+1}^*$ where $A_i \subseteq A$ and $a_i \in A$. It is *unambiguous* if every word $u \in P$ has a unique factorization $u = u_1 a_1 \cdots u_k a_k u_{k+1}$ with $u_i \in A_i^*$.

Theorem 3. *Let $L \subseteq A^*$. The following are equivalent:*

1. $L \in \mathcal{DA}(A^*)$ *is a right ideal.*
2. L *is a finite union of unambiguous monomials $A_1^* a_1 \cdots A_k^* a_k A^*$.*
3. L *is recognized by a complete flip one-pass po2dfa.*

Property (2) in Theorem 3 states that unambiguity of monomials and the ideal property can be achieved simultaneously which is non-trivial. A two-way automaton is *fully accepting* if all its states are final. As for one-way automata, this yields prefix-closed languages (at least for $\mathcal{DA}$). The following result for prefix-closed languages is easily deduced from Theorem 3.

Corollary 3. *Let $L \subseteq A^*$. The following are equivalent:*

1. $L \in \mathcal{DA}(A^*)$ *is prefix-closed.*
2. L *is recognized by a fully accepting one-pass po2dfa.*

We now turn to general one-pass po2dfa. A convenient intermediate step from languages in $\mathcal{DA}$ to automata are rankers. A *ranker* is a word in $\{X_a, Y_a \mid a \in A\}^*$. Intuitively, a ranker r represents a sequence of instructions X_a for "next a-position" and Y_a for "previous a-position" which is processed from left to right. That is, for a word $u = a_1 \cdots a_n$ with $a_j \in A$ and a position $i \in \{0, \ldots, n+1\}$ we set $\varepsilon(u, i) = i$ and

$$X_a r(u, i) = r(u, \min \{j > i \mid a_j = a\}),$$
$$Y_a r(u, i) = r(u, \max \{j < i \mid a_j = a\}).$$

If a nonempty ranker r starts with an X_a-modality, then we say that r is an X-*ranker*; and we define $r(u) = r(u, 0)$, *i.e.*, the evaluation of X-rankers starts at the beginning of the word u. Symmetrically, if r starts with Y_a, then $r(u) = r(u, n+1)$. As usual, $\min \emptyset$ and $\max \emptyset$ are undefined. Thus a nonempty ranker r either defines a unique position $r(u)$ in a word u or $r(u)$ is undefined. For example, $X_a Y_b X_c(bac) = 3$ whereas $X_a Y_b X_c(cba)$ is undefined. For a ranker r we set $L(r) = \{u \in A^* \mid r(u) \text{ is defined}\}$.

Theorem 4. *Let $L \subseteq A^*$. The following are equivalent:*

1. $L \in \mathcal{DA}(A^*)$ *is a Boolean combination of right ideals.*
2. L *is a finite union of unambiguous monomials $A_1^* a_1 \cdots A_k^* a_k A_{k+1}^*$ such that $\{a_i, \ldots, a_k\} \not\subseteq A_i$ for all $i \in \{1, \ldots, k\}$.*
3. L *is Boolean combination of languages $L(r)$ for X-rankers r.*
4. L *is recognized by a one-pass po2dfa.*

Right ideals are the finitary version of open sets in the Cantor topology over infinite words. It is therefore not surprising that a large part of Theorem 4 reduces to infinite words: The proof of the implication from (1) to (2) relies on a result of Diekert and Kufleitner [5, Theorem 6.6]. The step from (2) to (3) uses a characterization of X-rankers over infinite words [3, Theorem 3]. Showing the implication from (3) to (4) is the most technical part. In particular, one has to show that one-pass po2dfa are closed under union and intersection. Here, the respective result for po2-Büchi automata cannot be applied directly, but showing closure under union and intersection resembles techniques which were developed for deterministic po2-Büchi automata [8]. Finally, the step from (4) back to (1) easily follows by combining the characterization of po2dfa due to Schwentick, Thérien, and Vollmer [14, Theorem 3.1] with Theorem 1.

It is decidable whether a given regular language belongs to $\mathcal{DA}$. Therefore, using Proposition 1 and Theorem 1, it is decidable whether a regular language is recognized by an arbitrary (resp. flip, fully final) one-pass po2dfa. The temporal logic version of X-rankers is denoted $TL_X[X_a, Y_a]$, cf. [3]; it is a fragment of deterministic unary temporal logic $TL[X_a, Y_a]$ over the modalities X_a and Y_a. The logic $TL[X_a, Y_a]$ is expressively complete for $\mathcal{DA}$, and $TL_X[X_a, Y_a]$ defines the right ideals in $\mathcal{DA}$.

In analogy to Theorem 4, there is also an expressively complete two-way automaton model for Boolean combinations of ideals. A two-way automaton is *weak* if for every strongly connected component either all states are final or all states are non-final. Note that every partially ordered automaton is weak. The following result is our only general result for arbitrary (not partially ordered) deterministic two-way automata.

Proposition 3. *A regular language is a Boolean combination of right ideals if and only if it is recognized by a deterministic weak one-pass two-way automaton.*

Not every deterministic one-pass two-way automaton recognizing a Boolean combination of right ideals needs to be weak. Therefore, the equivalence of (1) and (4) in Theorem 4 does not follow from Proposition 3. Also note that the analogue of Proposition 3 does not work for right ideals (resp. prefix-closed languages) and deterministic flip (resp. fully accepting) one-pass two-way automata since deterministic two-way automata can also reject an input by an infinite cycle in its computation.

Remark 2. We use the shortcut "nfa" for *nondeterministic finite automaton*, and "po1" for *partially ordered one-way*. Using this notation, we have the following inclusions between language classes recognizable by partially ordered automata:

$$\text{po1dfa} \subsetneq \text{one-pass po2dfa} \subsetneq \text{po2dfa} \subsetneq \text{po2nfa} = \text{po1nfa}.$$

The following (very similar) languages show that the inclusions are strict. The language $\{a, c\}^* ab \{a, b, c\}^*$ is recognizable by some one-pass po2dfa but not by a po1dfa. The language $\{a, b, c\}^* ab \{b, c\}^*$ is recognizable by a po2dfa but not by any one-pass po2dfa. Finally, the language $\{a, b, c\}^* ab \{a, b, c\}^*$ is recognizable by some po1nfa but not by any po2dfa. The equivalence of po2nfa and po1nfa is due to Schwentick, Thérien, and Vollmer [14]. For each of the above language classes the membership problem is decidable: The class po1dfa corresponds to $\mathcal{R}$-trivial monoids [14], one-pass po2dfa correspond to $\mathcal{R}$-classes of monoids in **DA** (Theorem 1 and Theorem 4). The algebraic equivalent of po2dfa is the variety of finite monoids **DA** [14], and po2nfa are expressively complete for the level 3/2 of the Straubing-Thérien hierarchy [14] which is decidable by a result of Pin and Weil [13].   ◇

Acknowledgments. We thank the anonymous referees for several suggestions which helped to improve the presentation of the paper, and we are also grateful for bringing to our attention the works of Avgustinovich and Frid [1] and of Paz and Peleg [11].

References

1. Avgustinovich, S.V., Frid, A.E.: Canonical Decomposition of a Regular Factorial Language. In: Grigoriev, D., Harrison, J., Hirsch, E.A. (eds.) CSR 2006. LNCS, vol. 3967, pp. 18–22. Springer, Heidelberg (2006)
2. Beauquier, D., Pin, J.-É.: Languages and scanners. Theor. Comput. Sci. 84(1), 3–21 (1991)
3. Dartois, L., Kufleitner, M., Lauser, A.: Rankers over Infinite Words. In: Gao, Y., Lu, H., Seki, S., Yu, S. (eds.) DLT 2010. LNCS, vol. 6224, pp. 148–159. Springer, Heidelberg (2010)
4. Diekert, V., Gastin, P., Kufleitner, M.: A survey on small fragments of first-order logic over finite words. Int. J. Found. Comput. Sci. 19(3), 513–548 (2008)
5. Diekert, V., Kufleitner, M.: Fragments of first-order logic over infinite words. Theory Comput. Syst. 48, 486–516 (2011)
6. Gehrke, M., Grigorieff, S., Pin, J.-É.: Duality and Equational Theory of Regular Languages. In: Aceto, L., Damgård, I., Goldberg, L.A., Halldórsson, M.M., Ingólfsdóttir, A., Walukiewicz, I. (eds.) ICALP 2008, Part II. LNCS, vol. 5126, pp. 246–257. Springer, Heidelberg (2008)
7. Kufleitner, M., Lauser, A.: Around dot-depth one. In: Dömösi, P., Iván, S. (eds.) AFL 2011, pp. 255–269 (2011)
8. Kufleitner, M., Lauser, A.: Partially ordered two-way Büchi automata. Int. J. Found. Comput. Sci. 22(8), 1861–1876 (2011)
9. McNaughton, R., Papert, S.: Counter-Free Automata. The MIT Press (1971)
10. Muller, D.E., Saoudi, A., Schupp, P.E.: Alternating Automata, the Weak Monadic Theory of the Tree, and its Complexity. In: Kott, L. (ed.) ICALP 1986. LNCS, vol. 226, pp. 275–283. Springer, Heidelberg (1986)
11. Paz, A., Peleg, B.: Ultimate-definite and symmetric-definite events and automata. J. Assoc. Comput. Mach. 12(3), 399–410 (1965)
12. Perrin, D., Pin, J.-É.: Infinite words. Pure and Applied Mathematics, vol. 141. Elsevier (2004)
13. Pin, J.-É., Weil, P.: Polynomial closure and unambiguous product. Theory Comput. Syst. 30(4), 383–422 (1997)
14. Schwentick, T., Thérien, D., Vollmer, H.: Partially-Ordered Two-Way Automata: A New Characterization of DA. In: Kuich, W., Rozenberg, G., Salomaa, A. (eds.) DLT 2001. LNCS, vol. 2295, pp. 239–250. Springer, Heidelberg (2002)
15. Staiger, L.: ω-languages. In: Salomaa, A., Rozenberg, G. (eds.) Handbook of Formal Languages, vol. 3, pp. 339–387. Springer (1997)
16. Staiger, L., Wagner, K.W.: Automatentheoretische und automatenfreie Charakterisierungen topologischer Klassen regulärer Folgenmengen. Elektron. Inform.-verarb. Kybernetik 10(7), 379–392 (1974)
17. Tesson, P., Thérien, D.: Diamonds are forever: The variety DA. In: Gomes, G., et al. (eds.) Semigroups, Algorithms, Automata and Languages 2001, pp. 475–500. World Scientific (2002)
18. Thomas, W.: Automata on infinite objects. In: van Leeuwen, J. (ed.) Handbook of Theoretical Computer Science, ch. 4, pp. 133–191. Elsevier (1990)

Hyper-minimization
for Deterministic Tree Automata

Artur Jeż[1,*] and Andreas Maletti[2,**]

[1] Institute of Computer Science, University of Wrocław
ul. Joliot-Curie 15, 50-383 Wrocław, Poland
`aje@cs.uni.wroc.pl`
[2] Institute for Natural Language Processing, Universität Stuttgart
Pfaffenwaldring 5b, 70569 Stuttgart, Germany
`andreas.maletti@ims.uni-stuttgart.de`

Abstract. Hyper-minimization aims to reduce the size of the representation of a language beyond the limits imposed by classical minimization. To this end, the hyper-minimal representation can represent a language that has a finite difference to the original language. The first hyper-minimization algorithm is presented for (bottom-up) deterministic tree automata, which represent the recognizable tree languages. It runs in time $\mathcal{O}(\ell mn)$, where ℓ is the maximal rank of the input symbols, m is the number of transitions, and n is the number of states of the input tree automaton.

1 Introduction

Hyper-minimization for deterministic finite-state string automata (dfa) [17] allows us to reduce the size of a dfa at the expense of a finite number of errors. The original article [2] that introduced hyper-minimization and its theoretical foundations also presented the first hyper-minimization algorithm, which was subsequently improved to $\mathcal{O}(mn)$ [1] and to $\mathcal{O}(m \log n)$ [4,8], where m is the number of transitions and n is the number of states of the input dfa. Thus, the fastest hyper-minimization algorithms have the same asymptotic time complexity as the fastest algorithms for dfa minimization [9]. Since hyper-minimization trivially reduces to minimization [8], faster hyper-minimization algorithms would imply faster minimization algorithms, which have remained elusive. Hyper-minimization was already generalized to weighted dfa [13] and to dfa over infinite strings [15]. An overview of existing hyper-minimization algorithms can be found in [12].

We generalize hyper-minimization to deterministic tree automata (dta) [5,6], which have applications in XML processing [10] and natural language processing [11]. We faithfully generalize the existing definitions from dfa to dta. Thus,

---

* Financial support provided by the Polish National Science Centre (NCN) grant DEC-2011/01/D/ST6/07164.
** Financial support provided by the German Research Foundation (DFG) grant MA 4959/1-1.

N. Moreira and R. Reis (Eds.): CIAA 2012, LNCS 7381, pp. 217–228, 2012.

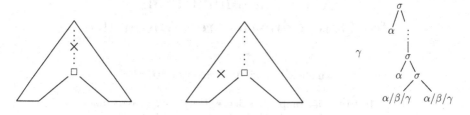

Fig. 1. Illustration of the difference locations in a context: along the path to the root (left) and off this path (middle), and illustration of the tree language $L(M_{ex})$ of Ex. 2

our hyper-minimization for dta is based on a congruence that is similar to the context-language equivalence used in dta minimization [3]. The fastest known algorithm for dta minimization [7] runs in time $\mathcal{O}(\ell m \log n)$, where ℓ is the maximal rank of the input symbols, m is the number of transitions, and n is the number of states of the input dta. The hyper-minimization algorithm that we present has the run-time complexity $\mathcal{O}(\ell m n)$, which is slightly worse than traditional minimization, but we believe that our algorithm can be improved using the standard techniques used in hyper-minimization of dfa. We sketch the improved version in Sect. 5.

Dta hyper-minimization is not a straightforward adjustment of dfa hyper-minimization. While they share the same principal structure, the actual properties used in the algorithms are different. The main reason for the differences is the location of the errors in the recognized context language. They can not only occur in the successor states (as for dfa) but can also occur in sibling states (see Fig. 1). This yields that several foundational results for dfa hyper-minimization [2] do not faithfully generalize to dta. Nevertheless, we borrow much of the surrounding infrastructure from the existing hyper-minimization algorithms [8,14] and despite the theoretical differences, we obtain an efficient hyper-minimization algorithm following the approach of [1].

2   Preliminaries

The set of all nonnegative integers is $\mathbb{N}$, and we let $[k] = \{i \in \mathbb{N} \mid 1 \leq i \leq k\}$ for every $k \in \mathbb{N}$. The cardinality of a finite set S is denoted by $|S|$. The symmetric difference $S \ominus T$ of sets S and T is $(S - T) \cup (T - S)$. If $S \ominus T$ is finite, then S and T are almost equal. A binary relation $\cong$ on S is an equivalence if it is reflexive, symmetric, and transitive. We often present them as partitions of S.

An alphabet Σ is a finite set, and a ranked alphabet (Σ, rk) consists of an alphabet Σ and a mapping $\mathrm{rk}\colon \Sigma \to \mathbb{N}$, which assigns a rank to each symbol of Σ. For every $k \in \mathbb{N}$, we let $\Sigma_k = \mathrm{rk}^{-1}(k)$ be the set of all symbols of rank k. In the following, we typically denote the ranked alphabet (Σ, rk) by just Σ. For a set T, we let $\Sigma(T) = \{\sigma(t_1, \ldots, t_k) \mid \sigma \in \Sigma_k, t_1, \ldots, t_k \in T\}$. The set $T_\Sigma(Q)$ of Σ-trees with states Q is the smallest set T such that $Q \cup \Sigma(T) \subseteq T$. We write T_Σ for $T_\Sigma(\emptyset)$. The height $\mathrm{ht}(t)$ of $t \in T_\Sigma(Q)$ is recursively defined as follows: $\mathrm{ht}(q) = 0$ for all $q \in Q$ and $\mathrm{ht}(\sigma(t_1, \ldots, t_k)) = 1 + \max\{\mathrm{ht}(t_i) \mid i \in [k]\}$ for all $\sigma \in \Sigma_k$ and $t_1, \ldots, t_k \in T_\Sigma(Q)$. The set $\mathrm{states}(t)$ is the minimal set Q such that $t \in T_\Sigma(Q)$. The set of positions of a tree $t \in T_\Sigma(Q)$ is denoted by $\mathrm{pos}(t)$, of which those that are labeled by $q \in Q$ form the set $\mathrm{pos}_q(t)$. Finally, for every $t \in T_\Sigma(Q)$, $q, q' \in Q$, and $w \in \mathrm{pos}_q(t)$, the tree $t[q']_w$ is obtained from t by relabeling the occurrence of q at w to q'.

A context c is a tree of $T_{\Sigma \cup \{\square\}}(Q)$, in which the special nullary symbol $\square$ occurs exactly once. The set of all such contexts is $C_\Sigma(Q)$, and we write C_Σ for $C_\Sigma(\emptyset)$. For every $c \in C_\Sigma(V)$ and $t \in T_{\Sigma \cup \{\square\}}(Q)$, the tree $c[t] \in T_{\Sigma \cup \{\square\}}(Q)$ denotes the tree obtained from c by replacing the unique occurrence of $\square$ by t. A tree $t' \in T_\Sigma(Q)$ is a subtree of $t \in T_\Sigma(Q)$ if there exists a context $c \in C_\Sigma(Q)$ such that $t = c[t']$. The subtree is strict if $t \neq t'$. The depth of a context $c \in C_\Sigma(Q)$ is recursively defined by $\mathrm{dp}(\square) = 0$ and

$$\mathrm{dp}(\sigma(t_1, \ldots, t_{i-1}, c, t_{i+1}, \ldots, t_k)) = 1 + \mathrm{dp}(c)$$

for every $\sigma \in \Sigma_k$, index $i \in [k]$, context $c \in C_\Sigma(Q)$, and $t_1, \ldots, t_k \in T_\Sigma(Q)$.

A deterministic tree automaton (dta) [5,6] is a tuple $M = (Q, \Sigma, \delta, F)$ where Q is a finite set of states, Σ is a ranked alphabet of input symbols, $\delta \colon \Sigma(Q) \to Q$ is a (partial) transition function, and $F \subseteq Q$ is a set of final states. The dta M is total if δ is total. The transition function δ extends to $\delta \colon T_\Sigma(Q) \to Q$ by $\delta(q) = q$ for every $q \in Q$ and $\delta(\sigma(t_1, \ldots, t_k)) = \delta(\sigma(\delta(t_1), \ldots, \delta(t_k)))$ for every $\sigma \in \Sigma_k$ and $t_1, \ldots, t_k \in T_\Sigma(Q)$. We let $L(M)_{q'}^q = \{c \in C_\Sigma \mid \delta(c[q']) = q\}$ for every $q, q' \in Q$. Moreover, $L(M)_{q'} = \bigcup_{f \in F} L(M)_{q'}^f$ contains all contexts c such that $c[q']$ takes M into a final state and $L(M)^q = \delta^{-1}(q) \cap T_\Sigma$ contains all (stateless) trees that take M into the state q. The dta M recognizes the tree language $L(M) = \bigcup_{f \in F} L(M)^f$. In the following, we assume that every considered dta M is trim (or equivalently: has only reachable states), which means that $L(M)^q \neq \emptyset$ for every $q \in Q$.

An equivalence $\cong$ on Q is a congruence (on the dta M) if we have that $\delta(\sigma(q_1, \ldots, q_k)) \cong \delta(\sigma(q'_1, \ldots, q'_k))$ for every $\sigma \in \Sigma_k$ and $q_1 \cong q'_1, \ldots, q_k \cong q'_k$. Two states $q, q' \in Q$ are equivalent, which is denoted by $q \equiv_M q'$ (or just $q \equiv q'$), if $L(M)_q = L(M)_{q'}$. We sometimes use those notions for states from different dta over the same ranked alphabet with the obvious meaning. Note that $\equiv_M$ is a congruence, and actually, the coarsest (i.e., least refined) congruence on M that respects F, which means that a final state cannot be equivalent to a nonfinal state. The dta M is minimal if there exists no equivalent dta with strictly fewer states. It is well-known that M is minimal if and only if it does not have two different, but equivalent states. For every dta M, an equivalent minimal dta can be computed efficiently using an adaptation [7] of HOPCROFT's algorithm [9], which runs in time $O(\ell m \log n)$ where $\ell = \max \mathrm{rk}(\Sigma)$ is the maximal rank of the input symbols, $m = |\mathrm{dom}(\delta)|$ is the number of transitions, and $n = |Q|$ is the number of states. From now on, let $M = (Q, \Sigma, \delta, F)$ be a minimal dta, which automatically yields that M is trim. Finally, we recall a central notion from [2] that will also be important in our setting. A state $q \in Q$ is a kernel state if $L(M)^q$ is infinite. Otherwise q is a preamble state. The sets of kernel and preamble states are denoted by $\mathrm{Ker}(M)$ and $\mathrm{Pre}(M)$, respectively.

3 Hyper-minimal Automata

The goal of hyper-minimization for a given dta M is the efficient computation of a dta that is as small as possible (measured by the number of states) and recognizes a tree language with finite difference to $L(M)$. A dta for which such a strictly smaller dta does not exist is called 'hyper-minimal', and we investigate the properties of these dta here. Before we can start, we need to introduce the main notions of this contribution and some essential properties.

For the rest of the section, we consider a minimal dta $M = (Q, \Sigma, \delta, F)$. To simplify the theoretical discussion, we assume that M is total. This can be achieved by adding a sink state $\bot$ as the target of all missing transitions of a partial dta. It should be noted that all properties of this section trivially extend to partial dta. The totality assumption made is purely a convenience.

A minimal dta is obtained by identifying and merging equivalent states. Accordingly, our goal is to obtain a hyper-minimal dta by identifying and merging almost equivalent states, where 'almost equivalent' has the usual mathematical meaning (i.e., equivalent up to a finite number of differences).

Definition 1 (cf. [2, Def. 2.2]). *The states $q, q' \in Q$ are almost equivalent, written $q \sim q'$, if $L(M)_q \ominus L(M)_{q'}$ is finite.*

Example 2. The running example dta M_{ex} is (Q, Σ, δ, F), where

- $Q = \{q_\alpha, q_\beta, q_\gamma, q_\sigma, \bot\}$,
- $\Sigma = \Sigma_0 \cup \Sigma_2$ with $\Sigma_0 = \{\alpha, \beta, \gamma\}$ and $\Sigma_2 = \{\sigma\}$,
- $F = \{q_\sigma, q_\gamma\}$, and
- δ returns $\bot$ except that for every $\sigma_0, \sigma_0' \in \Sigma_0$ we have

$$\delta(\sigma_0) = q_{\sigma_0} \qquad \delta(\sigma(q_{\sigma_0}, q_{\sigma_0'})) = q_\sigma \qquad \delta(\sigma(q_\alpha, q_\sigma)) = q_\sigma \ .$$

It recognizes the tree language $\{\gamma\} \cup \{c^n[\sigma(\sigma_0, \sigma_0')] \mid n \in \mathbb{N}, \sigma_0, \sigma_0' \in \Sigma_0\}$, where $c = \sigma(\alpha, \Box)$, $c^0 = \Box$, and $c^{n+1} = c^n[c]$ for every $n \in \mathbb{N}$ (see Fig. 1). Note that M_{ex} is minimal. However, q_β and q_γ are almost equivalent because

$$L(M_{\mathrm{ex}})_{q_\beta} = \{c^n[\sigma(\Box, \sigma_0)] \mid n \in \mathbb{N}, \sigma_0 \in \Sigma_0\} \cup \{c^n[\sigma(\sigma_0, \Box)] \mid n \in \mathbb{N}, \sigma_0 \in \Sigma_0\}$$
$$L(M_{\mathrm{ex}})_{q_\gamma} = \{\Box\} \cup L(M_{\mathrm{ex}})_{q_\beta} \ .$$

The state q_α is neither almost equivalent to q_β nor to q_σ.   ◇

We immediately observe that for all $q_1 \sim q_2$ there is an integer $k \in \mathbb{N}$ such that $\delta(c[q_1]) = \delta(c[q_2])$ for all $c \in C_\Sigma$ with $\mathrm{dp}(c) > k$. Since the difference $L(M)_{q_1} \ominus L(M)_{q_2}$ is finite, we can select k such that it is strictly larger than the depth of any context in the difference. For any context c of depth at least k we obtain that $\delta(c[q_1])$ and $\delta(c[q_2])$ are equivalent, and thus, equal by minimality.

In contrast to the string case, the converse of the previous statement is not true, which shows that the generalization is nontrivial. In a dta not only the successor, but also the sibling states determine the almost equivalence (see Fig. 1). Although q_α and q_β have the same successor states in Ex. 2, they are not almost equivalent as they expect different sibling states.

Clearly, almost equivalence is an equivalence on Q. Next, we show that it is even a congruence on M. In contrast to the context equivalence that respects F, the almost equivalence $\sim$ clearly need not respect F (see Ex. 2 where $q_\gamma \sim q_\beta$ but $q_\gamma \in F$ and $q_\beta \notin F$).

Lemma 3 (see [2, Lm. 2.10]). *For all $q \sim q'$ and contexts $c \in C_\Sigma$, we have $\delta(c[q]) \sim \delta(c[q'])$. In particular, $\sim$ is a congruence.*

Proof. For each context $c' \in L(M)_{\delta(c[q])} \ominus L(M)_{\delta(c[q'])}$ we have that the context $c'[c] \in L(M)_q \ominus L(M)_{q'}$. Clearly, different c' yield different contexts $c'[c]$, so there can only be finitely many such contexts c' because $L(M)_q \ominus L(M)_{q'}$ is finite, which proves that $\delta(c[q]) \sim \delta(c[q'])$. The latter property is a simple consequence of the former via particular contexts of depth 1 and the standard piecewise replacement. Let $\sigma \in \Sigma_k$ and $q_1 \sim q_1', \ldots, q_k \sim q_k'$ be almost equivalent states. Moreover, for each $q \in Q$, let $t_q \in L(M)^q$ be arbitrary. Then

$$\delta(\sigma(q_1, \ldots, q_k)) = \delta(\sigma(\Box, t_{q_2}, \ldots, t_{q_k})[q_1])$$
$$\sim \delta(\sigma(\Box, t_{q_2}, \ldots, t_{q_k})[q_1']) = \delta(\sigma(q_1', q_2, \ldots, q_k)) = \delta(\sigma(t_{q_1'}, \Box, t_{q_3}, \ldots, t_{q_k})[q_2])$$
$$\sim \ldots$$
$$\sim \delta(\sigma(t_{q_1'}, \ldots, t_{q_{k-1}'}, \Box)[q_k']) = \delta(\sigma(q_1', \ldots, q_k')) \ . \qquad \Box$$

To complete the essential definitions, two dta M and N are almost equivalent if $L(M)$ and $L(N)$ are almost equal. Naturally, this is an equivalence relation on dta. Next, we relate the states of almost equivalent dta in order to prepare our characterization of hyper-minimal dta.

Lemma 4. *Let $M = (Q, \Sigma, \delta, F)$ and $N = (P, \Sigma, \mu, G)$ be minimal dta that are almost equivalent. Then $L(M)_{\delta(t)}$ and $L(N)_{\mu(t)}$ are almost equal for all $t \in T_\Sigma$.*

Proof. For every $L \subseteq T_\Sigma$, let $t^{-1}L = \{c \in C_\Sigma \mid c[t] \in L\}$. Since $L(M)$ and $L(N)$ are almost equal, also $t^{-1}L(M)$ and $t^{-1}L(N)$ are almost equal. Together with $t^{-1}L(M) = L(M)_{\delta(t)}$ and $t^{-1}L(N) = L(N)_{\mu(t)}$, we proved the statement. $\qquad \Box$

Now we make hyper-minimality precise. The dta M is *hyper-minimal* if all almost equivalent dta are at least as large (i.e., have at least as many states). We already remarked that we want to obtain hyper-minimal dta with the help of merging. In a *merge* of $q \in Q$ into $q' \in Q$ we redirect all transitions leading to q into q'. Formally, for every two different states $q, q' \in Q$, the dta merge$(M, q \to q')$ is $(Q - \{q\}, \Sigma, \delta', F - \{q\})$ where for every $s \in \Sigma(Q - \{q\})$

$$\delta'(s) = \begin{cases} q' & \text{if } s \in \delta^{-1}(q) \\ \delta(s) & \text{otherwise.} \end{cases}$$

Lemma 5. *If $q \sim q'$ and q is a preamble state, then* merge($M, q \to q'$) *and M are almost equivalent.*

Proof. Let merge($M, q \to q'$) = $(Q', \Sigma, \delta', F')$. The set $D = L(M)_q \ominus L(M)_{q'}$ is finite because $q \sim q'$. We select ℓ with $\ell > \mathrm{ht}(c)$ for every $c \in D$. Let $t \in T_\Sigma$ be such that $\mathrm{ht}(t) \geq \ell + |Q|$. First we replace all subtrees $t' \in L(M)^q$ in t by just q. In this way, we obtain the tree u. Note that $\delta(t) = \delta(u)$ and $\mathrm{ht}(u) \geq \ell$ because $\mathrm{ht}(t') \leq |Q|$ for all $t' \in L(M)^q$ since q is a preamble state. Let $\mathrm{pos}_q(u) = \{w_1, \ldots, w_n\}$ with $w_1 < \cdots < w_n$ be the occurrences of q in u. For each $i \in [n]$, let $c_i = (u[q']_{w_1} \cdots [q']_{w_{i-1}})[\Box]_{w_i}$ be the context obtained from u by replacing the first $i - 1$ occurrences by q' and the occurrence w_i by $\Box$. Note that $\mathrm{ht}(c_i) = \mathrm{ht}(u) \geq \ell$, which allows us to obtain

$$\delta(t) = \delta(u) = \delta(c_1[q]) = \delta(c_1[q']) = \delta(c_2[q]) = \cdots = \delta(c_n[q']) \overset{\dagger}{=} \delta'(t) \ ,$$

where $\dagger$ holds because δ and δ' coincide on all transitions not involving q. Consequently, merge($M, q \to q'$) and M agree on all tall trees as desired. $\Box$

Example 6. Recall the dta $M_{\mathrm{ex}} = (Q, \Sigma, \delta, F)$ of Ex. 2. If we merge q_β into q_γ, then we obtain the dta merge($M_{\mathrm{ex}}, q_\beta \to q_\gamma$), which is $(Q - \{q_\beta\}, \Sigma, \delta', F)$ where δ' returns $\bot$ except that for every $\sigma_0, \sigma'_0 \in \Sigma_0$ we have

$$\delta'(\alpha) = q_\alpha \qquad\qquad \delta'(\beta) = q_\gamma \qquad\qquad \delta'(\gamma) = q_\gamma$$
$$\delta'(\sigma(q_{\sigma_0}, q_{\sigma'_0})) = q_\sigma \qquad\qquad \delta'(\sigma(q_\alpha, q_\sigma)) = q_\sigma \ . \qquad\qquad \diamond$$

Now we can characterize hyper-minimality [2]. The characterization allows us to easily determine whether M is hyper-minimal. Recall that a dta is minimal if and only if it does not have two different, but equivalent states. The condition for hyper-minimality is similar, but adds a restriction to preamble states.

Theorem 7. *The minimal dta M is hyper-minimal if and only if every pair of different, but almost equivalent states consists of only kernel states.*

Proof. We start with the "only if"-direction. Suppose that there exist two different, but almost equivalent states $q, q' \in Q$ such that q is a preamble state. Then M is not hyper-minimal because merge($M, q \to q'$) is strictly smaller and almost equivalent to M by Lm. 5. For the converse, let $N = (P, \Sigma, \mu, G)$ be a hyper-minimal dta that is strictly smaller (i.e., $|P| < |Q|$) and almost equivalent to M. The product dta $M' = (Q \times P, \Sigma, \delta \times \mu, F \times G)$ is given by

$$(\delta \times \mu)(\sigma(\langle q_1, p_1 \rangle, \ldots, \langle q_k, p_k \rangle)) = \langle \delta(\sigma(q_1, \ldots, q_k)), \mu(\sigma(p_1, \ldots, p_k)) \rangle$$

for every $\sigma \in \Sigma_k$ and $\langle q_1, p_1 \rangle, \ldots, \langle q_k, p_k \rangle \in Q \times P$. Since M is minimal, let $t_q \in L(M)^q$ for every $q \in Q$. If $q \in \mathrm{Ker}(M)$, then select t_q such that $\mathrm{ht}(t_q) \geq |Q|^2$. By the pigeon-hole principle with $|P| < |Q|$, there must exist different $q_1, q_2 \in Q$ and $p \in P$ such that $(\delta \times \mu)(t_q) = \langle q, p \rangle$ for $q \in \{q_1, q_2\}$. Consequently, $q_1 \sim q_2$ because $L(M)_{q_1}$ and $L(N)_p$ as well as $L(M)_{q_2}$ and $L(N)_p$ are almost equal by Lm. 4. This in turn yields that q_1 and q_2 are kernel states of M by assumption.

Algorithm 1. Structure of our dta hyper-minimization algorithm [2,8]

Require: a dta M
Return: an almost equivalent hyper-minimal dta

$\quad M \leftarrow \textsc{Minimize}(M)$ // complexity: $\mathcal{O}(\ell m \log n)$

2: $K \leftarrow \textsc{ComputeKernel}(M)$ // complexity: $\mathcal{O}(\ell m)$

$\quad \sim \;\leftarrow\; \{\langle q, q'\rangle \in Q^2 \mid L(M_\otimes)_{\langle q,q'\rangle} \text{ is finite}\}$ // see Sect. 4

4: **for all** $B \in (Q/{\sim})$ **do**

$\quad\quad$ select $q_B \in B$ such that $q_B \in K$ if possible

6: $\quad\quad$ **for all** $q \in B - K$ **do**

$\quad\quad\quad M \leftarrow \mathrm{merge}(M, q \rightarrow q_B)$ // complexity: $\mathcal{O}(1)$

8: **return** M

Moreover, $\langle q_1, p\rangle$ and $\langle q_2, p\rangle$ are kernel states of M' by the selection of the access trees with $\mathrm{ht}(t_{q_1}) \geq |Q|^2 \leq \mathrm{ht}(t_{q_2})$ (because the trees t_{q_1} and t_{q_2} can be pumped [5,6]). Now, for the sake of a contradiction, let $c \in L(M)_{q_1} \ominus L(N)_p$. Then $\{c[t] \mid t \in L(M')^{\langle q_1, p\rangle}\} \subseteq L(M) \ominus L(N)$. Since $\langle q_1, p\rangle$ is a kernel state of M', the set $L(M')^{\langle q_1, p\rangle}$ is infinite. This contradicts that M and N are almost equivalent, so consequently, $L(M)_{q_1} \ominus L(N)_p = \emptyset$, and $L(M)_{q_2} \ominus L(N)_p = \emptyset$ in the same manner. Thus, q_1 and q_2 are equivalent and $q_1 = q_2$ by the minimality of M, which shows that the dta N cannot exist. $\qquad\square$

Example 8. The dta M_{ex} of Ex. 2 is not hyper-minimal since $q_\beta \sim q_\gamma$ and both states are preamble states (see Ex. 2). However, with a little effort we can show that the dta $\mathrm{merge}(M_{\mathrm{ex}}, q_\beta \rightarrow q_\gamma)$ is hyper-minimal (see Ex. 6). $\qquad\diamond$

4 Hyper-minimization

The previous results suggest a hyper-minimization algorithm, which we sketch in Alg. 1. We work with the (potentially non-total) dta $M = (Q, \Sigma, \delta, F)$ now. In addition, we let $\ell = \max \mathrm{rk}(\Sigma)$, $m = |\mathrm{dom}(\delta)|$, and $n = |Q|$. Algorithm 1 simply determines the kernel states and the almost equivalence using methods that we describe later. It then merges states (by simply changing a reference) according to the conditions of Lm. 5, which guarantees that the result is almost equivalent. Finally, Thm. 7 shows that the obtained dta is hyper-minimal.

Corollary 9 (of Lm. 5 and Thm. 7). *Algorithm 1 returns a hyper-minimal dta that is almost equivalent to M.*

In Alg. 1 we use $\textsc{Minimize}$, which implements classical dta minimization [5,6,3] in time $O(\ell m \log n)$ using an adaptation of $\textsc{Hopcroft}$'s algorithm [9,7]. The procedure $\textsc{ComputeKernel}$ computes the kernel states of M using any fast algorithm for computing strongly connected components in a graph (e.g., $\textsc{Tarjan}$ [16]). The next proposition shows the trivial problem translation.

Proposition 10. $\mathrm{Ker}(M)$ *can be computed in time* $O(\ell m)$.

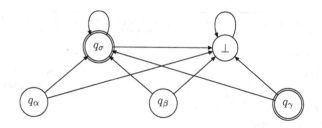

Fig. 2. The graph derived from the dta of Ex. 2

Proof. We turn our dta M into the graph (Q, E), where

$$E = \{(q, \delta(t)) \mid t \in \text{dom}(\delta), q \in \text{states}(t)\} \ .$$

It is simple to observe that $q \in \text{Ker}(M)$ if and only if it is reachable from a non-trivial strongly connected component of the graph (Q, E) [see [8] for details].   □

Example 11. The kernel states of the dta M_{ex} of Ex. 2 are $\{q_\sigma, \bot\}$, which is easily determined from the graph displayed in Fig. 2.

The final component is the identification of the almost equivalent states, which also determines the overall run-time of our hyper-minimization algorithm. For this final component, we use an adapted version of an algorithm from [1], which is simple but not the fastest. In the next section, we sketch how the currently fastest algorithms for dfa almost equivalence [4,8] can be adjusted to our dta setting.

To simplify the presentation, we assume that $\delta(s) = \bot$ for the special token $\bot \notin Q$ if $\delta(s)$ is undefined. Note that we do not add $\bot$ to Q, so we do not make M total. In contrast, we just introduce a notational convenience.

Definition 12. The *exclusive-or single-point self-product* of M is the dta

$$M_\otimes = (P \cup P^2, \Sigma, \delta \cup \delta', F')$$

such that $P = Q \cup \{\bot\}$ with $\bot \notin Q$,

- $F' = \{\langle q, q' \rangle \mid \text{either } q \in F \text{ or } q' \in F\}$, and
- for every $\sigma \in \Sigma_k$, $i \in [k]$ and $q, q', q_1, \ldots, q_k \in P$

$$\delta'(\sigma(c[\langle q, q' \rangle])) = \langle \delta(c[q]), \delta(c[q']) \rangle \ ,$$

where $c = \sigma(q_1, \ldots, q_{i-1}, \Box, q_{i+1}, \ldots, q_k)$ and at least one of the δ-entries has to be defined (i.e., $Q \cap \{\delta(c[q]), \delta(c[q'])\} \neq \emptyset$).
- δ' is undefined otherwise.   ◇

In other words, only the paired states in which one state is final and the other is nonfinal are now final states in $M_\otimes$. The transitions on pairs run componentwise with an explicit sink state component as long as at least one component is still a "normal" state. This special treatment is necessary to correctly handle partial dta.

Proposition 13. *We can construct $M_\otimes$ in time $\mathcal{O}(\ell m n)$.*

Proof. Clearly, we can create the states in time $\mathcal{O}(n^2)$. Since M is minimal, we have $n \leq m$, which yields $\mathcal{O}(n^2) \subseteq \mathcal{O}(mn)$. Clearly, for each transition in M, we construct at most $\mathcal{O}(\ell n)$ copies of that transition, which yields that we can construct all transitions in time $\mathcal{O}(\ell m n)$, where we assume that transition look-ups run in constant time. $\qquad\square$

Example 14. Now we can handle the dta M_{ex} of Ex. 2 as a partial dta. The dta $(M_{\text{ex}})_\otimes$ is $(Q \cup Q^2, \Sigma, \delta \cup \delta', F')$, where

- $F' = \{\langle q_\sigma, q_\alpha \rangle, \langle q_\sigma, q_\beta \rangle, \langle q_\sigma, \bot \rangle, \langle q_\gamma, q_\alpha \rangle, \langle q_\gamma, q_\beta \rangle, \langle q_\gamma, \bot \rangle\}^{\text{sym}}$, where L^{sym} is the symmetric closure of L, and
- some interesting transitions of δ' include

$$\delta'(\sigma(\langle q_\alpha, q_\beta \rangle, q_\sigma)) = \delta'(\sigma(q_\alpha, \langle q_\sigma, \bot \rangle)) = \langle q_\sigma, \bot \rangle$$
$$\delta'(\sigma(\langle q_\beta, q_\gamma \rangle, q_\alpha)) = \delta'(\sigma(\langle q_\beta, q_\gamma \rangle, q_\beta)) = \delta'(\sigma(\langle q_\beta, q_\gamma \rangle, q_\gamma)) = \langle q_\sigma, q_\sigma \rangle$$
$$\delta'(\sigma(q_\alpha, \langle q_\beta, q_\gamma \rangle)) = \delta'(\sigma(q_\beta, \langle q_\beta, q_\gamma \rangle)) = \delta'(\sigma(q_\gamma, \langle q_\beta, q_\gamma \rangle)) = \langle q_\sigma, q_\sigma \rangle \ . \ \diamond$$

For the sake of the next theorem, we assume that $M_\otimes$ is total to avoid a distinction between $\bot$ and undefinedness. It can easily be checked that the argument also works for partial dta.

Theorem 15. *$L(M_\otimes)_{\langle q, q' \rangle}$ is finite if and only if $q \sim q'$ for every $q, q' \in Q$.*

Proof. Let $M_\otimes = (Q', \Sigma, \delta', F')$. Clearly, $\delta'(c[\langle q, q' \rangle]) = \langle \delta(c[q]), \delta(c[q']) \rangle$ for every $c \in C_\Sigma$, which can be proven using standard induction. Now

$$c \in L(M_\otimes)_{\langle q, q' \rangle} \iff \delta'(c[\langle q, q' \rangle]) \in F' \iff \langle \delta(c[q]), \delta(c[q']) \rangle \in F'$$
$$\iff \text{ either } \delta(c[q]) \in F \text{ or } \delta(c[q']) \in F \iff c \in L(M)_q \ominus L(M)_{q'} \ .$$

This strong correspondence shows the statement because the finiteness of either set $(L(M_\otimes)_{\langle q, q' \rangle}$ or $L(M)_q \ominus L(M)_{q'})$ implies the finiteness of the other and $L(M)_q \ominus L(M)_{q'}$ is finite if and only if $q \sim q'$. $\qquad\square$

The finiteness of $L(M_\otimes)_{\langle q, q' \rangle}$ for all states $\langle q, q' \rangle$ can be determined easily (using standard algorithms) in linear time in the number of transitions of $M_\otimes$. Since the number of transitions of $M_\otimes$ is $\mathcal{O}(\ell m n)$, we can obtain $\sim$ in time $\mathcal{O}(\ell m n)$.

Example 16. In the dta M_{ex} of Ex. 2 we have $q_\alpha \not\sim q_\beta$ as demonstrated by the recursive transitions for $\langle q_\alpha, q_\beta \rangle$ in Ex. 14. Moreso, $q_\beta \sim q_\gamma$ because the language $L((M_{\text{ex}})_\otimes)_{\langle q_\beta, q_\gamma \rangle}$ is finite (it contains only $\square$). $\diamond$

Since we already proved that Alg. 1 is correct and have now established the run-time, we can state our main theorem.

Theorem 17. *Hyper-minimization of M can be performed in time $\mathcal{O}(\ell m n)$.*

Proof. We run Alg. 1, which runs in time $\mathcal{O}(\ell m n)$ because Line 3 runs in this time bound as demonstrated in this section. Finally, Cor. 9 proves the algorithm's correctness. $\qquad\square$

5 Discussion

In this section, we shortly discuss two minor issues. First, we demonstrate that dta minimization can be reduced in linear time to dta hyper-minimization. In the string case, this is achieved [8] with a new distinguished symbol that takes every state back to the initial state, thus making all states kernel states. Since we do not have a single initial state in a dta, we use a slightly different construction. Let $M = (Q, \Sigma, \delta, F)$ be a dta that is not necessarily minimal. For every $q \in \delta(\Sigma_0)$, let $\vec{q} \notin \Sigma$ be a new symbol of rank 1. Moreover, we use the two new symbols $\rightarrow$ and $\circlearrowleft$, which are of rank 0 and 1, respectively, and a new state $\imath \notin Q$. We construct the dta $M' = (Q \cup \{\imath\}, \Sigma', \delta', F)$ such that

- $\Sigma' = \Sigma \cup \{\vec{q} \mid q \in \delta(\Sigma_0)\} \cup \{\rightarrow, \circlearrowleft\}$,
- $\delta'(t) = \delta(t)$ for all $t \in \mathrm{dom}(\delta)$,
- $\delta'(\rightarrow) = \imath$ and $\delta'(\circlearrowleft(\imath)) = \imath$, and
- $\delta'(\vec{q}(\imath)) = q$ for all $q \in \delta(\Sigma_0)$.
- All remaining transitions are undefined.

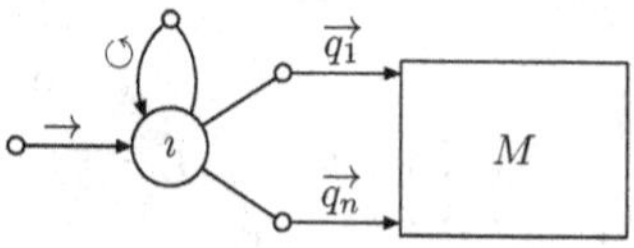

Fig. 3. Illustration

Clearly, M' can be constructed in linear time in the size of M. The construction is illustrated in Fig. 3. Clearly, all reachable states in M' are kernel states. It is easy to see that a dta in which all reachable states are kernel states is hyper-minimal if and only if it is minimal. Consequently, we can hyper-minimize M' to obtain a minimal dta M'' for $L(M')$. From M'' we can obtain a minimal dta for $L(M)$ by dropping all transitions involving the newly introduced symbols. Thus, we have reduced minimization to hyper-minimization, which shows that the complexity of dta minimization is a lower bound on the complexity of hyper-minimization.

Second, we sketch an improved version of our hyper-minimization algorithm, which uses the structure of the fastest dfa hyper-minimization algorithms [4,8]. First of all, we assume that M is total. We only present the computation of the almost equivalence because only this part needs to be improved to obtain the time bound $\mathcal{O}(\ell m \log n)$, which is also the time complexity of the fastest dta minimization algorithm [7]. Before we present the algorithm, we establish an auxiliary result.

Proposition 18. *Let M be minimal and $q, q' \in Q$. We have $q \sim q'$ if and only if for each context $c \in C_\Sigma(Q)$ we have*

- *$\delta(c[q]) \sim \delta(c[q'])$, and*
- *$\delta(c[q]) = \delta(c[q'])$ if $\mathrm{states}(c) \cap \mathrm{Ker}(M) \neq \emptyset$.*

Proof. The "only if" direction is a straightforward generalization of Lm. 3. For the converse, we simply take the trivial context $\square$. □

Proposition 18 shows that we need a completely new mechanism (compared to the string case) to compute the successor states. We define the successor states, where we keep two dta: (i) the original dta M_0 to enforce the equality constraints of the second item of Prop. 18 and (ii) a dta M obtained by successive merges to capture the almost equivalence.

Algorithm 2. Algorithm computing $\sim$.

Require: minimal dta $M = (Q, \Sigma, \delta, F)$
Return: the almost equivalence $\sim$ represented as a partition

  $M_0 \leftarrow M$ where $M_0 = (Q, \Sigma, \delta_0, F)$ // keep a copy of the input dta M
2: $\pi(q) \leftarrow \{q\}$ for all $q \in Q$ // trivial initial blocks
  $h \leftarrow \emptyset$ // empty hash map of type $h \colon Q^C \to Q$
4: $I \leftarrow Q; P \leftarrow Q$ // states that need to be considered and current states

     while $I \neq \emptyset$ **do**
6:      select $q \in I$ and remove it from I
     if $\mathrm{HasValue}(h, \mathrm{succ}_q^{M,M_0})$ **then**
8:      $q' \leftarrow \mathrm{Get}(h, \mathrm{succ}_q^{M,M_0})$ // retrieve state in bucket succ_q^{M,M_0} of h
     $\mathrm{Swap}(q', q)$ if $|\pi(q')| \geq |\pi(q)|$ // exchange roles of q' and q
10:      $P \leftarrow P - \{q'\}$ // state q' will be merged into q
     $I \leftarrow I \cup \{r \in P \mid t \in \delta^{-1}(q'), r \in \mathrm{states}(t)\}$ // add predecessors of q' in P
     to I
12:      $M \leftarrow \mathrm{merge}'(M, q' \to q)$ // merge state q' into q (do not remove q')
     $\pi(q) \leftarrow \pi(q) \cup \pi(q')$ // q' and q are almost equivalent
14:      $h \leftarrow \mathrm{Put}(h, \mathrm{succ}_q^{M,M_0}, q)$ // store q in h under key succ_q^{M,M_0}
     return π

Definition 19. *Let $M_0 = (Q, \Sigma, \delta_0, F)$ and M be dta. For every state $q \in Q$, let $\mathrm{succ}_q^{M,M_0} \colon C \to Q$ be the mapping such that for every $c \in C$*

$$\mathrm{succ}_q^{M,M_0}(c) = \begin{cases} \delta_0(c[q]) & \text{if } \mathrm{states}(c) \cap \mathrm{Ker}(M) \neq \emptyset \\ \delta(c[q]) & \text{otherwise,} \end{cases}$$

where $C = \{\sigma(q_1, \ldots, q_{i-1}, \square, q_{i+1}, \ldots, q_k) \mid \sigma \in \Sigma_k, i \in [k], q_1, \ldots, q_k \in Q\}$.

In other words, we compute with the original transition mapping δ_0 for all transition contexts containing a kernel state and use the current transition mapping δ for all other transition contexts. Let us attempt to explain Algorithm 2. Its overall structure is the same as in the string case [4,8]. We only changed the details to suit the new needs in the dta case. Roughly speaking, the algorithm first copies the input dta in order to have the original transition mapping available. Then it creates a block for each state. In I it keeps a set of states that need to be processed, and in P it stores the set of states that are still useful. Both are initially Q and we also create a hash map h of type $h \colon Q^C \to Q$, which initially has no entries. Clearly, the key set of this hash map is highly complex. The algorithm iteratively extracts a state q from I and computes its successors succ_q^{M,M_0}. It then looks succ_q^{M,M_0} up in the hash-map h, and simply stores them in h if they are so far unassociated. If the successors already have an entry in h, then the algorithm extracts the state with the same successors from h, compares the sizes of their respective blocks, and merges the state q' belonging to the smaller block into the one belonging to the bigger block. We use a variant of our merging procedure here, which does not delete the state q'. It also updates the blocks to reflect the merge, and it adds all states that have transitions leading to q' to I for processing because their successors have changed. The algorithm terminates when the set I is empty. The time complexity of this algorithm can be analyzed as in the string case [8]. Finally, its correctness still needs to be established.

References

1. Badr, A.: Hyper-minimization in $O(n^2)$. Int. J. Found. Comput. Sci. 20(4), 735–746 (2009)
2. Badr, A., Geffert, V., Shipman, I.: Hyper-minimizing minimized deterministic finite state automata. RAIRO Theor. Inf. Appl. 43(1), 69–94 (2009)
3. Comon, H., Dauchet, M., Gilleron, R., Löding, C., Jacquemard, F., Lugiez, D., Tison, S., Tommasi, M.: Tree automata: Techniques and applications (2007), http://tata.gforge.inria.fr/
4. Gawrychowski, P., Jeż, A.: Hyper-minimisation Made Efficient. In: Královič, R., Niwiński, D. (eds.) MFCS 2009. LNCS, vol. 5734, pp. 356–368. Springer, Heidelberg (2009)
5. Gécseg, F., Steinby, M.: Tree Automata. Akadémiai Kiadó, Budapest (1984)
6. Gécseg, F., Steinby, M.: Tree languages. In: Rozenberg, G., Salomaa, A. (eds.) Handbook of Formal Languages, vol. 3, ch. 1, pp. 1–68. Springer (1997)
7. Högberg, J., Maletti, A., May, J.: Backward and forward bisimulation minimization of tree automata. Theoret. Comput. Sci. 410(37), 3539–3552 (2009)
8. Holzer, M., Maletti, A.: An $n \log n$ algorithm for hyper-minimizing a (minimized) deterministic automaton. Theoret. Comput. Sci. 411(38-39), 3404–3413 (2010)
9. Hopcroft, J.E.: An $n \log n$ algorithm for minimizing states in a finite automaton. In: Kohavi, Z., Paz, A. (eds.) Theory of Machines and Computations, pp. 189–196. Academic Press (1971)
10. Hosoya, H.: Foundations of XML Processing: The Tree-Automata Approach. Cambridge University Press (2011)
11. Knight, K.: Capturing practical natural language transformations. Machine Translation 21(2), 121–133 (2007)
12. Maletti, A.: Notes on hyper-minimization. In: Proc. 13th Int. Conf. Automata and Formal Languages, pp. 34–49. Nyíregyháza College (2011)
13. Maletti, A., Quernheim, D.: Hyper-minimisation of deterministic weighted finite automata over semifields. In: Proc. 13th Int. Conf. Automata and Formal Languages, pp. 285–299. Nyíregyháza College (2011)
14. Maletti, A., Quernheim, D.: Optimal hyper-minimization. Int. J. Found. Comput. Sci. 22(8), 1877–1891 (2011)
15. Schewe, S.: Beyond hyper-minimisation — minimising DBAs and DPAs is NP-complete. In: Proc. 30th Int. Conf. Foundations of Software Technology and Theoretical Computer Science. LIPIcs, vol. 8, pp. 400–411. Schloss Dagstuhl - Leibniz-Zentrum für Informatik (2010)
16. Tarjan, R.E.: Depth-first search and linear graph algorithms. SIAM J. Comput. 1(2), 146–160 (1972)
17. Yu, S.: Regular languages. In: Rozenberg, G., Salomaa, A. (eds.) Handbook of Formal Languages, vol. 1, ch. 2, pp. 41–110. Springer (1997)

On the State and Computational Complexity of the Reverse of Acyclic Minimal DFAs

Galina Jirásková[1,*] and Tomáš Masopust[2,**]

[1] Mathematical Institute, Slovak Academy of Sciences,
Grešákova 6, 040 01 Košice, Slovak Republic
jiraskov@saske.sk
[2] Institute of Mathematics, Academy of Sciences of the Czech Republic
Žižkova 22, 616 62 Brno, Czech Republic
masopust@math.cas.cz

Abstract. We study the state complexity of the reverse of acyclic minimal deterministic finite automata, and the computational complexity of the following problem: Given an acyclic minimal DFA, is the minimal DFA for the reverse also acyclic? Note that we allow self-loops in acyclic automata. We show that there exists a language accepted by an acyclic minimal DFA such that the minimal DFA for its reverse is exponential with respect to the number of states, and we establish a tight bound on the state complexity of the reverse of acyclic DFAs. We also give a direct proof of the fact that the minimal DFA for the reverse is acyclic if and only if the original acyclic minimal DFA satisfies a certain structural property, which can be tested in quadratic time.

1 Introduction

The reverse of a machine or of a language is one of the classical operations in automata and formal language theory. However, in comparison with other operations, such as the boolean operations, the descriptive complexity of the reverse of regular languages is exponential in the worst case with respect to the number of states of minimal deterministic finite automata (DFAs). This paper demonstrates that this also holds true for a subclass of regular languages accepted by acyclic minimal DFAs. To prevent confusion with DFAs accepting only finite languages, it is important to explain here that we allow self-loops in acyclic automata. Thus, the notion of *acyclic* stands for automata without cycles of length two or more. This definition is adapted from the literature [7,15,16,18].

The first part of this paper studies the state complexity of the reverse of acyclic minimal DFAs, and proves that the tight bound for this subclass is 2^{n-1}, where n is the number of states of the input acyclic DFA. This bound can be met by an acyclic DFA over a ternary alphabet with a dead state, or by an acyclic DFA over a growing alphabet without the dead state. It remains open whether

* Research supported by VEGA grant 2/0183/11 and by grant APVV-0035-10.
** Research supported by the GAČR grant P202/11/P028, and by RVO: 67985840.

or not the upper bound can be met by an acyclic DFA over a binary alphabet independently on the presence of the dead state, as well as by an acyclic DFA over a fixed alphabet that has no dead state.

The exponential blow-up of states for this operation motivates the following computational complexity problem: Given an acyclic minimal DFA accepting a regular language, is the minimal DFA for the reverse of the language also acyclic? Surprisingly, the answer to this question depends only on a certain structural property of the input automaton which can be tested by a known algorithm with a quadratic-time complexity with respect to the size of the input automaton. This means that we do not need to compute the whole automaton for the reverse to answer the question. Although this result can be derived from other results concerning piecewise testable languages, as discussed in the conclusions, as far as the authors know it has never been proved directly in this context. Therefore, in the second part of this paper, we prefer to present a direct proof of the fact that the reverse is acyclic if and only if the original minimal acyclic automaton satisfies a structural property discussed below.

This problem can be generalized to many other operations and types of automata. It deserves attention especially in the case of operations that are of interest in practical applications and have exponential state complexity, such as projections or abstractions for DFAs [1,6,8,9].

2   Preliminaries and Definitions

The cardinality of a set Σ is denoted by $|\Sigma|$. An alphabet is a finite non-empty set. The free monoid generated by an alphabet Σ is denoted by Σ^*. A string over Σ is any element of Σ^*. The empty string (the identity of Σ^*) is denoted by ε. The length of a string w is denoted by $|w|$. A language over Σ is any subset of Σ^*.

A *nondeterministic finite automaton* (NFA) is a 5-tuple $N = (Q, \Sigma, \delta, Q_0, F)$, where Q is a finite non-empty set of states, Σ is an input alphabet, $Q_0 \subseteq Q$ is the set of initial states, $F \subseteq Q$ is the set of final states, and $\delta : Q \times \Sigma \to 2^Q$ is a transition function which can be inductively extended to the domain $2^Q \times \Sigma^*$. The language *accepted* by N is defined as the set $L(N) = \{w \in \Sigma^* \mid \delta(Q_0, w) \cap F \neq \emptyset\}$.

An NFA $N = (Q, \Sigma, \delta, Q_0, F)$ is a *complete deterministic finite automaton* (DFA) if $|Q_0| = 1$, and $|\delta(q, a)| = 1$ for each state q in Q and each input symbol a in Σ. In that case, we identify singleton sets of states with their elements, that is, we write q for a singleton set $\{q\}$. Moreover, we consider the transition function δ to be a total mapping from $Q \times \Sigma$ to Q that can be extended to the domain $Q \times \Sigma^*$.

Two states of a DFA are *distinguishable* if there exists a string w which is accepted from one of the states and rejected from the other one. Otherwise, the two states are *equivalent*. A DFA is *minimal* if all its states are reachable from the initial state, and no two different states are equivalent. A DFA is *acyclic* if all strongly connected components [4] of the directed graph of the DFA are trivial, that is, they consist only of one element [7,15,16,18]. Note that this definition allows self-loops.

The *subset automaton* corresponding to an NFA $N = (Q, \Sigma, \delta, Q_0, F)$ is the DFA $N' = (2^Q, \Sigma, \delta', Q_0, F')$, in which $F' = \{R \subseteq Q \mid R \cap F \neq \emptyset\}$ and $\delta'(R, a) = \delta(R, a)$ for each set R in 2^Q and each symbol a in Σ. The subset automaton N' accepts the same language as the automaton N, but it need not be minimal since some of its states may be unreachable or equivalent.

The *reverse* w^R of a *string* w is inductively defined as follows: $\varepsilon^R = \varepsilon$ and $(va)^R = av^R$ for a string v in Σ^* and a symbol a in Σ. The *reverse of a language* L is the language $L^R = \{w^R \mid w \in L\}$. The *reverse of a DFA* $M = (Q, \Sigma, \delta, q_0, F)$ is the NFA M^R obtained from M by reversing all the transitions and by swapping the role of the initial and final states, that is, $M^R = (Q, \Sigma, \delta^R, F, \{q_0\})$, where $\delta^R(q, a) = \{p \in Q \mid \delta(p, a) = q\}$. It is known that the states of the subset automaton corresponding to the reverse of a minimal DFA are pairwise distinguishable [2,3,11]. For the sake of completeness, we give a short proof of this fact here.

Lemma 1 ([2,3,11]). *All distinct states of the subset automaton corresponding to the reverse of a minimal DFA are pairwise distinguishable.*

Proof. Let M^R be the reverse of a minimal DFA M. Let q be an arbitrary state of the NFA M^R. Since state q is reachable in M, there exists a string w_q accepted by M^R from q. Furthermore, the string w_q is not accepted from any other state of M^R; otherwise, there would be two distinct computations of the DFA M on the string w_q^R. It follows that the states of the subset automaton corresponding to M^R are pairwise distinguishable since two distinct subsets of the state set of M^R must differ in a state q, and therefore the two subsets are distinguished by the string w_q.   □

3   Main Results

This section presents the main results of this paper. First, we show that the worst-case state complexity of the reverse of a language represented by a minimal acyclic DFA is exponential in the number of states of the DFA. As a consequence of this result, we get that the direct construction of the minimal automaton for the reverse may be computationally unfeasible. This motivates the study of structural properties that would be helpful in deciding the question whether or not the minimal DFA for the reverse of a language is acyclic, if the language is represented by a minimal acyclic DFA. We prove that the acyclicity of the minimal DFA for the reverse is equivalent to a structural property testable in quadratic time.

Recall that in the general case, the worst-case state complexity of the reverse of a language represented by an n-state DFA is 2^n [5,10,11,12,19]. Our next result shows that for acyclic DFAs, the upper bound on the state complexity of the reverse is 2^{n-1}.

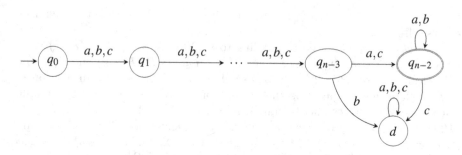

Fig. 1. The minimal acyclic DFA with the exponential reverse

Lemma 2. *Let M be an acyclic minimal DFA with n states. Then the minimal DFA accepting the reverse of the language $L(M)$ has no more than 2^{n-1} states.*

Proof. Let $M = (Q, \Sigma, \delta, q_0, F)$ be an n-state acyclic minimal DFA, and construct the NFA M^R for the reverse by swapping the role of the initial and final states, and by reversing all transitions. As M is acyclic, we can topologically order its states from left to right so that no transition goes from right to left. Let q be the rightmost state in this order. Since M is complete, q has self-loops under all symbols from Σ. If q is not final, it is the dead state of M, and we can remove it before constructing M^R, that is, the subset automaton corresponding to M^R has no more than 2^{n-1} states. On the other hand, if q is final, it appears because of the self-loops in all reachable states of the subset automaton corresponding to M^R. This again gives the upper bound 2^{n-1} on the number of states. The proof is complete. □

The following results show that the upper bound is tight.

Lemma 3. *There exists an acyclic minimal DFA M with n states over the alphabet $\{a, b, c\}$ such that the minimal DFA accepting the reverse of the language $L(M)$ has 2^{n-1} states.*

Proof. Consider the DFA shown in Fig. 1. To construct its reverse, omit the dead state d, make state q_{n-2} initial and state q_0 final, and reverse all the transitions. To simplify the proof, rename the states of the resulting NFA as shown in Fig. 2. We show that each subset of $\{0, 1, \ldots, n-2\}$ is reachable in the corresponding subset automaton.

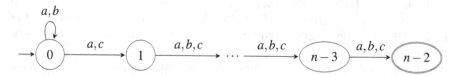

Fig. 2. The reverse of the DFA shown in Fig. 1; states renamed for the simplicity of the proof

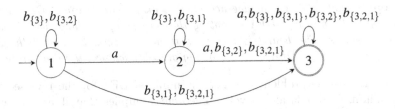

Fig. 3. The minimal acyclic DFA without the dead state with the exponential reverse

The proof is by induction on the size of subsets. Each singleton set $\{i\}$ is reached from the initial state $\{0\}$ by c^i. Each subset $\{i_1, i_2, \ldots, i_k\}$ of size k, where $2 \leq k \leq n-1$ and $0 \leq i_1 < i_2 < \cdots < i_k \leq n-2$, is reached from the set $\{0, i_3 - i_2, i_4 - i_2, \ldots, i_k - i_2\}$ of size $k-1$ by the string $ab^{i_2 - i_1 - 1}c^{i_1}$ since

$$\{0, i_3 - i_2, i_4 - i_2, \ldots, i_k - i_2\} \xrightarrow{a}$$

$$\{0, 1, i_3 - i_2 + 1, i_4 - i_2 + 1, \ldots, i_k - i_2 + 1\} \xrightarrow{b^{i_2 - i_1 - 1}}$$

$$\{0, i_2 - i_1, i_3 - i_1, i_4 - i_1, \ldots, i_k - i_1\} \xrightarrow{c^{i_1}} \{i_1, i_2, i_3, i_4, \ldots, i_k\}.$$

This gives 2^{n-1} reachable states of the subset automaton, which are all pairwise distinguishable by Lemma 1.   □

Note that the bound 2^{n-1} in the previous lemma follows naturally from the presence of the dead state, which is ignored in the construction of the reversed automaton. The next lemma shows, however, that the bound 2^{n-1} can also be met by an acyclic DFA without the dead state, but in this case we need an alphabet of exponential cardinality in comparison with the number of states, and it is not known whether the cardinality can be fixed.

Lemma 4. *There exists an acyclic minimal n-state DFA M without the dead state over a growing alphabet such that the minimal DFA accepting the reverse of the language $L(M)$ has 2^{n-1} states.*

Proof. Let $\Sigma_n = \{a\} \cup \{b_S \mid S \subseteq \{1, 2, \ldots, n\}$ and $n \in S\}$ be an alphabet consisting of a symbol a, and 2^{n-1} symbols b_S – one for each subset S of $\{1, 2, \ldots, n\}$ with $n \in S$.

Define an n-state acyclic DFA M over Σ_n with the state set $\{1, 2, \ldots, n\}$, where 1 is the initial state and n is the sole final state. By symbol a, state n goes to itself, and every other state i goes to state $i + 1$. By symbol b_S, every state in S goes to state n, and every other state goes to itself. Fig. 3 demonstrates this construction for $n = 3$.

In the subset automaton corresponding to the reverse of the DFA M, each subset S of $\{1, 2, \ldots, n\}$ containing state n is reached from the initial state $\{n\}$ by the symbol b_S. By Lemma 1, all these states are pairwise distinguishable, and the lemma follows.   □

As a consequence of the previous three lemmata we get the following result.

Theorem 1. *Let L be a language accepted by an acyclic minimal DFA with n states. Then the minimal DFA accepting the reverse of the language L has at most 2^{n-1} states. The bound is met by a ternary acyclic DFA with the dead state, or by an acyclic DFA over a growing alphabet without the dead state.*   □

Now we turn to the problem whether the minimal DFA for the reverse of an acyclic minimal DFA is also acyclic. Theorem 1 implies that it may be computationally unfeasible to directly construct the minimal DFA for the reverse. Therefore, we study structural properties of acyclic minimal DFAs to solve the problem. To this end, we need several definitions.

For two states p and q of a DFA $M = (Q, \Sigma, \delta, q_0, F)$, we write $p \prec q$ if $p \neq q$ and state q is reachable from state p, that is, there exists a string w in Σ^* such that $q = \delta(p, w)$. A state p is called *maximal* if there exists no state q such that $p \prec q$. Denote by $\Sigma(q)$ the set of all symbols appearing on the self-loops of state q, that is, $\Sigma(q) = \{a \in \Sigma \mid \delta(q, a) = q\}$.

Let $\Sigma_i \subseteq \Sigma$ and δ_i be the restriction of the transition function δ of the DFA M to the domain $Q \times \Sigma_i$. Denote by $\Gamma(\Sigma_i)$ the directed graph obtained from the deterministic automaton $(Q, \Sigma_i, \delta_i, q_0, F)$ by ignoring the labels of edges and eliminating the multi-edges. A *connected component* of the directed graph $\Gamma(\Sigma_i)$ with respect to a node q is the set of all nodes which are connected with q by a path disregarding the orientation of edges.

The following theorem characterizes the structural property which will be useful to derive the polynomial-time algorithm testing acyclicity of the reversed automaton. Although this result can be indirectly derived from other results concerning piecewise testable languages, as discussed in the conclusions, we prefer to give a direct proof of this fact here.

Theorem 2. *Let M be an acyclic minimal DFA. The minimal DFA accepting the reverse of the language $L(M)$ is acyclic if and only if for each state p of M, the connected component of the graph $\Gamma(\Sigma(p))$ containing state p has a unique maximal state with respect to the relation $\prec$.*

Proof. Let $M = (Q, \Sigma, \delta, q_0, F)$ be an acyclic minimal DFA and assume that the minimal DFA for the reverse, denoted by

$$M' = (Q', \Sigma, \delta', F, \{R \subseteq Q \mid q_0 \in R\}),$$

where $Q' \subseteq 2^Q$, is acyclic. The DFA M' is obtained from M by setting F to be the set of initial states, reversing all the transitions, converting the obtained NFA to a DFA, and minimizing the DFA. Each subset containing the initial state q_0 of M is set to be a final state of M'.

Assume that M' is acyclic. For the sake of contradiction, assume that there exists a state p in Q such that the connected component of the graph $\Gamma(\Sigma(p))$ containing state p has two distinct maximal states. Since state p is a maximal state of this component, there exists a state q in that component that is maximal and different from p. The DFA M is acyclic, thus either $p \nprec q$ or $q \nprec p$. Without loss of generality, we assume that $q \nprec p$. Then, there exist a state r in Q and two strings u, v in $\Sigma(p)^*$ such that $\delta(r, u) = p$ and $\delta(r, v) = q$. Since M is minimal, states p and q are distinguished by a string w in Σ^*. Let w be accepted from p and rejected from q as depicted in Fig. 4; the other case is symmetric.

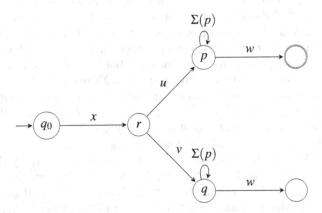

Fig. 4. Two maximal states p and q of the component $\Gamma(\Sigma(p))$ containing p

Consider the computation of M' on the string

$$w^R u^R v^R u^R v^R u^R v^R u^R v^R \ldots,$$

and let the computation be

$$F \xrightarrow{w^R} Z \xrightarrow{u^R} X_1 \xrightarrow{v^R} Y_1 \xrightarrow{u^R} X_2 \xrightarrow{v^R} Y_2 \cdots.$$

Since w is accepted by M from p but rejected from q, state p is in Z but q is not. Moreover, since p has a loop on each symbol in $\Sigma(p)$, it occurs in every X_i and Y_i. Now, consider the state r. It occurs in every set X_i since p goes to r by u^R in M'. However, r does not occur in any Y_i because otherwise we would have

$$r \xrightarrow{v} q \xrightarrow{uvuvuv\cdots uvu} q \xrightarrow{w} f$$

in M for a final state f of M; thus, string w would be accepted from state q, which is a contradiction. Now consider a sequence $X_1, Y_1, X_2, Y_2, \ldots$ of subsets of the states of M. Since we only have a finite number of such subsets, there exists a cycle in this sequence. Let X and X' be two consecutive subsets on this cycle. Then state r is in exactly one of X and X'. Without loss of generality, let $r \in X$. Since M is minimal, state r is reached in M from the initial state q_0 by a string $x \in \Sigma^*$. It follows that x^R is accepted from X in M'. On the other hand, since M is deterministic and $r \notin X'$, string x^R is not accepted from X' in M'. Thus X and X' are not equivalent, and therefore the cycle is not a self-loop. This contradicts our assumption that M' is acyclic.

To prove the converse implication, assume that for each state p of M, the connected component of the graph $\Gamma(\Sigma(p))$ containing p has a unique maximal state with respect to the relation $\prec$. For the sake of contradiction, assume that there exists a cycle of length at least two in the DFA M'. Let S and T be two different sets on this cycle. Without loss of generality, we can assume that there exists a state r in M with $r \notin S$ and $r \in T$. Assume that S goes to T by a string u, and T goes to S by a string v on the cycle in M', see Fig. 5. For $i \geq 0$, let $p_i = \delta(r, u^R(v^R u^R)^i)$ be the states of M reached from the state r by strings $u^R(v^R u^R)^i$. Then all the states p_i belongs to S. Since M is acyclic, there exists j such that p_j goes to itself on each symbol occurring in uv, denoted by $\Sigma(uv)$. Since p_j is in S and goes to itself on each symbol from $\Sigma(uv)$, it is also in T. Denote $p = p_j$. Then p is maximal with respect to $\Sigma(uv)$. Now the aim is to find another maximal state in the connected component of $\Gamma(\Sigma(uv))$ containing p.

To this aim, let $s_i = \delta(r, v^R(u^R v^R)^i)$ for $i \geq 0$. Since M is acyclic, there exists an index k such that s_k goes to itself on each symbol from $\Sigma(uv)$. Set $q = s_k$. State q is in the same connected component as p since both p and q are reached from r in M. We need to show that $q \neq p$. Assume to the contrary that $q = p$. Then state r is reached in M' from state p by the string $(vu)^k v$. Since state p is in T, state r is in S, which is a contradiction. Hence states p and q are distinct maximal states in the same connected component of the graph $\Gamma(\Sigma(uv))$. Since M is acyclic, either $p \not\prec q$ or $q \not\prec p$. Assume that $q \not\prec p$, and consider the graph $\Gamma(\Sigma(p))$. Then $\Gamma(\Sigma(uv)) \subseteq \Gamma(\Sigma(p))$, state p is maximal with respect to $\Sigma(p)$, and states p and q are connected in the graph $\Gamma(\Sigma(p))$. State q or a successor of q is maximal in the same connected component of $\Gamma(\Sigma(p))$, but it is different from p because $q \not\prec p$.   □

Now we demonstrate this technique on the following example.

Example 1. Consider the minimal DFA depicted in Fig. 6 (left). We have $\Sigma(1) = \{a, b\}$. Fig. 6 (right) shows the graph $\Gamma(\Sigma(1))$. The only connected component of $\Gamma(\Sigma(1))$ has two maximal states, namely 1 and d. By Theorem 2, the minimal DFA accepting the reverse of the language accepted by the DFA in Fig. 6 (left) has a cycle, as shown in Fig. 7.   □

Notice that this technique requires to consider complete minimal DFAs and it works neither for incomplete DFAs nor for complete DFAs that are not minimal. The previous example does not work if we ignore the dead state. In addition, in the case of non-minimal automata, we can have two different maximal accepting/non-accepting states that can be equivalent.

The condition whether for each state p of M, the connected component of $\Gamma(\Sigma(p))$ containing state p has a unique maximal state with respect to the

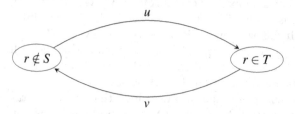

Fig. 5. A cycle in the minimal DFA M' for the reverse

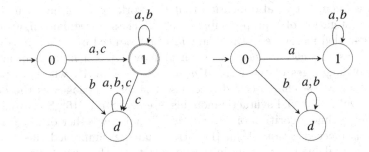

Fig. 6. An acyclic DFA and its graph $\Gamma(\Sigma(1))$

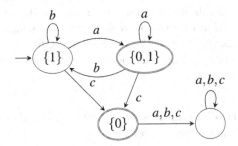

Fig. 7. The minimal DFA for the reverse of the language accepted by the DFA in Fig. 6 (left)

relation $\prec$ can be tested using the algorithm presented by Trahtman [18]. The algorithm runs in time $O(n^2)$, where n is the sum of the number of states and the number of transitions in M. As a consequence, we have the following theorem.

Theorem 3. *Let M be an acyclic minimal deterministic finite automaton with m states and k transitions. Let $n = mk$. There exists an algorithm solving the problem of acyclicity of the minimal deterministic automaton for the reverse of the language $L(M)$ in time $O(n^2)$.* □

4   Conclusions

We discussed the state complexity of acyclic minimal DFAs, and the problem of deciding whether or not the minimal DFA for the reverse of a language is acyclic if the language is represented by an acyclic minimal DFA. We showed that the minimal DFA for the reverse is acyclic if and only if the minimal acyclic DFA for the original language possesses a special structural property. This property can be tested in quadratic time using the result of Trahtman [18], even though the construction of the minimal DFA for the reverse may be exponential.

We could also ask the opposite question: Is there a structural property ensuring that the minimal DFA for the reverse of a language is acyclic if the language is represented by a minimal DFA with a cycle? As far as the authors know, this question is open. Let us also mention that the work by Trahtman is motivated by the investigation of a proper subclass of the class of regular languages, the class of so-called piecewise testable languages introduced by Simon in [13].

A *piecewise testable language* over an alphabet A is a finite boolean combination of languages of the form $A^*a_1A^*a_2A^*\ldots A^*a_kA^*$, where $k \geq 0$ and $a_i \in A$. Simon [14] characterized piecewise testable languages as the class of languages with $\mathcal{J}$-trivial syntactic monoids, see also Stern [15]. Stern suggested a polynomial-time algorithm of order $O(n^5)$ deciding whether or not a regular language is piecewise testable in [16]. Trahtman [18] improved this result by presenting an algorithm running in time quadratic in the size of the input, and provided a package TESTAS implementing the algorithm in [17].

Recently, Polák and Klíma [7] have mentioned another method for the verification of piecewise testability of a regular language. However, this method is based on the construction of a so-called *biautomaton*, which requires both the minimal DFA for a language and the minimal DFA for its reverse. According to Theorem 3, this construction may be unfeasible because of the complexity reasons.

Acknowledgements. The authors gratefully acknowledge useful suggestions and comments of the anonymous referees.

References

1. Boutin, O., Komenda, J., Masopust, T., Schmidt, K., van Schuppen, J.H.: Hierarchical control with partial observations: Sufficient conditions. In: Proc. of IEEE Conference on Decision and Control and European Control Conference (CDC-ECC 2011), Orlando, Florida, USA, pp. 1817–1822 (2011)
2. Brzozowski, J.A.: Canonical regular expressions and minimal state graphs for definite events. In: Proc. of the Symposium on Mathematical Theory of Automata. MRI Symposia Series, vol. 12, pp. 529–561. Polytechnic Institute of Brooklyn, New York (1963)
3. Champarnaud, J.M., Khorsi, A., Paranthoën, T.: Split and join for minimizing: Brzozowski's algorithm, http://jmc.feydakins.org/ps/c09psc02.ps
4. Cormen, T.H., Leiserson, C.E., Rivest, R.L., Stein, C.: Introduction to Algorithms, 3rd edn. The MIT Press (2009)
5. Jirásková, G., Šebej, J.: Note on Reversal of Binary Regular Languages. In: Holzer, M., Kutrib, M., Pighizzini, G. (eds.) DCFS 2011. LNCS, vol. 6808, pp. 212–221. Springer, Heidelberg (2011)
6. Jirásková, G., Masopust, T.: On a structural property in the state complexity of projected regular languages. Theoretical Computer Science (in press, 2012), doi:10.1016/j.tcs.2012.04.009
7. Klíma, O., Polák, L.: On biautomata. In: Proc. of NCMA 2011. books@ocg.at, vol. 282, pp. 153–164. Austrian Computer Society (2011)
8. Komenda, J., Masopust, T., van Schuppen, J.H.: Synthesis of controllable and normal sublanguages for discrete-event systems using a coordinator. Systems & Control Letters 60(7), 492–502 (2011)

9. Komenda, J., Masopust, T., van Schuppen, J.H.: Supervisory control synthesis of discrete-event systems using a coordination scheme. Automatica 48(2), 247–254 (2012)
10. Leiss, E.: Succinct representation of regular languages by boolean automata. Theoretical Computer Science 13, 323–330 (1981)
11. Mirkin, B.G.: On dual automata. Kibernetika 2, 7–10 (1966) (in Russian); English translation: Cybernetics 2, 6–9 (1966)
12. Salomaa, A., Wood, D., Yu, S.: On the state complexity of reversals of regular languages. Theoretical Computer Science 320, 315–329 (2004)
13. Simon, I.: Hierarchies of Events with Dot-Depth One. Ph.D. thesis, Dep. of Applied Analysis and Computer Science, University of Waterloo, Canada (1972)
14. Simon, I.: Piecewise Testable Events. In: Brakhage, H. (ed.) GI-Fachtagung 1975. LNCS, vol. 33, pp. 214–222. Springer, Heidelberg (1975)
15. Stern, J.: Characterizations of some classes of regular events. Theoretical Computer Science 35, 17–42 (1985)
16. Stern, J.: Complexity of some problems from the theory of automata. Information and Control 66(3), 163–176 (1985)
17. Trahtman, A.N.: A Package TESTAS for Checking Some Kinds of Testability. In: Champarnaud, J.-M., Maurel, D. (eds.) CIAA 2002. LNCS, vol. 2608, pp. 228–232. Springer, Heidelberg (2003)
18. Trahtman, A.N.: Piecewise and Local Threshold Testability of DFA. In: Freivalds, R. (ed.) FCT 2001. LNCS, vol. 2138, pp. 347–358. Springer, Heidelberg (2001)
19. Yu, S., Zhuang, Q., Salomaa, K.: The state complexities of some basic operations on regular languages. Theoretical Computer Science 125(2), 315–328 (1994)

Implementing Computations in Automaton (Semi)groups

Ines Klimann, Jean Mairesse, and Matthieu Picantin

Univ Paris Diderot, Sorbonne Paris Cité, LIAFA, UMR 7089 CNRS, Paris, France
{klimann,mairesse,picantin}@liafa.univ-paris-diderot.fr

Abstract. We consider the growth, order, and finiteness problems for automaton (semi)groups. We propose new implementations and compare them with the existing ones. As a result of extensive experimentations, we propose some conjectures on the order of finite automaton (semi)groups.

Keywords: automaton (semi)groups, growth, order, finiteness, minimization.

1 Introduction

Automaton (semi)groups — short for semigroups generated by Mealy automata or groups generated by invertible Mealy automata — were formally introduced a half century ago (for details, see [10,7] and references therein). Over the years, important results have started revealing their full potential. For instance, the article [9] constructs simple Mealy automata generating infinite torsion groups and so contributes to the Burnside problem, and, the article [5] produces Mealy automata generating the first examples of (semi)groups with intermediate growth and so answers the Milnor problem.

The classical decision problems have been investigated for such (semi)groups. The word problem is solvable using standard minimization techniques, while the conjugacy problem is undecidable [16]. Here we concentrate on the problems related to growth, order, and finiteness.

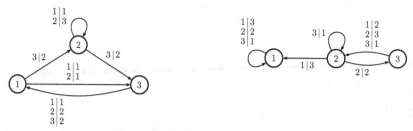

Fig. 1. A Mealy automaton and its dual

To illustrate, consider the two Mealy automata of Fig. 1. They are dual, that is, they can be obtained one from the other by exchanging the roles of stateset and alphabet. A (semi)group is associated in a natural way with each automaton (formally defined below). The two Mealy automata of Fig. 1 are associated with finite (semi)groups. Their orders are respectively: on the left a semigroup of order 234, on the right a group of order $1\,494\,186\,269\,970\,473\,680\,896 = 2^{64} \cdot 3^4 \approx 1.5 \times 10^{21}$.

Several points are illustrated by this example:

- An automaton and its dual generate (semi)groups which are either both finite or both infinite (see [12,2]).
- The order of a finite automaton (semi)group can be amazingly large. It makes a priori difficult to decide whether an automaton (semi)group is finite or not. Actually, the decidability of this question is open (see [10,2]).
- The order of the (semi)groups generated by a Mealy automaton and its dual can be strikingly different. It suggests to work with both automata together.

The contributions of the present paper are three-fold:

- We propose new implementations (in GAP [8]) of classical algorithms for the computation of the growth function; the computation of the order (if finite); the semidecision procedure for the finiteness.
- We compare the new implementations with the existing ones. Indeed, there exist two GAP packages dedicated to Mealy automata and their associated (semi) groups: FR by Bartholdi [4] and automgrp by Muntyan and Savchuk [11].
- We realize systematic experimentations on small Mealy automata as well as randomly chosen large Mealy automata. These serve as testbeds to some conjectures on the growth types of the associated (semi)groups, as well as on the order of a (semi)group.

The structure of the paper is the following. In Section 2, we present basic notions on Mealy automata and automaton (semi)groups. In Section 3, we give new implementations and compare them with the existing ones. Section 4 is dedicated to experimentations and to the resulting conjectures.

2   Automaton (Semi)groups

2.1   Mealy Automaton

If one forgets initial and final states, a *(finite, deterministic, and complete) automaton* $\mathcal{A}$ is a triple $\left(A, \Sigma, \delta = (\delta_i : A \to A)_{i \in \Sigma}\right)$, where the *set of states* A and the *alphabet* Σ are non-empty finite sets, and where the δ_i's are functions.

A *Mealy automaton* is a quadruple

$$\left(A, \Sigma, \delta = (\delta_i : A \to A)_{i \in \Sigma}, \rho = (\rho_x : \Sigma \to \Sigma)_{x \in A}\right),$$

such that both (A, Σ, δ) and (Σ, A, ρ) are automata. In other terms, a Mealy automaton is a letter-to-letter transducer with the same input and output alphabets. The transitions of a Mealy automaton are

$$x \xrightarrow{i \mid \rho_x(i)} \delta_i(x).$$

The graphical representation of a Mealy automaton is standard, see Fig. 1.

The notation $x \xrightarrow{\mathbf{u}|\mathbf{v}} y$ with $\mathbf{u} = u_1 \cdots u_n$, $\mathbf{v} = v_1 \cdots v_n$ is a shorthand for the existence of a path $x \xrightarrow{u_1|v_1} x_1 \xrightarrow{u_2|v_2} x_2 \longrightarrow \cdots \longrightarrow x_{n-1} \xrightarrow{u_n|v_n} y$ in $\mathcal{A}$.

In a Mealy automaton $(A, \Sigma, \delta, \rho)$, the sets A and Σ play dual roles. So we may consider the *dual (Mealy) automaton* defined by $\mathfrak{d}(\mathcal{A}) = (\Sigma, A, \rho, \delta)$, that is:

$$i \xrightarrow{x|y} j \in \mathfrak{d}(\mathcal{A}) \quad \Longleftrightarrow \quad x \xrightarrow{i|j} y \in \mathcal{A}.$$

It is pertinent to consider a Mealy automaton and its dual together, that is to work with the pair $\{\mathcal{A}, \mathfrak{d}(\mathcal{A})\}$, see an example in Fig. 1.

Let $\mathcal{A} = (A, \Sigma, \delta, \rho)$ and $\mathcal{B} = (B, \Sigma, \gamma, \pi)$ be two Mealy automata acting on the same alphabet; their *product* $\mathcal{A} \times \mathcal{B}$ is defined as the Mealy automaton with stateset $A \times B$, alphabet Σ, and transitions:

$$xy \xrightarrow{i|\pi_y(\rho_x(i))} \delta_i(x)\gamma_{\rho_x(i)}(y).$$

2.2   Generating (Semi)groups

Let $\mathcal{A} = (A, \Sigma, \delta, \rho)$ be a Mealy automaton. We view $\mathcal{A}$ as an automaton with an input and an output tape, thus defining mappings from input words over Σ to output words over Σ. Formally, for $x \in A$, the map $\rho_x : \Sigma^* \to \Sigma^*$, extending $\rho_x : \Sigma \to \Sigma$, is defined by:

$$\rho_x(\mathbf{u}) = \mathbf{v} \quad \text{if} \quad \exists y, \ x \xrightarrow{\mathbf{u}|\mathbf{v}} y.$$

By convention, the image of the empty word is itself. The mapping ρ_x is length-preserving and prefix-preserving. It satisfies

$$\forall u \in \Sigma, \ \forall \mathbf{v} \in \Sigma^*, \qquad \rho_x(u\mathbf{v}) = \rho_x(u)\rho_{\delta_u(x)}(\mathbf{v}).$$

We say that ρ_x is the *production function* associated with $(\mathcal{A}, x)$. For $\mathbf{x} = x_1 \cdots x_n \in A^n$ with $n > 0$, set $\rho_{\mathbf{x}} : \Sigma^* \to \Sigma^*, \rho_{\mathbf{x}} = \rho_{x_n} \circ \cdots \circ \rho_{x_1}$.

Denote dually by $\delta_i : A^* \to A^*, i \in \Sigma$, the production mappings associated with the dual Mealy automaton $\mathfrak{d}(\mathcal{A})$. For $\mathbf{v} = v_1 \cdots v_n \in \Sigma^n$ with $n > 0$, set $\delta_{\mathbf{v}} : A^* \to A^*, \ \delta_{\mathbf{v}} = \delta_{v_n} \circ \cdots \circ \delta_{v_1}$.

Definition 1. *Consider a Mealy automaton $\mathcal{A}$. The semigroup of mappings from Σ^* to Σ^* generated by $\rho_x, x \in A$, is called the semigroup generated by $\mathcal{A}$ and is denoted by $\langle \mathcal{A} \rangle_+$. A semigroup G is an automaton semigroup if there exists a Mealy automaton $\mathcal{A}$ such that $G = \langle \mathcal{A} \rangle_+$.*

A Mealy automaton $\mathcal{A} = (A, \Sigma, \delta, \rho)$ is *invertible* if all the mappings $\rho_x : \Sigma \to \Sigma$ are permutations. Then the production functions $\rho_x : \Sigma^* \to \Sigma^*$ are invertible.

Definition 2. *Let $\mathcal{A} = (A, \Sigma, \delta, \rho)$ be invertible. The group generated by $\mathcal{A}$ is the group generated by the mappings $\rho_x : \Sigma^* \to \Sigma^*, x \in A$. It is denoted by $\langle \mathcal{A} \rangle$.*

Let $\mathcal{A} = (A, \Sigma, \delta, \rho)$ be an invertible Mealy automaton. Its *inverse* is the Mealy automaton $\mathcal{A}^{-1}$ with stateset $A^{-1} = \{x^{-1}, x \in A\}$ and set of transitions

$$x^{-1} \xrightarrow{j|i} y^{-1} \in \mathcal{A}^{-1} \quad \Longleftrightarrow \quad x \xrightarrow{i|j} y \in \mathcal{A}.$$

A Mealy automaton is *reversible* if its dual is invertible. A Mealy automaton $\mathcal{A}$ is *bireversible* if both $\mathcal{A}$ and $\mathcal{A}^{-1}$ are invertible and reversible.

Theorem 1 ([2,12,13]). *The (semi)group generated by a Mealy automaton is finite if and only if the (semi)group generated by its dual is finite.*

2.3   Minimization and the Word Problem

Let $\mathcal{A} = (A, \Sigma, \delta, \rho)$ be a Mealy automaton. The *Nerode equivalence on A* is the limit of the sequence of increasingly finer equivalences ($\equiv_k$) recursively defined by:

$$\forall x, y \in A, \qquad x \equiv_0 y \iff \rho_x = \rho_y,$$
$$\forall k \geqslant 0,\, x \equiv_{k+1} y \iff x \equiv_k y \quad \text{and} \quad \forall i \in \Sigma,\, \delta_i(x) \equiv_k \delta_i(y).$$

Since the set A is finite, this sequence is ultimately constant; moreover if two consecutive equivalences are equal, the sequence remains constant from this point. The limit is therefore computable. For every element x in A, we denote by $[x]$ the class of x w.r.t. the Nerode equivalence.

Definition 3. *Let $\mathcal{A} = (A, \Sigma, \delta, \rho)$ be a Mealy automaton and let $\equiv$ be the Nerode equivalence on A. The minimization of $\mathcal{A}$ is the Mealy automaton $\mathrm{m}(\mathcal{A}) = (A/\equiv, \Sigma, \tilde{\delta}, \tilde{\rho})$, where for every (x, i) in $A \times \Sigma$, $\tilde{\delta}_i([x]) = [\delta_i(x)]$ and $\tilde{\rho}_{[x]} = \rho_x$.*

This definition is consistent with the standard minimization of "deterministic finite automata" where instead of considering the mappings $(\rho_x : \Sigma \to \Sigma)_x$, the computation is initiated by the separation between terminal and non-terminal states. Using Hopcroft algorithm, the time complexity of minization is $\mathcal{O}(\Sigma A \log A)$, see [1].

By construction, a Mealy automaton and its minimization generate the same semigroup. Indeed, two states of a Mealy automaton belong to the same class w.r.t the Nerode equivalence if and only if they represent the same element in the generated (semi)group.

Consider the *word problem*:

> **Input:** a Mealy automaton $(A, \Sigma, \delta, \rho)$; $\mathbf{x}, \mathbf{y} \in A^*$.
> **Question:** $(\rho_{\mathbf{x}} : \Sigma^* \to \Sigma^*) = (\rho_{\mathbf{y}} : \Sigma^* \to \Sigma^*)$?

The word problem is solvable by extending the above minimization procedure. FR uses this approach, while automgrp uses a method based on the wreath recursion [7].

3   Fully Exploiting the Minimization

Consider the following problems for the (semi)group given by a Mealy automaton: compute the growth function, compute the order (if finite), detect the finiteness. The packages FR and automgrp provide implementations of the three problems. Here we propose new implementations based on a simple idea which fully uses the automaton structure.

3.1   Growth

Consider a Mealy automaton $\mathcal{A} = (A, \Sigma, \delta, \rho)$ and an element $\mathbf{x} \in A^*$. The *length* of $\rho_{\mathbf{x}}$, denoted by $|\rho_{\mathbf{x}}|$, is defined as follows:

$$|\rho_{\mathbf{x}}| = \min\{n \mid \exists \mathbf{y} \in A^n, \rho_{\mathbf{x}} = \rho_{\mathbf{y}}\} .$$

The *growth series* of $\mathcal{A}$ is the formal power series given by

$$\sum_{g \in \langle \mathcal{A} \rangle_+} t^{|g|} = \sum_{n \in \mathbb{N}} \#\{g \in \langle \mathcal{A} \rangle_+ \; ; \; |g| = n\} \, t^n .$$

In words, the growth series enumerates the semigroup elements according to their length. This is an instanciation of the notion of spherical growth series for a finitely generated semigroup. Observe that the series is a polynomial if and only if the semigroup is finite.

Using the Generic Algorithm. Since the word problem is solvable, it is possible to compute an arbitrary but finite number of coefficients of the growth series. Indeed for each n, generate the set of elements of length n by multiplying elements of length $n - 1$ with generators and detecting-deleting duplicated elements by solving the word problem. The functions Growth from automgrp and WordGrowth from FR both follow this pattern. Therefore the structure of the underlying Mealy automaton is used only to get a solution to the word problem (in fact, both Growth and WordGrowth are generic, in the sense that they are applicable for any (semi)group with an implemented solution to the word problem).

New Implementation. We propose a new implementation based on a simple observation. Knowing the elements of length $n - 1$, Nerode minimization can be used in a global manner to obtain simultaneously the elements of length n. Concretely, with each integer $n \geq 1$ is associated a new Mealy automaton $\mathcal{A}_n$ defined recursively as follows:

$$\mathcal{A}_n = \mathfrak{m}(\mathcal{A}_{n-1} \times \mathfrak{m}(\mathcal{A})) \quad \text{and} \quad \mathcal{A}_1 = \mathfrak{m}(\mathcal{A}) .$$

Here, we assume, without real loss of generality, that the identity element is one of the generators (otherwise simply add a new state to the Mealy automaton coding the identity). This way, the elements of $\mathcal{A}_n$ are exactly the elements of length at most n.

```
AutomatonGrowth := function(arg)
local aut, radius, growth, sph, curr, next, r;
aut:=arg[1];   # Mealy automaton
if Length(arg)>1 then radius:=arg[2];
                else radius:=infinity;
fi;
r := 0;    curr := TrivialMealyMachine([1]);
next := Minimized(aut);
aut := Minimized(next+TrivialMealyMachine(Alphabet(aut)));
sph := aut!.nrstates - 1;  # number of non-trivial states
growth := [next!.nrstates-sph];
while sph>0 and r<radius
do  Add(growth,sph);
    r := r+1;    curr := next;
    next := Minimized(next*aut);
    sph := next!.nrstates-curr!.nrstates;
od;
return growth;
end;
```

Note that `AutomatonGrowth(aut)` computes the growth of the semigroup $\langle aut \rangle_+$, while `AutomatonGrowth(aut+aut^-1)` computes the growth of the group $\langle aut \rangle$.

Experimental Results. First we run `AutomatonGrowth` and FR's `WordGrowth` on the Grigorchuk automaton, a famous Mealy automaton generating an infinite group. For radius 10, `AutomatonGrowth` is much faster, 76 ms as opposed to 9912 ms[1]. The explanation is simple: `WordGrowth` calls the minimization procedure 57577 times while `AutomatonGrowth` calls it only 12 times. Here are the details.

```
gap> aut := GrigorchukMachine;; radius:= 10;;
gap> ProfileFunctions([Minimized]);
gap> WordGrowth(SCSemigroupNC(aut), radius); time;
[ 1, 4, 6, 12, 17, 28, 40, 68, 95, 156, 216 ]
9912
gap> DisplayProfile();
  count  self/ms  chld/ms  function
  57577     7712        0  Minimized
            7712           TOTAL
gap> ProfileFunctions([Minimized]);
gap> AutomatonGrowth(aut, radius); time;
[ 1, 4, 6, 12, 17, 28, 40, 68, 95, 156, 216 ]
76
gap> DisplayProfile();
  count  self/ms  chld/ms  function
     12       72        0  Minimized
              72           TOTAL
```

---

[1] All timings displayed in this paper have been obtained on an Intel Core 2 Duo computer with clock speed 3,06 GHz.

Now we compare the running times of the implementations for the computation of the first terms of the growth series for all 335 bireversible 3-letter 3-state Mealy automata (up to equivalence). In Tab. 1, some computations with FR's WordGrowth or with automgrp's Growth could not be completed in reasonable time for radius 7.

Table 1. Average time (in ms)

radius	1	2	3	4	5	6	7
FR's WordGrowth	3.4	29.0	555.0	8 616.5	131 091.4	2 530 170.3	?
automgrp's Growth	0.7	2.8	16.9	158.9	1 909.0	22 952.8	?
AutomatonGrowth	0.6	1.8	5.9	28.9	187.3	1 005.9	7 131.4

3.2 Order of the (Semi)group

Although the finiteness problem is still open, some semidecision procedures enable to find the order of an expected finite (semi)group. FR and automgrp use orthogonal approaches. Our new implementation refines the one of FR and remains orthogonal to the one of automgrp.

automgrp's *Implementation.* The GAP package automgrp provides the function LevelOfFaithfulAction, which allows to compute—very efficiently in some cases—the order of the generated group. The principle is the following. Let $\mathcal{A} = (A, \Sigma, \delta, \rho)$ be an invertible Mealy automaton and let G_k be the group generated by the restrictions of the production functions to Σ^k. If $\#G_k = \#G_{k+1}$ for some k, then $\langle \mathcal{A} \rangle$ is finite of order $\#G_k$. This function can be easily adapted to a non-invertible Mealy automaton.

Observe that LevelOfFaithfulAction cannot be used to compute the growth series. Indeed at each step a quotient of the (semi)group is computed. On the other hand LevelOfFaithfulAction is a good bypass strategy for the order computation. Furthermore, it takes advantage from the special ability of GAP to manipulate permutation groups.

FR's *Implementation and the New Implementation.* Any algorithm computing the growth series can be used to compute the order of the generated (semi)group if finite. It suffices to compute the growth series until finding a coefficient equal to zero. This is the approach followed by FR. Since we proposed, in the previous section, a new implementation to compute the growth series, we obtain as a byproduct a new procedure to compute the order. We call it AutomSGrOrder.

Experimental Results. The orthogonality of the two previous approaches can be simply illustrated by recalling the introductory example of Fig. 1. Neither FR's Order nor AutomSGrOrder are able to compute the order of the large group, while automgrp via LevelOfFaithfulAction succeeds in only 14 338 ms. Conversely, AutomSGrOrder computes the order of the small semigroup in 17 ms, while an adaptation of LevelOfFaithfulAction (to non-invertible Mealy automata) takes 2 193 ms.

3.3   Finiteness

There exist several criteria to detect the finiteness of an automaton (semi)group, see [2,3,6,14,15, ...]. But the decidability of the finiteness is still an open question. Each procedure to compute the order of a (semi)group yields a semidecision procedure for the finiteness problem. Both packages FR and automgrp apply a number of previously known criteria of (in)finiteness and then intend to conclude by ultimately using an order computation.

We propose an additional ingredient which uses minimization in a subtle way. Here, the semigroup to be tested is successively replaced by new ones which are finite if and only if the original one is finite. It is possible to incorporate this ingredient to get two new implementations, one in the spirit of FR and one in the spirit of automgrp. The new implementations are order of magnitudes better than the old ones. Both are useful since the fastest one depends on the cases.

3.3.1   mð-Reduction of Mealy Automata and Finiteness

The mð-reduction was introduced in [2] to give a sufficient condition of finiteness. The new semidecision procedures start with this reduction.

Definition 4. *A pair of dual Mealy automata is reduced if both automata are minimal. Recall that* m *(resp.* ð*) is the operation of minimization (resp. dualization). The* mð-*reduction of a Mealy automaton* A *consists in minimizing the automaton or its dual until the resulting pair of dual Mealy automata is reduced.*

The mð-reduction is well-defined: if both a Mealy automaton and its dual automaton are non-minimal, the reduction is confluent [2]. An example of mð-reduction is given in Fig. 2.

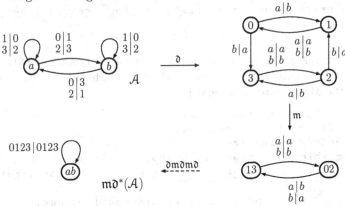

Fig. 2. The mð-reduction of a pair of dual Mealy automata

The sequence of minimization-dualization can be arbitrarily long: the minimization of a Mealy automaton with a minimal dual can make the dual automaton non-minimal.

If A is a Mealy automaton, we denote by mð*(A) the corresponding Mealy automaton after mð-reduction.

Theorem 2 ([2]). *A Mealy automaton $\mathcal{A}$ generates a finite (semi)group if and only if $\mathfrak{md}^*(\mathcal{A})$ generates a finite (semi)group.*

This is the starting point of the new implementations. We use an additional fact. We can prune a Mealy automaton by deleting the states which are not accessible from a cycle. This does not change the finiteness or infiniteness of the generated (semi)group [3].

3.3.2   The New Implementations

The design of procedure IsFinite1 is consistent with the one of AutomatonGrowth. Hence IsFinite1 is much closer to FR than to automgrp. Here we propose a version that works with the automaton and its dual in parallel.

```
IsFinite1 := function (aut, limit)
local radius, dual, curr1, next1, curr2, next2;
radius := 0;
aut := MDReduced(Prune(aut));     dual := DualMachine(aut);
curr1 := MealyMachine([[1]],[()]);     curr2 := curr1;
next1 := aut;     next2 := dual;
while curr2!.nrstates<>next2!.nrstates and radius<limit
do  radius := radius + 1;     curr1 := next1;
    next1 := Minimized(next1*aut);
    if curr1!.nrstates<>next1!.nrstates
    then  curr2 := next2;
          next2 := Minimized(next2*dual);
    else  return true;
    fi;
od;
if curr2!.nrstates = next2!.nrstates then   return true; fi;
return fail;
end;
```

The procedure IsFinite2 is a refinement of automgrp's LevelOfFaithfulAction: the minimization is called on the dual and can be enhanced again to work in parallel on the Mealy automaton and its dual.

```
IsFinite2 := function(aut,limit)
local f1, f2, next, cs, ns, lev;
aut := MDReduced(Prune(aut));
if IsInvertible(aut) then f1:=Group; f2:=PermList;
                     else f1:=Semigroup; f2:=Transformation;
fi;
lev := 0; cs := 1;   ns := Size(f1(List(aut!.output,f2)));
aut := DualMachine(aut);     next := aut;
while cs<ns and lev<limit
do lev := lev+1;   cs := ns;   next := Minimized(next*aut);
   ns := Size(f1(List(DualMachine(next)!.output,f2)));
od;
if cs=ns then return true; else return fail; fi;
end;
```

Experimental Results. Tab. 2 presents the average time to detect finiteness of (semi)groups generated by p-letter q-state invertible or reversible Mealy automata with $p + q \in \{5, 6\}$. To get a fair comparison of the implementations, what is given is the minimum of the running times for an automaton and its dual (see Theorem 1).

Table 2. Average time (in ms) to detect finiteness of (semi)groups

2- 3-				2- 4-				3- 3-			
FR	aut	Fin1	Fin2	FR	aut	Fin1	Fin2	FR	aut	Fin1	Fin2
0.68	0.81	0.49	0.49	36.36	1.79	0.52	0.62	1 342.12	3.78	0.61	0.70

FR: FR's `IsFinite`;   aut: automgrp's `IsFinite`;   Fin1: `IsFinite1`;   Fin2: `IsFinite2`

4   Conjectures

The efficiency of the new implementations enables to carry out extensive experimentations. We propose several conjectures supported by these experiments.

Recall the example given in the introduction. The (semi)groups generated by the Mealy automaton and its dual were strikingly different, with a very large one and a rather small one. This seems to be a general fact that we can state as an informal conjecture:

Whenever a Mealy automaton generates a finite (semi)group which is very large with respect to the number of states and letters of the automaton, then its dual generates a small one.

Observation: Any pair of finite (semi)groups can be generated by a pair of dual Mealy automata, see [2, Prop. 9]. The standard construction leads to automata whose sizes are related to the orders of the (semi)groups. Therefore it does not contradict the informal conjecture.

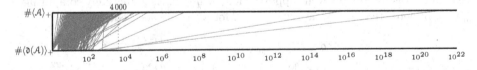

Fig. 3. Size of $\langle \mathcal{A} \rangle_+$ vs. size of $\langle \mathfrak{d}(\mathcal{A}) \rangle_+$

Fig. 3 illustrates this informal conjecture: for $\mathcal{A}$ covering the set of all 3-letter 3-state invertible Mealy automata, the endpoints of each segment represent respectively the order of $\langle \mathcal{A} \rangle_+$ and of $\langle \mathfrak{d}(\mathcal{A}) \rangle_+$, for all pairs detected as being finite.

To assess finiteness, the procedures `IsFinite1` and `IsFinite2` have been used. If the tested Mealy automaton and its dual were both found to have more than 4000 elements, the procedures were stopped, and the (semi)groups were supposed to be infinite. Based on the informal conjecture, we believe to have captured all finite groups. If true:

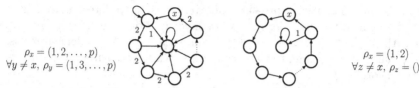

$\rho_x = (1, 2, \ldots, p)$
$\forall y \neq x,\ \rho_y = (1, 3, \ldots, p)$

4.1: among invertible automata: $\mathcal{M}_{p,q}$

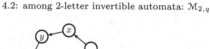

$\rho_x = (1, 2)$
$\forall z \neq x,\ \rho_z = ()$

4.2: among 2-letter invertible automata: $\mathcal{M}_{2,q}$

$\rho_{\bar{x}} = t(1, 2, \ldots, p)t^{-1}$
$\rho_y = (1, 3, \ldots, p)$
$t = \begin{cases} () & \text{for } p \text{ even} \\ (p, \frac{p+1}{2}) & \text{for } p \text{ odd} \end{cases}$

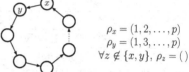

4.3: among 2-state invertible automata: $\mathcal{M}_{p,2}$

4.4: among bireversible automata: $\mathcal{B}_{p,q}$

$\rho_x = (1, 2, \ldots, p)$
$\rho_y = (1, 3, \ldots, p)$
$\forall z \notin \{x, y\},\ \rho_z = ()$

Fig. 4. Automata conjectured to generate the largest finite automaton groups

- There are 14 089 Mealy automata generating finite (semi)groups among the 233 339 invertible or reversible 3-letter 3-state Mealy automata;
- The group generated by Fig. 1-right is the largest finite group.

Our next conjectures are concerned with the largest finite groups that can be generated by automata of a given size.

Consider the family of p-letter q-state Mealy automata $(\mathcal{M}_{p,q})_{p+q>5}$ displayed on Fig. 4.1 for $p > 2$ and $q > 2$, while the specializations for $p = 2$ and $q = 2$ are displayed on Fig. 4.2 and Fig. 4.3. The example of Fig. 1-right is $\mathcal{M}_{3,3}$.

Conjecture 1. The group $\langle \mathcal{M}_{p,q} \rangle$ is finite. Every p-letter q-state invertible Mealy automaton generates a group which is either infinite or has an order smaller than $\#\langle \mathcal{M}_{p,q} \rangle$.

If true, Conjecture 1 implies the decidability of the finiteness problem for automaton groups. Without entering into the details of the experimentations, we consider that Conj. 1 is reasonably well supported for $p + q < 9$. As for actually computing $\#\langle \mathcal{M}_{p,q} \rangle$, here are the only cases with $q > 2$ for which we succeeded:

$$\forall q,\ 4 \leq q \leq 8, \qquad \#\langle \mathcal{M}_{2,q} \rangle = 2^{2^{q-1} + \frac{(q-2)(q-1)}{2}} - 2,$$

$$\#\langle \mathcal{M}_{3,3} \rangle = 2^{64} \cdot 3^4, \qquad \#\langle \mathcal{M}_{3,4} \rangle = 2^{325} \cdot 3^{13}, \qquad \#\langle \mathcal{M}_{4,3} \rangle = 2^{288} \cdot 3^{422}.$$

These groups are indeed huge. Incidentally, the finiteness of $\langle \mathcal{M}_{p,q} \rangle$ is checked for $p + q < 11$ and the informal conjecture is supported further by computing the order of the much smaller semigroups generated by the duals:

$\#\langle \partial(\mathcal{M}_{p,q}) \rangle_+$	2	3	4	5	6	7	8
2	–	–	219	1 759	13 135	94 143	656 831
3	–	238	1 552	8 140	37 786	162 202	$\cdots$
4	89	1 381	12 309	87 125	$\cdots$	$\cdots$	$\cdots$
5	131	6 056	67 906	602 656	$\cdots$	$\cdots$	$\cdots$
6	337	22 399	302 011	$\cdots$	$\cdots$	$\cdots$	$\cdots$
7	351	74 194	$\cdots$	$\cdots$	$\cdots$	$\cdots$	$\cdots$

Experimentally, the finite groups generated by bireversible Mealy automata seem to be much smaller. Consider the family of bireversible automata $(\mathcal{B}_{p,q})_{p,q}$ of Fig. 4.4. The group $\langle \mathcal{B}_{p,q} \rangle$ is isomorphic to $\mathfrak{S}_p^q$, while the group $\langle \mathfrak{d}(\mathcal{B}_{p,q}) \rangle$ is isomorphic to $\mathbb{Z}_q$. Again, the following is reasonably well supported for $p+q < 9$:

Conjecture 2. Every p-letter q-state bireversible Mealy automaton generates a group which is either infinite or has an order smaller than $\#\langle \mathcal{B}_{p,q} \rangle = p!^q$.

Our last conjecture is of a different nature and deals with the structure of infinite automaton semigroups.

Conjecture 3. Every 2-state reversible Mealy automaton generates a semigroup which is either finite or free of rank 2.

The conjecture has been tested and seems correct for reversible 2-state Mealy automata up to 6 letters. In the experiments, a semigroup generated by a p-letter automaton is conjectured to be free if its growth series coincides with $(2t)^n$ up to radius $p^2/2$ and if its dual generates a seemingly infinite group.

References

1. Aho, A., Hopcroft, J., Ullman, J.: The Design and Analysis of Computer Algorithms. Addison-Wesley (1974)
2. Akhavi, A., Klimann, I., Lombardy, S., Mairesse, J., Picantin, M.: On the finiteness problem for automaton (semi)groups. Internat. J. Algebra Comput. (accepted, 2012), arXiv:cs.FL/1105.4725
3. Antonenko, A.S.: On transition functions of Mealy automata of finite growth. Matematychni Studii 29(1), 3–17 (2008)
4. Bartholdi, L.: FR Functionally recursive groups — a GAP package, v.1.2.4.2 (2011), http://www.uni-math.gwdg.de/laurent/FR/
5. Bartholdi, L., Reznykov, I.I., Sushchanskii, V.I.: The smallest Mealy automaton of intermediate growth. J. Algebra 295(2), 387–414 (2006)
6. Bondarenko, I.V., Bondarenko, N.V., Sidki, S.N., Zapata, F.R.: On the conjugacy problem for finite-state automorphisms of regular rooted trees, ArXiv:math.GR/1011.2227
7. Cain, A.J.: Automaton semigroups. Theor. Comput. Sci. 410, 5022–5038 (2009)
8. The GAP Group: GAP – Groups, Algorithms, and Programming, v.4.4.12 (2008), http://www.gap-system.org
9. Grigorchuk, R.I.: On Burnside's problem on periodic groups. Funktsional. Anal. i Prilozhen. 14(1), 53–54 (1980)
10. Grigorchuk, R.I., Nekrashevich, V.V., Sushchanskii, V.I.: Automata, dynamical systems, and groups. Tr. Mat. Inst. Steklova 231, 134–214 (2000)
11. Muntyan, Y., Savchuk, D.: automgrp Automata Groups — a GAP package, v.1.1.4.1 (2008), http://finautom.sourceforge.net/
12. Nekrashevych, V.: Self-similar groups. Mathematical Surveys and Monographs, vol. 117. American Mathematical Society, Providence (2005)

13. Savchuk, D.M., Vorobets, Y.: Automata generating free products of groups of order 2. J. Algebra 336(1), 53–66 (2011)
14. Sidki, S.N.: Automorphisms of one-rooted trees: growth, circuit structure, and acyclicity. J. Math. Sci. (New York) 100(1), 1925–1943 (2000), algebra, 12
15. Silva, P.V., Steinberg, B.: On a class of automata groups generalizing lamplighter groups. Internat. J. Algebra Comput. 15(5-6), 1213–1234 (2005)
16. Šunik, Z., Ventura, E.: The conjugacy problem is not solvable in automaton groups (2010), arXiv:math.GR/1010.1993

On the Descriptional Complexity of the Window Size for Deterministic Restarting Automata

Martin Kutrib[1] and Friedrich Otto[2]

[1] Institut für Informatik, Universität Giessen
Arndtstr. 2, 35392 Giessen, Germany
kutrib@informatik.uni-giessen.de
[2] Fachbereich Elektrotechnik/Informatik, Universität Kassel
34109 Kassel, Germany
otto@theory.informatik.uni-kassel.de

Abstract. We investigate the descriptional complexity of restarting automata, an automaton model inspired from linguistics. More precisely, we study the impact of the window size. For window sizes of at least two, it is shown that between any two levels of the RW- and RRW-automata hierarchy, there are savings in the economy of description of the automata which cannot be bounded by any recursive function. This is true even if the automata are deterministic and/or stateless. The trade-off between window sizes two and one is recursive for deterministic devices. In addition, we establish polynomial upper bounds for the trade-offs between RRWW-automata with window sizes $k + 1$ and k for all $k \geq 2$.

Keywords: restarting automaton, window size, descriptional complexity, non-recursive trade-off.

1 Introduction

Restarting automata have been introduced in [4] in order to model the so-called "analysis by reduction," which is a technique used in linguistics to analyze sentences of natural languages that have free word order. This technique consists in a stepwise simplification of an extended sentence such that the (in)correctness of the sentence is not affected. A *restarting automaton* M consists of a flexible tape with end markers, a read/write window of a fixed size $k \geq 1$, and a finite-state control. It works in *cycles*, where each cycle begins with the window at the left end of the tape and M being in the initial state. During a cycle M scans its current tape contents from left to right and executes a single length-reducing rewrite step. The cycle ends with a restart that takes the window back to the left end of the tape and that resets M to the initial state. A computation is completed by a *tail computation* that is similar to a cycle but that ends with accepting or rejecting the input. In its rewrite steps M may introduce non-input symbols, so-called *auxiliary symbols*. This type of restarting automaton is called an RRWW-automaton. By placing certain restrictions on the definition, we obtain various subclasses of restarting automata (see, e.g., [16]).

N. Moreira and R. Reis (Eds.): CIAA 2012, LNCS 7381, pp. 253–264, 2012.

One of the most obvious parameters for restarting automata is the size of the read/write window. It is known that for restarting automata without auxiliary symbols, the size of the read/write window yields an infinite strict hierarchy of language classes [13]. Thus, it is natural to ask for the impact of this parameter on the size of the automata. Here we establish the following answers to this question. For RW- and RRW-automata and all $k \geq 2$, there are savings in the economy of description which cannot be bounded by any recursive function when changing from window size k to window size $k + 1$. This is true even if the automata are deterministic and/or stateless. This result is proved by reducing the halting problem for Turing machines on empty input to strictly monotone semi-Thue systems which, in turn, are simulated by restarting automata. Now a general result from [2,6] that is a slightly generalized and unified form of a result of Hartmanis [1] can be applied to obtain the intended non-recursive trade-offs. Interestingly, the trade-off between window size two and one is recursive for deterministic devices. In addition, we establish polynomial upper bounds for the trade-offs between RRWW-automata with window sizes $k + 1$ and k for $k \geq 2$.

Restarting automata in connection with descriptional complexity issues are also dealt with in [3,10,11]. Further results and references on both topics can be found, for example, in the surveys [2,16].

This paper is structured as follows. After recalling the necessary definitions on restarting automata we present some basics on descriptional complexity. Then we describe a semi-Thue system for Turing machine histories, which is the base for the reduction. In the subsequent section we derive our main results. We conclude with open and untouched questions for further investigations.

2 Notation and Definitions

For a finite alphabet Σ, Σ^* is the set of all words over Σ, and $\Sigma^+ = \Sigma^* \setminus \{\lambda\}$, where λ denotes the empty word. The length of a word w is written as $|w|$. We use $\subseteq$ for inclusions and $\subset$ for strict inclusions. The powerset of a set S is denoted by 2^S.

An RRWW-*automaton* M is given by an 8-tuple $M = (Q, \Sigma, \Gamma, \mathrm{c}, \$, q_0, k, \delta)$, where Q is a finite set of states, Σ is a finite input alphabet, Γ is a finite tape alphabet containing Σ, the symbols $\mathrm{c}, \$ \notin \Gamma$ serve as markers for the left and right border of the work tape, respectively, $q_0 \in Q$ is the initial state, $k \geq 1$ is the size of the *read/write window*, and δ is a *transition relation* that assigns finite sets of possible transitions to pairs of the form (q, u), where $q \in Q$ is a state and u is a possible content of the read/write window. There are four types of transition steps: *move-right steps* (MVR) that shift the window one position to the right and change the internal state, *rewrite steps* that replace the factor u contained in the window by a shorter word v, in this way also shortening the tape, change the internal state, and place the window immediately to the right of v, *restart steps* (Restart) that place the window over the left end of the tape and reset the internal state to q_0, and *accept steps* (Accept) that halt and accept. If $\delta(q, u) = \emptyset$ for some pair (q, u), then M necessarily halts, and we say that M *rejects* in this situation. The letters in $\Gamma \setminus \Sigma$ are called *auxiliary symbols*.

A *configuration* of M is a string $\alpha q \beta$, where $q \in Q$, and either $\alpha = \lambda$ and $\beta \in \{c\} \cdot \Gamma^* \cdot \{\$\}$ or $\alpha \in \{c\} \cdot \Gamma^*$ and $\beta \in \Gamma^* \cdot \{\$\}$. Here $q \in Q$ represents the current state, $\alpha\beta$ is the current content of the tape, and it is understood that the read/write window contains the first k symbols of β or all of β when $|\beta| \leq k$. A *restarting configuration* is of the form $q_0 c w\$$, where $w \in \Gamma^*$. If $w \in \Sigma^*$, then $q_0 c w\$$ is an *initial configuration*.

Each computation of M consists of a finite sequence of *cycles* that is followed by a *tail computation* (see Section 1). By $x \vdash_M^c y$ we denote the execution of a cycle that transforms the restarting configuration $q_0 c x\$$ into the restarting configuration $q_0 c y\$$. By $L(M) = \{ w \in \Sigma^* \mid M$ has an accepting computation on input $w \}$ we denote the *language* accepted by M.

As each cycle of a computation of M can be seen to consist of three phases, the transition relation of M can be described more compactly through so-called *meta-instructions* [15] of the form $(E_1, u \to v, E_2)$, where E_1 and E_2 are regular languages, and $u \to v$ stands for a rewrite step of M. On trying to execute this meta-instruction M will get stuck (and so reject) starting from the configuration $q_0 c w\$$, if w does not admit a factorization of the form $w = w_1 u w_2$ such that $c w_1 \in E_1$ and $w_2\$ \in E_2$. If, however, w does have factorizations of this form, then one such factorization is chosen nondeterministically, and $q_0 c w\$$ is transformed into $q_0 c w_1 v w_2\$$. In order to describe the tails of accepting computations we use meta-instructions of the form $(c \cdot E \cdot \$, \mathsf{Accept})$, which lead to acceptance starting from a configuration of the form $q_0 c w\$$ for $w \in E$.

Here we are also interested in certain restricted variants of RRWW-automata. Automaton M is an RWW-*automaton* if it must restart immediately after performing a rewrite operation. An RRWW-automaton is an RRW-*automaton*, if no auxiliary symbols are available, and it is an RR-*automaton* if each rewrite step $u \to v$ satisfies the property that v is a scattered subword of u. Analogously, we obtain RW- and R-automata from the RWW-automaton. We use the prefix det- to denote deterministic types of restarting automata, and for any $k \geq 1$ and any type X of restarting automaton, we use X(k) to denote the restarting automata of type X that have a read/write window of size k. Further, for any type X of automaton, $\mathscr{L}(\mathsf{X})$ denotes the class of languages that are accepted by automata of type X.

Concerning the influence of the size of the read/write window on the power of the various types of restarting automata, the following results are known.

Theorem 1 ([10,11,13]).
1. $\mathscr{L}(\mathsf{RWW}(1)) = \mathscr{L}(\mathsf{det\text{-}RRWW}(1)) = \mathsf{REG} \subset \mathscr{L}(\mathsf{RR}(1))$.
2. $\mathscr{L}(\mathsf{pref\text{-}X}(k)) \subset \mathscr{L}(\mathsf{pref\text{-}X}(k+1))$ *for all* $k \geq 1$, pref $\in \{\lambda, \mathsf{det}\}$, *and* $\mathsf{X} \in \{\mathsf{R}, \mathsf{RR}, \mathsf{RW}, \mathsf{RRW}\}$.

Each cycle C of an RRWW-automaton M contains a unique configuration of the form $c x q u y\$$ such that q is a state and $(q', v) \in \delta(q, u)$ is the rewrite step that is applied during this cycle. By $D_r(C)$ we denote the *right distance* $|uy\$|$ of this cycle. A sequence of cycles $C_1, C_2, \ldots, C_n$ is called *monotone* if $D_r(C_1) \geq D_r(C_2) \geq \cdots \geq D_r(C_n)$ holds. A computation of M is called *monotone* if the corresponding sequence of cycles is monotone, and the RRWW-automaton M itself is called *monotone* if each of its computations that starts from an initial configuration is monotone. To illustrate the way in which monotone RR(1)-automata can accept non-regular languages we present a simple example.

Example 2. Let M_0 be the RR(1)-automaton on $\Sigma = \{a, b\}$ that is given through the following meta-instructions (see above):

(1) $(\mathcal{c} \cdot (aa)^* \cdot a, b \to \lambda, (bb)^* \cdot \$)$,   (4) $(\mathcal{c} \cdot (aa)^*, a \to \lambda, (bb)^* \cdot \$)$,
(2) $(\mathcal{c} \cdot (aa)^+, b \to \lambda, (bb)^* \cdot b \cdot \$)$,   (5) $(\mathcal{c} \cdot (aa)^* \cdot a, a \to \lambda, (bb)^* \cdot b \cdot \$)$.
(3) $(\mathcal{c} \cdot \$, \mathsf{Accept})$,

The automaton M_0 processes inputs of the form $a^m b^n$. In fact, it alternately removes the first occurrence of the letter b and the last occurrence of the letter a. To distinguish between these two cases it uses the parity of the number of a's and the parity of the number of b's. Hence, M_0 is monotone, and it accepts the non-regular language $L(M_0) = \{\, a^m b^n \mid n \geq 0,\, m = n \text{ or } m = n + 1 \,\}$.   □

2.1 Descriptional Complexity

We recall some notation for descriptional complexity. Following [2] we say that a *descriptional system* S is a set of finite descriptors such that each $D \in S$ describes a formal language $L(D)$, and the underlying alphabet alph(D) over which D represents a language can be read off from D. The *family of languages represented* (or *described*) by S is $\mathscr{L}(S) = \{\, L(D) \mid D \in S \,\}$. For every language L, the set $S(L) = \{\, D \in S \mid L(D) = L \,\}$ is the set of its descriptors in S. A *complexity measure* for a descriptional system S is a total recursive mapping $c : S \to \mathbb{N}$.

Example 3. Monotone det-RRWW-automata can be encoded over some fixed alphabet such that their input alphabets can be extracted from the encodings. The set of these encodings is a descriptional system S, and $\mathscr{L}(S)$ is the family of deterministic context-free languages DCFL [5].

Examples for complexity measures for deterministic restarting automata are the total number of symbols, that is, the *length of the encoding* (length), or the product of the number of states and the number of possible window contents, that is, the *number of transitions* (trans).   □

Here we only use complexity measures that are recursively related to length. If there is a total recursive function $g : \mathbb{N} \times \mathbb{N} \to \mathbb{N}$ such that, for all $D \in S$, length(D) $\leq g(c(D), |\text{alph}(D)|)$, then c is said to be an *s-measure*. If, in addition, for any alphabet Σ, the set of descriptors in S describing languages over Σ is recursively enumerable in order of increasing size, then c is said to be an *sn-measure*. Clearly, length and trans are sn-measures for restarting automata.

Whenever we consider the relative succinctness of two descriptional systems S_1 and S_2, we assume the intersection $\mathscr{L}(S_1) \cap \mathscr{L}(S_2)$ to be non-empty. Let S_1 and S_2 be descriptional systems with complexity measures c_1 and c_2, respectively. A total function $f : \mathbb{N} \to \mathbb{N}$ is an *upper bound* for the increase in complexity when changing from a descriptor in S_1 to an equivalent descriptor in S_2, if for all $D_1 \in \mathcal{S}_1$ with $L(D_1) \in \mathcal{L}(\mathcal{S}_2)$, there exists a $D_2 \in \mathcal{S}_2(L(D_1))$ such that $c_2(D_2) \le f(c_1(D_1))$.

If there is no recursive upper bound, the *trade-off is said to be non-recursive*. Non-recursive trade-offs are independent of particular sn-measures. For establishing non-recursive trade-offs the following general result is useful that is a slightly generalized and unified form of a result of Hartmanis [1].

Theorem 4 ([2]). *Let $\mathcal{S}_1$ and $\mathcal{S}_2$ be two descriptional systems for recursive languages such that any descriptor D in $\mathcal{S}_1$ and $\mathcal{S}_2$ can effectively be converted into a Turing machine that decides $L(D)$, and let c_1 be a measure for $\mathcal{S}_1$ and c_2 be an sn-measure for $\mathcal{S}_2$. If there exists a descriptional system $\mathcal{S}_3$ and a property P that is not semi-decidable for descriptors from $\mathcal{S}_3$, such that, given an arbitrary $D_3 \in \mathcal{S}_3$, (i) there exists an effective procedure to construct a descriptor D_1 in $\mathcal{S}_1$, and (ii) D_1 has an equivalent descriptor in $\mathcal{S}_2$ if and only if D_3 does not have property P, then the trade-off between $\mathcal{S}_1$ and $\mathcal{S}_2$ is non-recursive.*

For deterministic RRWW- and RWW-automata, some non-recursive trade-offs have been obtained in [3]. In the following we show non-recursive trade-offs between restarting automata with window sizes $k+1$ and k by reduction of the halting problem for Turing machines on empty tape. In order to apply Theorem 4, we use the family of deterministic one-tape Turing machines as descriptional system $\mathcal{S}_3$. Property P is *not halting on empty input*. Next, given an arbitrary deterministic one-tape Turing machine M, that is, a descriptor $D_3 \in \mathcal{S}_3$, we must construct a det-RW($k + 1$)-automaton, that is, a descriptor D_1 in $\mathcal{S}_1$, that has an equivalent det-RW(k)-automaton, that is, a descriptor in $\mathcal{S}_2$, if and only if M halts on empty input.

3   A Semi-Thue System for Turing Machine Histories

Any deterministic one-tape Turing machine can be transformed into an equivalent one that never reenters its initial state, cannot print blanks, halts only in a particular halting state, and halts if and only if it accepts. So, we may assume without loss of generality that M is such a machine.

First we construct a semi-Thue system which generates strings that are encodings of the whole history of computations of M. Simulations of Turing machines by Thue systems have also been used, for example, in [7,14,17]. Afterwards a det-RW(3)-automaton is constructed from this semi-Thue system.

So, let Q be the state set of M, where q_0 is the initial state, q_f is the particular halting state, and Γ is the tape alphabet disjoint from Q containing the blank symbol. We assume that the transition function δ maps from $Q \times \Gamma$ to $Q \times \Gamma \times \{left, right\}$, that is, there are no stationary moves. A configuration of M is represented by a word from $\Gamma^* \cdot Q \cdot \Gamma^*$, where $x_1 \cdots x_i q x_{i+1} \cdots x_n$ is used to express that M is in state q, scanning tape symbol x_{i+1}, and $x_1 \cdots x_n$ is the support of the tape inscription. So, the initial configuration on empty tape is q_0.

The semi-Thue system $T = \langle S, P \rangle$ consists of an alphabet S and a finite binary relation $P \subseteq S^* \times S^*$. The pairs $(u, v) \in P$ are called *rules* and are written as

$u \rightarrow v$. These rules are extended to words over S^* as follows: $s \Rightarrow t$ iff there exist $x, y, u, v \in S^*$ such that $s = xuy$, $t = xvy$, and $(u \rightarrow v) \in P$. The reflexive and transitive closure of $\Rightarrow$ is denoted by $\Rightarrow^*$.

The semi-Thue system T will be used to encode the history of a computation of M in a single word. In order to track the whole history of a computation in a word, dummy symbols carrying some information are embedded into configurations. This information uniquely identifies the rule of T that was used to generate the corresponding dummy symbol. In this way the information carried by the dummy symbols in a word encoding the history of a computation of M allows us to uniquely reverse this computation. Moreover, to avoid ambiguity the symbols to the left of the state symbol are distinguished from those to the right of the state symbol, and the state symbol itself carries additional information about whether the tape symbol currently being scanned is to the left or to the right of the state symbol. The alphabet S of T is defined as

$$S = \{\triangleright, \triangleleft\} \cup \overrightarrow{Q} \cup \overleftarrow{Q} \cup \overrightarrow{\Gamma} \cup \overleftarrow{\Gamma} \cup D_L \cup D_R \cup H, \text{ where}$$
$$D_L = \{\, L_z \mid z \in (\overrightarrow{Q} \times (\{\triangleleft\} \cup \overleftarrow{\Gamma})) \cup ((\{\triangleright\} \cup \overrightarrow{\Gamma}) \times \overleftarrow{Q}) \,\},$$
$$D_R = \{\, R_z \mid z \in (\overrightarrow{Q} \times (\{\triangleleft\} \cup \overleftarrow{\Gamma})) \cup ((\{\triangleright\} \cup \overrightarrow{\Gamma}) \times \overleftarrow{Q}) \,\}, \text{ and}$$
$$H = \{\, I_z \mid z \in (\overrightarrow{Q} \times \{\triangleleft\}) \cup (\{\triangleright\} \times \overleftarrow{Q}) \,\}.$$

Here $\triangleright$ and $\triangleleft$ are the left and right endmarkers, $\overrightarrow{Q}$ and $\overleftarrow{Q}$ are disjoint copies of Q, where $\overrightarrow{q} \in \overrightarrow{Q}$ and $\overleftarrow{q} \in \overleftarrow{Q}$ mean state q such that the currently scanned tape symbol is the one to the right or the one to the left, and $\overrightarrow{\Gamma}$ and $\overleftarrow{\Gamma}$ are disjoint copies of Γ such that $\overrightarrow{\Gamma}$ and $\overleftarrow{\Gamma}$ indicate tape symbols such that the state symbol is to their right or to their left, respectively. The symbols L_z and R_z are dummy symbols to the left or to the right of the symbol representing the state, and the symbols I_z are intermediate symbols.

Next we define the rules of T. The histories of the computations of M are of the form $\triangleright (\overrightarrow{\Gamma} \cup D_L)^* \cdot (\overrightarrow{Q} \cup \overleftarrow{Q}) \cdot (\overleftarrow{\Gamma} \cup D_R)^* \triangleleft$. For each dummy symbol L_z and R_z and each state $q_i \in Q \setminus \{q_f\}$, we define the rules

$$\overrightarrow{q_i} R_z \rightarrow L_z L_z \overrightarrow{q_i} \quad \text{and} \quad L_z \overleftarrow{q_i} \rightarrow \overleftarrow{q_i} R_z R_z,$$

which are used to move the state symbol towards the currently scanned tape symbol. If the currently scanned symbol is non-blank, a transition $\delta(q_i, x_i) = (q_j, x_j, right)$ of M is simulated by the rules

$$\overrightarrow{q_i} \overleftarrow{x_i} \rightarrow L_{(\overrightarrow{q_i}, \overleftarrow{x_i})} \overrightarrow{x_j} \overrightarrow{q_j} \quad \text{and} \quad \overrightarrow{x_i} \overleftarrow{q_i} \rightarrow L_{(\overrightarrow{x_i}, \overleftarrow{q_i})} \overrightarrow{x_j} \overrightarrow{q_j}.$$

If x_i is the blank symbol, then the tape symbol is represented by an endmarker. So, we include the rules

$$\overrightarrow{q_i} \triangleleft \rightarrow L_{(\overrightarrow{q_i}, \triangleleft)} I_{(\overrightarrow{q_i}, \triangleleft)} \triangleleft \quad \text{and} \quad L_{(\overrightarrow{q_i}, \triangleleft)} I_{(\overrightarrow{q_i}, \triangleleft)} \rightarrow L_{(\overrightarrow{q_i}, \triangleleft)} \overrightarrow{x_j} \overrightarrow{q_j}, \text{ and}$$
$$\triangleright \overleftarrow{q_i} \rightarrow \triangleright L_{(\triangleright, \overleftarrow{q_i})} I_{(\triangleright, \overleftarrow{q_i})} \quad \text{and} \quad L_{(\triangleright, \overleftarrow{q_i})} I_{(\triangleright, \overleftarrow{q_i})} \rightarrow L_{(\triangleright, \overleftarrow{q_i})} \overrightarrow{x_j} \overrightarrow{q_j}.$$

Observe that the rules are uniquely determined by the dummy symbols L_z and I_z on the right-hand sides.

Symmetrically, a transition $\delta(q_i, x_i) = (q_j, x_j, left)$ of M is simulated by the rules

$$\overrightarrow{q_i}\overleftarrow{x_i} \to \overleftarrow{q_j}\overleftarrow{x_j}R_{(\overrightarrow{q_i},\overleftarrow{x_i})} \quad \text{and} \quad \overrightarrow{x_i}\overleftarrow{q_i} \to \overleftarrow{q_j}\overleftarrow{x_j}R_{(\overrightarrow{x_i},\overleftarrow{q_i})},$$
$$\overrightarrow{q_i}\lhd \to I_{(\overrightarrow{q_i},\lhd)}R_{(\overrightarrow{q_i},\lhd)}\lhd \quad \text{and} \quad I_{(\overrightarrow{q_i},\lhd)}R_{(\overrightarrow{q_i},\lhd)} \to \overleftarrow{q_j}\overleftarrow{x_j}R_{(\overrightarrow{q_i},\lhd)}, \quad \text{and}$$
$$\rhd\overleftarrow{q_i} \to \rhd I_{(\rhd,\overleftarrow{q_i})}R_{(\rhd,\overleftarrow{q_i})} \quad \text{and} \quad I_{(\rhd,\overleftarrow{q_i})}R_{(\rhd,\overleftarrow{q_i})} \to \overleftarrow{q_j}\overleftarrow{x_j}R_{(\rhd,\overleftarrow{q_i})}.$$

Also these rules are uniquely determined by the dummy symbols R_z and I_z on the right-hand sides. The semi-Thue system T can simulate computations of M. As on the left-hand side of each rule there is a state symbol for which δ is defined, or there is a corresponding symbol from H, and as δ is undefined for the halting state q_f, no further derivation step is possible for any encoding of a halting configuration of M. In the following, we consider the computation of M on empty input. So, $L_h(M) = \{\, w \mid \rhd\overrightarrow{q_0}\lhd \Rightarrow^* w \,\}$ is the set of histories that can be derived from the unique initial configuration of M on empty input in an arbitrary number of steps. It is worth mentioning that $L_h(M)$ is finite if and only if M halts on empty input.

Next we construct a det-RW(3)-automaton $M' = (Q', \Sigma, \Gamma, \mathord{\mathrm{c}}, \$, q_0', 3, \delta')$ that accepts exactly the words of $L_h(M)$ with endmarkers chopped off. To this end, let $Q' = \{q_0'\}$, and set $\Sigma = \Gamma = S \setminus \{\rhd, \lhd\}$. Basically, the idea is to let M' apply the rewriting steps of T in reversed manner, that is, whenever T applies a rule $s \to t$ to expand a string, M' reduces the string by rewriting the substring t by s. Whenever an endmarker is involved in a rule, M' can apply the same rule where the endmarkers are replaced by its own endmarkers. So, let K be the set of right-hand sides of rules of T where the endmarkers $\rhd$ and $\lhd$ have been replaced by the delimiters $\mathord{\mathrm{c}}$ and $\$$, respectively, and for each rule $(s \to t) \in P$, let s' (and t') be obtained from s (and t) by replacing any endmarker accordingly. Then we define

$$\delta'(q_0', z) = (q_0', \mathsf{MVR}), \text{ if } z \notin K, \text{ and}$$
$$\delta'(q_0', t') = (q_0', s'), \text{ if } t' \in K \text{ and } (s \to t) \in P.$$

Finally, M' accepts if and only if an input can be reduced to the encoding of the initial configuration of M on empty input, that is, $\delta'(q_0', \mathord{\mathrm{c}}\overrightarrow{q_0}\$) = \mathsf{Accept}$.

All rules of T are strictly increasing. Therefore, each rewrite step of M' shortens the tape. Moreover, any rewrite step occurs with a state symbol or a symbol from H in the read/write window. Whenever a state symbol occurs that is labeled by a right (left) arrow, then the two symbols to the left (to the right) of it determine uniquely the rule to be applied. Further, if a symbol from H occurs, then this symbol uniquely determines the rule to be applied. Thus, M' is deterministic. As M' only executes an accept step on the encoding of the initial configuration of M on empty input, it follows that M' accepts exactly the words of $L_h(M)$ with endmarkers chopped off. So, we obtain the following proposition.

Proposition 5. *From a deterministic one-tape Turing machine M, one can effectively construct a deterministic RW(3)-automaton for the language $L_h(M)$.*

4   The Descriptional Impact of the Window Size

In order to generalize the above approach to arbitrary window sizes, let $\tilde{\Sigma} = \{\tilde{x} \mid x \in \Sigma\}$ be a disjoint copy of Σ, and let $\varphi_k : \Sigma^* \to 2^{(\Sigma \cup \tilde{\Sigma})^*}$ be the finite substitution defined by $\varphi_k(x) = \{x, \tilde{x}^k\}$ $(x \in \Sigma)$.

Proposition 6. *Let $k \geq 3$. From a deterministic one-tape Turing machine M, one can construct a det-RW(k)-automaton for the language $\varphi_k(L_h(M))$.*

Proof. Each word $w \in \varphi_k(L_h(M))$ is obtained from a word $w' \in L_h(M)$ by replacing $s \geq 0$ symbols, where a symbol is replaced by k consecutive marked copies of itself. A det-RW(k)-automaton M'' accepting $\varphi_k(L_h(M))$ simulates the det-RW(3)-automaton for $L_h(M)$ as long as no symbol $\tilde{x}$ from $\tilde{\Sigma}$ appears in the window. Whenever this happens, M'' moves the window to the right until it contains k copies of $\tilde{x}$. If this is not possible, M'' halts and rejects. Otherwise, it rewrites $\tilde{x}^k$ into x and restarts. This construction is effective.   □

In order to complete the proof of the non-recursive trade-offs we modify the Turing machines as follows. First a deterministic one-tape Turing machine M is transformed into a machine $\hat{M}$ having three tracks on its single tape. The first track is used to simulate M on empty input. On the second track a unary counter is maintained, which is initially set to one. Then $\hat{M}$ starts to simulate M on empty input. After every simulation step, $\hat{M}$ marks the current tape square, remembers the current state, moves to the counter, and increases it by one. Afterwards it computes n^2 in unary on the third track, where n is the value of the counter. Now it changes to a special *test state* for one time step, and moves across the n^2 tape squares marked on the third track. The set of test states can simply be implemented as a disjoint copy of the state set. Finally, $\hat{M}$ changes again to a *test state* for one time step, clears the third track, returns to the tape position at which the simulation of M has been interrupted, and changes to the state remembered. It then simulates the next step of M and so on. Clearly, the language $L(\hat{M})$ accepted by $\hat{M}$ is finite if and only if $L(M)$ is finite.

Theorem 7. *For $X \in \{\text{det-RW}, \text{det-RRW}\}$ and all $k \geq 3$, the trade-off from $X(k)$-automata to $X(k-1)$-automata is non-recursive.*

Proof. As mentioned above, this theorem is proved using Theorem 4. Given a deterministic one-tape Turing machine M, we first construct the Turing machine $\hat{M}$ as explained above. Then, a det-RW(k)-automaton M'' for the language $\varphi_k(L_h(\hat{M}))$ is obtained according to Proposition 6. Clearly, M'' can also be seen as a det-RRW(k)-automaton. It remains to be shown that $\varphi_k(L_h(\hat{M}))$ is accepted by an X-automaton with window size $k-1$ iff M halts on empty input.

If M halts on empty input, then so does $\hat{M}$. Hence, $L_h(\hat{M})$ and $\varphi_k(L_h(\hat{M}))$ are finite. Thus, $\varphi_k(L_h(\hat{M}))$ is regular, and so it is accepted by X-automata with window size $k-1$. On the other hand, let M' be an X-automaton with window size $k-1$, and let $w \in \varphi_k(L_h(\hat{M}))$ such that w only contains symbols from $\tilde{\Sigma}$. If $w \vdash_{M'}^c w'$, then $w' \notin \varphi_k(L_h(\hat{M}))$, as in a single cycle M' shortens its tape content by at least one symbol, rewriting a substring of length at most $k - 1$. Therefore, if M' accepts the language $\varphi_k(L_h(\hat{M}))$, then it can do so only by accepting tail computations, which means that $\varphi_k(L_h(\hat{M}))$ is regular.

Let $R \subset \Sigma^*$ be the regular language of all words over Σ that contain an even number of test state symbols (see above). Then $\varphi_k(L_h(\hat{M})) \cap R$ is also regular. However, since in every step during the simulation of $\hat{M}$ by the semi-Thue system a dummy symbol is inserted, the lengths of the words in $\varphi_k(L_h(\hat{M})) \cap \Sigma^*$ increase with every step of $\hat{M}$. In particular, the lengths increase at least by n^2 when $\hat{M}$ moves across the n^2 marked squares on its third track after having simulated the nth step of M. Before $\hat{M}$ starts these moves, it has changed to a test state for an odd number of times. At the end of these moves it again changes to a test state. Therefore, all words in $\varphi_k(L_h(\hat{M})) \cap \Sigma^*$ that are encodings of the configurations during these moves are filtered out by the intersection with R. So, there are gaps of arbitrary sizes in the sequence of the lengths of the words in $\varphi_k(L_h(\hat{M})) \cap R$. It follows that $\varphi_k(L_h(\hat{M}))$ is not context-free if it is infinite. Since we know already that it is regular, it must be finite. As $\varphi_k(L_h(\hat{M}))$ is finite if and only if M halts on empty input, the theorem follows.   □

The computational power of stateless restarting automata has been studied in [8,9]. The construction of a det-RW(3)-automaton for the language $L_h(M)$ preceding Proposition 5 provides, in fact, a stateless automaton. Moreover, the generalization to $\varphi_k(L_h(M))$ shown in Proposition 6 does not require additional states. Furthermore, the proof of Theorem 7 reveals that nondeterminism cannot help to accept $\varphi_k(L_h(\hat{M}))$ by an RW- or RRW-automaton with window size $k - 1$. This implies the following theorem, where we use the prefix stl- to denote stateless types of restarting automata.

Theorem 8. *For all* X ∈ {stl-det-RW, stl-det-RRW, det-RW, det-RRW, stl-RW, stl-RRW, RW, RRW} *and all* $k \geq 3$, *the trade-off from* X(k)-*automata to* X(k−1)-*automata is non-recursive.*

From the reduction of the halting problem for Turing machines on empty tape some undecidability results are obtained.

Theorem 9. *For all* X ∈ {stl-det-RW, stl-det-RRW, det-RW, det-RRW, stl-RW, stl-RRW, RW, RRW} *and all* $k \geq 3$, *infiniteness is not even semi-decidable, and finiteness, regularity, and context-freeness are undecidable for* X(k)-*automata.*

Proof. Given an arbitrary deterministic one-tape Turing machine M, an X(k)-automaton accepting the language $\varphi_k(L_h(M))$ or $\varphi_k(L_h(\hat{M}))$ can effectively be constructed by Proposition 6. Since $\varphi_k(L_h(M))$ is finite if and only if M halts on empty input, the non-semi-decidability of infiniteness and the undecidability of finiteness follow immediately.

The proof of Theorem 7 shows that the language $\varphi_k(L_h(\hat{M}))$ is context-free if and only if M halts on empty input. This implies the undecidability of regularity and context-freeness.   □

So far, we obtained trade-offs between window sizes k and $k-1$ only for $k \geq 3$. In order to investigate the trade-offs between window size one and two, we recall from [10,11,13] that $\mathscr{L}(\mathrm{R}(1)) = \mathscr{L}(\mathrm{RW}(1)) = \mathscr{L}(\mathrm{RWW}(1)) = \mathrm{REG}$ and $\mathscr{L}(\mathrm{det}\text{-}\mathrm{RR}(1)) = \mathscr{L}(\mathrm{det}\text{-}\mathrm{RRW}(1)) = \mathscr{L}(\mathrm{det}\text{-}\mathrm{RRWW}(1)) = \mathrm{REG} \subset \mathscr{L}(\mathrm{RR}(1)) = \mathscr{L}(\mathrm{RRW}(1)) = \mathscr{L}(\mathrm{RRWW}(1))$ (see Theorem 1). Interestingly, the trade-off between window size two and one is recursive for deterministic devices.

Lemma 10. *Every deterministic restarting automaton with window size two is monotone.*

Proof. Assume that a deterministic restarting automaton M with window size two performs a rewrite step $\mathord{\mathrm{c}}xquy\$ \vdash \mathord{\mathrm{c}}xvq'y\$$. The right distance of this cycle is $|uy\$|$, and $|v| \leq 1$. Due to the deterministic behavior, the rewrite step of the next cycle cannot occur before the read/write window contains the first symbol following x. So, the right distance of the next cycle is at most $1 + |vy\$| \leq |uy\$|$. Therefore, M is monotone. $\qquad\square$

Theorem 11. *For* $\mathrm{X} \in \{\mathrm{det}\text{-}\mathrm{RW}, \mathrm{det}\text{-}\mathrm{RRW}\}$, *the trade-off from* $\mathrm{X}(2)$-*automata to* $\mathrm{X}(1)$-*automata is recursive.*

Proof. One fundamental result in [5] shows that monotone det-R- and monotone det-RRWW-automata accept exactly the class DCFL. In particular, a construction is presented that transforms a deterministic and monotone restarting automaton into an equivalent deterministic pushdown automaton. Another fundamental result of [19,20] is the decidability of regularity for deterministic pushdown automata. The effective procedure reveals a recursive upper bound for the number of states of an equivalent deterministic finite automaton. $\qquad\square$

The precise trade-off for the conversion from deterministic restarting automata with window size two to ones with window size one is a challenging task for further investigations. One point is to determine a good upper bound. The upper bound for the DPDA to DFA conversion from [20] reads as follows: Let M be a deterministic pushdown automaton with n states, t stack symbols, and h is the length of the longest word pushed in a single transition. If $L(M)$ is regular, then $2^{2^{O(n^2 \log n + \log t + \log h)}}$ states are sufficient for a DFA to accept $L(M)$. The second point is to determine matching or at least good lower bounds. The following lower bound for the DPDA to DFA conversion can be found in [2,12]. Let $n \geq 1$ be an integer. Then there is a language L_n that is accepted by a deterministic pushdown automaton of size $O(n^3)$, and each equivalent DFA has at least 2^{2^n} states.

Concerning RRWW-automata, it has been shown in [18] that, for all $k \geq 2$, $\mathscr{L}(\mathrm{RRWW}(k+1)) = \mathscr{L}(\mathrm{RRWW}(k))$. In fact, one can effectively construct an equivalent RRWW(k)-automaton $M_2 = (Q_2, \Sigma, \Gamma_2, \mathord{\mathrm{c}}, \$, q_0, k, \delta_2)$ from a given RRWW($k+1$)-automaton $M_1 = (Q_1, \Sigma, \Gamma_1, \mathord{\mathrm{c}}, \$, q_0, k+1, \delta_1)$. Actually, the construction presented in [18] satisfies the inequalities

$$|Q_2| \in O(|Q_1|^3 \cdot |\Gamma_1|^{2k+2}) \quad \text{and} \quad |\Gamma_2| \in O(|Q_1|^2 \cdot |\Gamma_1|^{2k+3}).$$

When taking the complexity sn-measure trans, that is, the product of the number of states and the number of possible window contents, then we obtain $|Q_2| \in O(\text{trans}(M_1)^3)$ and $|\Gamma_2| \in O(\text{trans}(M_1)^3)$ and, hence, $\text{trans}(M_2) \in O(\text{trans}(M_1)^3 \cdot \text{trans}(M_1)^{3k}) = O(\text{trans}(M_1)^{3k+3})$.

Corollary 12. *For all $k \geq 2$, the trade-off between $\text{RRWW}(k+1)$-automata and $\text{RRWW}(k)$-automata is at most polynomial, where the degree of the polynomial depends on the window size k.*

5 Conclusion

It remains to determine the trade-off between nondeterministic RW- and RRW-automata with window size two and those of window size one. Here the situation is different from that for deterministic machines. The proof of Theorem 11 does not generalize for at least two reasons. First, regularity is undecidable for context-free languages and, second, in [13] it is shown that there is a nondeterministic RR(1)-automaton that accepts a non-context-free language.

The impact of the window size on the descriptional complexity of (non)deterministic R-, RR-, and RWW-automata is also a promising field for further investigations.

References

1. Hartmanis, J.: On Gödel speed-up and succinctness of language representations. Theoret. Comput. Sci. 26, 335–342 (1983)
2. Holzer, M., Kutrib, M.: Descriptional complexity – An introductory survey. In: Scientific Applications of Language Methods, pp. 1–58. Imperial College Press (2010)
3. Holzer, M., Kutrib, M., Reimann, J.: Non-recursive trade-offs for deterministic restarting automata. J. Autom. Lang. Comb. 12, 195–213 (2007)
4. Jančar, P., Mráz, F., Plátek, M., Vogel, J.: Restarting Automata. In: Reichel, H. (ed.) FCT 1995. LNCS, vol. 965, pp. 283–292. Springer, Heidelberg (1995)
5. Jančar, P., Mráz, F., Plátek, M., Vogel, J.: On monotonic automata with a restart operation. J. Autom. Lang. Comb. 4, 287–311 (1999)
6. Kutrib, M.: The phenomenon of non-recursive trade-offs. Int. J. Found. Comput. Sci. 16, 957–973 (2005)
7. Kutrib, M., Malcher, A.: When Church-Rosser becomes context-free. Int. J. Found. Comput. Sci. 18, 1293–1302 (2007)
8. Kutrib, M., Messerschmidt, H., Otto, F.: On stateless deterministic restarting automata. Acta Inform. 47, 391–412 (2010)
9. Kutrib, M., Messerschmidt, H., Otto, F.: On stateless two-pushdown automata and restarting automata. Int. J. Found. Comput. Sci. 21, 781–798 (2010)
10. Kutrib, M., Reimann, J.: Optimal simulations of weak restarting automata. Int. J. Found. Comput. Sci. 19, 795–811 (2008)
11. Kutrib, M., Reimann, J.: Succinct description of regular languages by weak restarting automata. Inform. Comput. 206, 1152–1160 (2008)

12. Meyer, A.R., Fischer, M.J.: Economy of description by automata, grammars, and formal systems. In: Symposium on Switching and Automata Theory (SWAT 1971), pp. 188–191. IEEE (1971)
13. Mráz, F.: Lookahead hierarchies of restarting automata. J. Autom. Lang. Comb. 6, 493–506 (2001)
14. Narendran, P., Ó'Dúnlaing, C., Rolletschek, H.: Complexity of certain decision problems about congruential languages. J. Comput. System Sci. 30, 343–358 (1985)
15. Niemann, G., Otto, F.: On the power of RRWW-automata. In: Ito, M., Păun, G., Yu, S. (eds.) Words, Semigroups, and Transductions - Festschrift in Honor of Gabriel Thierrin, pp. 341–355. World Scientific (2001)
16. Otto, F.: Restarting Automata. In: Ésik, Z., Martin-Vide, C., Mitrana, V. (eds.) Recent Advances in Formal Languages and Applications. SCI, vol. 25, pp. 269–303. Springer, Heidelberg (2006)
17. Post, E.L.: Recursive unsolvability of a problem of Thue. J. Symbolic Logic 12, 1–11 (1947)
18. Schluter, N.: Restarting Automata with Auxiliary Symbols and Small Lookahead. In: Dediu, A.-H., Inenaga, S., Martín-Vide, C. (eds.) LATA 2011. LNCS, vol. 6638, pp. 499–510. Springer, Heidelberg (2011)
19. Stearns, R.E.: A regularity test for pushdown machines. Inform. Control 11, 323–340 (1967)
20. Valiant, L.G.: Regularity and related problems for deterministic pushdown automata. J. ACM 22, 1–10 (1975)

A Disambiguation Algorithm
for Finite Automata and Functional Transducers

Mehryar Mohri

Courant Institute of Mathematical Sciences and Google Research
251 Mercer Street,
New York, NY 10012, USA

Abstract. We present a new disambiguation algorithm for finite automata and functional finite-state transducers. We give a full description of the algorithm, including a detailed pseudocode and analysis, and several illustrating examples. Our algorithm is often more efficient and the result dramatically smaller than the one obtained using determinization for finite automata or an existing disambiguation algorithm for transducers based on a construction of Schützenberger. In a variety of cases, the size of the unambiguous transducer returned by our algorithm is only linear in that of the input transducer while the transducer given by the construction of Schützenberger is exponentially larger. Our algorithm can be used effectively in many applications to make automata and transducers more efficient to use.

1 Introduction

Finite automata and transducers are used in a variety of applications in text and speech processing [10,13], bioinformatics [8], image processing [1], optical character recognition [6], and many others. In these applications, automata and transducers are often the result of various complex operations and in general are not efficient to use. Some optimization algorithms such as determinization can make their use more time-efficient. However, the result of determinization is sometimes prohibitively large and not all finite-state transducers are determinizable [7,11].

This paper presents and analyzes an alternative optimization algorithm, *disambiguation*, which in practice can have efficiency benefits similar to determinization. Our disambiguation algorithm is novel and applies to finite automata, including automata with ϵ-transitions, and to *functional finite-state transducers*, that is those representing a partial function. Disambiguation returns an automaton or transducer equivalent to the input that is *unambiguous*, that is one that admits no two accepting paths labeled with the same (input) string. In many instances, the absence of ambiguity can be useful to make search more efficient by reducing the number of paths to explore for very large automata or transducers with several hundred thousand or millions of transitions in text and speech processing or in bioinformatics, and there are many other critical needs for the disambiguation of automata and transducers.

N. Moreira and R. Reis (Eds.): CIAA 2012, LNCS 7381, pp. 265–277, 2012.
© Springer-Verlag Berlin Heidelberg 2012

For finite automata, one way to proceed to obtain an unambiguous and equivalent automaton is simply to apply the standard determinization algorithm. But, as we shall see, for some input automata our algorithm can take exponentially less time than determinization and return an equivalent unambiguous automaton exponentially smaller than the one obtained by using determinization.

For finite-state transducers, disambiguation applies to a broader set of transducers than those that can be determinized using the algorithm described in [11], it applies to any functional transducer. In contrast, it was shown by [3] that a functional transducer is determinizable if and only if it additionally verifies the *twins property* [7,11,2]. Our disambiguation algorithm is also often dramatically more efficient and results in substantially smaller transducers than those obtained using a disambiguation algorithm based on a construction of Schützenberger [16,15], also described by E. Roche and Y. Schabes in the introductory chapter of [14]. In particular, when the input transducer is unambiguous, our algorithm simply returns the same transducer, while the result of the algorithm presented in [14] can be exponentially larger.

The remainder of this paper is organized as follows. In Section 2, we introduce the notation and basic concepts needed for the presentation and analysis of our algorithm. In Section 3, we present our disambiguation algorithm for finite automata in detail, including the proof of its correctness and a brief description of its extension to finite automata with ϵ-transitions. In Section 4, we show how the algorithm can be be used to disambiguate functional transducers and illustrate it with several examples.

2   Preliminaries

We will denote by ϵ the empty string. A finite automaton A with ϵ-transitions is a system (Σ, Q, I, F, E) where Σ is a finite alphabet, Q a finite set of states, $I \subseteq Q$ the set of initial states, $F \subseteq Q$ the set of final states, and E a finite multiset of transitions, which are elements of $Q \times (\Sigma \cup \{\epsilon\}) \times Q$. We denote by $|A| = |Q| + |E|$ the *size of an automaton* A, that is the sum of the number states and transitions defining A.

A path π of an automaton is an element of E^* with consecutive transitions. The label of a path is the string obtained by concatenation of the labels of its constituent transitions. We denote by $P(p, x, q)$ the set of paths from p to q labeled with x or, more generally, by $P(R, x, R')$ the set of paths labeled with x from some set of states R to some set of states R'. We also denote by $P(R, R')$ the set of all paths from R to R'. An *accepting path* is an element of $P(I, F)$. The *language accepted by an automaton* A is the set of strings labeling its accepting paths and is denoted by $L(A)$. Two automata A and B are said to be equivalent when $L(A) = L(B)$.

We will say that a state p can be reached by a string x when there exists a path from an initial state to p labeled with x. When two states can be reached by the same string, we say that they are *co-reachable*. We will also say that two states p and q *share a common future* when they admit a common string x to reach a final state, that is when there exists a string x such that $P(p, x, F) \cap P(q, x, F) \neq \emptyset$. For any subset $s \subseteq Q$ and $x \in \Sigma^*$, we will denote by $\delta(s, x)$ the set of states that can be reached from the states in s by a path labeled with x.

A *finite-state transducer* is a finite automaton in which each transition is augmented with an output label, which is an element of $(\Delta \cup \{\epsilon\})$, where Δ is a finite alphabet. For any transducer T, we denote by T^{-1} its *inverse*, that is the transducer obtained from T by swapping the input and output label of each transition.

We will use the standard algorithm to compute the intersection $A \cap A'$ of two automata A and A' [12], whose states are pairs formed by a state of A and a state of A', and whose transitions are of the form $((p, q), a, (p', q'))$, where (p, a, q) is a transition in A and (p', a, q') in A'.

An automaton A is said to be *trim* if all of its states lie on some accepting path. It is said to be *unambiguous* if no string $x \in \Sigma^*$ labels two distinct accepting paths, *finitely ambiguous* if there exists $k \in \mathbb{N}$ such that no string labels more than k accepting paths, *polynomially ambiguous* if there exists a polynomial P with coefficients in $\mathbb{N}$ such that no string x labels more than $P(|x|)$ accepting paths. The finite, polynomial, and exponential ambiguity of an automaton with ϵ-transitions can be tested in polynomial time [4].

3   Disambiguation Algorithm for Finite Automata

In this section, we describe in detail our disambiguation algorithm for finite automata. The algorithm is first described for automata without ϵ-transitions. The extension to the case of automata with ϵ-transitions is discussed later. Our algorithm in general does not require a full determinization. In fact, in some cases where the determinization creates 2^n states where n is the number of states of the input automaton, the cost of our new algorithm or the size of its output is only in $O(n)$.

3.1   Description

Figure 1 gives the pseudocode of the algorithm. The first step of the algorithm consists of computing the automaton $A \cap A$ and of trimming it by removing non-coaccessible states (line 1). The cost of this computation is in $O(|A|^2)$ since the complexity of intersection is quadratic and since trimming can be done in linear time. The automaton B thereby constructed can be used to determine in constant time if two states q and r of A that can be reached from I via the same string share a common future simply by checking if (q, r) is a state of B. Indeed, by definition of intersection, this property holds iff (q, r) is a state of B. As shown by the following proposition, the automaton B is in fact directly related to the ambiguity of A.

Proposition 1 ([4]). *Let A be a trim finite automaton with no ϵ-transition. A is unambiguous iff no coaccessible state in $A \cap A$ is of the form (p, q) with $p \neq q$.*

DISAMBIGUATION(A)
1   $B \leftarrow$ TRIM$(A \cap A)$
2   **for** each $i \in I$ **do**
3       $s \leftarrow \{i' : i' \in I \wedge (i, i') \in B\}$
4       $I' \leftarrow Q' \leftarrow Q' \cup \{(i, s)\}$
5   ENQUEUE$(\mathcal{Q}, (i, s))$
6   **for** each $(u, u') \in I'^2$ **do**
7       $R \leftarrow R \cup \{(u, u')\}$
8   **while** $\mathcal{Q} \neq \emptyset$ **do**
9       $(p, s) \leftarrow$ HEAD$(\mathcal{Q})$
10   DEQUEUE$(\mathcal{Q})$
11   **if** $((p \in F)$ **and** $(\not\exists (p', s') \in F'$ with $(p', s') \, R \, (p, s)))$ **then**
12          $F' \leftarrow F' \cup \{(p, s)\}$
13   **for** each $(p, a, q) \in E$ **do**
14          $t \leftarrow \{r \in \delta(s, a) : (q, r) \in B\}$
15   **if** $\big(\not\exists ((p', s'), a, (q, t)) \in E'$ with $(p', s') \, R \, (p, s)\big)$ **then**
16   **if** $((q, t) \notin Q')$ **then**
17                $Q' \leftarrow Q' \cup \{(q, t)\}$
18   ENQUEUE$(\mathcal{Q}, (q, t))$
19             $E' \leftarrow E' \cup \{((p, s), a, (q, t))\}$
20   **for** each (p', s') s.t. $((p', s') R \, (p, s))$ **and** $((p', s'), a, (q', t')) \in E'$ **do**
21                $R \leftarrow R \cup \{(q, t), (q', t')\}$
22 **return** A'

Fig. 1. New disambiguation algorithm for finite automata

Proof. Since A is trim, the states of $A \cap A$ are all accessible by construction. Thus, a state (p, q) in $A \cap A$ is coaccessible iff it lies on an accepting path, that is by definition of intersection, iff there are two paths $\pi = \pi_1 \pi_2 \in P(I, F)$ and $\pi' = \pi_1' \pi_2' \in P(I, F)$ with $\pi_1 \in P(I, p)$ and $\pi_1' \in P(I, q)$, with π_1 and π_1' sharing the same label and π_2 and π_2' also sharing the same label. Thus, A is unambiguous iff $p = q$. ☐

The algorithm constructs an unambiguous automaton $A' = (Q', E', I', F')$. The set of states Q' are of the form (p, s) where p is a state of A and s a subset of the states of A. Line 2 defines the initial states which are of the form (i, s) with $i \in I$ and s a subset of the states in I sharing a common future with i. The algorithm maintains a relation R such that two states of A' are in relation via R iff they can be reached by the same string from the initial states. In particular, since all initial states are reachable by ϵ, any two pair of initial states are in relation via R (lines 6-7).

The algorithm also maintains a queue $\mathcal{Q}$ containing the set of states (p, s) of Q' left to examine and for which the outgoing transitions are to be determined. The queue discipline, that is the order in which states are added or extracted from $\mathcal{Q}$ is arbitrary and does not affect the correctness of the algorithm. However, different orderings can result in different but equivalent resulting automata.

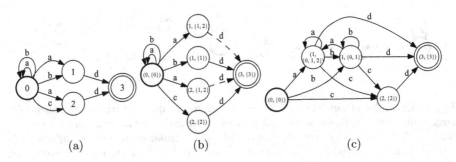

Fig. 2. Illustration of the disambiguation algorithm. (a) Automaton A. (b) Result of disambiguation algorithm applied to A. One of the two dashed transitions is disallowed by the algorithm. (c) Result of determinization applied to A.

At each execution of the loop of lines 8-21, a new state (p, s) is extracted from Q (lines 9-10). To avoid an ambiguity due to finality, state (p, s) is made final only if there is no final state $(p', s') \in F'$ in relation with (p, s) (lines 11-12).

Each outgoing transition (p, a, q) of p is then examined. Line 14 defines t to be the subset of the states of A that can be reached from a state of s by reading x but excludes states q' that do not share a common future with q. This is because the subsets are used to detect ambiguities. If q and q' do not share a common future even though there are paths with the same label x reaching them, these paths cannot be completed to reach a final state with the same label. Thus, if X is the set of strings leading to a state (p, s) of Q', the subset s contains exactly the set of states r of A that can be reached via X from I and that share a common future with p.

To avoid creating two paths from I' to (q, t) with the same labels, the transition from (p, s) to (q, t) with label q is not created if there exists already one from (p', s') to (q, t) for a state (p', s') that can reached by a string also reaching (p, s) (condition of line 15). Note that if (p, s) is extracted from Q before a state (p', s') with $(p', s')R(p, s)$, then the transition from (p, s) to (q, t) is created first and the one from (p', s') to (q, t) not created. This is how the queue discipline directs the choice of the transitions created.

Lines 16-18 add (q, t) to Q' when it is not already in Q' and line 19 adds the new transition defined to E'. After creation of this transition, the destination state (q, t) is then put in relation with all states (q', t') reached by a transition labeled with $a \in \Sigma$ from a state (p', s') that is in relation with (p, s).

Figure 2 illustrates the application of the algorithm in a simple case. Observe that states 1 or 2 are not included in the subset of $(0, \{0\})$ in the automaton of Figure 2(b) since 0 does not share a common future with 1 or 2. Figure 2 also shows the result of the application of determinization to the same example. As can be seen from this example, in some instances, determinization creates more transitions than disambiguation. Some states created by the disambiguation algorithm may be non-coaccessible, that is, they may admit no transition to a final state because their output transitions were not constructed to avoid

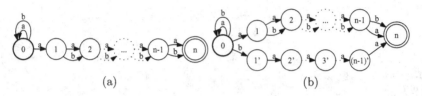

Fig. 3. Examples of automata A for which determinization returns an exponentially larger automaton while our algorithm returns A (for (a)) or an automaton whose size is linear in A (for (b)). (a) Automaton representing the regular expression $(a+b)^*a(a+b)^n$, whose minimal deterministic equivalent has size $\Omega(2^n)$. (b) Automaton representing the regular expression $(a + b)^*(a(a + b)^n + ba^n)$, whose determinization results in an automaton with $\Omega(2^n)$ states.

generating ambiguity. These states and the transitions leading to them can be removed in linear time using a standard trimming algorithm. In the case of the automaton of Figure 2(b), the state whose dashed transition is not constructed can be trimmed.

More generally, note that when the input automaton is unambiguous, the subsets created by our algorithm are reduced to singletons: by Proposition 1, a subset cannot contain two distinct states in that case. In such cases, our algorithm simply returns the same automaton A. The work done after computation of B is also linear in $|A|$. In contrast, the determinization of A may lead to a blow-up, even when the automaton is unambiguous. In particular, for the standard case of the non-deterministic automaton of Figure 3(a) representing the regular expression $(a + b)^*a(a + b)^n$, it is known that determinization creates $2^{n+1} - 1$ states. However, this automaton is unambiguous and our algorithm returns the same automaton unchanged. The automaton of Figure 3(b) is similar but is ambiguous. Nevertheless, it is not hard to see that again the size of the automaton returned by determinization is exponential and that that of the automaton output by our algorithm is only linear.

3.2 Analysis

The termination of the algorithm is guaranteed by the fact that the number of states and transitions created must be finite. This is because the number of possible subsets s of states of A is finite, thereby also the number of pairs (p, s) created by the algorithm where p is a state of A and s a subset. Also, the number of transitions created at a state (p, s) is at most equal to the number of states leaving p in A. In the worst case, the algorithm may create exponentially many subsets and thus the computational complexity of the algorithm is exponential. In many practical cases, however, this worst case behavior is not observed. In particular, the automaton returned by our disambiguation algorithm is substantially smaller than the one obtained by application of determinization.

We will now show that the automaton returned by the algorithm is unambiguous using the following lemma.

Lemma 1. *Let (q, t) and (q', t') be two states constructed by algorithm* DISAM-BIGUATION *run on input automaton A, then (q, t) R (q', t') iff (q, t) and (q', t') are co-reachable.*

Proof. We will show by induction on the length of strings x that if two states (q, t) and (q', t') are both reachable by x, then (p, s) R (q', t'). The steps of lines 6-7 ensure that (q, t) R (q', t') when both states are initial, that is, when they are reachable by ϵ. Assume that it holds for all strings x of length less than or equal to n. Let $x = x'a$ be a string of length $n + 1$ with $x' \in \Sigma^*$ and $a \in \Sigma$ and assume that (q, t) and (q', t') are both reachable by x. Then, there exists a state (p, s) reachable by x' and admitting a transition labeled with a leading to (q, t) and similarly a state (p', s') reachable by x' and admitting a transition labeled with a leading to (q', t'). Then, by the induction hypothesis, we have (p, s) R (p', s'), thus (q, t) R (q', t') is guaranteed by execution of the steps of lines 20-21. This proves the implication corresponding to one side. The converse holds straightforwardly by construction (lines 6-7 and 20-21). □

Proposition 2. *The automaton A' returned by algorithm* DISAMBIGUATION *run on input automaton A is unambiguous.*

Proof. Let π_1 and π_2 be two paths in A' from I' to F' with the same label $x \in \Sigma^*$. If $x = \epsilon$, π_1 is a path from some initial state (i_1, s_1) to (i_1, s_1) and similarly π_2 a path from some initial state (i_2, s_2) to (i_2, s_2). All initial states are in relation (lines 6-7), therefore at most one can be made final (lines 11-12). This implies that $(i_1, s_1) = (i_2, s_2)$ and $\pi_1 = \pi_2$. Let (q_1, t_1) be the destination state of π_1 and (q_2, t_2) the destination state of π_2. Since (q_1, t_1) and (q_2, t_2) are both reachable by x, by Lemma 1, we have (q_1, t_1) R (q_2, t_2). Since no two distinct equivalent states can be made final (lines 11-12), we must have $(q_1, t_1) = (q_2, t_2)$.

If $x = \epsilon$, this implies that the two paths π_1 and π_2 coincide. If $x \neq \epsilon$, x can be written as $x = x'a$ with $x' \in \Sigma^*$ and $a \in \Sigma$ and π_1 and π_2 can be decomposed as $\pi_1 = \pi_1'e_1$ and $\pi_2 = \pi_2'e_2$ with e_1 and e_2 transitions labeled with a leading to (q_1, t_1). Let (p_1, s_1) be the destination state of π_1' and (p_2, s_2) the destination state of π_2'. Since π_1' and π_2' are both labeled with x', by Lemma 1, we have (p_1, s_1) R (p_1', s_1'). By the condition of line 15, if $(p_1, s_1) \neq (p_1', s_1')$, (p_1, s_1) and (p_1', s_1') cannot both admit a transition labeled with a and leading to the same state (q_1, t_1). Thus, we must have $(p_1, s_1) = (p_1', s_1')$. Proceeding in the same way with π_1' and π_2' and so on shows that the paths π_1 and π_2 coincide, which concludes the proof. □

The following lemmas will be used to show the equivalence between the automaton returned by the algorithm and the input automaton.

Lemma 2. *Let (p, s) be a state constructed by algorithm* DISAMBIGUATION *run on input automaton A. If (p, s) is reachable by the strings u and v in A', then the set of states reachable by u in A and sharing a common future with p coincides with the set of states reachable by v in A and sharing a common future with p .*

Proof. We show by recurrence on the length of u that if state (p, s) is reachable by u in A', then s is the set of states reachable by u and sharing a common future with p. This property holds straightforwardly for $u = \epsilon$ by the construction of lines 2-5. Assume now that it holds for all u of length less than or equal to n. Let $u = u'a$ with $u' \in \Sigma^*$ of length n and $a \in \Sigma$. If (p, s) is reachable by u, there must exist some state (p', s') reachable by u' and admitting a transition labeled with a leading to (p, s). By the induction hypothesis, s' is the set of states reachable by u' and sharing a common future with p'. By definition of s (line 14), $s = \{q \in \delta(s', a) : (q, p) \in B\}$, thus the states in s are all reachable by u and share a common future with p. Conversely, let q be a state reachable by u and sharing future with p. There is a transition labeled with a from some state q' reachable by u'. Since q' admits a transition to q labeled with a and p' admits a transition labeled with a to p, and p and q share a common future, p' and q' must also share a common future. By the induction hypothesis, s' is the set of states reachable by u' and sharing a common future with p', therefore q' is in s'. Since $q \in \delta(q', a)$ and q shares a common future with p, this implies that q is in s. This shows that the states in s are those reachable by u and sharing a common future with p. $\square$

Lemma 3. *Let A' be the automaton returned by algorithm* DISAMBIGUATION *run on input automaton A. Let q be a state reachable in A by string x. Then, there exists a state (q, t) in A' for some subset t such that (q, t) is reachable by x in A'.*

Proof. We will prove the property by induction on the length of x. The property straightforwardly holds for $x = \epsilon$ by the construction steps of lines 2-5. Assume now that it holds for all strings of length less than or equal to n and let $x = ua$ with u a string of length n and $a \in \Sigma$. If q is reachable by string x in A, then there exists a state p_0 in A reachable by u and admitting a transition labeled with a leading to q. By the induction hypothesis, there exists a state (p_0, s_0) in A' reachable by u. Now, the property clearly holds for (q, t_0) if the transition labeled with a leaving (p_0, s_0) is constructed at lines 15-19, with t_0 defined at line 14. Otherwise, by the test of line 15, there must exist in A' a distinct state (p_1, s_0') admitting a transition labeled with a leading to (q, t_0) with $(p_1, s_0') R (p_0, s_0)$. Note that we cannot have $p_1 = p_0$, since the same string cannot reach two distinct states (p_0, s_0) and (p_0, s_1). Now, since (p_1, s_0') admits a transition labeled with a leading to (q, t_0), p_1 must admit a transition labeled with a and leading to q. Thus, p_1 and p_0 share a common future in A. Since $(p_1, s_0') R (p_0, s_0)$, by Lemma 1, they are reachable by a common string v. Thus, both u and v reach (p_0, s_0). By Lemma 2, this implies that the set of states in A reachable by u and v and sharing a common future with p_0 are the same. Since p_1 and p_0 share a common future in A and v reaches both p_0 and p_1, u must also reach p_1 in A.

If u reaches (p_1, s_0'), then (q, t_0) can be reached by x since (p_1, s_0') admits a transition labeled with a leading to (q, t). Otherwise, by the induction hypothesis, there must exist a distinct state (p_1, s_1) in A' reachable by u, with p_1 admitting a transition labeled with a to q. Reapplying the argument already presented for

(p_0, s_0) to (p_1, s_1), either we find a path in A' labeled with x to a state (q, t_1), or there exists a state (p_2, s_2) in A' with the same property as (p_0, s_0) with p_2 distinct from p_1 and p_0. Since the number of distinct such states is finite, reiterating this process guarantees finding a path in A' labeled with x to a state (q, t_k) after some finite number of times k. Thus, the property holds in all cases.

□

Lemma 4. *Let A' be the automaton returned by algorithm* DISAMBIGUATION *run on input automaton A, then $L(A') \subseteq L(A)$.*

Proof. The proof argument is similar to that of Lemma 3. Let x be a string reaching a final state $q_0 \in F$ in A. By Lemma 3, there exists a state (q_0, t_0) in A' reachable by x. If state (q_0, t_0) is made final (lines 11-12), this shows that x is accepted by A'. Otherwise, there must exist a final state (q_1, t_0') with $(q_1, t_0') \, R \, (q_0, t_0)$. Note that this implies that q_1 is final. Note also that we have $q_1 \neq q_0$ since two states (q_0, t_0) and (q_0, t_0') cannot be co-reachable with $t_0' \neq t_0$. Since $(q_1, t_0') \, R \, (q_0, t_0)$, there exists a string x_1 reaching both states. Since (q_0, t_0) is reachable by both x and x_1, by Lemma 2, the set of states in A reachable by x and sharing a common future with q_0 and those reachable by x_1 and sharing a common future with q_0 are the same. q_1 shares a common future with q_0 since both states are final and q_1 is reachable by x_1, therefore q_1 is reachable by x.

Now, if x reaches (q_1, t_0'), this shows that x is accepted by A'. Otherwise, by Lemma 3, there exists a state (q_1, t_1) in A' reachable by x. We can reapply to (q_1, t_1) the same argument as for (q_0, t_0) since q_1 is a final state. Doing so, we either find a final state in A' reachable by x or a state (q_2, t_2) in A' with the same properties as (q_0, t_0) with q_0, q_1, and q_2 all distinct. Since the number of states of A' is finite, reiterating this process guarantees finding a final state reachable by x. This concludes the proof.   □

Proposition 3. *The automaton A' returned by algorithm* DISAMBIGUATION *run on input automaton A is equivalent to A.*

Proof. By construction, a path $((p_1, s_1), a_1, (p_2, s_2)) \cdots ((p_k, s_k), a_k, (p_{k+1}, s_{k+1}))$ is created in A' only if the path $(p_1, a_1, p_2) \cdots (p_k, a_k, p_{k+1})$ exists in A, and a state (p, s) is made final in A' only if p is final in A. Thus, if a string $x = a_1 \cdots a_k$ is accepted by A' it is also accepted by A, which shows that $L(A') \subseteq L(A)$. the reverse inclusion holds by Lemma 4.

The following theorem follows directly by Propositions 2 and 3.

Theorem 1. *The automaton A' returned by algorithm* DISAMBIGUATION *run on input automaton A is an unambiguous automaton equivalent to A.*

Note that the states disallowed via the condition of our algorithm are the minimal ones that can be safely removed from the subsets to check the presence of ambiguities.

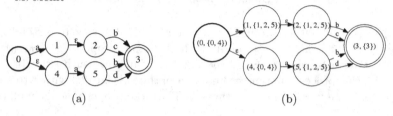

Fig. 4. (a) Automaton A with ϵ-transitions. (b) Unambiguous automaton equivalent to A returned by our disambiguation algorithm. The dashed transition is disallowed by the algorithm.

3.3 Disambiguation of Automata with ϵ-Transitions

Our algorithm can also be extended to the case of automata with ϵ-transitions. We briefly describe that extension. Let A be an input automaton with ϵ-transitions. Here, the automaton B used to determine pairs of states sharing the same future is obtained similarly by computing the intersection $A \cap A$ by using an ϵ-filter [12] and by trimming the result by removing non-coaccessible states and transitions. For any set R of states of A, let $\epsilon[R]$ denote the ϵ-closure of R, that is the set of states reachable from states of R via paths labeled with ϵ.

To extend the algorithm to cover the case of automata with ϵ-transitions, it suffices to proceed as follows. The initial states are defined by the set of (i, s) with $i \in I$ and $s = \{q \in \epsilon[I] \colon (i, q) \in B\}$. At line 14, $\delta(s, a)$ is defined as the set of states reachable from s by reading a, including via ϵ-transitions. Finally, the relation R is extended to ϵ-transitions as follows: for each (p', s') such that $(p', s') \, R \, (p, s)$ and $((p, s), \epsilon, (q', t')) \in E'$, (p', s') is put in relation with (q', t'). Figure 4 illustrates the application of our algorithm in that case.

4 Disambiguation of Finite-State Transducers

In this section, we consider the problem of determining an unambiguous transducer equivalent to a given *functional finite-state transducer*, that is a finite-state transducer representing a (partial) rational function, or equivalently one associating at most one output string to any input string. The functionality of a finite-state transducer T can be tested efficiently from the transducer $T \circ T^{-1}$ as shown by [2].

Theorem 2 ([2]). *There exists an algorithm for testing the functionality of a finite-state transducer T with output alphabet Δ in time $O(|E|^2 + |\Delta| \, |Q|^2)$.*

One possible algorithm for finding an unambiguous transducer equivalent to a functional transducer is determinization [11], however, as discussed earlier, not all functional transducers admit an equivalent deterministic transducer. Figure 5(a) shows an example of such a functional transducer which in fact is unambiguous. A trim functional transducer is determinizable iff it admits the twins property [3].

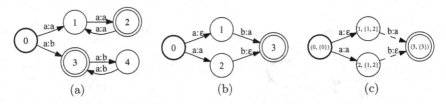

Fig. 5. (a) Unambiguous finite-state transducer admitting no sequential or deterministic equivalent. (a) Functional transducer T. (b) Disambiguated transducer equivalent to T returned by our algorithm. One of the two dashed transitions is disallowed by the algorithm.

We will describe instead a disambiguation algorithm does not require that additional property. It is known that any functional transducer can be represented by an unambiguous transducer [9,5]. For a functional transducer, by definition, two accepting paths with the same input label have the same output labels. Thus, for disambiguating a functional transducer, only input labels matter and our automata disambiguation can be readily applied to create an unambiguous transducer equivalent to an input functional transducer. Our disambiguation algorithm gives a constructive proof of the existence of an equivalent unambiguous transducer for a rational function. The different possible *cross-sections* of the construction of [9] correspond to different orders in which transitions are visited and disallowed by our algorithm. Figure 5(b)-(c) illustrates the application of the algorithm in the case of a simple functional transducer.

As already pointed out, our algorithm compares favorably with the existing disambiguation algorithm for finite-state transducers of Schützenberger [16,15]. That construction can be concisely described as follows. Let D be a deterministic automaton obtained by determinization of the input automaton A of the functional transducer T, that is the automaton obtained by removing the output labels of T. Then, the algorithm consists of composing D with T using the standard composition algorithm for finite-state transducers while disallowing finality of two composition states (p, s) and (q, s) with the same determinization subset s and distinct states p and q of T, and similarly disallowing all but one transition labeled with a from two states (p, s) and (q, s) to the same state, to avoid generating ambiguities. As can be seen from this description, the algorithm requires the determinization of A. This is implicit in the description of this construction in [14].

In contrast, our disambiguation algorithm that does not require the determinization of A and as seen in the previous sections can return exponentially smaller automata than those returned by determinization is some cases. Consider for example the finite-state transducers defined as the automata of Figure 3 with each transition augmented with an output label identical to its output label. The construction of Schützenberger requires for those transducers the determinization of the input automata, thus its cost as well as the size of the result are exponential with respect to the size of the output as already discussed in Section 3. Unlike that construction, as in the automata case, our algorithm returns the same transducer or returns one whose size is only linear in that of the input.

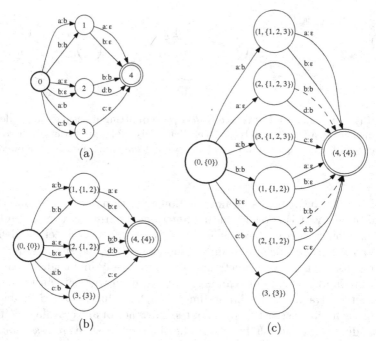

Fig. 6. Disambiguation of functional transducers. (a) Functional transducer T. (b) Unambiguous transducer equivalent to T returned by our algorithm. The dashed transitions are disallowed by the algorithm. (c) Unambiguous transducer returned by the disambiguation construction of Schützenberger [16,15].

The subsets defined by our disambiguation algorithm are never larger than those defined in the subset construction of determinization. This is because for a state (p, s) constructed in the algorithm, only states sharing a common future with p are kept in the subset s. In addition to making the size of the subsets shorter, this also reduces the number of states created: two possible states (p, s') and $(p, s")$ in the construction of Schützenberger are reduced to the same (p, s) after removal from s' and $s"$ of the states not sharing a common future with p. This leads in many cases to transducers exponentially smaller than those generated by the construction of Schützenberger and similar improvements in time efficiency.

The observation just emphasized can be illustrated by the simple example of Figure 6. The transducer T of Figure 6(a) is functional but is not unambiguous. Figure 6(b) shows the result of our disambiguation algorithm which is an unambiguous transducer equivalent to T with the same number of states. In contrast, the transducer created by the construction of Schützenberger (Figure 6(c)) has several more states and transitions and some larger subsets.

5   Conclusion

We presented a new and often more efficient algorithm for the disambiguation of finite automata and functional transducers. This algorithm is of great practical importance in a variety of applications including text and speech processing, bioinformatics, and in many other applications where they can be used to increase search efficiency. We have also designed a natural extension of these algorithms to some broad families of weighted automata and transducers defined over different semirings. We will present these extensions as well as their theoretical analysis in a longer version of this paper.

Acknowledgments. I thank Cyril Allauzen and Michael Riley for discussions about this work. This research was supported by a Google Research Award.

References

1. Albert, J., Kari, J.: Digital image compression. In: Handbook of Weighted Automata. Springer (2009)
2. Allauzen, C., Mohri, M.: Efficient algorithms for testing the twins property. Journal of Automata, Languages and Combinatorics 8(2), 117–144 (2003)
3. Allauzen, C., Mohri, M.: Finitely subsequential transducers. International Journal of Foundations of Computer Science 14(6), 983–994 (2003)
4. Allauzen, C., Mohri, M., Rastogi, A.: General algorithms for testing the ambiguity of finite automata and the double-tape ambiguity of finite-state transducers. Int. J. Found. Comput. Sci. 22(4), 883–904 (2011)
5. Berstel, J.: Transductions and Context-Free Languages. Teubner Studienbucher (1979)
6. Breuel, T.M.: The OCRopus open source OCR system. In: Proceedings of IS&T/SPIE 20th Annual Symposium (2008)
7. Choffrut, C.: Contributions à l'étude de quelques familles remarquables de fonctions rationnelles. Ph.D. thesis, Université Paris 7, LITP: Paris, France (1978)
8. Durbin, R., Eddy, S.R., Krogh, A., Mitchison, G.J.: Biological Sequence Analysis: Probabilistic Models of Proteins and Nucleic Acids. Cambridge University Press (1998)
9. Eilenberg, S.: Automata, Languages and Machines, vol. A. Academic Press (1974)
10. Kaplan, R.M., Kay, M.: Regular models of phonological rule systems. Computational Linguistics 20(3) (1994)
11. Mohri, M.: Finite-state transducers in language and speech processing. Computational Linguistics 23(2), 269–311 (1997)
12. Mohri, M.: Weighted automata algorithms. In: Handbook of Weighted Automata, pp. 213–254. Springer (2009)
13. Mohri, M., Pereira, F.C.N., Riley, M.: Speech recognition with weighted finite-state transducers. In: Handbook on Speech Processing and Speech Communication. Springer (2008)
14. Roche, E., Schabes, Y. (eds.): Finite-State Language Processing. MIT Press (1997)
15. Sakarovitch, J.: A construction on finite automata that has remained hidden. Theor. Comput. Sci. 204(1-2), 205–231 (1998)
16. Schützenberger, M.P.: Sur les relations rationnelles entre monoides libres. Theor. Comput. Sci. 3(2), 243–259 (1976)

Synchronization of Automata
with One Undefined or Ambiguous Transition

Pavel V. Martyugin

Ural Federal University,
620083 Ekaterinburg, Russia
martugin@mail.ru

Abstract. We consider the careful synchronization of partial automata
with only one undefined transition and the generalized synchronization
of nondeterministic automata with only one ambiguous transition. For
each of the two cases we prove that the problem of checking whether
or not a given automaton is synchronizable is PSPACE-complete. The
restrictions of these problems to 2-letter automata are also PSPACE-
complete.

Keywords: Synchronizing words, Careful synchronization, Nondeterministic automata, Computational Complexity.

1 Introduction

A *deterministic finite automaton (DFA)* is a triple $\mathscr{A} = (Q, \Sigma, \delta)$, where Q is
a finite set of states, Σ is a finite alphabet and δ is a totally defined transition
function. Denote by Σ^* the free Σ-generated monoid with the empty word λ. The
function δ extends in a natural way to the action $Q \times \Sigma^* \to Q$. This extension is
also denoted by δ. A DFA $\mathscr{A} = (Q, \Sigma, \delta)$ is called *synchronizing* if there exists
a word $w \in \Sigma^*$ whose action *resets* $\mathscr{A}$, that is, leaves the automaton in one
particular state no matter at which state in Q it started: $\delta(q, w) = \delta(q', w)$ for
all $q, q' \in Q$. Any word w with this property is said to be a *reset* or *synchronizing*
word for the automaton $\mathscr{A}$.

A conjecture proposed by Černý in [2] states that every synchronizing DFA
with n states can be synchronized by a word of length at most $(n-1)^2$. There
have been many attempts to prove it but they all have failed so far. The best
upper bound known up to date is $n(7n^2 + 6n + 16)/48$, see [14]. Černý in [2]
describes an algorithm that checks whether a given DFA with n states and
k letters is synchronizing or not. This algorithm takes $O(n^2k)$ time, i.e. it is
polynomial in n and k. A survey of results concerning synchronizing words can
be found in [15].

The notion of a synchronizing word can be generalized to the case of automata
with a partial transition function (PFA) and to the case of nondeterministic finite
automata (NFA). A *partial finite automaton (PFA)* is a triple $\mathscr{A} = (Q, \Sigma, \delta)$,
where Q is a finite set of states, Σ is a finite alphabet and δ is a partial function

N. Moreira and R. Reis (Eds.): CIAA 2012, LNCS 7381, pp. 278–288, 2012.

from $Q \times \Sigma$ to Q. The function δ can be undefined on some pairs from the set $Q \times \Sigma$. Denote by 2^Q the set of all subsets of Q. The function δ can be naturally extended to $2^Q \times \Sigma^*$ as follows. We put $\delta(q, \lambda) = q$ for every $q \in Q$. Let $q \in Q$, $a \in \Sigma$, $w \in \Sigma^*$. If both states $p = \delta(q, w)$ and $\delta(p, a)$ are defined, then we put $\delta(q, wa) = \delta(p, a)$. If $S \subseteq Q$, $w \in \Sigma^*$ and the values $\delta(q, w)$ are defined for all states $q \in S$, then we put $\delta(S, w) = \{\delta(q, w) | q \in S\}$.

A PFA $\mathscr{A} = (Q, \Sigma, \delta)$ is called *carefully synchronizing*, if there is a word $w \in \Sigma^*$ such that the value $\delta(Q, w)$ is defined and $|\delta(Q, w)| = 1$. We say that such a word w is a *carefully synchronizing word (c.s.w.)* for the automaton $\mathscr{A}$. A c.s.w. for some PFA synchronizes it and does not "break" it in the sense that no undefined transition is ever used.

Clearly, as each DFA is also a PFA, c.s.w. for a DFA is also synchronizing for it. Therefore, thee careful synchronization of PFA is a natural generalization of the synchronization of DFA.

The Černý-type problem can be also considered for PFA. Let $c(n)$ be the maximal length of the shortest c.s.w. among all carefully synchronizing PFA with n states. It follows from [12] and [5] that $\Omega(3^{n/3}) \leq c(n) \leq O(n^2 \cdot 4^{n/3})$. This means that the length of the shortest c.s.w. may not be polynomial in n. This length may not be polynomial in n even for 2-letter PFA, see [11]. Checking whether a given PFA is carefully synchronizing is harder then checking whether a given DFA is synchronizing. Indeed, synchronizablity of a DFA can be checked in polynomial time but checking whether a given PFA is carefully synchronizing is PSPACE-complete. The restriction of this problem to the class of 2-letter PFA is also PSPACE-complete, see [13].

The synchronization of DFA is fast and easy to check, the careful synchronization for an arbitrary PFA is slow and hard to check. We have a problem: where does the precise border between simplicity and hardness lie? In this paper we consider a subclass of the class of all PFA which are very close to DFA, namely, PFA with only one undefined transition. It turns out that already in this simple case checking whether a given PFA is carefully synchronizing is PSPACE-complete. We will also prove the PSPACE-completeness of the same problem for PFA with a binary alphabet.

There is another point of view to the definition of PFA. We can consider an undefined transition not as a forbidden transition, but as a transition which maps a state to some unknown state. Such transition is equivalent in some sense to a nondeterministic mapping which maps the state to the set of all states. Thus, it is natural to consider synchronization for nondeterministic automata.

A *nondeterministic finite automaton (NFA)* is the triple $\mathscr{A} = (Q, \Sigma, \delta)$ such that Q is a finite set of states, Σ is a finite alphabet, and δ is a function from $Q \times \Sigma$ to 2^Q. The function δ can be naturally extended to the set $2^Q \times \Sigma^*$. Let $S \subseteq Q$, $a \in \Sigma$, then we put $\delta(S, a) = \bigcup_{q \in S} \delta(q, a)$. We also put $\delta(S, \lambda) = S$. Let $S \subseteq Q$, $w \in \Sigma^*$, $w = ua$, $a \in \Sigma$ and the set $\delta(S, u)$ is defined, then we put $\delta(S, w) = \delta(\delta(S, u), a)$.

Let $\mathscr{A} = (Q, \Sigma, \delta)$ be an NFA and $w \in \Sigma^*$. The word w is D_1-*directing (D_1-d.w.)* (or D_1-*synchronizing*) if $\delta(q, w) \neq \emptyset$ for all $q \in Q$ and $|\delta(Q, w)| = 1$. The word w is D_2-*directing* (D_2-*d.w.*) if $\delta(q, w) = \delta(Q, w)$ for all $q \in Q$. The word w is D_3-*directing* (D_3-*d.w.*) if $\bigcap_{q \in Q} \delta(q, w) \neq \emptyset$. The NFA $\mathscr{A}$ is called D_1, D_2 or D_3-*directable* if there is a D_1, D_2 or D_3-directing word for it. D_1-directability was first studied in [1]. The definitions in the form presented here were introduced and D_2- and D_3-directability were first studied in [6].

The D_1, D_2 and D_3-directability reduce to the ordinary synchronization when the NFA under consideration is a DFA. The D_1 and D_3-directability reduce to the careful synchronization in the case when an NFA is a PFA. A state z in an automaton is called a *zero state* if every letter of the automaton is defined on the state z and maps z to itself. Directable NFA with a zero state were investigated in [7]. The D_2-directability is a generalization of the careful synchronization for automata with a zero state.

Let $d_1(n)$, $d_2(n)$ and $d_3(n)$ be the maximal lengths of the shortest D_1, D_2 or D_3-directing words for NFA with n states, respectively. The following lower and upper bounds of the length of the shortest directing words were obtained in [5, 8, 9, 12]:

$$2^n - n \leq d_1(n) \leq \Theta(2^n);$$

$$2^n - n - 1 \leq d_2(n) \leq \Theta(2^n);$$

$$\Omega(3^{n/3}) \leq d_3(n) \leq \Theta(n^2 \cdot 4^{n/3}).$$

It is proved in [13] that the problems of checking whether a given NFA is D_1, D_2, or D_3-directable are PSPACE-complete even for 2-letter automata.

For NFA, we can ask the the same question as for PFA: where is the precise border between simplicity and hardness. In this paper we consider a class of NFA with a totally defined transition relation such that there exists only one ambiguous transition which maps one state to a set of two states. In such NFA $\mathscr{A} = (Q, \Sigma, \delta)$ for any $q \in Q, a \in \Sigma$ we have $|\delta(q, a)| \in \{1, 2\}$ and $|\delta(q, a)| = 2$ for only one pair (q, a). The class of such NFA is very close to the class of all DFA, but we prove that the problem of checking the D_1 and D_2-directability are PSPACE-complete for NFA with one ambiguous transition even if only 2-letter automata are considered.

What about D_3-synchronization? The NFA $\mathscr{A} = (Q, \Sigma, \delta)$ is called *complete* if for any state $q \in Q$ and for any letter $a \in \Sigma$ it follows that $\delta(q, a) \neq \emptyset$. The class of complete NFA contains the class of NFA with only one ambiguous transition. Every D_3-directable complete NFA with n states has a D_3-d.w. of length at most $1 + n(n-1)(n-2)/2$, see [6,9]. A polynomial time algorithm that checks whether a given NFA is D_3-directable is described in [6]. That is why the D_3-directability of complete NFA is much simpler than D_1 and D_2-directability.

Let us give some auxiliary notation. Let $w \in \Sigma^*$. Denote by $|w|$ the length of the word w. Let $i, j \in \{1, \ldots, |w|\}$. Let $w[i]$ denote the i-th letter of the word w. Denote the word $w[i]w[i+1] \cdots w[k]$ by $w[i, k]$. For an automaton $\mathscr{A} = (Q, \Sigma, \delta)$ and $q \in Q, a \in \Sigma$, we denote $\delta(q, a)$ by $q.a$ if there is no chance of confusion. We also denote $\delta(S, w)$ by $S.w$ for any subset $S \subseteq Q$ and for any word $w \in \Sigma^*$.

Moreover, we denote singleton $\{q\}$ as q without brackets. In particular, if we have $\delta(q_1, a) = \{q_2\}$ in some NFA, we denote $\delta(q_1, a) = q_2$.

2 Classes of Automata and Computational Problems

The first class that we consider is a class of automata with only one undefined transition. The PFA $\mathscr{A} = (Q, \Sigma, \delta)$ is a PFA *with one undefined transition* if $\delta(q_0, a_0)$ is undefined for some $q_0 \in Q, a_0 \in \Sigma$ and $\delta(q, a)$ is defined for all $(q, a) \in Q \times \Sigma \setminus \{(q_0, a_0)\}$ (this means that $|\delta(q, a)| = 1$). Every PFA is a special case of NFA in the sense that if $\delta(q, a)$ is undefined in PFA, then $\delta(q, a) = \emptyset$ in the corresponding NFA. Therefore, we can define D_1, D_2 or D_3-d.w. for PFA.

Lemma 1. *Let $\mathscr{A} = (Q, \Sigma, \delta)$ be a PFA with the zero state $z \in Q$. Then the set of all c.s.w., the set of all D_1-d.w., the set of all D_2-d.w. and the set of all D_3-d.w. are equal for $\mathscr{A}$.*

Proof. Let w be a c.s.w. for $\mathscr{A}$. In this case for any state $q \in Q$ we have $|q.w| = 1$. Moreover, $z.w = z$. Therefore, $q.w = z \neq \emptyset$. Therefore, $Q.w = z = q.w$ for any $q \in Q$. Moreover, $\bigcap_{q \in Q} q.w = z \neq \emptyset$. Thus, the word w is D_1, D_2 and D_3-directing.

Let w be a D_1-d.w. for $\mathscr{A}$, then $|Q.w| = 1$ and $q.w \neq \emptyset$ for any $q \in Q$. Thus w is a c.s.w.

Let w be a D_2-d.w. for $\mathscr{A}$, then for any state $q \in Q$ we have $q.w = Q.w$. The state z is a zero state. Therefore, $z = z.w = Q.w = q.w \neq \emptyset$ for any state $q \in Q$. Thus w is a c.s.w.

Let w be a D_3-d.w. for $\mathscr{A}$, then $\bigcap_{q \in Q} q.w \neq \emptyset$. Let $q \in Q$. We have, $z.w = z$. Therefore, $z \in q.w$. $\mathscr{A}$ is a PFA. Therefore, $|q.w| \leq 1$. Therefore, $q.w = z$. Thus, w is a c.s.w.

For D_1 and D_2-d.w. a stronger fact has been proved in [7].

Lemma 2. *Let $\mathscr{A} = (Q, \Sigma, \delta)$ be an NFA with a zero state $z \in Q$. Then the set of all D_1-d.w. for $\mathscr{A}$ is equal to the set of all D_2-d.w. for $\mathscr{A}$.*

Let $\mathscr{A} = (Q, \Sigma, \delta)$ be a PFA with one undefined transition $\delta(q, a)$. In previous lemma we forbid to use the mapping $\delta(q, a)$. But we can treat this situation in a different way. We can imagine that the transition $\delta(q, a)$ is not forbidden but maps the state q to an unknown state. Thus, in this case we have an NFA with only one ambiguous value $\delta(q, a)$ which is equal to the set of all states. So this is not a "simplest" nondeterministic case because we can consider an NFA $B = (Q, \Sigma, \delta)$ such that for any pair (q, a) except some (q_0, a_0) it follows $|\delta(q, a)| = 1$, but $|\delta(q_0, a_0))| = 2$. The class of such NFA is very close to the class of all DFA. We call such an automaton an NFA *with one ambiguous transition*.

We study the computational complexity of careful synchronization and directability of automata. Namely we consider the following computational problems.

Problem: CARSYN
Input: A PFA $\mathscr{A} = (Q, \Sigma, \delta)$.
Question: Is the automaton $\mathscr{A}$ carefully synchronizing?

Problem: D1DIR (D2DIR, D3DIR)
Input: An NFA $\mathscr{A} = (Q, \Sigma, \delta)$.
Question: Is the automaton $\mathscr{A}$ D_1 (D_2, D_3)-directable?

Let PROBLEM be any problem in {CARSYN, D1DIR, D2DIR, D3DIR}. If we consider the restriction of PROBLEM to all possible automata over an alphabet of size $\leq k$ for some fixed k, then we call such a problem k-PROBLEM (for example, 2-CARSYN or 3-D1DIR). We denote by PROBLEM ONE UNDEF the restriction of PROBLEM to the class of all automata with one undefined transition. We denote by PROBLEM ONE AMBIG the restriction of PROBLEM to the class of all automata with one ambiguous transition. We use the polynomial-time reducibility $\leq_p$ of problems. The following lemma is trivial. The square brackets in the next lemma denote an optional expression.

Lemma 3. *Let PROBLEM* $\in$ {*CARSYN, D1DIR, D2DIR, D3DIR*} *and $k \geq 2$ be an integer. Then*

1. *PROBLEM ONE UNDEF* $\leq_p$ *PROBLEM;*
2. *[$k-$]PROBLEM ONE UNDEF* $\leq_p$ *[$k-$]PROBLEM;*
3. *PROBLEM ONE AMBIG* $\leq_p$ *PROBLEM;*
4. *[$k-$]PROBLEM ONE AMBIG* $\leq_p$ *[$k-$]PROBLEM;*
5. *2-PROBLEM ONE UNDEF* $\leq_p$ *[$k-$]PROBLEM ONE UNDEF;*
6. *2-PROBLEM ONE AMBIG* $\leq_p$ *[$k-$]PROBLEM ONE AMBIG.*

We show that almost all problems listed above are PSPACE-complete. At first, we prove that these problems belong to PSPACE.

Lemma 4. *Let PROBLEM* $\in$ {*CARSYN, D1DIR, D2DIR, D3DIR*} *and let $k \geq 2$ be an integer. Then all possible problems [$k-$] PROBLEM [ONE UNDEF] and [$k-$] PROBLEM [ONE AMBIG] belong to PSPACE.*

Proof. It is already proved in [13] that the problems CARSYN, D1DIR, D2DIR, D3DIR and the corresponding k-problems for $k \geq 2$ are PSPACE-complete. Therefore, from Lemma 3 we obtain that all considered problems are in PSPACE. The proof of this fact is very simple. It is a consequence of Savitch's theorem (which states that PSPACE=NPSPACE) because any carefully synchronizing or directing word can be nondeterministically applied to a given PFA or NFA using $O(n^2)$ bits of memory where n is the number of states.

In Section 3 we prove that all ONE UNDEF problems are PSPACE-complete. In Section 4 we prove that problems D1DIR and D2DIR ONE AMBIG are PSPACE-complete. In Section 5 we reduce all problems to 2-letter alphabet and prove that all considered 2-problems (except of D3DIR ONE AMBIG) are PSPACE-complete.

3   Automata with One Undefined Transition

Theorem 1. *The problems CARSYN(D1DIR, D2DIR, D3DIR) ONE UNDEF are PSPACE-complete.*

Proof. From Lemma 4 we have that the problems are in PSPACE. To prove the PSPACE-hardness, we use the reduction of the classical PSPACE-complete problem FINITE AUTOMATA INTERSECTION to our problems. We reduce the instance of initial problem to an instance of problem CARSYN ONE UNDEF. The input of this instance shall be a PFA $\mathscr{B}$ with a zero state. Therefore, due to Lemma 1 this instance shall be also a required instance for the problems D1DIR, D2DIR and D3DIR ONE UNDEF.

We use the PSPACE-complete problem FINITE AUTOMATA INTERSECTION (see [10] and [4]). We consider deterministic finite automata of the form $\mathscr{A} = (Q, \Sigma, \delta, s, F)$ as *recognizers*, where Q is a set of states, Σ is an alphabet, δ is a totally-defined transition function, $s \in Q$ is an initial state and $F \subseteq Q$ is a set of final states. Let w be a word in Σ^*. The automaton $\mathscr{A}$ *accepts* the word w if and only if $\delta(s, w) \in F$.

Problem: FINITE AUTOMATA INTERSECTION
Input: The number $k \geq 2$ and the recognizers $\mathscr{A}_0 = (Q_0, \Sigma, \delta_0, s_0, F_0), \ldots$ $\mathscr{A}_{k-1} = (Q_{k-1}, \Sigma, \delta_{k-1}, s_{k-1}, F_{k-1})$.
Question: Is there a word $w \in \Sigma^*$ such that $\delta_0(s_0, w) \in F_0, \ldots, \delta_{k-1}(s_{k-1}, w) \in F_{k-1}$?

We reduce this problem to the problem CARSYN ONE UNDEF. Let recognizers $\mathscr{A}_0 = (Q_0, \Sigma', \delta_0, s_0, F_0), \ldots \ldots, \mathscr{A}_{k-1} = (Q_{k-1}, \Sigma', \delta_{k-1}, s_{k-1}, F_{k-1})$ be an input of the problem FINITE AUTOMATA INTERSECTION. We construct a PFA with one undefined transition $\mathscr{B} = (Q, \Sigma, \delta)$ such that there is a c.s.w. for the automaton $\mathscr{B}$ if and only if the recognizers $\mathscr{A}_0, \ldots, \mathscr{A}_{k-1}$ have a common accepting word.

Let $\Sigma = \Sigma' \cup \{b, c\}$, $\Sigma' \cap \{b, c\} = \emptyset$ and $Q = Q_0 \cup \cdots \cup Q_{k-1} \cup T \cup U$, where $T = \{t_0, \ldots, t_{k-1}\}$ and $U = \{u_{-k}, u_{-k+1}, \ldots u_{k-2}, u_{k-1}\}$ and the sets $Q_0, Q_1, \ldots, Q_{k-1}, T$, and U are pairwise disjoint. Let us define the function δ. If $q \in Q_i$ for some $i \in \{0, \ldots, k-1\}$ and $a \in \Sigma'$, then

$$\delta(q, a) = \delta_i(q, a), \delta(q, b) = s_i, \delta(q, c) = \begin{cases} u_i, & \text{if } q \in F_i; \\ t_i, & \text{if } q \in Q_i \setminus F_i. \end{cases}$$

Let $i \in \{0, \ldots, k-1\}$ and $a \in \Sigma'$, then

$$\delta(t_i, a) = t_i, \delta(t_i, b) = s_i, \delta(t_i, c) = \begin{cases} t_{i-1}, & \text{if } i > 0; \\ \text{undefined}, & \text{if } i = 0. \end{cases}$$

Let $i \in \{-k, \ldots, k-1\}$ and $a \in \Sigma'$, then

$$\delta(u_i, a) = u_i, \delta(u_i, b) = \begin{cases} s_{i \bmod k}, & \text{if } i > -k; \\ u_{-k}, & \text{if } i = -k. \end{cases}, \delta(u_i, c) = \begin{cases} u_{i-1}, & \text{if } i > -k; \\ u_{-k}, & \text{if } i = -k. \end{cases}$$

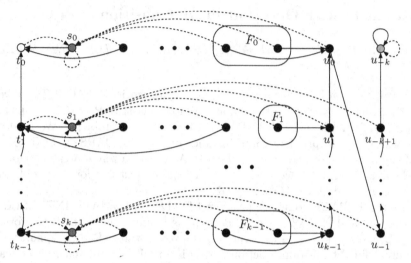

Fig. 1. Automaton $\mathscr{B}$

Only the transition $\delta(t_0, c)$ is undefined in the PFA $\mathscr{B}$. The state u_{-k} is a zero state. Figure 1 represents the automaton $\mathscr{B}$. Solid and dotted lines stand for the action of c and b, respectively. The actions of letters from Σ' are not drawn.

Let $v \in \Sigma'^*$ be a word accepted by all the recognizers $\mathscr{A}_0, \ldots, \mathscr{A}_{k-1}$. Let us prove that the word bvc^{2k} is a c.s.w. for the PFA $\mathscr{B}$. Indeed, $\delta(Q, b)$ is defined and $\delta(Q, b) = \{s_0, \ldots, s_{k-1}, u_{-k}\}$. All recognizers $\mathscr{A}_0, \ldots, \mathscr{A}_{k-1}$ accept the word $v \in \Sigma'^*$. Hence, $\delta_0(s_0, v) \in F_0, \ldots, \delta_{k-1}(s_{k-1}, v) \in F_{k-1}$. Therefore, the function $\delta(\bullet, v)$ is defined on the set $\{s_1, \ldots, s_k\}$ and $\delta(\{s_0, \ldots, s_{k-1}\}, v) \subseteq F_0 \cup \ldots \cup F_{k-1}$. Hence $\delta(Q, bvc) \subseteq \{u_{-1}, u_0, \ldots, u_{k-1}\}$. Therefore, $\delta(Q, bvc^{2k}) = u_{-k}$. Thus the word bvc^{2k} is a c.s.w. for the PFA $\mathscr{B}$.

Otherwise, let $\mathscr{B}$ be carefully synchronizing and let w be the shortest c.s.w. for $\mathscr{B}$. Note that $w[1] \neq c$ because $Q.c$ is not defined. Let $w' \in \Sigma'^*$. The value $Q.w'c$ is also undefined. Furthermore, $Q.w'b = \{s_0, \cdots, s_{k-1}, u_{-k}\} = Q.b$ and the word w cannot start with w'. Therefore, $w[1] = b$. We have $Q.w[1] = Q.b = \{s_0, \cdots, s_{k-1}, u_{-k}\}$. For every $i \in \{0, \ldots, k-1\}$ the set $Q.b$ contains exactly one state from the set Q_i. For every state from Q_i, each letter from Σ' moves it to some state from Q_i. Therefore, for every $i \in \{0, \ldots, k-1\}$ and for every word $v \in \Sigma'^*$, we have $|Q.bv \cap Q_i| = 1$. Moreover, $s_i.vb \subseteq Q_i.vb = s_i$. Hence $Q.bvb = Q.b$, i.e. the word bvb cannot be a prefix of the shortest c.s.w. Therefore, the word w starts with bvc for some $v \in \Sigma'^*$.

The letters from Σ' do not move states from the sets T and U. Therefore, there is no reason to use letters from Σ' after bvc before using the letter b. Note that $s_i.vc \in \{t_i, u_i\}$. Hence, for every $y \in \{0, \ldots, k-1\}$, if the state $s_i.vc^{y+1}$ exists, then it belongs to $\{t_{i-y}, u_{i-y}\}$. If the set $Q.bvc^{y+1}$ exists, then it contains the state u_{-k} and exactly one state from every pair $\{t_{i-y}, u_{i-y}\}$ for $i \in \{0, \ldots, k-1\}$. We have $t_{i-y}.b = u_{i-y}.b = s_{(i-y) \bmod k}$. Therefore, $Q.bvc^{y+1}b = Q.b$. Hence, the word w cannot start with $bvc^{y+1}b$ for $y < k$. Therefore, the word w starts with bvc^{k+1}.

If there exists $i \in \{0, \ldots, k-1\}$ such that $s_i.vc = t_i$, then the state $s_i.vc^{k+1}$ is undefined and we obtain a contradiction. Hence $s_i.vc = u_i$ for any i. Hence, for any $i \in \{0, \ldots, k-1\}$ the word v maps the state s_i to a state from F_i. Thus, v is a common accepting word for the recognizers $\mathscr{A}_1, \ldots, \mathscr{A}_k$.

Therefore, the PFA $\mathscr{B}$ is carefully synchronizing if and only if there exists a common accepting word for $\mathscr{A}_0, \ldots, \mathscr{A}_{k-1}$. Now the statement of the theorem follows from Lemma 1, because every shortest c.s.w. is the shortest $D_1(D_2, D_3)$-d.w. for PFA with a zero state.

4   Automata with One Ambiguous Transition

Theorem 2. *The problems D1DIR ONE AMBIG and D2DIR ONE AMBIG are PSPACE-complete.*

Proof. From Lemma 4 we have that the problems are in PSPACE. To prove the PSPACE-hardness we use the problem FINITE AUTOMATA INTERSECTION.

We reduce every instance of this problem to an instance of the problem D1DIR. The input of the latter instance will be an NFA $\mathscr{C}$ with a zero state. Let $k \geq 2$ and recognizers $\mathscr{A}_0 = (Q_0, \Sigma', \delta_0, s_0, F_0), \ldots,$ $\mathscr{A}_{k-1} = (Q_{k-1}, \Sigma', \delta_{k-1}, s_{k-1}, F_{k-1})$ be an input of the problem FINITE AUTOMATA INTERSECTION. We construct an NFA $\mathscr{C} = (Q, \Sigma, \delta)$ with one ambiguous transition such that there is a D_1-d.w. for the automaton $\mathscr{C}$ if and only if the recognizers $\mathscr{A}_0, \ldots, \mathscr{A}_{k-1}$ have a common accepting word.

Let $\Sigma = \Sigma' \cup \{b, c\}$, $\Sigma' \cap \{b, c\} = \emptyset$ and $Q = Q_0 \cup \cdots \cup Q_{k-1} \cup T \cup U$, where $T = \{t_0, \ldots, t_{k-1}\}$ and $U = \{u_{-2k+1}, u_{-2k+2}, \ldots u_{k-2}, u_{k-1}\}$ and the sets $Q_0, Q_1, \ldots, Q_{k-1}, T$, and U are pairwise disjoint.

Let us define the function δ.

If $q \in Q_i$ for some $i \in \{0, \ldots, k-1\}$ and $a \in \Sigma'$, then

$$\delta(q, a) = \delta_i(q, a), \delta(q, b) = s_i, \delta(q, c) = \begin{cases} u_i, & \text{if } q \in F_i; \\ t_i, & \text{if } q \in Q_i \setminus F_i. \end{cases}$$

Let $i \in \{0, \ldots, k-1\}$ and $a \in \Sigma'$, then

$$\delta(t_i, a) = t_i, \delta(t_i, b) = s_i, \delta(t_i, c) = \begin{cases} t_{i-1}, & \text{if } i > 0; \\ \{t_0, t_{k-1}\}, & \text{if } i = 0. \end{cases}$$

Let $i \in \{-2k+1, \ldots, k-1\}$ and $a \in \Sigma'$, then $\delta(u_i, a) = u_i$,

$$\delta(u_i, b) = \begin{cases} s_{i \bmod k}, & \text{if } i > -2k+1; \\ u_{-2k+1}, & \text{if } i = -2k+1. \end{cases}, \delta(u_i, c) = \begin{cases} u_{i-1}, & \text{if } i > -2k+1; \\ u_{-2k+1}, & \text{if } i = -2k+1. \end{cases}$$

Only the transition $\delta(t_0, c)$ in this NFA is ambiguous. The state u_{-2k+1} is a zero state. The construction of the NFA $\mathscr{C}$ is very similar to the construction of the PFA $\mathscr{B}$ from Theorem 1. The differences are that the transition $t_0.c$ is defined but it is ambiguous and the number of states in the set U is $3k - 1$ instead of

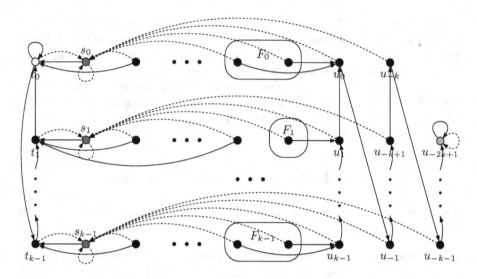

Fig. 2. Automaton $\mathscr{C}$

$2k$. Figure 2 represents the automaton $\mathscr{C}$, where solid and dotted lines stand for the action of respectively c and b.

Similarly to the proof of Theorem 1, we obtain that if $v \in \Sigma'^*$ is a word accepted by all the recognizers $\mathscr{A}_0, \dots, \mathscr{A}_{k-1}$, then the word bvc^{3k-1} is a D_1-d.w. for the NFA $\mathscr{C}$. Otherwise, let w be the shortest D_1-d.w. for $\mathscr{C}$. It can be proved that the word w starts with bvc^{2k} for some $v \in \Sigma'^*$. The proof is very similar to the proof of the same fact for the word bvc^{k+1} in Theorem 1.

Aiming at a contradiction, suppose that there exists $i \in \{0, \dots, k-1\}$ such that $s_i.v \notin F_i$ and $s_i.vc = t_i$. In this case, $t_0 \in s_i.vc^k$. Therefore, $s_i.vc^{2k-1} \supseteq t_0.c^{k-1} = T$. We have $T.a = T$ for $a \neq b$ and $u_{-2k+1} \notin T$. Hence we need at least one letter b after the word bvc^{k+1} to obtain the D_1-d.w. But $(T \cup u_{-2k+1}).b = \{s_0, \dots, s_{k-1}, u_{-2k+1}\} = Q.b$. We obtain a contradiction with the minimality of w. Therefore, $s_i.v \in F_i$ for any $i \in \{0, \dots, k-1\}$. Thus, v is a common accepting word for recognizers $\mathscr{A}_0, \dots, \mathscr{A}_{k-1}$.

Now the statement of the theorem follows from Lemma 2, because every D_1-d.w. is the D_2-d.w. for NFA with a zero state. The theorem is proved.

5   Automata with Binary Alphabet

The automata $\mathscr{B}$ and $\mathscr{C}$ from the proof of Theorem 1 and 2 can have arbitrary number of letters. What about automata with binary alphabet?

Lemma 5. *Let $\mathscr{A}$ be a PFA with one undefined transition (NFA with one ambiguous transition) with n states and m letters. Then there exists a 2-letter PFA with one undefined transition (NFA with one ambiguous transition) $F_2(\mathscr{A})$ with nm states such that the automaton $\mathscr{A}$ is carefully synchronizing (D_1, D_2, D_3-directable) iff the automaton $F_2(\mathscr{A})$ is carefully synchronizing (D_1, D_2, D_3-directable).*

Proof. If $\mathscr{A} = \{Q, \Sigma, \delta\}$ is a PFA with one undefined transition (NFA with one ambiguous transition), $Q = \{q_0, \ldots, q_{n-1}\}$ $(n > 1)$ and $\Sigma = \{a_0, \ldots, a_{m-1}\}$ $(m > 2)$ we can suppose without loss of generality that $|\delta(q_i, a_j)| = 1$ for each $j > 0$. Define a reduction F_2 as follows: $F_2(\mathscr{A}) = (Q', \{x, y\}, \gamma)$ where $Q' = Q \times \Sigma$ and $\gamma((q_i, a_j), x) = (q_i, a_{\min\{j+1, m-1\}})$ and $\gamma((q_i, a_j), y) = \delta(q_i, a_j) \times \{a_0\}$. Then obviously $F_2(\mathscr{A})$ is a PFA with one undefined transition (NFA with one ambiguous transition) too.

Define $f : \{x, y\}^* y \to \Sigma^+$ recursively: $f(x^i y) = a_{\min\{i, m-1\}}$, and $f(vyx^i y) = f(vy) f(x^i y)$ for $v \in \{x, y\}^*$. It can be easy proved that for any $q \in Q, a \in \Sigma$ and $\gamma((q, a_0), f^{-1}(a)) = (\delta(q, a), a_0)$.

Let $w \in \Sigma^*$ be a c.s.w.(D_1, D_2 or D_3-d.w.) for $\mathscr{A}$ and let $w' \in f^{-1}(w)$. We have by our assumption $|\delta(q_i, a_j)| = 1$ for each $j > 0$, particularly for $j = 1$. Hence, $|q.xy| = 1$ for each $q \in Q'$ and $Q'.xy \subseteq Q \times \{a_0\}$. Hence, $\gamma(Q', xyw') \subseteq \gamma(Q \times \{a_0\}, w') = (\delta(Q, w) \times \{a_0\})$. Therefore xyw' is a c.s.w.(D_1, D_2 or D_3-d.w.) for $F_2(\mathscr{A})$.

Let u be a c.s.w. (D_1, D_2 or D_3-d.w.) for $F_2(\mathscr{A})$. In this case, $|\gamma(Q', uxy)| = 1$ and $\gamma(Q', uxy) = (\delta(Q, f(uxy)), a_0)$. Therefore, the word $f(uxy)$ is a c.s.w. (D_1, D_2 or D_3-d.w.) for $\mathscr{A}$.

Theorem 3. *1. The problems 2-CARSYN (D1DIR,D2DIR,D3DIR) ONE UNDEF are PSPACE-complete;*
2. The problems 2-D1DIR(D2DIR) ONE AMBIG are PSPACE-complete.

Proof. We consider the problem 2-CARSYN ONE UNDEF. Let us use Lemma 5 to automata $\mathscr{B} = (Q, \Sigma, \delta)$ from Theorem 1. We obtain $F_2(\mathscr{B})$ is a PFA with two letters and one undefined transition. We also obtain that $\mathscr{B}$ is carefully synchronizing iff $F_2(\mathscr{B})$ is carefully synchronizing. The PFA $\mathscr{B}$ is an input of an instance of CARSYN ONE UNDEF, the PFA $F_2(\mathscr{B})$ is an input of an instance of 2-CARSYN ONE UNDEF. The PFA $F_2(\mathscr{B})$ has $|\Sigma| \cdot |Q|$ states. Thus, we have a polynomial in time reduction. Therefore, the problem 2-CARSYN ONE UNDEF is PSPACE-complete. The PSPACE-completeness of other problems can be proved similarly using Theorems 1 and 2 and Lemma 5.

Acknowledgement. The author acknowledges support from the Presidential Programm for young researchers, grant MK-266.2012.1.

References

1. Burkhard, H.V.: Zum Längenproblem homogener Experimente an determinierten und nicht-deterministischen Automaten. Elektronische Informationsverarbeitung und Kybernetik 12, 301–306 (1976)
2. Černý, J.: Poznámka k homogénnym experimentom s konečnými automatmi. Mat.-Fyz. Cas. Slovensk. Akad. Vied. 14, 208–216 (1964) (in Slovak)
3. Eppstein, D.: Reset sequences for monotonic automata. SIAM J. Comput. 19, 500–510 (1990)
4. Garey, M.R., Johnson, D.S.: Computers and Intractability: A Guide to the Theory of NP-Completeness. W. H. Freeman (1979)

5. Gazdag, Z., Ivan, S., Nagy-Gyorgy, J.: Improved upper bounds on synchronizing nondeterministic automata. Information Processing Letters 109(17), 986–990 (2009)
6. Imreh, B., Steinby, M.: Directable nondeterministic automata. Acta Cybernetica 14, 105–115 (1999)
7. Imreh, B., Imreh, C., Ito, M.: On directable nondeterministic trapped automata. Acta Cybernetica 16, 37–45 (2003)
8. Ito, M., Shikishima-Tsuji, K.: Some Results on Directable Automata. In: Karhumäki, J., Maurer, H., Păun, G., Rozenberg, G. (eds.) Theory Is Forever (Salomaa Festschrift). LNCS, vol. 3113, pp. 125–133. Springer, Heidelberg (2004)
9. Ito, M.: Algebraic Theory of Automata and Languages. World Scientific, Singapore (2004)
10. Kozen, D.: Lower bounds for natural proof systems. In: Proceedings of the 18th Annual Symposium on Foundations of Computer Science, pp. 254–266 (1977)
11. Martyugin, P.: Lower bounds for the length of the shortest carefully synchronizing words for two- and three-letter partial automata. Diskretn. Anal. Issled. Oper. 15(4), 44–56 (2008)
12. Martyugin, P.V.: A Lower Bound for the Length of the Shortest Carefully Synchronizing Words. Russian Mathematics (Iz. VUZ) 54(1), 46–54 (2010)
13. Martyugin, P.V.: Complexity of Problems Concerning Carefully Synchronizing Words for PFA and Directing Words for NFA. In: Ablayev, F., Mayr, E.W. (eds.) CSR 2010. LNCS, vol. 6072, pp. 288–302. Springer, Heidelberg (2010)
14. Trahtman, A.N.: Modifying the Upper Bound on the Length of Minimal Synchronizing Word. In: Owe, O., Steffen, M., Telle, J.A. (eds.) FCT 2011. LNCS, vol. 6914, pp. 173–180. Springer, Heidelberg (2011)
15. Volkov, M.V.: Synchronizing Automata and the Černý Conjecture. In: Martín-Vide, C., Otto, F., Fernau, H. (eds.) LATA 2008. LNCS, vol. 5196, pp. 11–27. Springer, Heidelberg (2008)

Restarting Tiling Automata[*]

Daniel Průša[1] and František Mráz[2]

[1] Czech Technical University, Faculty of Electrical Engineering
Karlovo náměstí 13, 121 35 Prague 2, Czech Republic
prusapa1@cmp.felk.cvut.cz
[2] Charles University, Faculty of Mathematics and Physics
Malostranské nám. 25, 118 25 Prague 1, Czech Republic
frantisek.mraz@mff.cuni.cz

Abstract. We present a new model of a two-dimensional computing device called restarting tiling automaton. The automaton defines a set of tile-rewriting, weight-reducing rules and a scanning strategy by which a tile to rewrite is being searched. We investigate properties of the induced families of picture languages. Special attention is paid to picture languages that can be accepted independently of the scanning strategy. We show that this family strictly includes REC and exhibits similar closure properties. Moreover, we prove that its intersection with the set of one-row languages coincides with the regular languages.

Keywords: two-dimensional languages, tiling systems, restarting automata.

1 Introduction

Recently there were developed several theoretical models for picture languages, which simultaneously can describe many "naturally" specified picture languages and have some desirable theoretical properties like closure properties under several regular operations [5].

A prominent place belongs to the class of recognizable picture languages (REC), which is based on tiling systems. Several automata models can accept REC. Among others online tessellation automata (2OTA) based on cellular automata, tiling automata (TA) and Wang automata (WTA).

Anselmo, Giammarresi and Madonia [1] introduced tiling automata (TA) as a generalization of one-dimensional automata into 2D, which in general has the same descriptional power as 2OTA and recognize exactly the class of picture languages REC. The idea is to equip a tiling system with a scanning strategy – an order in which all symbols of an input picture are scanned with several properties: each symbol of the input picture is visited exactly once, and at the time a position is visited, all the positions needed for computing the resulting

[*] The authors were supported by the Grant Agency of the Czech Republic: the first author under the project P103/10/0783 and the second author under the projects P103/10/0783 and P202/10/1333.

N. Moreira and R. Reis (Eds.): CIAA 2012, LNCS 7381, pp. 289–300, 2012.

symbol (according to the rules of the given tiling system) have been already visited.

Such a scanning strategy enables them to define deterministic TA. Nevertheless, deterministic TAs are weaker than their nondeterministic variant and they recognize a class of picture languages incomparable by inclusion with the class of languages accepted by the 4-way finite automata (4FA) [3].

Wang automata (WTA) introduced by Lonati and Pradella [7] are based on Wang tiling systems, which use tiles (square pictures of size 2×2) with colors on their edges. Wang automata use also a kind of scanning strategy. In contrast to [1], the authors limit the scanning strategies to "polite" strategies, which do not enable the automaton "to jump" between not adjacent positions. Moreover, the next visited position should be computable locally from the symbol at the current position and its neighborhood without knowing the size of the scanned picture. Once again, the nondeterministic Wang automata recognize REC and their deterministic variant is weaker.

Here we introduce a new automata model called two-dimensional restarting tiling automaton (2RTA), which is a generalization of tiling automata [1] and simultaneously borrows stepwise reductions from the so-called restarting automata [9]. A 2RTA can rewrite pictures according to a given finite set of tile-rewriting rules. Each rewriting changes exactly one symbol and it is required that the symbol is replaced by another symbol with lower weight. One rewriting consists of scanning the current picture according to the given scanning strategy until a place (tile) is found which can be rewritten according to some of the rules of the automaton. Then, one of the possible rewriting rules is applied and the automaton restarts its computation. The automaton repeats the rewriting until it cannot rewrite anymore. If the resulting picture belongs to a given local language, the automaton accepts, otherwise it rejects. In contrast to a restarting automaton, a 2RTA has no states. It is easy to see that 2RTAs are at least as strong as tiling systems.

Barták [2] studied a version of 2RTA (called two-dimensional restarting automaton) with a fixed scanning strategy, which scans the picture row by row from the top to the bottom and left-to-right in each row.

The structure of the paper is the following. Section 2 recalls basic notions from the theory of picture languages. Next, in Section 3 we introduce scanning strategies, 2RTAs and show several closure properties for the class of languages accepted by 2RTAs with given scanning strategy. Moreover, we introduce a new class of languages which can be accepted by 2RTAs equipped with any strategy – strategy independent – si-2RTL. Further we show that si-2RTL is closed under intersection, union, mirroring and projection. Section 4 presents a limited version of two-dimensional Turing machines, which are used to show that si-2RTL is a proper superset of REC and that the class of picture languages accepted by 4FA is included into deterministic si-2RTL. Section 5 shows that si-2RTL restricted to one-dimensional pictures equals to the regular (string) languages. In Section 6, we conclude by summarizing why it is important to study the new classes of automata and picture languages.

2   Basic Definitions

This section gives basic definitions on two-dimensional (2D) objects, languages and 2D specific operations.

Definition 1. *A* picture *over a finite alphabet Σ is a two-dimensional matrix whose elements are from Σ. The set of all pictures over Σ is denoted by $\Sigma^{*,*}$. A* picture language *over Σ is a subset of $\Sigma^{*,*}$.*

Let P be a picture over Σ. We denote the number of rows and columns of P by rows(P) and cols(P), respectively. The pair $(\text{rows}(P), \text{cols}(P))$ is called the *size* of P. The *empty picture* Λ is defined as the only picture of size $(0,0)$. The set of all pictures of size (m,n) over Σ is denoted by $\Sigma^{m,n}$. Assuming $1 \leq i \leq \text{rows}(P)$ and $1 \leq j \leq \text{cols}(P)$, $P(i,j)$ (or shortly $P_{i,j}$) identifies the symbol located in the i-th row and the j-th column in P.

Two (partial) binary operations are used to concatenate pictures. Let P and Q be pictures over Σ of sizes (k,l) and (m,n), respectively. The *column concatenation* $P \oplus Q$ is defined iff $k = m$, the *row concatenation* $P \ominus Q$ is defined iff $l = n$. The products are specified by the following schemes:

$$P \oplus Q = \begin{matrix} P_{1,1} & \cdots & P_{1,l} & Q_{1,1} & \cdots & Q_{1,n} \\ \vdots & \ddots & \vdots & \vdots & \ddots & \vdots \\ P_{k,1} & \cdots & P_{k,l} & Q_{m,1} & \cdots & Q_{m,n} \end{matrix} \qquad P \ominus Q = \begin{matrix} P_{1,1} & \cdots & P_{1,l} \\ \vdots & \ddots & \vdots \\ P_{k,1} & \cdots & P_{k,l} \\ Q_{1,1} & \cdots & Q_{1,n} \\ \vdots & \ddots & \vdots \\ Q_{m,1} & \cdots & Q_{m,n} \end{matrix}$$

We also define $\Lambda \ominus P = P \ominus \Lambda = \Lambda \oplus P = P \oplus \Lambda = P$ for any picture P.

In addition, we introduce the *clockwise rotation* P^{R}, *vertical mirroring* P^{VM} and *horizontal mirroring* P^{HM}.

$$P^{\mathrm{R}} = \begin{matrix} P_{m,1} & \cdots & P_{1,1} \\ \vdots & \ddots & \vdots \\ P_{m,n} & \cdots & P_{1,n} \end{matrix} \qquad P^{\mathrm{VM}} = \begin{matrix} P_{1,n} & \cdots & P_{1,1} \\ \vdots & \ddots & \vdots \\ P_{m,n} & \cdots & P_{m,1} \end{matrix} \qquad P^{\mathrm{HM}} = \begin{matrix} P_{m,1} & \cdots & P_{m,n} \\ \vdots & \ddots & \vdots \\ P_{1,1} & \cdots & P_{1,n} \end{matrix}$$

Let $\pi : \Sigma \to \Gamma$ be a mapping between two alphabets. The projection by π of $P \in \Sigma^{m,n}$ is the picture $P' \in \Gamma^{m,n}$ such that $P'(i,j) = \pi(P(i,j))$ for all $1 \leq i \leq m$, $1 \leq j \leq n$. Note that each of the above introduced operations can be naturally extended to languages.

Let $\mathcal{S} = \{\vdash, \dashv, \top, \bot, \#\}$ be a set of special markers (*sentinels*). In the text we always implicitly assume that $\Sigma \cap \mathcal{S} = \emptyset$ for any alphabet Σ. For $P \in \Sigma^{m,n}$, we define a *boundary picture* $\widehat{P}$ over $\Sigma \cup \mathcal{S}$ of size $(m+2, n+2)$. The content is given by the following scheme.

#	⊤ ⊤	⋯	⊤ ⊤	#

$$\begin{array}{|c|c c c c c|c|} \hline \# & \top\ \top & \cdots & \top\ \top & \# \\ \hline \vdash & & & & \dashv \\ \vdots & & P & & \vdots \\ \vdash & & & & \dashv \\ \hline \# & \bot\ \bot & \cdots & \bot\ \bot & \# \\ \hline \end{array}$$

Usually, only # is used to distinguish the border. Our version simplifies the definition of bounded computations, keeping the recognition abilities unchanged.

3 Two-Dimensional Restarting Tiling Automata

Our model of a two-dimensional restarting automaton is based on the following two concepts. The first one is the tiling automaton [1], which nondeterministically guesses an inverse image of an input picture according to rewriting rules of a given tiling system and checks whether the inverse image belongs to a given local language. The second one is the (one-dimensional) restarting automaton (namely shrinking restarting automaton [6]) that iteratively simplifies the input string until either a correct "simple" string is obtained and the input is accepted or until the simplification gets stuck, in which case the input is rejected. During each simplification, the automaton scans the current string, performs a single rewriting operation and restarts. Each rewriting must be shrinking, i.e., weight-decreasing. A variant of the following model has already been studied by Barták in [2]. His model called two-dimensional restarting automaton used a fixed scanning strategy.

A tiling automaton [1] is a tiling system equipped with a scanning strategy, which determines the order in which all the positions of an input picture are visited. A scanning strategy must satisfy several requirements: each position must be visited exactly once, when a position is scanned, all three neighbors of the position needed for computing inverse image of the respective tile must have been already visited (and stored in a suitable data structure).

Cherubini and Pradella [4] added some more requirements to the scanning strategy and later they studied such scanning strategies together with another model of automata which use tiles with colors on their border, the so-called Wang tiles [7,8]. Nevertheless, but we will follow the more general approach based on [1]. A *scanning strategy* is represented by a single starting position in the picture, which will be one of its corners, and a partial computable function $f : \mathbb{N}^4 \to \mathbb{N}^2$ where $f(i, j, m, n)$ is the next scanned position after scanning the position (i, j) in a picture of size (m, n), $\mathbb{N}$ denotes the set of all positive integers. The starting position will be represented simply as an integer $c_s \in \{1, 2, 3, 4\}$ with the meaning 1 for the top-left corner, 2 for the top-right corner, 3 for the bottom-right corner and 4 for the bottom-left corner of a picture. In contrast to [1] and [4], we will not consider any additional restrictions on the scanning strategy only that if (i_0, j_0) is a starting position on a picture P of size (m, n) (for any $m, n > 0$), then the sequence of positions visited according to this strategy $(i_0, j_0), (i_1, j_1), \ldots, (i_{(m+1)(n+1)-1}, j_{(m+1)(n+1)-1})$ is a permutation of the set of positions of all tiles of size $(2,2)$ in the extended picture $\widehat{P}$. A scanning strategy will be represented as a pair $\nu = (c_s, f)$.

Barták [2] used the following fixed scanning strategy $\nu_{\text{row}} = (1, f_{\text{row}})$ scanning a picture row by row from left to right with the starting position $(1,1)$ and the next position function

$$f_{\text{row}}(i,j,m,n) = \begin{cases} (i, j+1) & \text{if } j < n+1; \\ (i+1, 1) & \text{if } j = n+1, i < m. \end{cases}$$

Definition 2. *A* two-dimensional restarting tiling automaton, *referred to as* 2RTA, *is a 6-tuple* $\mathcal{M} = (\Sigma, \Gamma, \Theta_f, \delta, \nu, \mu)$, *where* Σ *is a finite input alphabet,* Γ *is a finite working alphabet* $(\Gamma \supseteq \Sigma)$, $\Theta_f \subseteq (\Gamma \cup \mathcal{S})^{2,2}$ *is a set of accepting tiles,* $\nu = (c_s, f)$ *is a scanning strategy,* $\mu : \Gamma \to \mathbb{N}$ *is a weight function and* $\delta \subseteq \{(U \to V) \mid U, V \in (\Gamma \cup \mathcal{S})^{2,2}\}$ *is a set of rewriting rules satisfying the condition that in every rule* $u \to v$ *only a single position of* u, *containing a symbol* a *from* Γ, *is changed into some* $b \in \Gamma$ *such that* $\mu(b) < \mu(a)$.

Symbols from $\Gamma \setminus \Sigma$ are called auxiliary symbols and must not be contained in any input word. On a given input picture $P \in \Sigma^{m,n}$, the automaton $\mathcal{M}$ works in phases. Each phase starts in a starting position (i_0, j_0) corresponding to the corner c_s of the boundary picture $\widehat{P}$, that is,

- $(1,1)$ for $c_s = 1$,
- $(1, n+1)$ for $c_s = 2$,
- $(m+1, n+1)$ for $c_s = 3$, and
- $(m+1, 1)$ for $c_s = 4$.

Then $\mathcal{M}$ scans the current picture using a window of size $(2,2)$ which is moved according to the scanning strategy ν. Let $\nu(r, m, n)$ denote the sequence of the first r positions of a picture of size (m,n) visited according to the scanning strategy ν. When $\mathcal{M}$ finds a tile for which a rewriting rule is defined it performs one of the possible rewritings and finishes the phase by restart (i.e., it goes to the to the corner indicated by c_s again). When no rewriting rule can be executed for the whole picture, the automaton verifies whether the picture belongs to $L(\Theta_f)$ – the local language defined by Θ_f – and if so, it accepts. Formally:

Definition 3. *Let* $\mathcal{M} = (\Sigma, \Gamma, \Theta_f, \delta, \nu, \mu)$ *be a two-dimensional restarting tiling automaton,* P_1, P_2 *be two pictures over the alphabet* Γ *of the same size* (m,n) *and* $\nu((m+1)(n+1), m, n) = (i_0, j_0), \ldots, (i_{(m+1)(n+1)-1}, j_{(m+1)(n+1)-1})$. *We say that the picture* P_1 *can be* directly reduced *to picture* P_2, *denoted by* $P_1 \vdash_{\mathcal{M}} P_2$, *if there exists an integer* s, $0 \leq s < (m+1)(n+1)$, *such that* $\widehat{P_1}(k,l) = \widehat{P_2}(k,l)$ *for all pairs of the indices* k, l, *where* $1 \leq k \leq \text{rows}(\widehat{P_1})$, $1 \leq l \leq \text{cols}(\widehat{P_1})$ *except the pairs* (i_s, j_s), $(i_s, j_s + 1)$, $(i_s + 1, j_s)$, $(i_s + 1, j_s + 1)$ *and there exists a rule*

$$\begin{array}{|c|c|}\hline P_1(i_s, j_s) & P_1(i_s, j_s+1) \\\hline P_1(i_s+1, j_s) & P_1(i_s+1, j_s+1) \\\hline\end{array} \rightarrow \begin{array}{|c|c|}\hline P_2(i_s, j_s) & P_2(i_s, j_s+1) \\\hline P_2(i_s+1, j_s) & P_2(i_s+1, j_s+1) \\\hline\end{array} \quad \text{in } \delta.$$

Moreover, there is no rule in δ that could be applied to any tile

$P_1(i_r, j_r)$	$P_1(i_r, j_r+1)$
$P_1(i_r+1, j_r)$	$P_1(i_r+1, j_r+1)$

, where $0 \leq r < s$. We say that P_1 can be reduced to P_2 (denoted by $P_1 \vdash_{\mathcal{M}}^* P_2$) if there exists a sequence of reductions $Q_1 \vdash_{\mathcal{M}} Q_2, Q_2 \vdash_{\mathcal{M}} Q_3, \ldots, Q_{n-1} \vdash_{\mathcal{M}} Q_n$, where $n \geq 1$, $Q_1 = P_1$ and $Q_n = P_2$. Obviously, the relation $\vdash_{\mathcal{M}}^*$ is the reflexive and transitive closure of the relation $\vdash_{\mathcal{M}}$.

Let $\mathcal{M} = (\Sigma, \Gamma, \Theta_f, \delta, \nu, \mu)$ be a 2RTA. The language accepted by $\mathcal{M}$ is the set $L(\mathcal{M}) = \{P \in \Sigma^{*,*} \mid \exists Q \in \Gamma^{*,*} : P \vdash_{\mathcal{M}}^* Q \text{ and } Q \in L(\Theta_f)\}$.

2RTA is by definition nondeterministic, as on a tile one of several rewriting rules with the same left-hand side can be applied. We define a deterministic version of two-dimensional restarting automata (2DRTA) by requiring that no two rewriting rules can have the same left-hand side.

Definition 4. *A deterministic two-dimensional restarting automaton, referred to as 2DRTA, is a 2RTA $\mathcal{M} = (\Sigma, \Gamma, \Theta_f, \delta, \nu, \mu)$ with the set of rewriting rules δ satisfying one additional condition that for every tile $T \in (\Gamma \cup \mathcal{S})^{2,2}$ there exists at most one rule with the left-hand side T in δ.*

To illustrate our definition we present a deterministic 2RTA accepting the language of all squares over a one-letter alphabet

$$L = \{P \in a^{n,n} \mid n \geq 0\}.$$

Example 1. Let $\mathcal{M} = (\Sigma, \Gamma, \Theta_f, \delta, \mu)$ be a deterministic 2RTA with $\Sigma = \{a\}$, $\Gamma = \{a, 1\}$,

$$\delta = \left\{ \begin{array}{cc} \boxed{\begin{array}{cc} \# & \top \\ \vdash & a \end{array}} \rightarrow \boxed{\begin{array}{cc} \# & \top \\ \vdash & 1 \end{array}} , \boxed{\begin{array}{cc} 1 & a \\ a & a \end{array}} \rightarrow \boxed{\begin{array}{cc} 1 & a \\ a & 1 \end{array}} \end{array} \right\}, \quad \Theta_f = \left\{ \boxed{\begin{array}{cc} \# & \top \\ \vdash & 1 \end{array}} , \boxed{\begin{array}{cc} \top & \top \\ 1 & a \end{array}} , \right.$$

$$\boxed{\begin{array}{cc} \top & \top \\ a & a \end{array}} \boxed{\begin{array}{cc} \top & \# \\ a & \dashv \end{array}} , \boxed{\begin{array}{cc} a & \dashv \\ a & \dashv \end{array}} \boxed{\begin{array}{cc} a & \dashv \\ 1 & \dashv \end{array}} , \boxed{\begin{array}{cc} 1 & \dashv \\ 1 & \dashv \end{array}} \boxed{\begin{array}{cc} a & 1 \\ \bot & \# \end{array}} , \boxed{\begin{array}{cc} a & a \\ \bot & \bot \end{array}} \boxed{\begin{array}{cc} a & a \\ \bot & \bot \end{array}} \boxed{\begin{array}{cc} \vdash & a \\ \# & \bot \end{array}} , \boxed{\begin{array}{cc} \vdash & a \\ \vdash & a \end{array}} \boxed{\begin{array}{cc} \vdash & 1 \\ \vdash & a \end{array}} ,$$

$$\boxed{\begin{array}{cc} 1 & a \\ a & 1 \end{array}} \boxed{\begin{array}{cc} a & 1 \\ a & a \end{array}} , \boxed{\begin{array}{cc} a & a \\ 1 & a \end{array}} \boxed{\begin{array}{cc} a & a \\ a & a \end{array}} , \boxed{\begin{array}{cc} \# & \# \\ \# & \# \end{array}} \right\}.$$

The automaton has the single auxiliary symbol 1, which is used to mark the main diagonal of the picture in order to check that the input is a square picture. The weight function μ is defined in a way that the only allowed rewriting is from a to 1, e.g. $\mu(a) = 2$ and $\mu(1) = 1$.

Barták [2] showed several closure properties of his two-dimensional restarting automata and related them to other classes of picture languages like 2OTA, REC and tiling systems. Unfortunately several of his results are conjectures only. The class 2RTA with the scanning strategy ν will be denoted as ν-2RTA. Next we will show that for any scanning strategy ν, the class of languages recognizable by ν-2RTA (denoted as $\mathcal{L}(\nu$-2RTA)) is closed under projection.

Lemma 1. *For each scanning strategy ν, $\mathcal{L}(\nu$-2RTA) is closed under projection.*

Proof. Let Σ_1, Σ_2 be two finite alphabets and let $\varphi : \Sigma_1 \rightarrow \Sigma_2$ be a projection. We will prove that if a picture language $L_1 \subseteq \Sigma_1^{*,*}$ is accepted by a ν-2RTA $\mathcal{A}_1$, then $L_2 = \varphi(L_1)$ is accepted by a ν-2RTA $\mathcal{A}_2$.

First, the automaton $\mathcal{A}_2$ starts to "reverse" the projection φ on the input picture starting from the lower right corner of the picture and then simulates a computation of $\mathcal{A}_1$. The process of reversing the projection and the simulation of $\mathcal{A}_1$ can be mixed. Nevertheless, the picture is accepted only when all symbols of the picture are "reversed" [10].   □

Actually the technique used in the proof of Lemma 1 can be easily extended in such a way that each symbol x of the input picture is first replaced by the pair (x, x) and then two computations of two different 2RTAs can be simulated – the first one operating on the first elements of the pairs and the second one operating on the second elements of the pairs. Clearly, in this way it is possible to construct 2RTA for accepting union and intersection of any two given 2RTAs. Hence we have the following.

Corollary 1. *For any scanning strategy ν, $\mathcal{L}(\nu\text{-2RTA})$ is closed under union and intersection.*

The above results concerning 2RTAs indicate that for some properties the scanning strategy employed by an 2RTA does not put any restriction on its power. In the following we will use the class of languages which can be accepted using each scanning strategy. Formally, a language L is *strategy independent*, if for each scanning strategy ν, there exists a ν-2RTA $\mathcal{M}$ such that $L(\mathcal{M}) = L$. The class of all strategy independent languages will be denoted as si-2RTL. Similarly we can define deterministic strategy independent languages accepted by deterministic 2RTA. The respective class of languages will be denoted as si-2DRTL.

Corollary 2. *Both si-2RTL and si-2DRTL are closed under projection, union and intersection.*

Theorem 1. *Both si-2RTL and si-2DRTL are closed under vertical and horizontal mirroring and rotation.*

Proof. Let L be a picture language in si-2RTL and let $\nu = (c, f)$ be any scanning strategy. We will prove, e.g., $L^{VM} \in \mathcal{L}(\nu\text{-2RTA})$. Define $\nu^{VM} = (c^{VM}, f^{VM})$ as follows: $c^{VM} = 1$ if $c = 2$, $c^{VM} = 2$ if $c = 1$, $c^{VM} = 3$ if $c = 4$, $c^{VM} = 4$ if $c = 3$ and
$$f^{VM}(i, j, m, n) = (i_1, n + 2 - j_1) \text{ where } f(i, j, m, n) = (i_1, j_1).$$
There is a ν^{VM}-2RTA $\mathcal{M} = (\Sigma, \Gamma, \Theta_f, \delta, \nu^{VM}, \mu)$ such that $L = L(\mathcal{M})$. Define a mapping $\pi : \Gamma \cup \mathcal{S} \to \Gamma \cup \mathcal{S}$ fulfilling $\pi(\vdash) = \dashv$, $\pi(\dashv) = \vdash$ and $\pi(a) = a$ for all $a \in (\Gamma \cup \{\top, \bot\})$. Modify $\mathcal{M}$ to a ν-2RTA $\mathcal{M}' = (\Sigma, \Gamma, \Theta'_f, \delta', \nu, \mu)$, where $\Theta'_f = \pi(\Theta_f^{VM})$, and for each rewriting rule $U \to V$ in δ, the mirrored rule $\pi(U^{VM}) \to \pi(V^{VM})$ is inserted into δ'. Then, $\mathcal{M}'$ accepts P iff $\mathcal{M}$ accepts P^{VM}, implying $L(\mathcal{M}') = L^{VM}$. If $\mathcal{M}$ is deterministic, $\mathcal{M}'$ is deterministic too.   □

Theorem 2. *Let ν be a scanning strategy such that the last position scanned according to ν on pictures of all (positive) sizes is always the same corner (containing the sentinel $\#$). Then the class $\mathcal{L}(\nu\text{-2DRTA})$ is closed under complement.*

Proof. A deterministic 2RTA $\mathcal{M} = (\Sigma, \Gamma, \Theta_f, \delta, \nu, \mu)$ rejects an input picture P, when $P \vdash_{\mathcal{M}}^* Q$, no rule from δ can be applied on any subpicture of Q and $Q \notin L(\Theta_f)$. It is possible deterministically detect the situation when $\mathcal{M}$ has no applicable rewriting rule – when it scans the last position according to ν and it has no applicable rule in δ. At this moment, an ν-2DRTA $\mathcal{M}' = (\Sigma, \Gamma', \Theta_f', \delta', \nu, \mu')$ accepting the complement of $L(\mathcal{M})$ can start additional verification whether $\mathcal{M}$ accepts the picture. Then $\mathcal{M}'$ accepts iff $\mathcal{M}$ rejects. For the full construction see [10]. $\qquad\qquad\square$

Using similar techniques as above, it is possible to prove closure on both column and row concatenations for $\mathcal{L}(\nu_{\text{row}}\text{-2RTA})$. Unfortunately, we do not know whether si-2RTL is closed under column and row concatenations.

4   Two-Dimensional Bounded Turing Machines and 2RTAs

For technical reasons, we introduce a limited version of Turing machine working on pictures. Let $\mathcal{H} = \{R, L, D, U, Z\}$ be the set of *head movements*. The first four elements denote directions: left, right, up, down. Z stands for zero (none) movement. Let $\upsilon : \mathcal{S} \to \mathcal{H}$ be a mapping such that $\upsilon(\vdash) = R$, $\upsilon(\dashv) = L$, $\upsilon(\top) = D$, $\upsilon(\bot) = U$ and $\upsilon(\#) = Z$.

Definition 5. *A (non-deterministic) two-dimensional bounded Turing machine, 2BTM for short, $\mathcal{A} = (Q, \Sigma, \Gamma, \delta, q_0, Q_F)$ is a Turing machine working on a two-dimensional tape with a finite set of states Q containing the initial state q_0, an input alphabet Σ, a working alphabet Γ ($\supseteq \Sigma$), where $\Gamma \cap \mathcal{S} = \emptyset$, a set of final states $Q_F \subseteq Q$ and a transition relation $\delta : (Q \backslash Q_F) \times (\Gamma \cup \mathcal{S}) \to 2^{Q \times (\Gamma \cup \mathcal{S}) \times \mathcal{H}}$ satisfying for any pair $(q, a) \in (Q \backslash Q_F \times (\Gamma \cup \mathcal{S})$ and each element $(q', a', d) \in \delta(q, a)$:*

- *$a \in \mathcal{S}$ implies $d = \upsilon(a)$ and $a' = a$, and*
- *$a \notin \mathcal{S}$ implies $a' \notin \mathcal{S}$.*

We say that $\mathcal{A}$ is a deterministic 2BTM (2DBTM), if for for each $q \in Q$ and $a \in \Gamma \cup \mathcal{S}$ it holds $|\delta(q, a)| \leq 1$.

In the initial configuration of $\mathcal{A}$ on an input picture $P \in \Sigma^{*,*}$, its tape contains $\widehat{P}$, its control unit is in state q_0 and the head scans the top-left corner of P. When $P = \Lambda$, the head scans the bottom-right corner of $\widehat{P}$ containing $\#$. The machine accepts P iff there is a computation of $\mathcal{A}$ starting in the initial configuration on P and finishing in an accepting state from Q_F.

Definition 6. *Let $k \in \mathbb{N}$ be an integer such that during each computation of M over any picture from $\Sigma^{*,*}$ each tape field is visited at most k times. Then, we say that M is a* constant-visit-2BTM *(cv-2BTM, for short). We will denote deterministic* cv-2BTM *by* cv-2DBTM.

In contrast to 2BTM, all computations of an cv-2BTM are finite and have time complexity $t(m, n) = O(mn)$ for pictures of size (m, n). It is easily seen that cv-2BTM and cv-2DBTM can simulate 2OTA and 2DOTA, respectively. Thus we get the following.

Theorem 3. REC *is included in* $\mathcal{L}$(cv-2BTM), DREC *is included in* $\mathcal{L}$(cv-2DBTM).

For each 4FA $\mathcal{A}$ there exists a constant k such that in any accepting computation of $\mathcal{A}$ on an input picture P, the head of $\mathcal{A}$ visits any symbol of P at most k times. It is possible to show that even deterministic cv-2BTM can simulate any (nondeterministic) 4FA.

Theorem 4 ([10]). $\mathcal{L}$(4FA) *is included in* $\mathcal{L}$(cv-2DBTM).

In [10], we showed how to construct a language not in REC, but accepted by a cv-2DBTM. Define L_D over $\Sigma = \{0, 1, 2\}$ consisting of pictures $P = Q \textoslash C \textoslash Q^{\text{VM}}$ where $C \in \{2\}^{*,1}$, each row of Q contains exactly one symbol 1 and each diagonal of Q parallel to the minor diagonal contains at most one symbol 1. An example of such a picture follows.

0	1	0	0	2	0	0	1	0
0	1	0	0	2	0	0	1	0
0	0	0	1	2	1	0	0	0
1	0	0	0	2	0	0	0	1

Proposition 1 ([10]). L_D *is in* $\mathcal{L}$(cv-2DBTM) \ REC.

Further we will relate cv-2BTMs and 2RTAs

Theorem 5. *For any scanning strategy* ν, $\mathcal{L}$(cv-2BTM) *is included in* $\mathcal{L}(\nu\text{-2RTA})$ *and* $\mathcal{L}$(cv-2DBTM) *is included in* $\mathcal{L}(\nu\text{-2DRTA})$.

Proof. Let $\mathcal{M} = (Q, \Sigma, \Gamma, \delta, q_0, Q_F)$ be a cv-2BTM. We describe a ν-2RTA $\mathcal{T} = (\Sigma, \Gamma', \Theta_f, \delta', \nu, \mu')$ such that $\mathcal{L}(\mathcal{T}) = \mathcal{L}(\mathcal{M})$. The idea is to simulate a computation of $\mathcal{M}$ over any input $P \in \Sigma^{*,*}$. If $\mathcal{M}$ scans the tape field f and the control unit is in state q, then $\mathcal{T}$ stores q into f. A set of rewriting rules is designed for changing the current configuration of $\mathcal{M}$ into a configuration after a single step of $\mathcal{M}$.

Let $k = |\Gamma|$, $m = \max\{k, 5\}$ and $I = \{0, \ldots, k\}$. Elements in Γ' are of five types:

1. $a \in (\Sigma \cup \mathcal{S})$, $\mu'(a) = mk + 4$, is an initial input symbol,
2. $(i, a) \in (I \times \Gamma)$, $\mu'((i, a)) = mi + 3$, represents a field containing a, the head of $\mathcal{M}$ is not placed here, at most i instructions of $\mathcal{M}$ can be performed over the field,
3. $(i, a, q) \in (I \times \Gamma \times Q)$, $\mu'((i, a, q)) = mi + 1$, the same meaning of a and i as above, moreover, the head of $\mathcal{M}$ scans this field and the control unit is in the state q,

4. $(i, a, q, d) \in (I \times \Gamma \times Q \times \mathcal{H})$, $\mu'((i,a,q,d)) = mi$, the same meaning of a, i and q as above, moreover, $\mathcal{M}$ will move from this field in the direction d,
5. $(i, a, q, b) \in (I \times \Gamma \times Q \times \Gamma)$, $\mu((i,a,q,b)) = mi + 2$, an auxiliary symbol with the meaning: $\mathcal{M}$ moved to this field containing a in the state q from a neighboring field by an instruction which writes b.

There is no loss of generality in assuming that $\mathcal{M}$ moves the head in each computation step and enters a state in Q_F only when it scans the bottom right corner of P. The specification of rewriting rules and Θ_f follows. For each $a \in \Sigma$, there is a rule creating the representation of the initial configuration:

$$\begin{array}{|c|c|} \hline \# & \top \\ \hline \vdash & a \\ \hline \end{array} \rightarrow \begin{array}{|c|c|} \hline \# & \top \\ \hline \vdash & (k, a, q_0) \\ \hline \end{array} \ .$$

For each instruction $(q, a) \rightarrow (q', a', R)$ in δ, rules matching the following patterns are added:

$$\begin{array}{|c|c|} \hline (i,a,q) & (j,b) \\ \hline s & t \\ \hline \end{array} \rightarrow \begin{array}{|c|c|} \hline (i,a,q,R) & (j,b) \\ \hline s & t \\ \hline \end{array}, \quad \begin{array}{|c|c|} \hline (i,a,q) & b \\ \hline s & t \\ \hline \end{array} \rightarrow \begin{array}{|c|c|} \hline (i,a,q,R) & b \\ \hline s & t \\ \hline \end{array},$$

$$\begin{array}{|c|c|} \hline (i,a,q,R) & (j,b) \\ \hline s & t \\ \hline \end{array} \rightarrow \begin{array}{|c|c|} \hline (i,a,q,R) & (j,b,q',a') \\ \hline s & t \\ \hline \end{array},$$

$$\begin{array}{|c|c|} \hline (i,a,q,R) & b \\ \hline s & t \\ \hline \end{array} \rightarrow \begin{array}{|c|c|} \hline (i,a,q,R) & (0,b,q',a') \\ \hline s & t \\ \hline \end{array},$$

$$\begin{array}{|c|c|} \hline (i,a,q,R) & (j,b,q',a') \\ \hline s & t \\ \hline \end{array} \rightarrow \begin{array}{|c|c|} \hline (i-1,a') & (j,b,q',a') \\ \hline s & t \\ \hline \end{array},$$

$$\begin{array}{|c|c|} \hline (i-1,a') & (j,b,q',a') \\ \hline s & t \\ \hline \end{array} \rightarrow \begin{array}{|c|c|} \hline (i-1,a') & (j,b,q') \\ \hline s & t \\ \hline \end{array},$$

where $i, j \in I$, $0 < i \le k$, $b \in \Gamma \setminus \mathcal{S}$, $s, t \in \Sigma \cup \mathcal{S} \cup (I \times \Gamma)$. By writing an auxiliary symbol of the form (i, a, q, d) on the tape, we ensure that $\mathcal{T}$ cannot start to simulate simultaneously two instructions of $\mathcal{M}$ moving from a field in different (e.g. opposite) directions.

A special attention has to be paid to the situation, when the head of $\mathcal{M}$ moves outside P. We do not represent such a configuration, but rather the following configuration reached by applying a next instruction of the form $(q', \dashv) \rightarrow (q'', \dashv, L)$. Thus, the set of rules is completed by

$$\begin{array}{|c|c|} \hline (i,a,q) & \dashv \\ \hline s & t \\ \hline \end{array} \rightarrow \begin{array}{|c|c|} \hline (i-1,a',q'') & \dashv \\ \hline s & t \\ \hline \end{array} \ .$$

The weight function μ has been defined to conform the rules. Similar rules are added also for the remaining instructions which move the head of $\mathcal{M}$ left, up or down.

Θ_f contains all the tiles of the form $\begin{array}{|c|c|}\hline (i,a,q_f) & \dashv \\ \hline \bot & \# \\ \hline \end{array}$, for all $a \in \Gamma$, $i \in I$, $q_f \in Q_F$

and all the tiles of the form $\begin{array}{|c|c|}\hline a & b \\ \hline c & d \\ \hline \end{array}$, where $a, b, c, d \in (\Sigma \cup S) \cup (I \times \Gamma)$ except

the tiles of the form $\begin{array}{|c|c|}\hline a & \dashv \\ \hline \bot & \# \\ \hline \end{array}$, where $a \in (\Sigma \cup S) \cup (I \times \Gamma)$. This ensures the

rewriting process finishes only if $\mathcal{M}$ reaches an accepting state (with its head at

the bottom right corner). If $P = \Lambda$, then $\widehat{P} = \begin{array}{|c|c|}\hline \# & \# \\ \hline \# & \# \\ \hline \end{array}$. This tile is in Θ_f iff $\mathcal{M}$

accepts Λ. To finish the proof, it is easy to see that when $\mathcal{M}$ is deterministic, $\mathcal{T}$
is deterministic as well.   □

Note that the simulation of a cv-2BTM by a 2RTA does not depend on the
employed scanning strategy. Using Theorem 3, Theorem 4 and Proposition 1 we
obtain the following.

Corollary 3. REC *is a proper subclass of* si-2RTL, DREC *is a a proper subclass
of* si-2DRTL, $\mathcal{L}(\text{4FA}) \subset$ si-2DRTL.

5   2RTA Working over One-Row Pictures

These automata can be quite powerful when we have freedom to design a suitable
scanning strategy. They can accept also non-regular (one-dimensional) languages
[10]. On the other hand, a simple strategy ν_{row} leads to the recognition of regular
languages only.

Lemma 2 ([10]). *Let* $\mathcal{A} = (\Sigma, \Gamma, \Theta_f, \delta, \nu_{\text{row}}, \mu)$ *be a* ν_{row}*-2RTA accepting a
one-dimensional picture language* $(L(\mathcal{A}) \subseteq \Sigma^{1,*} = \Sigma^*)$. *There is a* cv-2BTM $\mathcal{M}$
such that $L(\mathcal{M}) = L(\mathcal{A})$.

Each cv-2BTM accepting only one-row pictures recognizes a regular language
([10]). This gives the following consequences.

Proposition 2. *If a* ν_{row}*-2RTA accepts a one-dimensional picture language,
then it is a regular language.*

Corollary 4. *Both* $\mathcal{L}(\nu_{\text{row}}$*-2RTA) and* si-2RTL *restricted to one-dimensional
picture languages are equal to the class of regular languages.*

6   Conclusions

We have introduced a new two-dimensional model of restarting tiling automaton. An arbitrary scanning strategy makes it quite powerful. However, when we
restrict to languages recognizable independently on the strategy, we obtain a
family (si-2RTL) exhibiting good properties. It is a proper superset of REC and
has nearly the same closure properties. Namely, we showed that si-2RTL is closed

under union, intersection, projection, mirroring and rotation. When considering one-row inputs only, the family collapses to the class of regular languages.

Deterministic 2RTAs lead to family si-2DRTL, which is an extension of DREC. Beside DREC, it includes also $\mathcal{L}(4FA)$. These facts demonstrate the significance of the model and entitle us to see si-2DRTL as an important family of deterministically recognizable picture languages.

References

1. Anselmo, M., Giammarresi, D., Madonia, M.: A computational model for tiling recognizable two-dimensional languages. Theoretical Computer Science 410, 3520–3529 (2009)
2. Barták, J.: Recognition of picture languages. Master thesis, Faculty of Mathematics and Physics, Charles University, Prague (2008)
3. Blum, M., Hewitt, C.: Automata on a 2-dimensional tape. In: Proceedings of the 8th Annual Symposium on Switching and Automata Theory (SWAT 1967), pp. 155–160. IEEE Computer Society, Washington, DC (1967)
4. Cherubini, A., Pradella, M.: Picture Languages: From Wang Tiles to 2D Grammars. In: Bozapalidis, S., Rahonis, G. (eds.) CAI 2009. LNCS, vol. 5725, pp. 13–46. Springer, Heidelberg (2009)
5. Giammarresi, D., Restivo, A.: Two-dimensional languages. In: Rozenberg, G., Salomaa, A. (eds.) Handbook of Formal Languages, vol. 3, pp. 215–267. Springer-Verlag New York, Inc., New York (1997)
6. Jurdziński, T., Otto, F.: Shrinking restarting automata. Int. J. Found. Comput. Sci. 18(2), 361–385 (2007)
7. Lonati, V., Pradella, M.: Picture Recognizability with Automata Based on Wang Tiles. In: van Leeuwen, J., Muscholl, A., Peleg, D., Pokorný, J., Rumpe, B. (eds.) SOFSEM 2010. LNCS, vol. 5901, pp. 576–587. Springer, Heidelberg (2010)
8. Lonati, V., Pradella, M.: Towards More Expressive 2D Deterministic Automata. In: Bouchou-Markhoff, B., Caron, P., Champarnaud, J.-M., Maurel, D. (eds.) CIAA 2011. LNCS, vol. 6807, pp. 225–237. Springer, Heidelberg (2011)
9. Otto, F.: Restarting Automata. In: Ésik, Z., Martín-Vide, C., Mitrana, V. (eds.) Recent Advances in Formal Languages and Applications. SCI, vol. 25, pp. 269–303. Springer, Heidelberg (2006)
10. Průša, D., Mráz, F.: New models for recognition of picture languages: Sgraffito and restarting tiling automata. Research Report CTU–CMP–2012–08, Center for Machine Perception, K13133 FEE Czech Technical University, Prague, Czech Republic (March 2012)

Crossing the Syntactic Barrier:
Hom-Disequalities for $\mathcal{H}_1$-Clauses

Andreas Reuß[*] and Helmut Seidl

Technische Universität München
{a.reuss,seidl}@in.tum.de

Abstract. We extend $\mathcal{H}_1$-clauses with disequalities between images of
terms under a tree homomorphism (hom-disequalities). This extension
allows to test whether two terms are distinct modulo a semantic inter-
pretation, allowing, e.g., to neglect information that is not considered
relevant for the intended comparison. We prove that $\mathcal{H}_1$-clauses with
hom-disequalities are more expressive than $\mathcal{H}_1$-clauses with ordinary
term disequalities, and that they are incomparable with $\mathcal{H}_1$-clauses with
disequalities between paths. Our main result is that $\mathcal{H}_1$-clauses with
this new type of constraints can be normalized into an equivalent tree
automaton with hom-disequalities. Since emptiness for that class of automata turns out to be decidable, we conclude that satisfiability is decidable for positive Boolean combinations of queries to predicates defined
by $\mathcal{H}_1$-clauses with hom-disequalities.

1 Introduction

Analyses of tree-manipulating programs can nicely be specified by means of
Horn clauses. This approach has successfully been applied to such different
kinds of programming formalisms as Prolog programs and cryptographic protocols [4,1,9,8]. Pure Horn clauses have difficulties, though, to express negative
information such as that two values must be different. In order to compensate
for this deficiency, we have extended Horn clauses with disequality constraints.
It turns out that when extending the decidable class of $\mathcal{H}_1$-clauses [14], again
a decidable class is obtained [12,13]. The class $\mathcal{H}_1$ differs from general Horn
clauses in that the terms in the heads may contain at most one constructor and
no variable may occur twice in a head.

Disequality on terms or subterms, however, may be too imprecise in the presence of semantic interpretations because syntactically distinct terms may represent the same value. One of the simplest forms of such interpretations are
homomorphisms. Homomorphisms allow, e.g., to relabel nodes, to select specific
subtrees (depending on labels), or permute subtrees. In the tree representation
of a tuple of trees, a suitable homomorphism thus allows to select individual
components. Perhaps most useful in the context of cryptographic protocols is
the possibility of disequalities modulo homomorphisms to compare messages

---

[*] The author was supported by the DFG Graduiertenkolleg 1480 (PUMA).

N. Moreira and R. Reis (Eds.): CIAA 2012, LNCS 7381, pp. 301–312, 2012.

while disregarding irrelevant information such as random padding or session keys. Analyses of anonymity violation or non-interference [6,2,3] may search for values which are independent of sender identities or secret subparts, respectively.

Example 1. Consider the following example where the predicates p_u, p_v model the set of states reaching stages u, v of a protocol. For simplicity, assume that the value at stage u is obtained from the value at stage v by combining the value at v with a secret value under a data constructor f, where the secret value is taken from some set *input*. This can be formalized by the following clauses:

$$p_u(f(Z, Y)) \quad \Leftarrow high(Z), p_v(Y)$$
$$high(secret(X)) \Leftarrow input(X)$$

where the value Y at stage v may contain secrets as well. Now assume that we want to verify that the *public view* of values at stage u is independent of the secrets included into the values. Here, the public view of a value is realized by a homomorphism H which maps the constructor *secret* (along with its respective subtrees) to some constant $\square$ and is the identity for the remaining constructors. Then a potential violation of the independence could be expressed by:

$$error \Leftarrow p_u(X), p_u(Y), \ X \neq_H Y$$

where $\neq_H$ applies disequality to the images under H. $\square$

In [12,13], we extended the normalization procedure for $\mathcal{H}_1$ from [9,7] to clauses with term disequality constraints and disequality constraints between paths. This procedure transforms every finite set of $\mathcal{H}_1$-clauses with term disequality constraints into an equivalent finite set of *automata clauses* with term disequality constraints [10] and thus allows to decide whether or not a given query is satisfiable. Our goal here is to make this approach work also in presence of homomorphisms. The first step in this direction is to introduce finite tree automata with disequality constraints modulo homomorphisms and to prove that k-finiteness is decidable for these. In order to do so, we build on techniques provided in [10]. Related automata techniques have been proposed in [5] which do not only apply to term disequalities but also to disequalities between paths.

In the second step, we then indicate how, based on these new automata, a similar normalization procedure can be realized as has been applied in [9,12,13]. Particular care for this extension is required at the splitting rule. This rule allows to remove variables from a clause which do not occur in the head. Also, a refined argument must be applied to prove termination of the procedure.

2   Preliminaries

Terms and Constraints. We consider ordered ranked trees made up of symbols from a ranked alphabet (Σ, ar) where Σ denotes a set of symbols and $ar : \Sigma \to \mathbb{N} \cup \{0\}$ specifies each symbol's arity. If the arities of symbols are understood, then the ranked alphabet is denoted by Σ alone. For a ranked alphabet Σ and a set $\mathbf{V} = \{X_1, X_2, \ldots\}$ of *variables*, the set $\mathcal{T}_\Sigma(\mathbf{V})$ of (finite ordered) trees over Σ and $\mathbf{V}$ consists of all terms t given by the grammar: $t ::= X_i \mid a \mid b(t_1, \ldots, t_k)$ where $a, b \in \Sigma$, and a has arity 0, while b has arity $k > 0$. For a tree t, the expression vars(t) denotes the set of variables occurring in t. The tree t is called *ground* if t does not contain any variable $X \in \mathbf{V}$. The set of ground terms is also denoted by $\mathcal{T}_\Sigma$. A *literal* A is an expression of the form $p(t)$ where p is a unary predicate and $t \in \mathcal{T}_\Sigma(\mathbf{V})$. Non-unary predicates can be integrated in our framework, too, by equipping their arguments with an *implicit* constructor of the same arity as the predicate. A *term constraint* is a conjunction of disequalities $s \neq t$ for terms s, t. A substitution θ is a mapping $\theta : \mathbf{V} \to \mathcal{T}_\Sigma(\mathbf{V})$. We write θt and θA for the result of applying θ to the term t and the literal A, respectively. θ is called *ground* if θX_i is ground for all i. The substitution θ satisfies the term constraint ϕ (denoted by: $\theta \models \phi$) if it satisfies each constraint occurring in ϕ. The substitution θ satisfies the constraint $s \neq t$ if $\theta s \neq \theta t$.

Tree Homomorphisms. For a given ranked alphabet Σ, a tree homomorphism H maps each symbol $f \in \Sigma$ of arity $k \geq 0$ to a term $t(X_1, \ldots, X_k)$, i.e., a term containing only variables from the set $\{X_1, \ldots, X_k\}$. The mapping $\mathrm{H}^* : \mathcal{T}_\Sigma(\mathbf{V}) \mapsto \mathcal{T}_\Sigma(\mathbf{V})$ is then recursively defined as $\mathrm{H}^*(X) = X$ for $X \in \mathbf{V}$, and $\mathrm{H}^*(f(t_1, \ldots, t_k)) = \mathrm{H}(f)(\mathrm{H}^*(t_1), \ldots, \mathrm{H}^*(t_k)) = t[\mathrm{H}^*(t_1)/X_1, \ldots, \mathrm{H}^*(t_k)/X_k]$. Here, $\mathrm{H}^*(s)$ is ground whenever s is ground. The function H^{-1} reverses the effect of applying H^*, i.e., $\mathrm{H}^{-1}s = \{t \mid \mathrm{H}^*t = s\}$. We extend H^* and H^{-1} for sets of terms T by defining $\mathrm{H}^*T = \{\mathrm{H}^*t \mid t \in T\}$ and $\mathrm{H}^{-1}T = \{t \mid \mathrm{H}^*t \in T\}$. For notational convenience, we may also write $\mathrm{H}e$ instead of H^*e, meaning that the mapping H^* is applied to the expression e.

A *hom-disequality* is an expression $s \neq_\mathrm{H} t$ where $s, t \in \mathcal{T}_\Sigma(\mathbf{V})$. In case that both s and t are ground terms, $s \neq_\mathrm{H} t$ is equivalent to $\mathrm{H}^*s \neq \mathrm{H}^*t$. A ground substitution θ satisfies a hom-disequality $s \neq_\mathrm{H} t$ if (and only if) $\theta s \neq_\mathrm{H} \theta t$, i.e., $\mathrm{H}^*(\theta s) \neq \mathrm{H}^*(\theta t)$. For a substitution θ (not necessarily ground) and a homomorphism H, let θ_H denote the substitution given by: $\theta_\mathrm{H} X = \mathrm{H}^*(\theta X)$. Then it holds that $\mathrm{H}^* \circ \theta = \theta_\mathrm{H} \circ \mathrm{H}^*$, i.e., for every term t, we have: $\mathrm{H}^*(\theta t) = \theta_\mathrm{H}(\mathrm{H}^*t)$.

For monotone Boolean combinations of term disequalities and hom-disequalities, satisfiability is recursively defined by:

$$\theta \models (\phi_1 \wedge \phi_2) \text{ iff } (\theta \models \phi_1) \wedge (\theta \models \phi_2)$$
$$\theta \models (\phi_1 \vee \phi_2) \text{ iff } (\theta \models \phi_1) \vee (\theta \models \phi_2)$$

For convenience, we also define the *hom-equality* $s =_\mathrm{H} t \Leftrightarrow \neg(s \neq_\mathrm{H} t)$.

Horn Clauses with Constraints. A *constrained Horn clause* c is given by

$$B_0 \Leftarrow B_1, \ldots, B_m, \phi$$

where $B_0, \ldots, B_m$ are literals and ϕ is a either a conjunction of term disequalities or a conjunction of hom-disequalities. The left-hand side B_0 is the *head* of the clause c while the sequence $B_1, \ldots, B_m, \phi$ denotes the *body* or *precondition* of c. The constraint ϕ imposes an additional restriction on the applicability of the clause. A constraint ϕ which is always true can be omitted. Assume that we are given a finite set $\mathcal{C}$ of constrained Horn clauses. Then the *least* model $\mathcal{M_C}$ of $\mathcal{C}$ is the least set M of ground facts $p(t), t \in \mathcal{T_\Sigma}$, such that $\mathcal{M_C} \supseteq \mathcal{T_C}(\mathcal{M_C})$. Here, the operator $\mathcal{T_C}$ is defined as follows. Assume that M is any set of ground facts $p(t)$. Then $\mathcal{T_C}(M)$ is the set of all ground facts θB_0 where θ is a ground substitution, $B_0 \Leftarrow B_1, \ldots, B_m, \phi$ is in $\mathcal{C}$, $\theta B_1, \ldots, \theta B_m \in M$, and $\theta \models \phi$. The language $\{t \in \mathcal{T_\Sigma} \mid p(t) \in \mathcal{M_C}\}$ of p is also denoted by $[\![p]\!]_\mathcal{C}$. For convenience, we also consider the set $[\![p]\!]_\mathcal{C}^i = \{t \in \mathcal{T_\Sigma} \mid p(t) \in \mathcal{T_C}^i(\emptyset)\}$ which consists of all trees t where the fact $p(t)$ can be derived by at most i rounds of fixpoint iteration.

$\mathcal{H}_1$-Clauses, Normal Clauses, Automata Clauses. Let us briefly introduce the subclasses of Horn clauses which we consider here. Essentially, these classes are obtained from the classes considered in [12] by replacing constraints consisting of disequalities between terms with constraints consisting of disequalities between terms modulo a given tree homomorphism. Thus, a Horn clause is an $\mathcal{H}_1$-clause if the term t in the head $p(t)$ contains at most one constructor, and no variable occurs twice in t. For convenience, we adopt the convention that the variables in the heads of $\mathcal{H}_1$-clauses are enumerated $X_1, \ldots, X_k$, i.e., t either equals X_1 or is of the form $f(X_1, \ldots, X_k)$ for a constructor of arity k where the case of atoms is subsumed by choosing $k = 0$. Moreover for a distinction, variables *not* occurring in the head will be denoted $Y, Y_1, \ldots$

The Horn clause is a *normal clause* if it is of the form:

$$p(f(X_1, \ldots, X_k)) \Leftarrow p_1(X_{i_1}), \ldots, p_r(X_{i_r}), \phi$$

where all variables occurring in the body of the clause also occur in the head. Moreover, the Horn clause is an *automata clause* if additionally each variable X_i occurring in the head occurs exactly once in the literals occurring in the body and the head contains exactly one constructor, i.e., the clause has the form:

$$p(f(X_1, \ldots, X_k)) \Leftarrow p_1(X_1), \ldots, p_k(X_k), \phi \ .$$

In particular, each normal clause as well as each automata clause is an $\mathcal{H}_1$-clause. In the following, we do not differentiate between tree automata and (finite) sets of automata clauses. The predicates and clauses of a set of automata clauses correspond to the states and transition rules of the corresponding tree automaton and vice versa.

As with ordinary Horn clauses or Horn clauses with term [12] or *path* [13] constraints, every set $\mathcal{N}$ of normal clauses can be transformed to an equivalent set $\mathcal{A}$ of automata clauses whose predicates correspond to conjunctions $p_1 \cap \ldots \cap p_j, j \geq 0$, of original predicates p_i from $\mathcal{N}$.

Lemma 1. *For every finite set $\mathcal{N}$ of normal clauses, a finite set $\mathcal{A}$ of automata clauses can be constructed with $[\![p]\!]_\mathcal{N} = [\![p]\!]_\mathcal{A}$ for each predicate p of $\mathcal{N}$.*   □

Effects of Tree Homomorphisms. Tree homomorphisms are rather expressive. E.g., a homomorphism may

- rename constructors: $H^*(g(X_1)) = h(X_1)$
- delete constructors: $H^*(g(X_1)) = X_1$
- delete subtrees: $H^*(f(X_1, X_2)) = g(X_2)$
- add constructors: $H^*(f(X_1, X_2)) = f(g(X_1), h(g(X_2)))$
- copy subtrees: $H^*(g(X_1)) = f(X_1, X_1)$
- permute subtrees: $H^*(f(X_1, X_2)) = f(X_2, X_1)$
- combine two or more of these features.

3 Expressiveness

This section compares automata classes extended with term, path, and hom-disequalities. (Unlabeled) path disequalities are expressions $X.\pi \neq Y.\pi'$ where X, Y are variables and π, π' are *paths* specifying subterms of trees as (possibly empty) sequences of numbers. E.g., $t.1.2$ denotes the second child of the first child of t if it exists, and is undefined otherwise. The ground substitution θ satisfies the disequality $X.\pi \neq Y.\pi'$ if either of $(\theta X).\pi$ or $(\theta Y).\pi'$ is undefined, or both are defined but the resulting terms differ.

First we show that hom-disequalities cannot be simulated by term disequalities, by presenting a specific language defined through a set of automata clauses which cannot be defined by a finite set of automata clauses with path disequalities only. Intuitively, path constraints can only express disequalities between subterms of at most a certain depth d as specified as part of the corresponding path expression. Hom-disequalities, however, may disregard an unbounded number of constructors on top of the tree.

Let $\Sigma = \{a, s, f\}$ where a, s and f have arities 0, 1 and 2, respectively. Let H be the homomorphism defined by: $Hs = X_1$ while terms rooted a or f are not changed by H. Consider the language $L = \{f(t_1, t_2) \mid t_1 \neq_H t_2\}$. The following automaton with hom-disequalities accepts L through predicate p.

$$
\begin{aligned}
\top(a) &\Leftarrow \\
\top(s(X_1)) &\Leftarrow \top(X_1) \\
\top(f(X_1, X_2)) &\Leftarrow \top(X_1), \top(X_2) \\
p(f(X_1, X_2)) &\Leftarrow \top(X_1), \top(X_2), X_1 \neq_H X_2
\end{aligned}
$$

Lemma 2. *There is no tree automaton for L with path disequalities only.* $\square$

Proof. Assume for a contradiction that an automaton $\mathcal{A}$ with path disequalities exists which accepts L through a predicate p. It is known [11,5,13] that to $\mathcal{A}$, a complement automaton $\mathcal{B}$ with *path equalities* only can be constructed such that there is a predicate $\bar{p}$ which accepts the complement language $\bar{L}$ given by: $\bar{L} = \mathcal{T}_\Sigma \setminus L = \{f(t_1, t_2) \mid t_1 =_H t_2\} \cup \{a\} \cup \{s(t) \mid t \in \mathcal{T}_\Sigma\}$. Let d be the maximal depth of a path occurring in $\mathcal{B}$. For a ground term t, let $r_{t1}, r_{t2}, \ldots$ denote the infinite sequence of terms defined by: $r_{ti} = s^d(f(t, s^i(a)))$. Then for $t \neq_H t'$, it holds that for all i, j, $r_{ti} =_H r_{tj}$ and $r_{ti} \neq_H r_{t'j}$, but for all paths π occurring in

$\mathcal{B}$, $r_{ti}.\pi \neq r_{tj}.\pi$ if $i \neq j$. As there are infinitely many sequences (r_{ti}) but only finitely many clauses, $\mathcal{B}$ has a clause

$$\overline{p}(f(X_1, X_2)) \Leftarrow q_1(X_1), q_2(X_2), \phi$$

such that for two terms t, t' with $t \neq_H t'$, there are two terms t_1, t_2 from the sequence (r_{ti}) and two terms t_3, t_4 from the sequence $(r_{t'i})$ such that both $f(t_1, t_2)$ and $f(t_3, t_4)$ are in $[\![\overline{p}]\!]_{\mathcal{B}}$ by application of this clause. Especially, $t_1, t_3 \in [\![q_1]\!]_{\mathcal{B}}$ and $t_2, t_4 \in [\![q_2]\!]_{\mathcal{B}}$. Since ϕ is a conjunction of path equalities, ϕ must be equivalent to true because $t_1.\pi \neq t_2.\pi$ for all paths π occurring in $\mathcal{B}$. But then the clause also accepts the term $f(t_1, t_4)$ for $\overline{p}$ — contradiction.   □

Now consider the set $\mathcal{C}$:

$$\begin{aligned} \top(a) &\Leftarrow \\ \top(f(X_1, X_2)) &\Leftarrow \top(X_1), \top(X_2) \\ p(f(X_1, X_2)) &\Leftarrow \top(X_1), \top(X_2), X_1 \neq X_2.1 \end{aligned}$$

for $\Sigma = \{a, f\}$. The language $[\![p]\!]_{\mathcal{C}}$ is not accepted by any automaton with hom-disequalities since hom-disequalities cannot directly access arbitrary subtrees independent of the labels in the tree. Let us denote by T the language $[\![p]\!]_{\mathcal{C}}$ of p w.r.t. $\mathcal{C}$. We have $T = \{f(t, a) \mid t \in \mathcal{T}_\Sigma\} \cup \{f(t, f(t_1, t_2)) \mid t, t_1, t_2 \in \mathcal{T}_\Sigma, t \neq t_1\}$.

Lemma 3. *There is no tree automaton $\mathcal{A}$ with hom-disequalities only that defines a predicate p with $[\![p]\!]_{\mathcal{A}} = T$.*   □

Proof. This example is the *unlabeled-path* variant of the corresponding example from [13] which provides the language $[\![p]\!]_{\mathcal{C}}$ that is accepted by an automaton with path disequalities but not by any automaton with term disequalities only.

Assume for a contradiction that an automaton $\mathcal{A}$ with hom-disequalities exists which accepts T through a predicate p. As in Lemma 2 we construct the complement automaton $\mathcal{B}$ with *hom-equalities* only, containing a predicate $\overline{p}$ which accepts the language $[\![\overline{p}]\!]_{\mathcal{B}} = \overline{T} = \mathcal{T}_\Sigma \setminus T = \{a\} \cup \{f(t, f(t, s)) \mid s, t \in \mathcal{T}_\Sigma\}$.

Case 1:   If both X_1 and X_2 occur in Hf, then it holds for each equality of $\mathcal{B}$ that $l =_H r$ if and only if $l = r$, and we refer to the proof in [13] that for $\overline{T}$ no automaton with term equalities exists.

Case 2:   If $Hf \in \{X_1, X_2, g\}$ for a ground term g, then each equality $l =_H r$ either is vacuously true or false (in case $g \neq Ha$), so that $\mathcal{B}$ may be considered an automaton with term equalities only, and as in Case 1, the proof in [13] applies.

Case 3:   Assume therefore that Hf equals a term $t = f(t_1, t_2)$ where t only contains the variable X_1 (respectively X_2). Then we define $d(t)$ to be the maximal $i \geq 0$ so that the path 1^i (respectively 2^i) is defined for t. Then $l =_H r$ iff $d(l) = d(r)$. The contradiction now follows from an argument analogous to Lemma 2 based on the fact that there are infinitely many values $d(t)$ but only finitely many clauses in $\mathcal{B}$.   □

Automata with term disequalities can be simulated by path-disequality automata [13]. Thus, automata with hom-disequalities are incomparable to path-disequality automata, while both classes are more expressive than automata with term disequalities only.

4   Automata with Hom-Disequalities

In this section we first show that to every set of automata clauses $\mathcal{A}$ with hom-disequality constraints, a *generalized* automaton with (ordinary) term disequalities $\mathcal{A}_H$ can be constructed such that $H[\![p]\!]_{\mathcal{A}} = [\![p]\!]_{\mathcal{A}_H}$. Secondly, we show that it is decidable for $\mathcal{A}_H$, every predicate p and number k whether $|[\![p]\!]_{\mathcal{A}_H}| < k$ — and thus, whether $|H[\![p]\!]_{\mathcal{A}}| < k$.

The class HDA (Hom-Disequality-Automata) of general automata with disequality term constraints consists of finite sets of clauses of the form:

$$p(t) \Leftarrow p_1(X_1), \ldots, p_k(X_k), \phi \qquad (k \geq 0)$$

where $p, p_1, \ldots, p_k$ are unary predicates, t is a term with vars$(t) \subseteq \{X_1, \ldots, X_k\}$, and ϕ is a conjunction of disequalities $t_i \neq t_j$ which may only mention variables from $\{X_1, \ldots, X_k\}$. The set vars(t) of variables occurring in the head of such a clause c is denoted hv(c) while its complement with respect to $\{X_1, \ldots, X_k\}$, i.e., the set of variables occurring only in the body of c, is denoted bv(c). The number of disequalities in a clause c is denoted $dc(c)$.

Lemma 4. *Let $\mathcal{A}$ be a finite set of automata clauses with hom-disequalities $t_1 \neq_H t_2$ for an arbitrary tree homomorphism H. Then an HDA $\mathcal{A}_H$ can be constructed such that for every predicate p and ground term t, $[\![p]\!]_{\mathcal{A}_H} = H[\![p]\!]_{\mathcal{A}}$.*

Proof. Note that an analogous lemma has been provided by Godoy et al. for tree automata with disequalities between paths [5]. $\mathcal{A}_H$ is obtained from $\mathcal{A}$ by transforming each clause c of the form:

$$p(f(X_1, \ldots, X_k)) \Leftarrow p_1(X_1), \ldots, p_k(X_k), \phi$$

with $\phi = l_1 \neq_H r_1 \wedge \ldots \wedge l_m \neq_H r_m$, $m \geq 0$ to the new clause c':

$$p(H^*(f(X_1, \ldots, X_k))) \Leftarrow p_1(X_1), \ldots, p_k(X_k), \phi'$$

with $\phi' = H^* l_1 \neq H^* r_1 \wedge \ldots \wedge H^* l_m \neq H^* r_m$. Instead of $[\![q]\!]_{\mathcal{A}_H} = H[\![q]\!]_{\mathcal{A}}$ for all predicates q, we prove by induction that $[\![q]\!]_{\mathcal{A}_H}^i = H[\![q]\!]_{\mathcal{A}}^i$ for all $i \geq 0$ and all q, with the base case $[\![q]\!]_{\mathcal{A}_H}^0 = H[\![q]\!]_{\mathcal{A}}^0 = \emptyset$. Recall that $H^* \theta s = \theta_H H^* s$ for terms s, where $\theta_H = \{X_i \mapsto H^* \theta X_i\}$. For a collection of terms $t_1, \ldots, t_k$, let $\theta = \{X_i \mapsto t_i\}$ (hence $\theta_H = \{X_i \mapsto H^* t_i\}$). Now assume that $f(t_1, \ldots, t_k) \in [\![p]\!]_{\mathcal{A}}^{i+1}$ by application of the clause c, i.e., $t_j \in [\![p_j]\!]_{\mathcal{A}}^i$ for all $1 \leq j \leq k$ and $\theta \models l_j \neq_H r_j$ for all $1 \leq j \leq m$. By induction hypothesis, the first condition holds iff $H^* t_j \in [\![p_j]\!]_{\mathcal{A}_H}^i \; \forall 1 \leq j \leq k$. The latter condition means that for all $1 \leq j \leq m$ we have $H^* \theta l_j \neq H^* \theta r_j$, which is equivalent to $\theta_H H^* l_j \neq \theta_H H^* r_j$ (for all j). Thus, $f(t_1, \ldots, t_k) \in [\![p]\!]_{\mathcal{A}}^{i+1}$ through application of c if and only if $H^* t_j \in [\![p_j]\!]_{\mathcal{A}_H}^i \; \forall 1 \leq j \leq k$ and $\theta_H \models H^* l_j \neq H^* r_j \; \forall 1 \leq j \leq m$, which is equivalent to $H^* f(t_1, \ldots, t_k) \in [\![p]\!]_{\mathcal{A}_H}^{i+1}$ through application of c', with the ground substitution θ_H, since $H(f)(H^* t_1, \ldots, H^* t_k) = H^* f(t_1, \ldots, t_k)$. $\square$

Example 2. Consider the homomorphism:

$$H = \{b \mapsto a, f(X_1, X_2) \mapsto g(X_1, g(X_1, a))\}$$

where all other constructors are preserved. Then the set of automata clauses:

$$p(b) \quad \Leftarrow$$
$$p(f(X_1, X_2)) \Leftarrow p(X_1), p(X_2), X_1 \neq_H f(X_2, X_2)$$

is transformed into the following set of clauses:

$$p(a) \quad \Leftarrow$$
$$p(g(X_1, g(X_1, a))) \Leftarrow p(X_1), p(X_2), X_1 \neq g(X_2, g(X_2, a))$$

Note that the variable X_2 no longer occurs in the head of the second clause, while the variable X_1 occurs more than once.   □

Deciding k-Finiteness of HDA. We proceed along the lines of deciding k-finiteness of automata with term disequalities in [12]. This base algorithm, though, must be extended as now heads are no longer just single constructor applications. Moreover, not all variables occurring in preconditions necessarily also occur in the head of a clause. Again, we start with a semi-algorithm that decides for a given HDA $\mathcal{A}$, a predicate p, and a number $k \geq 1$ whether $|[\![p]\!]_{\mathcal{A}}| \geq k$, by computing in every round i, $i \geq 1$, the sets $[\![q]\!]^i_{\mathcal{A}}$ for all predicates q until $|[\![p]\!]^i_{\mathcal{A}}| \geq k$ after some round i. Here, the sets $[\![q]\!]^i_{\mathcal{A}}$ can be computed from the sets $[\![q']\!]^{i-1}_{\mathcal{A}}$ by applying the implications $c \in \mathcal{A}$ — starting with $[\![q]\!]^0_{\mathcal{A}} = \emptyset$ for all q.

In order to obtain an algorithm, we establish an upper bound for the number of rounds which are needed for deciding HDA-k-finiteness. By a counting argument (analogous to [12]), it suffices to increase the sets $[\![q]\!]^i_{\mathcal{A}}$ only up to $k + \sum_{c \in \mathcal{A}} dc(c)$ trees for each predicate $q \neq p$. Our claim is based on the following lemma which says that each term constraint ϕ of a clause c "filters out" no more than $dc(c)$ trees. More precisely, if a clause $q_1(t) \Leftarrow \alpha_1, q_2(X), \alpha_2, \phi$ can produce a tree for predicate q_1 in round i, and $X \in vars(t)$, then the clause can produce at least $|[\![q_2]\!]^{i-1}_{\mathcal{A}}| - |\phi|_X$ trees until round i, where $|\phi|_X$ denotes the number of disequalities in ϕ which mention X.

Lemma 5. *Let $\mathcal{A}$ be an HDA, and $c \in \mathcal{A}$ a clause $q(t) \Leftarrow p_1(X_1), \ldots, p_k(X_k), \phi$. Assume that we are given a ground substitution $\theta \models \phi$ with $X_i\theta \in [\![p_i]\!]^d_{\mathcal{A}} \ \forall \ i \in \{1, \ldots, k\}$ for some $d \geq 0$. Then $|[\![q]\!]^{d+1}_{\mathcal{A}}| \geq \max\{|[\![p_i]\!]^d_{\mathcal{A}}| - dc(c) \mid X_i \in hv(c)\}$.*

Proof. Let $X_j \in hv(c)$ and $\phi \equiv C_1 \wedge \ldots \wedge C_m$, $m = dc(c)$. Reorder the C_i s. t. X_j is mentioned exactly in $C_1, \ldots, C_l$, $0 \leq l \leq m$. Choose θ s.t. $\theta \models C_{l+1} \wedge \ldots \wedge C_m$ and $X_i\theta \in [\![p_i]\!]^d_{\mathcal{A}}$ for all $i \in \{1, \ldots, k\} \setminus \{j\}$. Making $C_1, \ldots, C_l$ true by choosing $\theta(X_j)$ can be considered as an instance of the pigeonhole principle implying that there are at least $|[\![p_j]\!]^d_{\mathcal{A}}| - l \geq |[\![p_j]\!]^d_{\mathcal{A}}| - m$ different trees in $[\![p_j]\!]^d_{\mathcal{A}}$ which satisfy all C_i, $1 \leq i \leq l$. Since $X_j \in hv(c)$, each of them can be used in combination with the trees $X_i\theta, i \neq j$, to produce one tree for $[\![q]\!]^{d+1}_{\mathcal{A}}$.   □

Our main theorem considers a procedure which iteratively constructs all facts $p(t)$ with a proof depth less than or equal to some $m \geq 0$ which depends only on the number k, the number of predicates occurring in $\mathcal{A}$, and the total number of disequalities in $\mathcal{A}$. The theorem generalizes the one in [12] in that now variables have to be taken into consideration which only occur in the precondition of a clause but not in the head, and the heads of clauses are not restricted to terms with exactly one constructor.

Theorem 1. *Let $\mathcal{A}$ be an HDA with n predicates and $d = \sum_{c \in \mathcal{A}} dc(c)$ disequality constraints. Let k be a positive number. Then for all predicates p, it holds that $|[\![p]\!]_{\mathcal{A}}| < k$ if and only if $|[\![p]\!]_{\mathcal{A}}^{n(d+k)}| < k$.* $\square$

Corollary 1. *Let $\mathcal{A}$ be an HDA. Then for all predicates p and numbers $k \geq 0$, it can be effectively decided whether $|[\![p]\!]_{\mathcal{A}}| \leq k$. Moreover, if $|[\![p]\!]_{\mathcal{A}}| \leq k$, then $[\![p]\!]_{\mathcal{A}}$ can be effectively computed.* $\square$

From Lemma 4 and Corollary 1, we conclude:

Corollary 2. *Let $\mathcal{A}$ be a finite set of automata clauses (with hom-disequalities). Then for all predicates p it can be effectively decided whether $[\![p]\!]_{\mathcal{A}} = \emptyset$.* $\square$

Corollary 3. *Let $\mathcal{A}$ be a finite set of automata clauses. Then for all predicates p and numbers $k > 0$ it can be effectively decided whether $|\mathrm{H}[\![p]\!]_{\mathcal{A}}| < k$. Moreover, in case that $|\mathrm{H}[\![p]\!]_{\mathcal{A}}| = m < k$, a sequence $t_1, \ldots, t_m \in [\![p]\!]_{\mathcal{A}}$ can be effectively constructed such that the terms $\mathrm{H}t_i$, $i = 1, \ldots, m$, are pairwise distinct.* $\square$

5 $\mathcal{H}_1$-Normalization

In this section, we describe the normalization procedure which constructs for every finite set $\mathcal{C}$ of $\mathcal{H}_1$-clauses with hom-disequalities a finite set $\mathcal{N}$ of normal clauses with hom-disequalities which is equivalent to $\mathcal{C}$. Thus, this general procedure is quite in-line with the normalization procedures for unconstrained $\mathcal{H}_1$-clauses [9,7] or $\mathcal{H}_1$-clauses with term disequalities [12] or path disequalities [13]. The normalization procedure consists of three rules, *resolution*, *splitting*, and *propagation*, each of which adds finitely many simpler clauses which are implied by the current set of clauses. These rules are repeatedly applied until the set of clauses becomes saturated. The following paragraphs briefly collect the three types of normalization rules. A significant modification w.r.t. [12] only is required when it comes to *splitting*. We refer to the current set of all implied clauses (whether originally present or added during normalization) as $\mathcal{C}$, while $\mathcal{N} \subseteq \mathcal{C}$ denotes the current subset of normal clauses in $\mathcal{C}$.

Resolution: Complex queries in preconditions are simplified by a resolution step with a *normal clause*. Assume that $\mathcal{C}$ contains a clause $h \Leftarrow \alpha_1, p(t), \alpha_2, \psi$. If $\mathcal{N}$ has a clause $p(f(X_1, \ldots, X_k)) \Leftarrow \beta, \phi$, and $t = f(t_1, \ldots, t_k)$, then

$$h \Leftarrow \alpha_1, \alpha', \alpha_2, \psi \wedge \psi'$$

is added with $\alpha' = \beta[t_1/X_1, \ldots, t_k/X_k]$ and likewise, $\psi' = \phi[t_1/X_1, \ldots, t_k/X_k]$.

Splitting: *Splitting* removes variables that are not contained in the head of a clause. Assume that $\mathcal{C}$ contains a clause $h \Leftarrow \alpha, \psi$ and Y is a variable which occurs in the precondition α, ψ but neither occurs in h nor in any literal $q(t)$ with $t \neq Y$ within α. Then we can rearrange α into a sequence $\alpha', q_1(Y), \ldots, q_r(Y)$ where α' does not contain Y. Let ψ contain n disequalities involving Y.

In the case of term disequalities, the key issue is to decide whether the conjunction $[\![q_1]\!]_\mathcal{N} \cap \ldots \cap [\![q_r]\!]_\mathcal{N}$ contains less than $n + 1$ terms — and if so, to provide all terms of this set. In presence of the homomorphism H however, this is no longer sufficient. Instead, we must refer to the number of *images* of terms from $[\![q_1]\!]_\mathcal{N} \cap \ldots \cap [\![q_r]\!]_\mathcal{N}$ under H. In order to do so, we apply Lemma 1 from Section 4 and construct for $\mathcal{N}$ an HDA $\mathcal{A}$ such that $[\![p]\!]_\mathcal{A} = \mathrm{H}[\![p]\!]_\mathcal{N}$ for all predicates p of $\mathcal{N}$. By Corollary 3, we can decide k-finiteness (choosing $k = n + 1$) of the conjunction of the q_i with respect to this automaton. If only $n' < n + 1$ terms are in the set $\mathrm{H}([\![q_1]\!]_\mathcal{N} \cap \ldots \cap [\![q_r]\!]_\mathcal{N})$, the corollary provides us with n' witnesses in the set $[\![q_1]\!]_\mathcal{N} \cap \ldots \cap [\![q_r]\!]_\mathcal{N}$ whose images under H are pairwise distinct.

Let $\mathrm{H}([\![q_1]\!]_\mathcal{N} \cap \ldots \cap [\![q_r]\!]_\mathcal{N})$ contain m terms. If $m > n$, then we add the clause $h \Leftarrow \alpha', \psi'$ to the set $\mathcal{C}$ where ψ' is obtained from ψ by removing all disequalities that mention Y. If $m \leq n$, let $t_1, \ldots, t_m$ be the terms as provided by Corollary 3. Then we add to $\mathcal{C}$ all clauses

$$h \Leftarrow \alpha', \psi[t_i/Y], \quad i = 1, \ldots, m$$

Example 3. Consider again the clause

$$error \Leftarrow p_u(X), p_u(Y), X \neq_\mathrm{H} Y$$

from the example in the introduction (here, we use $X, Y, Z \ldots$ as variable names), and assume that $\mathrm{H}[\![p_u]\!]_\mathcal{N} = \{t\}$ for some ground term $t = f(\Box, b)$, where $\mathcal{N}$ denotes the whole (current) subset of normal clauses. One potential pre-image of t then is the term $t' = f(secret(a), b)$. Applying splitting for variable Y (and assuming $t' \in [\![p_u]\!]_\mathcal{N}$), we obtain the new clause

$$error \Leftarrow p_u(X), X \neq_\mathrm{H} f(secret(a), b)$$

Now applying splitting for variable X results in

$$error \Leftarrow f(secret(a'), b) \neq_\mathrm{H} f(secret(a), b)$$

for some (possibly different) pre-image $f(secret(a'), b) \in [\![p_u]\!]_\mathcal{N}$ of t. The disequality of the clause turns out to be false, which is due to the fact that p_u does not accept two or more terms that are *different modulo* H.   □

Propagation: Consider clauses $p(X_1) \Leftarrow q_1(X_1), \ldots, q_r(X_1), \psi$ where ψ only contains the variable X_1 (or none). Assume that $r > 0$, and $\mathcal{N}$ contains normal clauses $q_j(f(X_1, \ldots, X_k)) \Leftarrow \alpha_j, \psi_j$ for $j = 1, \ldots, r$. Then

$$p(f(X_1, \ldots, X_k)) \Leftarrow \alpha_1, \ldots, \alpha_r, \psi_1 \wedge \ldots \wedge \psi_r \wedge \psi'$$

is added where $\psi' = \psi[f(X_1, \ldots, X_k)/X_1]$.

The correctness of the construction can be proven along the lines in [12]:

Theorem 2. *Let $\mathcal{C}$ denote a finite set of $\mathcal{H}_1$-clauses. Let $\overline{\mathcal{C}}$ denote the set of all clauses obtained from $\mathcal{C}$ by adding all clauses according to the resolution, splitting and propagation rules. Then the subset $\mathcal{N}$ of all normal clauses in $\overline{\mathcal{C}}$ is equivalent to $\mathcal{C}$, i.e., $[\![p]\!]_\mathcal{C} = [\![p]\!]_\mathcal{N}$ for every predicate p occurring in $\mathcal{C}$.*   □

Termination of $\mathcal{H}_1$-normalization is achieved by avoiding to add certain clauses that are *subsumed* by the current set of clauses. Two clauses $h_i \Leftarrow \alpha_i, \phi_i, i = 1, 2$, are said to belong to the same *family* if they agree in their heads h_i and their preconditions α_i consist of the same set of literals. The two clauses still may differ in their disequality constraints ϕ_i. A clause $h \Leftarrow \alpha, \phi$ is subsumed by a set of clauses $h \Leftarrow \alpha_i, \phi_i, i = 1, \ldots, n$, from the same family, if ϕ implies the disjunction $\bigvee_{i=1}^{n} \phi_i$. Subsumed clauses can be omitted as they do not contribute new facts to the least model of a set of clauses.

Theorem 3. *Let $\mathcal{C}$ denote a finite set of $\mathcal{H}_1$-clauses. Let $\overline{\mathcal{C}}$ denote the set of clauses obtained from $\mathcal{C}$ by adding all clauses according to resolution, splitting and propagation, that are not subsumed. Then $\overline{\mathcal{C}}$ is finite.*

Proof. Since the number of predicates and constructors is finite, there are only finitely many distinct heads of clauses. The number of literals occurring in preconditions is bounded since new literals $p(t)$ are only added for *subterms* t of terms already present in the original set $\mathcal{C}$ of clauses. Therefore, the number of families of clauses occurring during normalization is finite. For each family f let $\psi_{\mathcal{C},f}$ denote the (possibly empty) disjunction of constraints of clauses of $\mathcal{C}$ which belong to f. Each clause that is added to $\mathcal{C}$ extends one of the finitely many constraints $\psi_{\mathcal{C},f}$ to $\psi_{\mathcal{C},f} \vee \phi$ for a conjunction of disequalities ϕ. The number of variables in each constraint $\psi_{\mathcal{C},f}$ is bounded since neither resolution with normal clauses nor splitting does introduce new variables, while propagation steps may introduce fresh variables, but directly produces normal clauses.

Now consider a sequence $\psi_i, i \geq 1$, of conjunctions of hom-disequalities. It remains to show that the disjunction $\bigvee_{i=1}^{m} \psi_i, m \geq 1$, eventually becomes *stable*, i.e., there exists some M such that $\bigvee_{i=1}^{m} \psi_i = \bigvee_{i=1}^{M} \psi_i$ for all $m \geq M$. In order to construct such an M consider the sequence $\psi_{\mathrm{H},i}, i \geq 1$, of ordinary term disequalities where $\psi_{\mathrm{H},i}$ is obtained from ψ_i by replacing each hom-disequality $s \neq_{\mathrm{H}} t$ with $\mathrm{H}(s) \neq \mathrm{H}(t)$. Then θ is a solution to ψ_i, iff $\mathrm{H} \circ \theta$ is a solution to $\psi_{\mathrm{H},i}$. In [12] we have shown that disjunctions of sequences of conjunctions of ordinary term disequalities are ultimately stable. Therefore, there exists an M' such that $\bigvee_{i=1}^{m} \psi_{\mathrm{H},i} = \bigvee_{i=1}^{M'} \psi_{\mathrm{H},i}$ for all $m \geq M'$. Then we choose the constant M as M'. In order to prove that the sequence $\bigvee_{i=1}^{m} \psi_i$ for $m \geq M'$ is implied by $\bigvee_{i=1}^{M'} \psi_i$, assume that θ is a solution of $\bigvee_{i=1}^{m} \psi_i$ for some $m \geq M'$. Then $\mathrm{H} \circ \theta$ is a solution of $\bigvee_{i=1}^{m} \psi_{\mathrm{H},i}$ and therefore also of $\bigvee_{i=1}^{M'} \psi_{\mathrm{H},i}$. Consequently, θ must also be a solution of $\bigvee_{i=1}^{M'} \psi_i$. Therefore, we conclude that also $\bigvee_{i=1}^{m} \psi_i = \bigvee_{i=1}^{M'} \psi_i$ for all $m \geq M$. This implies that eventually all clauses that can be added are subsumed. Therefore, the normalization procedure terminates. $\square$

According to Theorem 2 and Theorem 3, for every finite set $\mathcal{C}$ of $\mathcal{H}_1$-clauses with hom-disequalities an equivalent finite set $\mathcal{N}$ of normal clauses can be constructed. By Lemma 1, $\mathcal{N}$ can then be transformed into an equivalent finite set $\mathcal{A}$ of automata clauses. Finally, by Corollary 2, emptiness is decidable for every predicate defined by $\mathcal{A}$. Altogether, we obtain:

Theorem 4. *To every finite set C of $\mathcal{H}_1$-clauses with hom-disequality constraints, a finite set $\mathcal{A}$ of automata clauses can be effectively constructed such that for every predicate p of C, $[\![p]\!]_C = [\![p]\!]_\mathcal{A}$. In particular, emptiness is decidable.* □

6   Conclusion

We have shown that finite sets of $\mathcal{H}_1$-clauses with hom-disequalities, i.e., disequalities between images of terms under a given tree homomorphism, can be effectively transformed into finite tree automata with hom-disequalities. Since emptiness is decidable for these automata, we have provided a procedure to decide arbitrary conjunctive or disjunctive queries to predicates defined by such clauses. It remains for future work to explore how Horn clauses, extended with hom-disequalities can be applied to the verification of security properties of protocols, such as anonymity or non-interference [6,2].

References

1. Blanchet, B.: An efficient cryptographic protocol verifier based on prolog rules. In: CSFW, pp. 82–96 (2001)
2. Bugliesi, M., Rossi, S.: Non-interference proof techniques for the analysis of cryptographic protocols. Journal of Computer Security 13(1), 87–113 (2005)
3. Chatzikokolakis, K.: Probabilistic and Information-Theoretic Approaches to Anonymity. Ph.D. thesis, École polytechnique (2007)
4. Frühwirth, T.W., Shapiro, E.Y., Vardi, M.Y., Yardeni, E.: Logic programs as types for logic programs. In: LICS, pp. 314–328 (1991)
5. Godoy, G., Giménez, O., Ramos, L., Àlvarez, C.: The hom problem is decidable. In: STOC, pp. 485–494. ACM (2010)
6. Goguen, J.A., Meseguer, J.: Security policies and security models. In: IEEE Symposium on Security and Privacy, pp. 11–20 (1982)
7. Goubault-Larrecq, J.: Deciding H1 by resolution. IPL 95(3), 401–408 (2005)
8. Goubault-Larrecq, J., Parrennes, F.: Cryptographic Protocol Analysis on Real C Code. In: Cousot, R. (ed.) VMCAI 2005. LNCS, vol. 3385, pp. 363–379. Springer, Heidelberg (2005)
9. Nielson, F., Riis Nielson, H., Seidl, H.: Normalizable Horn Clauses, Strongly Recognizable Relations, and Spi. In: Hermenegildo, M.V., Puebla, G. (eds.) SAS 2002. LNCS, vol. 2477, pp. 20–35. Springer, Heidelberg (2002)
10. Reuß, A., Seidl, H.: Bottom-Up Tree Automata with Term Constraints. In: Fermüller, C.G., Voronkov, A. (eds.) LPAR-17. LNCS, vol. 6397, pp. 581–593. Springer, Heidelberg (2010)
11. Seidl, H., Neumann, A.: On Guarding Nested Fixpoints. In: Flum, J., Rodríguez-Artalejo, M. (eds.) CSL 1999. LNCS, vol. 1683, pp. 484–498. Springer, Heidelberg (1999)
12. Seidl, H., Reuß, A.: Extending H1-clauses with disequalities. IPL 111(20), 1007–1013 (2011)
13. Seidl, H., Reuß, A.: Extending $\mathcal{H}_1$-Clauses with Path Disequalities. In: Birkedal, L. (ed.) FOSSACS 2012. LNCS, vol. 7213, pp. 165–179. Springer, Heidelberg (2012)
14. Weidenbach, C.: Towards an Automatic Analysis of Security Protocols in First-Order Logic. In: Ganzinger, H. (ed.) CADE 1999. LNCS (LNAI), vol. 1632, pp. 314–328. Springer, Heidelberg (1999)

Factor and Subsequence Kernels
and Signatures of Rational Languages

Ahmed Amarni and Sylvain Lombardy

Laboratoire d'informatique Gaspard-Monge
University Paris-Est Marne-la-Vallée
{Ahmed.Amarni,Sylvain.Lombardy}@univ-mlv.fr

Abstract. The kernels are popular methods to measure the similarity between words for classification and learning. We generalize the definition of rational kernels in order to apply kernels to the comparison of languages. We study this generalization for factor and subsequence kernels and prove that these kernels are defined for parameters chosen in an appropriate interval. We give different methods to build weighted transducers which compute these kernels.

1 Introduction

In classification and learning, kernel methods, like *support vector machines*, are widely used ([1–3]). In many domains, like speech and handwritten document recognition, or computational biology, the kernel methods offer a simple and efficient answer for classification and pattern matching.

In this paper we consider *rational kernels*, introduced in [4] and extensively studied in [5]. We generalize the constructions introduced in [5] and define the kernels between rational languages. Kernels are usually defined between words and measure the similarity between a word and a witness. We consider here the comparison between two languages: a word (or even a language) can be evaluated with respect to a known corpus. We will define this kernel in such a way that the value of the word is higher if it is closer to small words of the corpus.

In Section 2, we present basic notions on weighted automata and transducers. In Section 3, we define the rational kernels of languages. For positive definite symmetric kernels, we define rational signature of languages as the behaviour of weighted transducers. We focus in the last sections on two specific rational kernels (and signatures). First, in Section 4, we consider factor kernels, second, in Section 5, we consider sequence kernels. For each of these kernels, we show how it corresponds to the behaviour of weighted automata. We also study, for applications, which values of the parameters involved in these kernels allow us to evaluate them. On top of this study, we provide an efficient construction for the subsequence signature of rational languages.

2 Basic Notions and Definitions

The definitions of weighted automata and weighted transducers given in this part follow the classical definitions (*cf.* [6] or [7]).

N. Moreira and R. Reis (Eds.): CIAA 2012, LNCS 7381, pp. 313–320, 2012.

2.1   Weighted Automata

Let A be an alphabet. A $\mathbb{R}$-automaton $\mathcal{A}$ over A^* is an automaton where each transition, initial arrow and final arrow is endowed with a weight in $\mathbb{R}$. The weight of a computation is the product of weights along this computation, and the weight of a word accepted by such an automaton is the sum of the weights of accepting computations. If a word is not accepted by $\mathcal{A}$ then its weight is set to zero. A $\mathbb{R}$-automaton $\mathcal{A}$ thus realizes a mapping from words into $\mathbb{R}$; this mapping can be seen as a formal power series. This series is called the *behaviour* of the $\mathbb{R}$-automaton, and $\mathbb{R}$-automata are equivalent if their behaviour is the same. For every word w in A^*, we denote by $\mathcal{A}(w)$ the weight associated to w by $\mathcal{A}$.

2.2   Weighted Transducers

Let A and B be two alphabets. Like $\mathbb{R}$-automata, $\mathbb{R}$-transducers over $A^* \times B^*$ are transducers endowed with weights in $\mathbb{R}$. The transition labels of our $\mathbb{R}$-transducers are in $(A \cup \{\varepsilon\}) \times (B \cup \{\varepsilon\})$ (ε is the empty word). There are two ways to consider a $\mathbb{R}$-transducer.

First, as a weighted acceptor: each computation is labeled by a pair of words in $A^* \times B^*$; the weight of a pair of words is therefore the sum of the weights of all accepting computations for this pair. The $\mathbb{R}$-transducer $\mathcal{T}$ realizes a mapping from $A^* \times B^*$ into $\mathbb{R}$, and for every pair (u, v) in $A^* \times B^*$, $\mathcal{T}(u, v)$ is the weight in $\mathbb{R}$ associated to (u, v). A weighted transducer realizing a rational kernel is such an acceptor.

Second, as a translator: if (u, v) is the label of a computation with weight k, this computation reads the input u and outputs the word v with weight k. Then, the image of a word u by the transducer is the sum of the weighted outputs of all transitions with input u. This image is therefore a polynomial over B^* with coefficients in $\mathbb{R}$, or even a formal power series, if there are infinitely many computations with input u; we denote by $\mathcal{T}(u)$ the image of u by $\mathcal{T}$. For the realization of signatures, the weighted transducers are seen as translators.

3   Rational Kernel of Languages and Signatures

In this paper, the kernels we consider are particular cases of rational kernels studied in [4].

Definition 1. *Let $\mathcal{T}$ be a $\mathbb{R}$-transducer over $A^* \times A^*$. The rational kernel K induced by $\mathcal{T}$ is the application of $A^* \times A^*$ into $\mathbb{R}$ realized by $\mathcal{T}$: $K(u, v) = \mathcal{T}(u, v)$, for every pair of words u and v.*

We address in this paper the extension of kernels defined for words to kernels for rational languages. Since the kernels are defined in $\mathbb{R}$, it is natural to consider linear extensions. Contrary to the extension proposed in [5], on the one hand, we deal with infinite languages and thus our automata are not acyclic, on the other hand, we consider languages through their characteristic series in $\mathbb{R}$, which is not a *closed semiring*, and where summation issues arise.

Definition 2. *Let K be a rational kernel, and let L and H be two languages. The rational kernel of parameter μ applied to L and H is defined by*

$$K_\mu(L, H) = \sum_{u \in L, v \in H} K(u, v) \mu^{|u| + |v|}. \tag{1}$$

If μ is a formal parameter, $K_\mu(L, H)$ is a formal power series. It is defined as soon as the rational kernel is defined for each pair of words. For applications, it may be interesting to evaluate this series. Then, the radius of convergence depends on the nature of the kernel.

We focus in this paper on positive definite symmetric rational kernels (*cf.* [5]). These kernels are defined as scalars product of rational *signatures*.

Definition 3. *Let A and B be two alphabets. Let $\mathcal{T}$ be an $\mathbb{R}$-transducer over $A^* \times B^*$. The $\mathcal{T}$-signature of a word w in A^* is $\mathcal{T}(w)$. The $\mathcal{T}$-signature of parameter μ of a language L included in A^* is*

$$\sigma_\mu(L) = \sum_{w \in L} \mathcal{T}(w) \mu^{|w|}. \tag{2}$$

Proposition 1. *If $\mathcal{T}$ is an $\mathbb{R}$-transducer, the $\mathcal{T}$-signature with parameter μ of a language is realized by a transducer $\mathcal{T}_\mu$.*

The $\mathcal{T}$-signature of a language L with parameter μ is therefore a power series over B^* whose coefficients are series in $\mathbb{R}[[\mu]]$. For each value of μ in the radius of convergence, the $\mathcal{T}$-signature of a rational language L is a rational power series over B^*.

If L is a rational language, the characteristic series of L is realized by any unambiguous automaton $\mathcal{A}$ recognizing L (for instance the minimal automaton of L), seen as an automaton with multiplicities, and the $\mathcal{T}$-signature of L is realized by the application of the transducer $\mathcal{T}_\mu$ (Proposition 1) on $\mathcal{A}$.

The componentwise product (Hadamard product) of the $\mathcal{T}$-signatures of two languages L and H is called the $\mathcal{T}$-*indicator series* $I_\mu(L, H)$ of L and H.

The $\mathcal{T}$-kernel of rank k and the $\mathcal{T}$-kernel of two languages L and H are respectively defined as

$$K_\mu^{(k)}(L, H) = \sum_{w \in B^k} \langle \sigma_\mu(L), w \rangle \langle \sigma_\mu(H), w \rangle, \tag{3}$$

$$K_\mu(L, H) = \sum_{k=0}^{\infty} K_\mu^{(k)}(L, H) = \sum_{w \in B^*} \langle \sigma_\mu(L), w \rangle \langle \sigma_\mu(H), w \rangle. \tag{4}$$

The $\mathcal{T}$-norm of a language L is defined as $\|L\|_\mu = \sqrt{K_\mu(L, L)}$ and can be evaluated if μ belongs to the radius of convergence.

Proposition 2. *Let $\mathcal{T}$ be a $\mathbb{R}$-transducer over $A^* \times B^*$, and let $\mathcal{T}_\mu$ be the transducer that realizes the $\mathcal{T}$-signature with parameter μ. Then, the $\mathcal{T}$-kernel with parameter μ is realized by the transducer $\mathcal{T}_\mu^{-1} \circ \mathcal{T}_\mu$.*

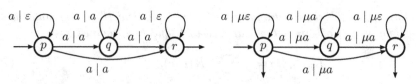

(a) The transducer computing χ (b) The transducer computing Ψ_μ

Fig. 1. Transducer extracting factors. Every transition is valid with any letter in input: transitions $a \mid a$ copy the input on the output, transitions $a \mid \varepsilon$ have no output.

4   Factor Signature

The similarity between two words can be evaluated by the *factor kernel*. we first extend this to languages. As symmetric rational kernels, factor kernels are characterized by signatures.

4.1   Factor Signature of Words

Let w be a word of length n in A^*. Let (i, l) be in $[0; n-1] \times [1; n-i]$; the factor of w with offset i and length l is $f(w, i, l) = w_{i+1} \ldots w_{i+l}$. Let $\mathsf{fact}(w)$ be the set of admissible pairs (i, l) describing a factor of w.

Definition 4. *Let w be a word. The factor signature of w is the linear combination defined by:*

$$\chi(w) = \sum_{(i,l) \in \mathsf{fact}(w)} f(w, i, l). \tag{5}$$

Proposition 3. *Let $\mathcal{T}$ be the $\mathbb{N}$-transducer of Figure 1(a). $\mathcal{T}$ computes the factor signature.*

4.2   Factor Signature of Languages

The factor signature of a language L with parameter μ is the series defined as

$$\Psi_\mu(L) = \sum_{w \in L} \chi(w)\mu^{|w|}. \tag{6}$$

The signature Ψ_μ is realized by the transducer of Figure 1(b). By metonymy, in the sequel, this transducer is called Ψ_μ.

If L is a language, let $\mathsf{Fact}_n(L)$ be the number of factors of length n in L, we consider the entropy of L, $\mathsf{E}(L)$:

$$\mathsf{E}(L) = \limsup_{n \to \infty} \frac{\log_2(\mathsf{Fact}_n(L))}{n}. \tag{7}$$

Proposition 4. *Let L be a rational language. If μ belongs to $[0; \frac{1}{2^{\mathsf{E}(L)}}[$, the factor signature $\Psi_\mu(L)$ is defined.*

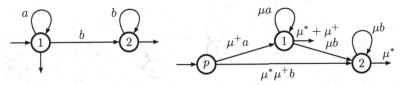

(a) The minimal automaton of L_2 (b) An automaton realizing $\Psi_\mu(L_2)$

Fig. 2. The factor signature of $L_2 = a^*b^*$

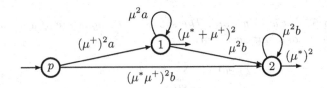

Fig. 3. The factor indicator series $I_\mu(L_2, L_2)$

Notice that $2^{\mathsf{E}(L)}$ is in $[1; |A|]$, where A is the alphabet; therefore if μ is smaller than $1/|A|$, $\Psi_\mu(L)$ is defined.

Example 1. $L_2 = a^*b^*$ is recognized by the deterministic automaton $\mathcal{A}_2$ of Figure 2(a). The application of Ψ_μ to $\mathcal{A}_2$, followed by the ε-transition removal gives the automaton of Figure 2(b).

Proposition 5. *If μ is in $[0; \frac{1}{|A|}[$, where A is the alphabet, for every pair of languages L and H, the factor kernel $K_\mu(L, H)$ is defined.*

Example 2. If $L = a^*b^*$, the factor indicator series $I_\mu(L_2, L_2)$ is realized by the automaton of Figure 3. This automaton is the square of the automaton of Figure 2(b).

5   Subsequence Signature and Kernel

Subsequences kernels, also called gappy n-gram kernels are rational kernels (*cf.* [8]) which involve a decay factor λ. We extend them to languages and obtain therefore kernels (described by signatures) with two parameters.

5.1   Subsequence Signature of Words

Let w be a word in A^*, for every k in $[1; n]$, every increasing sequence s of length k with values in $[1; |w|]$ leads to a word v of length k such that $v_i = w_{s_i}$: v is the subsequence of w indexed by s and is denoted by $\sigma(w, s)$. We denote the length of s by $|s|$ and we define the width of s as $\ell(s) = s_k - s_1 + 1$. Let $S(w)$ be the set of increasing sequences in $[1; |w|]$.

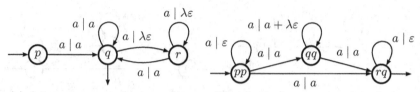

(a) A transducer erasing inner letters. (b) A transducer computing the subsequence signature.

Fig. 4. From factors to subsequences

Definition 5. *Let w be a word. The subsequence signature of w with parameter λ is the linear combination defined as:*

$$\varphi_\lambda(w) = \sum_{s \in S(w)} \lambda^{\ell(s) - |s|} \sigma(w, s). \tag{8}$$

The signature $\varphi_\lambda(w)$ can be evaluated for any value of λ in $]0; 1]$. If $\lambda = 1$, the coefficient of a word v in $\varphi_\lambda(w)$ is the number of occurences of v as a subsequence in w, and $\varphi_1(\lambda)$ is known as the *Magnus transformation (cf. [9])* of w; if $\lambda = 0$ every subsequence with a hole is discarded, and $\varphi_\lambda(w) = \chi(w)$.

Proposition 6. *There exists a $\mathbb{N}[\lambda]$-rational transduction τ_λ such that $\varphi_\lambda = \tau_\lambda \circ \chi$.*

As a consequence φ_λ is the $\mathbb{N}[\lambda]$-rational relation realized by the transducer of Figure 4(b) which is the composition of transducers of Figure 1(a) and Figure 4(a).

This definition of the subsequence signature leads to a kernel which is slightly different than the one given in [5], which is equal to $K_\lambda^{(k)}(u, v)\lambda^{2k}$, where $K_\lambda^{(k)}$ is the kernel defined in our framework.

5.2   Subsequence Signature of Languages

The subsequence signature of a language L is the bivariate series defined by:

$$\Phi_{\lambda,\mu}(L) = \sum_{w \in L} \varphi_\lambda(w)\mu^{|w|}. \tag{9}$$

For numerical applications, this signature can be evaluated. The following proposition gives values of parameters for which the signature is defined.

Proposition 7. *Let L be a rational language. For all μ in $[0; \frac{1}{2^{E(L)}}[$ and λ in $[0; \frac{1}{\mu^2 2^{E(L)}}[$, $\Phi_{\lambda,\mu}(L)$ is defined.*

Proposition 8. *The function $\Phi_{\lambda,\mu}$ which maps a language to its subsequence signature is a rational function.*

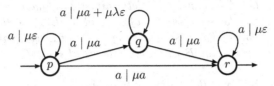

Fig. 5. The subsequence signature of a language: the transducer $\Phi_{\lambda,\mu} = \tau_\lambda \circ \Psi_\mu$

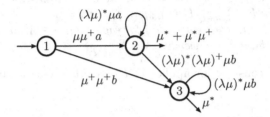

Fig. 6. An automaton realizing the subsequence signature of $L_2 = a^*b^*$

Example 3. Let $L_2 = a^*b^*$. The minimal automaton of L_2 is given on Figure 2(a). The subsequence signature of L_2 is obtained by the application of $\Phi_{\lambda,\mu}$ on this automaton. After ε-removal and state merging we get the automaton of Figure 6.

The subsequence signature allows to define a parametrized norm for rational languages. This norm is an indicator about the richness of subsequences in the language. The subsequence norm of L with parameter (λ, μ) is:

$$\|L\|_{\mu,\lambda} = \sqrt{\sum_{u \in A^*} \langle \Phi_{\lambda,\mu}(L), u \rangle^2}. \tag{10}$$

Proposition 9. *Let L be a rational language over an alphabet A. If λ and μ are two non negative reals such that μ is in $[0; \frac{1}{|A|}[$ and λ is in $[0; \frac{1}{\mu|A|} - 1[$, then $\|L\|_{\mu,\lambda}$ is defined.*

5.3 Direct Computation of the Automaton Realizing the Subsequence Signature

We focus in this part on the computation of the realization of the subsequence signature of a rational language. (A similar construction exists for factor signatures). We consider a rational language L recognized by standard deterministic automaton $\mathcal{A}$: there is no incoming transition to the initial state of $\mathcal{A}$.

Let Q be the set of states of $\mathcal{A}$. We define the $Q \times Q$ matrix $E(x)$ as follows. For every pair of states (p, q), $E_{p,q}(x) = kx$, where k is the number of transitions from p to q. $E_{p,q}^n(x) = kx^n$ if there are exactly k paths of length n from p to q, and $E_{p,q}^*(x)$ is the series that counts the number of paths of each length between p and q.

Proposition 10. *Let $\mathcal{A}$ be a standard deterministic automaton which accepts a language L. Let Q be the set of states of $\mathcal{A}$, and let $E(x)$ be the parametrized matrix defined as above. The subsequence signature of L with parameters λ and μ is realized by the automaton $\mathcal{B}$ with the same set of states Q as $\mathcal{A}$, and defined by:*

– The initial state i of $\mathcal{B}$ is the initial state of $\mathcal{A}$.
– For each state $p \neq i$, p is final in $\mathcal{B}$ with weight $k = \sum_{q \in T} E^*(\mu)_{p,q}$ if $k \neq 0$.
– For each state r, for each letter $a \in A$, there is a transition from i to r with label a and weight $k = \sum_{q \in Q} E^*(\mu)_{i,q} \mu \delta(a)_{q,r}$ if $k \neq 0$.
– For each state $p \neq i$, each state $r \neq i$ for each letter $a \in A$, there is a transition from p to r with label a and weight : $k = \sum_{q \in Q} E^*_{p,q}(\lambda\mu)\mu\delta(a)_{q,r}$ if $k \neq 0$.

Example 4. This construction applied to L_2 directly gives the automaton drawn in Figure 6.

References

1. Burges, C.: A tutorial on support vector machines for pattern recognition. Data Mining Knowledge Discovery 2, 121–167 (1998)
2. Lodhi, H., Saunders, C., Shawe-Taylor, J., Cristianini, N., Watkins, C.: Text classification using string kernels. Journal of Machine Learning Research 2, 419–444 (2002)
3. Kontorovich, L., Cortes, C., Mohri, M.: Kernel methods for learning languages. Theoretical Computer Science 405, 223–236 (2008)
4. Cortes, C., Haffner, P., Mohri, M.: Rational kernels. In: Becker, S., Thrun, S., Obermayer, K. (eds.) NIPS, pp. 601–608. MIT Press (2002)
5. Cortes, C., Haffner, P., Mohri, M.: Rational kernels: Theory and algorithms. Journal of Machine Learning Research 5, 1035–1062 (2004)
6. Kuich, W., Salomaa, A.: Semirings, Automata, Languages. Monographs in Theoretical Computer Science. An EATCS Series, vol. 5. Springer (1986)
7. Sakarovitch, J.: Elements of Automata Theory. Cambridge University Press (2009)
8. Cortes, C., Haffner, P., Mohri, M.: Positive Definite Rational Kernels. In: Schölkopf, B., Warmuth, M.K. (eds.) COLT/Kernel 2003. LNCS (LNAI), vol. 2777, pp. 41–56. Springer, Heidelberg (2003)
9. Sakarovitch, J., Simon, I.: Subwords. In: Combinatorics on Words. Encyclopedia of Mathematics and its Applications, vol. 17, pp. 104–144. Addison-Wesley, Reading (1983)

Multi-Tilde-Bar Derivatives

Pascal Caron, Jean-Marc Champarnaud, and Ludovic Mignot

LITIS, Université de Rouen, 76801 Saint-Étienne du Rouvray Cedex, France
{pascal.caron,jean-marc.champarnaud,ludovic.mignot}@univ-rouen.fr

Abstract. Multi-tilde-bar operators allow us to extend regular expressions. The associated extended expressions are compatible with the structure of Glushkov automata and they provide a more succinct representation than standard expressions. The aim of this paper is to examine the derivation of multi-tilde-bar expressions. Two types of computation are investigated: Brzozowski derivation and Antimirov derivation, as well as the construction of the associated automata.

1 Introduction

Regular expression word derivatives have been introduced in [5] by Brzozowski in order to compute language quotients via expression derivatives: for any word w, the language denoted by the derivative of a regular expression E w.r.t. w is the left quotient of the language denoted by E w.r.t. w. Regular expression derivation plays a fundamental role in theory of automata. In particular, under the assumption that the set D of all the derivatives of a regular expression E is finite, it is possible to construct a FA (finite automaton) with D as a set of states that recognizes the language denoted by E.

Word derivatives handle unrestricted regular expressions; they are themselves expressions and they provide a DFA (deterministic finite automaton), as far as the ACI (associativity, commutativity and idempotence) properties of the sum of two expressions are used. Alternative types of derivation have been designed since Brzozowski's seminal work. Partial derivatives, due to Antimirov [2], only address simple regular expressions; they are sets of expressions and they provide both a DFA and a NFA (non-deterministic finite automaton). Antimirov derivatives have been recently extended to unrestricted regular expressions [10]; extended partial derivatives are sets of sets of expressions and they provide a DFA, a NFA and an AFA (alternating finite automaton) [11]. Some derivations are based on the linearization of the (simple) input expression: let us cite the continuations of Berry and Sethi [4], the c-continuations of Champarnaud and Ziadi [14] and the derivatives of Ilie and Yu [18]. Let us mention that Antimirov derivation has been extended to the case of weighted rational expressions [21,13].

As reported in [2], the concept of derivation has been successfully used to investigate the properties of regular expressions [17,15,7,20,3,1]. More recently, Brzozowski introduced a new approach for studying the state complexity of regular languages, based on the counting of their quotients (or of their derivatives) [6].

N. Moreira and R. Reis (Eds.): CIAA 2012, LNCS 7381, pp. 321–328, 2012.

Moreover, derivatives provide a useful tool to implement regular matching algorithms [23,16], or scanner generators as reported in [22].

A close topic is the derivation of new operators that extend regular expressions. For example, the computation of the derivatives of an approximative regular expression (that denotes a languages at a bounded distance from a given language) has been presented in [12]. The aim of this paper is to investigate the derivation of the multi-tilde-bar expressions introduced in [8,9]. These expressions are built upon simple operators and multi-tilde-bar operators and their main interest is that they are compatible with the structure of Glushkov automata and more succinct than standard expressions. We provide formulae for the computation of word and partial derivatives of multi-tilde-bar expressions and investigate the properties of these derivatives.

The next section gathers classical notions concerning regular languages, regular expressions and finite automata; it also recalls the definition and main properties of multi-tilde-bar operators. The definition of the quotient of the language of an extended to multi-tilde-bar expression is introduced in Section 3. Section 4 is devoted to the computation of the Brzozowski derivatives of an extended expression and Section 5 to the computation of the Antimirov derivatives. In both cases, the construction of the associated automaton is provided.

2   Preliminaries

We recall some definitions and notation concerning regular languages, regular expressions , finite automata and multi-tilde-bar expressions. For further details about these topics, we refer to classical books such as [24].

Languages, Regular Expressions and Automata
An *alphabet* is a finite set of symbols. Given an alphabet Σ, any subset of Σ^* is a *language over* Σ. The *set of regular languages* over Σ is denoted by $\mathrm{Reg}(\Sigma^*)$ and is defined as the smallest family of languages containing $\emptyset$ and $\{a\}$ for every symbol a in Σ and closed under union, catenation and Kleene star. A *regular expression* E over an alphabet Σ is inductively defined by $E = 0$, $E = 1$, $E = a$, $E = (F + G)$, $E = (F \cdot G)$, $E = (F^*)$ with a a symbol in Σ, and F and G two regular expressions over Σ. The *language denoted by* a regular expression is inductively defined by $L(0) = \emptyset$, $L(1) = \{\varepsilon\}$, $L(a) = \{a\}$, $L(F+G) = L(F) \cup L(G)$, $L(F \cdot G) = L(F) \cdot L(G)$ and $L(F^*) = L(F)^*$, with a a symbol in Σ, and F and G two regular expressions over Σ. By construction, the language denoted by a regular expression is regular. The *alphabetic width* $|E|$ of E is the number of occurrences of symbols of Σ appearing in E. A *finite automaton* A is a 5-tuple $(\Sigma, Q, I, F, \delta)$ where Σ is an alphabet, Q is a finite set of states, $I \subset Q$ a set of *initial states*, $F \subset Q$ a set of *final states* and $\delta \subset Q \times \Sigma \times Q$ a set of *transitions*. The set δ can be seen as a function from $Q \times \Sigma$ to 2^Q defined by $q' \in \delta(q, a) \Leftrightarrow (q, a, q') \in \delta$. The domain of the function δ can be extended to $2^Q \times \Sigma^*$ by setting, for all $Q' \subset Q$, $\delta(Q', \varepsilon) = Q'$, $\delta(Q', a) = \bigcup_{q \in Q'} \delta(q, a)$, $\delta(Q', a \cdot w) = \delta(\delta(Q', a), w)$ for all word w in Σ^* . The language *recognized* by the automaton A is the set $L(A) = \{w \in \Sigma^* \mid \delta(I, w) \cap F \neq \emptyset\}$. A language

L is *recognizable* if there exists an automaton that recognizes it. The *set of recognizable languages* over Σ is denoted by $\mathrm{Rec}(\Sigma^*)$. Kleene theorem [19] asserts that $\mathrm{Reg}(\Sigma^*) = \mathrm{Rec}(\Sigma^*)$. Consequently , for every regular language L, there exist an automaton A and an expression E such that $L = L(E) = L(A)$.

The Multi-tilde-Bar Operators [8,9]

The unary operators *tilde*, denoted by $\sim$, and *bar*, denoted by $-$ are defined for every expression E by $L(\widetilde{E}) = L(E) \cup \{\varepsilon\}$ and $L(\overline{E}) = L(E) \setminus \{\varepsilon\}$. They are extended to *multi-tilde-bar operators*, which are applied to a list of expressions, according to the following definitions.

Let n be a positive integer. For convenience, the list $(E_1, \ldots, E_n)$ of expressions is denoted by $E_{1,n}$. Similarly, a catenation $E_1 \cdots E_n$ is denoted by $E_{1 \cdots n}$. The set of integers $\{1, \ldots, n\}$ is denoted by $[\![1, n]\!]$. The subset of pairs (i, j) such that if $1 \leq i \leq j \leq n$ is denoted by $[\![1, n]\!]^2_{\leq}$. The set of finite lists of pairs in $[\![1, n]\!]^2_{\leq}$ is denoted by $\mathcal{S}_n$.

Let S be a list in $\mathcal{S}_n$. Let k be in $[\![1, n]\!]$. The list $S_{\leq k}$ (resp. $S_{\geq k}$) is defined by $S_{\leq k} = ((i, f) \in S \mid f \leq k)$ (resp. $S_{\geq k} = ((i - k + 1, f - k + 1) \in S \mid i \geq k))$. Let us notice that a renumbering is performed for the computation of $S_{\geq k}$. A list S is said to be *free* if for all pairs $(i, f), (i', f')$ in S such that $(i, f) \neq (i', f')$, $[\![i, f]\!] \cap [\![i', f']\!] = \emptyset$. Let $L_1, \ldots, L_n$ be n nonempty regular languages over Σ and w be a word in $L_1 \cdots L_n$. A sequence $(w_1, \ldots, w_n)$ satisfying $w_1 \cdots w_n = w \wedge \forall k \in [\![1, n]\!], w_k \in L_k$ is said to be a *split up* of w over $(L_1, \ldots, L_n)$.

Multi-tilde-bar operators are a natural combination of multi-tilde and multi-bar operators [9]. The respective role of tildes and bars is explicited in the two following definitions.

Definition 1. *Let $(w_1, \ldots, w_n)$ be a split up of a word w over a list of languages $(L_1 \cup \{\varepsilon\}, \ldots, L_n \cup \{\varepsilon\})$. Let T be a free list in $\mathcal{S}_n$. The sequence $(w_1, \ldots, w_n)$ is generated by the list T if it holds: $w_k = \varepsilon$ if $k \in \bigcup_{(i,f) \in T} [\![i, f]\!]$ and $w_k \in L_k$ otherwise.*

Bars are used to forbid some combinations of tildes. Consequently, the satisfaction of a bar by a sequence has to be defined with a list of tildes as a context.

Definition 2. *Let $E_{1,n}$ be a list of n expressions. Let $(w_1, \ldots, w_n)$ be a split up of a word w over $(L(E_1) \cup \{\varepsilon\}, \ldots, L(E_n) \cup \{\varepsilon\})$ generated by a free list T in $\mathcal{S}_n$. Let $b = (i, f)$ be a pair in $[\![1, n]\!]^2_{\leq} \setminus T$. The bar b is said to be satisfied by $(w_1, \ldots, w_n)$ w.r.t. T if at least one of the three following conditions is satisfied:*
   (1) there exists a pair t in T such that t overlaps b,
   (2) there exists a pair t in T such that b is included in t,
   (3) $w_i \cdots w_f \neq \varepsilon$.

According to the two previous definitions, the language denoted by a multi-tilde-bar can be expressed as follows:

Definition 3 ([8]). *Let $E_{1,n}$ be a list of expressions over an alphabet Σ and L' be the list $(L(E_1) \cup \{\varepsilon\}, \ldots, L(E_n) \cup \{\varepsilon\})$ of languages. Let B and T be two lists* in S_n such that $B \cap T = \emptyset$. The multi-tilde-bar $E = \overset{\approx}{}_{T;B}(E_{1,n})$ denotes the language

$$L(E) = \left\{ w \in \Sigma^* \;\middle|\; \begin{array}{l} \text{there exists a split up of } w \text{ over } L' \text{ generated by a free} \\ \text{sublist } T' \text{ of } T \text{ satisfying every bar in } B \text{ w.r.t. } T'. \end{array} \right\}$$

Example 1. Let us consider the EMRE E_1 defined by

$$E_1 = \overset{\approx}{}_{(1,1),(2,2);(1,2)}\big((a^*b),(b^*a)\big) \cdot a^* \quad (\textit{i.e. } \overbrace{(\overset{\approx}{a^*b})(\overset{\approx}{b^*a})} \cdot a^*).$$

The language denoted by E_1 is the set

$$L(E_1) = (((L(a^*b) \cup \{\varepsilon\}) \cdot (L(b^*a) \cup \{\varepsilon\})) \setminus \{\varepsilon\}) \cdot L(a^*).$$

Definition 4. *Let Σ be an alphabet. An* Extended to multi-tilde-bar Regular Expression *(**EMRE**) over Σ is inductively defined by:*

$$E = 0, \quad E = 1, \quad E = a,$$
$$E = E_1 + E_2, \quad E = E_1 \cdot E_2, \quad E = E_1^*, \quad E = \overset{\approx}{}_{T;B}(E_{1,n}),$$

where $E_1, \ldots, E_n$ are any n EMREs over an alphabet Σ, a is any symbol in Σ and T and B are any two disjoint lists in S_n.

Definition 5. *An EMRE is said to be* total *if and only if for any of its multi-tilde-bar subexpressions $\overset{\approx}{}_{T;B}(E_{1,n})$ it holds $T \cup B = [\![1,n]\!]_\leq^2$.*

Lemma 1 ([8]). *Any EMRE admits an equivalent total one.*

3   Quotient Formulae

We now recall the inductive computation of the quotient $w^{-1}(L)$ of a language L w.r.t. a word w in Σ^*, that is the set $\{w' \in \Sigma^* \mid ww' \in L\}$.

Lemma 2. *Let L be language in $\mathrm{Reg}(\Sigma^*)$ and w be a word in Σ^*. The quotient $w^{-1}(L)$ of L w.r.t. w is inductively computed as follows:*

$$\varepsilon^{-1}(L) = L, \quad (aw')^{-1}(L) = w'^{-1}(a^{-1}(L)),$$
$$a^{-1}(\emptyset) = a^{-1}(\{\varepsilon\}) = a^{-1}(\{b\}) = \emptyset, \quad a^{-1}(a) = \{\varepsilon\},$$
$$a^{-1}(L_1 \cup L_2) = a^{-1}(L_1) \cup a^{-1}(L_2), \quad a^{-1}(L_1^*) = a^{-1}(L_1) \cdot L_1^*,$$
$$a^{-1}(L_1 \cdot L_2) = \begin{cases} a^{-1}(L_1) \cdot L_2 \cup a^{-1}(L_2) & \text{if } \varepsilon \in L_1, \\ a^{-1}(L_1) \cdot L_2 & \text{otherwise.} \end{cases}$$

where L_1 and L_2 are any two languages in $\mathrm{Reg}(\Sigma^*)$, a and b are any two distincts symbols in Σ and w' is any word in Σ^*.

Lemma 3. *Let $E = \overset{\approx}{}_{T;B}(E_{1,n})$ be a total EMRE over an alphabet Σ. Then:*

$$L(E) = \left(\begin{array}{l} \{\varepsilon \mid (1,n) \in T\} \cup (L(E_1) \setminus \{\varepsilon\}) \cdot L(\overset{\approx}{}_{T_{\geq 2};B_{\geq 2}}(E_{2,n})) \\ \cup \bigcup_{(1,k-1) \in T} (L(E_k) \setminus \{\varepsilon\}) \cdot L(\overset{\approx}{}_{T_{\geq k+1};B_{\geq k+1}}(E_{k+1,n})) \end{array} \right).$$

Corollary 1. *Let $E = \overset{\approx}{}_{T;B}(E_{1,n})$ be a total EMRE over an alphabet Σ and let a be a symbol in Σ. Then:*

$$a^{-1}(L(E)) = \left(\begin{array}{l} a^{-1}(L(E_1)) \cdot L(\overset{\approx}{}_{T_{\geq 2};B_{\geq 2}}(E_{2,n})) \\ \cup \bigcup_{(1,k-1) \in T} a^{-1}(L(E_k)) \cdot L(\overset{\approx}{}_{T_{\geq k+1};B_{\geq k+1}}(E_{k+1,n})) \end{array} \right)$$

4 Word Derivatives of an EMRE

The set of all the word derivatives of a regular expression can be infinite. However Brzozowski derivation yields a finite set of derivatives (called *dissimilar derivatives*) based on the use of the $+_{ACI}$ operator that is associative, commutative and idempotent. We extend these results to the case of EMREs and give the construction of the dissimilar derivative DFA of an EMRE.

Definition 6. *Let E be regular expression over the alphabet Σ and w be a word in Σ^*. The dissimilar derivative $\frac{d}{d_a}(E)$ of E w.r.t. w is inductively computed as*

$$\frac{d}{d_\varepsilon}(E) = E, \quad \frac{d}{d_{aw'}}(E) = \frac{d}{d_{w'}}\left(\frac{d}{d_a}(E)\right),$$

$$\frac{d}{d_a}(0) = \frac{d}{d_a}(1) = \frac{d}{d_a}(b) = 0, \quad \frac{d}{d_a}(a) = 1,$$

$$\frac{d}{d_a}(F+G) = \frac{d}{d_a}(F) + \frac{d}{d_a}(G), \quad \frac{d}{d_a}(F^*) = \frac{d}{d_a}(F) \cdot F^*,$$

$$\frac{d}{d_a}(F \cdot G) = \begin{cases} \frac{d}{d_a}(F) \cdot G +_{ACI} \frac{d}{d_a}(G) & \text{if } \varepsilon \in L(F), \\ \frac{d}{d_a}(F) \cdot G & \text{otherwise.} \end{cases}$$

where F and G are any two regular expressions over the alphabet Σ, a and b are any two distincts symbols of Σ and w' is any word in Σ^*.

Definition 7. *Let $E = \widetilde{}_{T;B}(E_{1,n})$ be a total EMRE over an alphabet Σ, let a be a symbol in Σ and w be a word in Σ^*. Then:*

$$\frac{d}{d_a}(E) = \left(\begin{array}{l} \frac{d}{d_a}(E_1) \cdot \widetilde{}_{T_{\geq 2};B_{\geq 2}}(E_{2,n}) \\ +_{ACI} \sum_{ACI(1,k-1) \in T} \frac{d}{d_a}(E_k) \cdot \widetilde{}_{T_{\geq k+1};B_{\geq k+1}}(E_{k+1,n}) \end{array} \right),$$

$$\frac{d}{d_w}(E) = \begin{cases} E & \text{if } w = \varepsilon, \\ \frac{d}{d_{w'}}\left(\frac{d}{d_b}(E)\right) & \text{if } w = b \cdot w' \wedge b \in \Sigma \wedge w' \in \Sigma^*. \end{cases}$$

Proposition 1. *The derivative of an EMRE E w.r.t. a word w denotes the set $w^{-1}(L(E))$.*

Proposition 2. *The set of dissimilar derivatives of an EMRE is finite.*

Definition 8. *Let E be an EMRE over an alphabet Σ and $\mathcal{D}_E$ be the set of the dissimilar derivatives of E. Let $A = (\Sigma, Q, I, F, \delta)$ be the automaton defined by $Q = \mathcal{D}_E$, $I = \{E\}$, $F = \{E' \in Q \mid \varepsilon \in L(E')\}$, $\forall E' \in Q$, $\forall a \in \Sigma$, $\delta(E', a) = \{\frac{d}{d_a}(E')\}$. The automaton A is the* dissimilar derivative DFA *of E.*

Proposition 3. *The dissimilar derivative DFA of an EMRE E recognizes $L(E)$.*

Example 2. Let us consider the total EMRE $E_1 = \widetilde{(a^*b)}\widetilde{(b^*a)} \cdot a^*$ defined in Example 1. Successive dissimilar derivatives of E are computed as follows:

$$\frac{d}{d_a}(E_1) = a^*b \cdot \widetilde{(b^*a)} \cdot a^* + a^* = E_2 \qquad \frac{d}{d_a}(E_4) = a^* = E_5$$

$$\frac{d}{d_b}(E_1) = \widetilde{(b^*a)} \cdot a^* + b^*a \cdot a^* = E_3 \qquad \frac{d}{d_b}(E_4) = b^*a \cdot a^* = E_6$$

$$\frac{d}{d_a}(E_2) = a^*b \cdot \widetilde{(b^*a)} \cdot a^* + a^* = E_2 \qquad \frac{d}{d_a}(E_5) = a^* = E_5$$

$$\frac{d}{d_b}(E_2) = \widetilde{(b^*a)} \cdot a^* = E_4 \qquad \frac{d}{d_b}(E_5) = 0$$

$$\frac{d}{d_a}(E_3) = a^* = E_5 \qquad \frac{d}{d_a}(E_6) = a^* = E_5$$

$$\frac{d}{d_b}(E_3) = b^*a \cdot a^* = E_6 \qquad \frac{d}{d_b}(E_6) = b^*a \cdot a^* = E_6$$

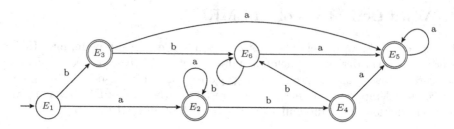

Fig. 1. The Dissimilar Derivative DFA of E_1

5   Partial Derivatives of an EMRE

Partial derivatives [2] of a regular expression are defined as follows.

Definition 9. *The partial derivative of a regular expression E w.r.t. a word w is the set $\frac{\partial}{\partial_a}(E)$ of expressions inductively computed as follows:*

$$\frac{\partial}{\partial_\varepsilon}(E) = E, \quad \frac{\partial}{\partial_{aw'}}(E) = \frac{\partial}{\partial_{w'}}(\frac{\partial}{\partial_a}(E)),$$

$$\frac{\partial}{\partial_a}(0) = \frac{\partial}{\partial_a}(1) = \frac{\partial}{\partial_a}(b) = \emptyset, \quad \frac{\partial}{\partial_a}(a) = \{1\},$$

$$\frac{\partial}{\partial_a}(F + G) = \frac{\partial}{\partial_a}(F) \cup \frac{\partial}{\partial_a}(G), \quad \frac{\partial}{\partial_a}(F^*) = \frac{\partial}{\partial_a}(F) \cdot F^*,$$

$$\frac{\partial}{\partial_a}(F \cdot G) = \begin{cases} \frac{\partial}{\partial_a}(F) \cdot G \cup \frac{\partial}{\partial_a}(G) & \text{if } \varepsilon \in L(F), \\ \frac{\partial}{\partial_a}(F) \cdot G & \text{otherwise.} \end{cases}$$

where: F and G are any two regular expressions over the alphabet Σ, a and b are any two distincts symbols of Σ and w' is any word in Σ^* and for any set of expressions $\mathcal{E}$, $\frac{\partial}{\partial_a}(\mathcal{E}) = \bigcup_{E \in \mathcal{E}} \frac{\partial}{\partial_a}(E)$, $L(\mathcal{E}) = \bigcup_{E \in \mathcal{E}} L(E)$.

We now define the partial derivatives of a total EMRE.

Definition 10. *Let $E = \overset{\frown}{}_{T;B}(E_{1,n})$ be a total EMRE over an alphabet Σ, let a be a symbol in Σ and w be a word in Σ^*. Then:*

$$\frac{\partial}{\partial_a}(E) = \left(\begin{array}{c} \frac{\partial}{\partial_a}(E_1) \cdot \overset{\frown}{}_{T_{\geq 2};B_{\geq 2}}(E_{2,n}) \\ \cup \bigcup_{(1,k-1) \in T} \frac{\partial}{\partial_a}(E_k) \cdot \overset{\frown}{}_{T_{\geq k+1};B_{\geq k+1}}(E_{k+1,n}) \end{array} \right),$$

$$\frac{\partial}{\partial_w}(E) = \begin{cases} \{E\} & \text{if } w = \varepsilon, \\ \frac{\partial}{\partial_{w'}}(\frac{\partial}{\partial_b}(E)) & \text{if } w = b \cdot w' \wedge b \in \Sigma \wedge w' \in \Sigma^*. \end{cases}$$

Proposition 4. *Let $E = \overset{\frown}{}_{T;B}(E_{1,n})$ be a total EMRE over an alphabet Σ and w be a word in Σ^*. Then $L(\frac{\partial}{\partial_w}(E)) = w^{-1}(L(E))$.*

By definition, a partial derivative of an expression E is a set of expressions and each of these expressions is called a *derivated term* of E. We show that the set $\mathcal{D}'_E$ of all the derivated terms of an EMRE E is finite and we give the construction of the derivated term NFA.

Lemma 4. *Let $E = \overset{\frown}{}_{T;B}(E_{1,n})$ be a total EMRE over an alphabet Σ and let w be a word in Σ^+. Then:*

$$\frac{\partial}{\partial_w}(E) \subset \bigcup_{w=uv \wedge v \neq \varepsilon} \bigcup_{k=1}^{n} \frac{\partial}{\partial_v}(E_k) \cdot \overset{\frown}{}_{T_{\geq k+1};B_{\geq k+1}}(E_{k+1,n}).$$

Proposition 5. *Let E be a total EMRE . Then:* $(\#\mathcal{D}'_E) \leq |E| + 1$.

Definition 11. *Let E be an EMRE over an alphabet Σ . Let $A = (\Sigma, Q, I, F, \delta)$ be the automaton defined by $Q = \mathcal{D}'_E$, $I = \{E\}$, $F = \{E' \in Q \mid \varepsilon \in L(E')\}$, for any expression $E' \in Q$, for any symbol a in Σ, $\delta(E', a) = \frac{\partial}{\partial_a}(E')$. The automaton A is the* derivated term NFA *of E.*

Proposition 6. *The derivated term automaton of an EMRE E recognizes $L(E)$.*

Example 3. Let us consider the total EMRE $E_1 = \overline{(\overset{\lower1mm\hbox{$\sim\sim$}}{a^*b})(\overset{\lower1mm\hbox{$\sim\sim$}}{b^*a})} \cdot a^*$ defined in Example 2. Successive derivated terms of E are computed as follows:

$$\frac{\partial}{\partial_a}(E_1) = \{a^*b(\overset{\sim\sim}{b^*a}) \cdot a^*, a^*)\}$$
$$= \{E'_2, E'_3\}$$
$$\frac{\partial}{\partial_b}(E_1) = \{(\overset{\sim\sim}{b^*a}) \cdot a^*, b^*a \cdot a^*\}$$
$$= \{E'_4, E'_5\}$$
$$\frac{\partial}{\partial_a}(E'_2) = \{a^*b(\overset{\sim\sim}{b^*a}) \cdot a^*\} = \{E'_2\}$$
$$\frac{\partial}{\partial_b}(E'_2) = \{(\overset{\sim\sim}{b^*a}) \cdot a^*\} = \{E'_4\}$$

$$\frac{\partial}{\partial_a}(E'_3) = \{a^*\} = \{E'_3\}$$
$$\frac{\partial}{\partial_b}(E'_3) = \emptyset$$
$$\frac{\partial}{\partial_a}(E'_4) = \{a^*\} = \{E'_3\}$$
$$\frac{\partial}{\partial_b}(E'_4) = \{b^*a \cdot a^*\} = \{E'_5\}$$
$$\frac{\partial}{\partial_a}(E'_5) = \{a^* = \{E'_3\}$$
$$\frac{\partial}{\partial_b}(E'_5) = \{b^*a \cdot a^*\} = \{E'_5\}$$

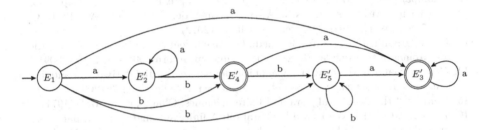

Fig. 2. The Derivated Term NFA of E_1

6 Conclusion

We have shown how the Brzozowski derivation and the Antimirov one can be applied to the case of (simple) regular expressions extended to multi-tilde-bar operators. The computation of the c-continuations for such expressions has been already investigated even though it is not presented here. The main interest of c-continuations is that they allow us to efficiently implement Glushkov and Antimirov NFAs. We also intend to generalize these derivations to the case of unrestricted regular expressions extended to multi-tilde-bar operators.

References

1. Almeida, M., Moreira, N., Reis, R.: Antimirov and Mosses's rewrite system revisited. Int. J. Found. Comput. Sci. 20(4), 669–684 (2009)
2. Antimirov, V.: Partial derivatives of regular expressions and finite automaton constructions. Theoret. Comput. Sci. 155, 291–319 (1996)

3. Antimirov, V.M., Mosses, P.D.: Rewriting extended regular expressions. Theor. Comput. Sci. 143(1), 51–72 (1995)
4. Berry, G., Sethi, R.: From regular expressions to deterministic automata. Theoret. Comput. Sci. 48(1), 117–126 (1986)
5. Brzozowski, J.A.: Derivatives of regular expressions. J. Assoc. Comput. Mach. 11(4), 481–494 (1964)
6. Brzozowski, J.A.: Quotient complexity of regular languages. Journal of Automata, Languages and Combinatorics 15(1/2), 71–89 (2010)
7. Brzozowski, J.A., Leiss, E.L.: On equations for regular languages, finite automata, and sequential networks. Theor. Comput. Sci. 10, 19–35 (1980)
8. Caron, P., Champarnaud, J.M., Mignot, L.: Erratum to "acyclic automata and small expressions using multi-tilde-bar operators". [Theoret. Comput. Sci. 411(38-39), 3423–3435] (2010); Theor. Comput. Sci. 412(29), 3795–3796 (2011)
9. Caron, P., Champarnaud, J.M., Mignot, L.: Multi-bar and multi-tilde regular operators. Journal of Automata, Languages and Combinatorics 16(1), 11–26 (2011)
10. Caron, P., Champarnaud, J.-M., Mignot, L.: Partial Derivatives of an Extended Regular Expression. In: Dediu, A.-H., Inenaga, S., Martín-Vide, C. (eds.) LATA 2011. LNCS, vol. 6638, pp. 179–191. Springer, Heidelberg (2011)
11. Caron, P., Champarnaud, J.M., Mignot, L.: A general frame for the derivation of regular expressions (submitted, 2012)
12. Champarnaud, J.-M., Jeanne, H., Mignot, L.: Approximate Regular Expressions and Their Derivatives. In: Dediu, A.-H., Martín-Vide, C. (eds.) LATA 2012. LNCS, vol. 7183, pp. 179–191. Springer, Heidelberg (2012)
13. Champarnaud, J.M., Ouardi, F., Ziadi, D.: An efficient computation of the equation $\mathbb{K}$-automaton of a regular $\mathbb{K}$-expression. Fundam. Inform. 90(1-2), 1–16 (2009)
14. Champarnaud, J.M., Ziadi, D.: Canonical derivatives, partial derivatives, and finite automaton constructions. Theoret. Comput. Sci. 239(1), 137–163 (2002)
15. Conway, J.H.: Regular algebra and finite machines. Chapman and Hall (1971)
16. Frishert, M.: FIRE Works & FIRE Station: A finite automata and regular expression playground. Ph.D. thesis, Eindhoven University, Netherlands (2005)
17. Ginzburg, A.: A procedure for checking equality of regular expressions. J. ACM 14(2), 355–362 (1967)
18. Ilie, L., Yu, S.: Follow automata. Inf. Comput. 186(1), 140–162 (2003)
19. Kleene, S.: Representation of events in nerve nets and finite automata. Automata Studies Ann. Math. Studies 34, 3–41 (1956)
20. Krob, D.: Differentation of K-rational expressions. Internat. J. Algebra Comput. 2(1), 57–87 (1992)
21. Lombardy, S., Sakarovitch, J.: Derivatives of rational expressions with multiplicity. Theor. Comput. Sci. 332(1-3), 141–177 (2005)
22. Owens, S., Reppy, J.H., Turon, A.: Regular-expression derivatives re-examined. J. Funct. Program. 19(2), 173–190 (2009)
23. Sulzmann, M., Lu, K.: Partial derivative regular expression pattern matching (December 2007) (manuscript)
24. Yu, S.: Regular languages. In: Rozenberg, G., Salomaa, A. (eds.) Handbook of Formal Languages. Word, Language, Grammar, vol. I, pp. 41–110. Springer, Berlin (1997)

On Positive TAGED
with a Bounded Number of Constraints

Pierre-Cyrille Héam[*], Vincent Hugot[**], and Olga Kouchnarenko

FEMTO-ST CNRS 6174, University of Franche-Comté & INRIA/CASSIS, France
{pierre-cyrille.heam,vincent.hugot,olga.kouchnarenko}@inria.fr

Abstract. Tree Automata With Global Equality Constraints (aka. positive TAGED, or TAGE) are a variety of Bottom-Up Tree Automata, with added expressive power. While there is interest in using this formalism to extend existing regular model-checking frameworks – built on vanilla tree automata – such a project can only be practical if the algorithmic complexity of common decision problems is kept tractable. Unfortunately, useful TAGE decision problems sport very high complexities: Membership is NP-complete, Emptiness and Finiteness are both EXPTIME-complete, Universality and Inclusion are undecidable. It is well-known that restricting the *kind* of equality constraints can have a dramatic effect on complexity, as evidenced by Rigid Tree Automata. However, the influence of the *number* of constraints on complexity has yet to be examined. In this paper, we focus on three common decision problems: Emptiness, Finiteness and Membership, and study their algorithmic complexity under a bounded number of equality constraints.

1 Introduction

Tree Automata are a pervasive tool of contemporary Computer Science, with applications running the gamut from XML processing [8] to program verification. Since their original introduction in the fifties, they have spawned an ever-growing family of variants, each with its own characteristics of expressiveness and decision complexity. Notable among them is the sub-family of Tree Automata With Constraints, which increases the expressiveness of vanilla tree automata by providing some means of comparing subtrees. Examples of such devices are are Automata With Equality and Disequality Constraints [4], Automata with Constraints on Brothers [2], and Visibly Tree Automata with Memory and Constraints [3]. In this paper, we focus on one of the latest strains: Tree Automata With Global Equality Constraints (TAGE) [6,5]. Their increased expressiveness is well paid for in terms of algorithmic complexity: Membership is NP-complete, Emptiness and Finiteness are both EXPTIME-complete [1], Universality and Inclusion are undecidable. While those complexities are fairly prohibitive, restrictions on the constraints can dramatically simplify some problems — for instance Rigid Tree

---

[*] This author is supported by the project ANR 2010 BLAN 0202 02 FREC.
[**] This author is supported by the French DGA (Direction Générale de l'Armement).

N. Moreira and R. Reis (Eds.): CIAA 2012, LNCS 7381, pp. 329–336, 2012.

Automata (RTA) [9], a more restrictive class of TAGE, enjoy a trivial, linear-time decision procedure for Emptiness. An application of TAGE of particular interest is the extension of regular model-checking techniques, where the increased expressiveness permits a wider range of applications. For such extensions to be practical, algorithmic complexities must be kept tractable — for instance RTA achieve that for Emptiness by restricting the *kind* of equality constraints which may be taken; in contrast, the present paper studies how bounding the *number* of constraints influences the complexity of three common decision problems: Emptiness and Finiteness (Sec. 3) are shown to be in PTIME for one constraint, and EXPTIME-complete for two or more; Membership (Sec. 4) is shown to stay in PTIME, regardless of how high the bound is.

2   Preliminaries

Relations and Intervals. Let $R \subseteq Q^2$ be a binary relation on a set Q; we denote by R^+, R^* and $R^{\equiv}$ its transitive, reflexive-transitive, and equivalence closure (symmetric-reflexive-transitive), respectively. Unless explicitly stated otherwise, reflexive closures are taken on $\mathrm{dom}(R) = \{\, x \mid \exists y : xRy \text{ or } yRx \,\}$, even if R has been introduced as a relation on the larger set Q. The integer interval $[n, m] \cap \mathbb{Z}$ is written $[\![n, m]\!]$.

Trees. We denote by $\mathbb{N}^*$ the set of words over $\mathbb{N}$; if $v, w \in \mathbb{N}^*$, then $v.w$ stands for the concatenation of the words v and w. A *ranked alphabet* is a finite set Σ of symbols, equipped with an arity function $\mathsf{arity} : \Sigma \to \mathbb{N}$. The subset of symbols of Σ with arity n is denoted by Σ_n, and the notation σ/n is shorthand for "σ, with arity $\sigma = n$". The set $\mathcal{T}(\Sigma)$ of trees over Σ is defined inductively as the smallest set such that $\Sigma_0 \subseteq \mathcal{T}(\Sigma)$ and, if $n \geqslant 1$, $\sigma \in \Sigma_n$ and $u_1, \ldots, u_n \in \mathcal{T}(\Sigma)$, then $\sigma(u_1, \ldots, u_n) \in \mathcal{T}(\Sigma)$. If t is a tree, then the set of *positions* (or *nodes*) $Pos(t) \subseteq \mathbb{N}^*$ is defined inductively by $Pos(t) = \{\varepsilon\}$ if $t \in \Sigma_0$ and $Pos(\sigma(u_1, \ldots, u_n)) = \{\varepsilon\} \cup \{\, k.\alpha_k \mid k \in [\![1, n]\!],\ \alpha_k \in Pos(u_{k+1}) \,\}$ otherwise, where n is the arity of σ. We see a tree t as a function $t : Pos(t) \to \Sigma$ which maps a position to the symbol at that position in t. Positions are equipped with a non-strict (resp. strict) partial order $\trianglelefteq$ (resp. $\lhd$), such that $\alpha \trianglelefteq \beta$ iff β is a prefix of α (resp. $\alpha \trianglelefteq \beta$ and $\alpha \neq \beta$). The *subtree of a tree* t *at position* $\alpha \in Pos(t)$ is the tree $t|_\alpha$ such that $Pos(t|_\alpha) = \{\, \beta \mid \alpha.\beta \in Pos(t) \,\}$ and $\forall \beta \in Pos(t|_\alpha)$, $t|_\alpha(\beta) = t(\alpha.\beta)$. Subterms are ordered by the relations $u \trianglelefteq v \iff \exists \alpha \in Pos(v) : v|_\alpha = u$ and $u \lhd v \iff u \trianglelefteq v \wedge u \neq v$. Note that $\alpha \trianglelefteq \beta \implies t|_\alpha \trianglelefteq t|_\beta$. Two positions α and β are *incomparable*, written $\alpha \curlywedge \beta$, if neither $\alpha \trianglelefteq \beta$ nor $\beta \trianglelefteq \alpha$. The *size* of a tree t is denoted by $\|t\|$ and defined by $\|t\| = |Pos(t)|$.

Tree Automata. Let Q be a finite set of symbols of arity 0, called *states*, such that $Q \cap \Sigma = \varnothing$. A *transition* is a rewrite rule $\sigma(q_1, \ldots, q_n) \to q$, where $q_1, \ldots, q_n, q \in Q$ and $\sigma \in \Sigma_n$. A *bottom-up non-deterministic finite tree automaton* (tree automaton, or TA for short) over Σ is a tuple $\mathcal{A} = \langle \Sigma, Q, F, \Delta \rangle$, such that $F \subseteq Q$ and Δ is a finite set of transitions. A *run* of $\mathcal{A}$ on a term $t \in \mathcal{T}(\Sigma)$ is a tree $\rho : Pos(t) \to Q$ such that for all $\alpha \in Pos(t)$, $t(\alpha)(\rho(\alpha.1), \ldots, \rho(\alpha.n)) \to \rho(\alpha) \in \Delta$.

A run ρ is a q-run if $\rho(\varepsilon) = q$, and it is called *accepting* (or *successful*) if $\rho(\varepsilon) \in F$. A set of trees is called a *language*. The set of all trees on which there exists a q-run of $\mathcal{A}$ is written $\mathcal{L}^q(\mathcal{A})$, and the set of trees on which there exists an accepting run is denoted by $\mathcal{L}(\mathcal{A}) = \bigcup_{q \in F} \mathcal{L}^q(\mathcal{A})$, and called the *language recognised* (or *accepted*) by $\mathcal{A}$.

Tree Automata with Equality Constraints: TAGE. A TAGE, or "positive TAGED" [6] is a tuple $\mathcal{A} = \langle \Sigma, Q, F, \Delta, \approx \rangle$, where $\langle \Sigma, Q, F, \Delta \rangle$ is a tree automaton over Σ and $\approx \subseteq Q^2$ is a binary relation on Q. The *underlying tree automaton* $\langle \Sigma, Q, F, \Delta \rangle$ is denoted by $\mathsf{ta}(\mathcal{A})$. A run of a TAGE $\mathcal{A}$ on a tree t is a run of $\mathsf{ta}(\mathcal{A})$ on t *satisfying* the equality constraints of $\approx$, which is to say: for all positions $\alpha, \beta \in \mathcal{P}os(t)$, if $\rho(\alpha) \approx \rho(\beta)$ then $t|_\alpha = t|_\beta$. An *accepting* run of $\mathcal{A}$ is a run of $\mathcal{A}$ which is accepting for $\mathsf{ta}(\mathcal{A})$; accepted languages are defined similarly to TA. The Membership problem for TAGE is NP-complete [6]. Emptiness and Finiteness are EXPTIME-complete, whereas Universality and Inclusion are undecidable [9, Table 1]. Following the respective definitions of runs, it is straightforward that for every TAGE $\mathcal{A}$, $\mathcal{L}(\mathcal{A}) \subseteq \mathcal{L}(\mathsf{ta}(\mathcal{A}))$. A TAGE $\mathcal{A}$ is said to be *rigid* (i.e. a RTA) if $\approx \subseteq id_Q$, i.e. if every constraint is of the form $p \approx p$. The standard *disjoint union* of two TAGE $\mathcal{A}$ and $\mathcal{B}$ is a TAGE $\mathcal{A} \uplus \mathcal{B}$, such that $\mathcal{L}(\mathcal{A} \uplus \mathcal{B}) = \mathcal{L}(\mathcal{A}) \cup \mathcal{L}(\mathcal{B})$ [6]. Two TAGE $\mathcal{A}$ and $\mathcal{B}$ are said to be *equivalent* if $\mathcal{L}(\mathcal{A}) = \mathcal{L}(\mathcal{B})$.

TAGE-Specific Notations. Throughout this paper, any TAGE $\mathcal{X}$ will be assumed to have attributes of the form $\langle \mathcal{X}{:}\Sigma, \mathcal{X}{:}Q, \mathcal{X}{:}F, \mathcal{X}{:}\Delta, \mathcal{X}{:}\approx \rangle$. In addition, $\mathcal{A}$ will simply be assumed to be $\langle \Sigma, Q, F, \Delta, \approx \rangle$. We write the modification of an existing TAGE as $\{\mathcal{X} \mid <modifs>\}$, where $<modifs>$ is a comma-separated list of attribute changes. For brevity, within the scope of $\{\mathcal{X} \mid \cdots \}$ any unqualified attribute x stands for $\mathcal{X}{:}x$ — this takes precedence over the $\mathcal{A}{:}x$ convention. For instance, $\{\mathcal{X} \mid \approx := \varnothing\}$ is the bare tree automaton associated with $\mathcal{X}$, or $\mathsf{ta}(\mathcal{X})$. Modifications of the form "$x := f(x)$" will just be written "$f(x)$"; for instance $\{\mathcal{X} \mid Q \setminus \{q\}\}$ is $\mathcal{X}$ from which the state q has been removed, as with "$Q := Q \setminus \{q\}$" (or even "$\mathcal{X}{:}Q := \mathcal{X}{:}Q \setminus \{q\}$"). Of course in this example the modification "$F \setminus \{q\}$" is completely omitted, as it is implied by "$Q \setminus \{q\}$", given that by definition $\mathcal{X}{:}F \subseteq \mathcal{X}{:}Q$. The same goes for the removal of all the rules of $\mathcal{X}{:}\Delta$ and constraints of $\mathcal{X}{:}\approx$ that used q.

Tree Automata With Bounded Equality Constraints: TAGE$_k$. A TAGE$_k$, where $k \in \mathbb{N}$, is a TAGE whose number of constraints is at most k. In other words, a TAGE$_k$ $\mathcal{A}$ is such that $\mathrm{Card}(\approx) \leqslant k$. By extension, we also denote by TAGE$_k$ the set of all automata which are TAGE$_k$. Note that trivially TAGE$_k \subseteq$ TAGE$_{k+1} \subseteq$ TAGE.

3   The Emptiness and Finiteness Problems

Lemma 1 (*Incomparable Positions*). *Let $\mathcal{A}$ be a TAGE with the constraint $p \approx q$, and ρ an accepting run of $\mathcal{A}$ on a tree t. Assume that both those states are involved in the run: $\{p, q\} \subseteq \mathrm{ran}\, \rho$; then any two distinct positions $\alpha, \beta \in \rho^{-1}(\{p, q\})$, $\alpha \neq \beta$, are incomparable: $\alpha \perp \beta$.*

Proof. Since $\alpha, \beta \in \rho^{-1}(\{p, q\})$ and $\{p, q\} \subseteq \operatorname{ran} \rho$ and $p \cong q$, we have $t|_\alpha = t|_\beta$. Suppose wlog. that $\alpha \lhd \beta$, then $t|_\alpha \lhd t|_\beta$; this is absurd since $t|_\beta$ cannot be structurally equal to one of its own strict subterms. Therefore $\alpha \curlywedge \beta$.   □

Lemma 2 (*Rigidification*). *For every* TAGE$_1$ $\mathcal{A}$*, there exists an equivalent* RTA $\mathcal{B}$ *whose size is at most quadratic in that of* $\mathcal{A}$.[1]

Proof. If $\mathcal{A}$ has no constraints, or a rigid constraint ($p \cong p$), then $\mathcal{B} = \mathcal{A}$. Assume $\mathcal{A}$ has a constraint of the form $p \cong q$, with $p \neq q$, and suppose wlog. that $p, q \notin F$. BUILDING BLOCKS. We let $\mathcal{B}_p^{\neg} = \langle \mathcal{A} \mid Q \setminus \{p\} \rangle$, $\mathcal{B}_q^{\neg} = \langle \mathcal{A} \mid Q \setminus \{q\} \rangle$, $\mathcal{B}_p = \langle \mathcal{B}_q^{\neg} \mid F := \{p\}, \Delta := \Delta_p \rangle$ —where Δ_p is $\mathcal{B}_q^{\neg} : \Delta$ from which all rules where p appears in the left-hand side have been removed, and $\mathcal{B}_q$, which is defined symmetrically to $\mathcal{B}_p$. Lastly, $\mathcal{B}_{pq}$ is built to accept the intersection of the languages of $\mathcal{B}_p$ and $\mathcal{B}_q$; using the standard product algorithm, it has a single final state $q_f = (p, q)$. Note that they are all vanilla tree automata. CONSTRUCTION. We let

$$\mathcal{B} = \mathcal{B}_p^{\neg} \uplus \mathcal{B}_q^{\neg} \uplus \langle \mathcal{A} \mid Q', \Delta', q_f \cong q_f \rangle, \text{ with } \begin{cases} Q' = (Q \setminus \{p, q\}) \uplus (\mathcal{B}_{pq} : Q) \\ \Delta' = \Delta_{pq}^{q_f} \uplus (\mathcal{B}_{pq} : \Delta) \end{cases},$$

where $\Delta_{pq}^{q_f}$ is $\mathcal{A} : \Delta$ from which all left-hand side occurrences of p or q have been replaced by q_f. EQUIVALENCE. Let $t \in \mathcal{L}(\mathcal{A})$, accepted through a run ρ; one of the following is true: (**1**) neither p nor q appears in ρ (**2**) p appears, and q does not (**3**) q appears, and p does not (**4**) both p and q appear. In the three first cases, the constraints are not involved, and t is accepted by: (1) both $\mathcal{B}_p^{\neg}$ and $\mathcal{B}_q^{\neg}$ (2) $\mathcal{B}_q^{\neg}$ (3) $\mathcal{B}_p^{\neg}$. In case (4), a subterm evaluating to p will belong to $\mathcal{L}^p(\mathcal{A})$ by definition, and also to $\mathcal{L}^q(\mathcal{A})$ as it needs to be equal to another extant subterm evaluating to q. Furthermore, p and q can only appear at the root of each subruns, lest $p \cong q$ be trivially violated. Therefore, a successful run of $\mathcal{B}$ can be constructed by simply substituting all p and q subruns by q_f-runs of $\mathcal{B}_{pq}$. Thus $t \in \mathcal{L}(\mathcal{B})$. Conversely, let $t \in \mathcal{L}(\mathcal{B})$; it is immediately seen by construction that $\mathcal{L}(\mathcal{B}_p^{\neg}) \subseteq \mathcal{L}(\mathcal{A})$ and $\mathcal{L}(\mathcal{B}_q^{\neg}) \subseteq \mathcal{L}(\mathcal{A})$. Suppose that t is accepted through a run of the third and last part of $\mathcal{B}$ (namely $\langle \mathcal{A} \mid \cdots \rangle$); then every q_f-subrun can be replaced by either a p-run or a q-run of $\mathcal{A}$. The result of this operation is trivially an accepting run of $\operatorname{ta}(\mathcal{A})$; there remains to observe that it satisfies $p \cong q$, because the corresponding subtrees must be equal given the constraint $(q_f, q_f) \in \mathcal{B} :\cong$. Thus $t \in \mathcal{L}(\mathcal{A})$. SIZE & TIME. All building blocks are of size $O(\|\mathcal{A}\|)$, except $\mathcal{B}_{pq}$, which is of size $O(\|\mathcal{A}\|^2)$. Globally, the size of $\mathcal{B}$ is at most quadratic in that of $\mathcal{A}$. The construction is also straightforwardly done in quadratic time.   □

Proposition 3 (*Emptiness*). *The Emptiness problem is in* PTIME *for* TAGE$_1$*, and* EXPTIME-*complete for* TAGE$_2$.

Proof. TAGE$_1$. Emptiness is testable in linear time for RTA [9], therefore the emptiness of $\mathcal{A}$ is testable in quadratic time using the construction of Lemma 2. TAGE$_2$.

---

[1] Note that the general construction for TAGE is exponential [6, Thm. 10].

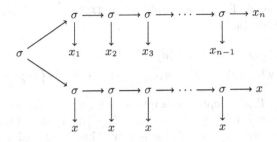

Fig. 1. Language L

OVERVIEW. We reduce the test of the emptiness of the intersection of n tree automata $\mathcal{A}_1, \ldots, \mathcal{A}_n$, which is an EXPTIME-complete problem, to the emptiness of a TAGE$_2$ $\mathcal{A}$. This is similar to the arguments of [5, Thm. 1], the major difference being that we can only use two constraints instead of an unbounded number of constraints. The idea is to take advantage of the fact that an explicit equality constraint between two positions effectively enforces an arbitrary number of implicit equality constraints on the sub-positions. ASSUMPTIONS. It is assumed wlog. that $n \geqslant 2$ and the sets of states of the $\mathcal{A}_i$ are pairwise disjoint; that is to say, $\forall i, j \in [\![1, n]\!]$, $i \neq j \Rightarrow (\mathcal{A}_i : Q) \cap (\mathcal{A}_j : Q) = \varnothing$. Furthermore, it can be assumed that each $\mathcal{A}_i$ has exactly one final state q_{f_i}. If that is not the case, then $\mathcal{A}_i$ can be modified to be so, which results in its size doubling in the worst case. LANGUAGE. We define the language L as the set of trees of the form given in Figure 1$_{[p333]}$, where σ is a fresh binary symbol and for all i, $x_i \in \mathcal{L}(\mathcal{A}_i)$ and $x = x_i$. Note that this implies that $x \in \bigcap_i \mathcal{L}(\mathcal{A}_i)$, and therefore L is empty iff $\bigcap_i \mathcal{L}(\mathcal{A}_i)$ is empty. AUTOMATON. We build a TAGE$_2$ $\mathcal{A}$ that accepts L, by first building a universal tree automaton $\mathcal{U}$, of final state q^{u}. Then, we let $\mathcal{A} = \langle \Sigma, Q, F, \Delta, \approxeq \rangle$, where

$$Q = (\biguplus_i \mathcal{A}_i : Q) \uplus (\mathcal{U} : Q) \uplus \{ q_1^{u}, \ldots, q_{n-1}^{u}, q_1^{v}, \ldots, q_{n-1}^{v} \} \uplus \{ q_f \}$$

$$F = \{ q_f \} \qquad q^{u} \approxeq q^{u}, \qquad q_1^{u} \approxeq q_1^{v} \qquad \Sigma = (\textstyle\bigcup_i \mathcal{A}_i : \Sigma) \uplus \{ \sigma / 2 \}$$

$$\Delta = \{ \sigma(q_1^{u}, q_1^{v}) \to q_f \} \cup (\textstyle\bigcup_i \mathcal{A}_i : \Delta) \cup (\mathcal{U} : \Delta) \cup$$
$$\{ \sigma(q^{u}, q_{k+1}^{u}) \to q_k^{u} \mid k \in [\![1, n-2]\!] \} \cup \{ \sigma(q^{u}, q^{u}) \to q_{n-1}^{u} \} \cup$$
$$\{ \sigma(q_{fk}, q_{k+1}^{v}) \to q_k^{v} \mid k \in [\![1, n-2]\!] \} \cup \{ \sigma(q_{fn-1}, q_{fn}) \to q_{n-1}^{v} \} \ .$$

Note that we have $\mathcal{L}(\mathcal{A}) = L$ and $\|\mathcal{A}\| = O\left(\sum_{k=1}^{n} \|\mathcal{A}_i\|\right)$, which concludes the proof. $\square$

Proposition 4 (Finiteness). *The Finiteness problem is in* PTIME *for* TAGE$_1$, *and* EXPTIME-*complete for* TAGE$_2$.

Proof. TAGE$_1$. Finiteness is testable in linear time for RTA [9], therefore the finiteness of $\mathcal{A}$ is testable in quadratic time using the construction of Lemma 2.

TAGE₂. We reduce the Emptiness problem for TAGE₂ to the Finiteness problem. Given a TAGE₂ $\mathcal{A}$, we build

$$\mathcal{A}' = \left\{ \mathcal{A} \mid Q \uplus \{p\}, F := \{p\}, \Sigma \uplus \{\sigma/_1\}, \Delta' \right\}$$
$$\text{where } \Delta' = \Delta \cup \{ \sigma(q_f) \to p \mid q_f \in F \} \cup \{ \sigma(p) \to p \} .$$

$\mathcal{A}'$ is also a TAGE₂. If $\mathcal{A}$ accepts the empty language, then so does $\mathcal{A}'$. Conversely, if $t \in \mathcal{L}(\mathcal{A})$, then $\sigma^*(t) \subseteq \mathcal{L}(\mathcal{A}')$, and thus $\mathcal{L}(\mathcal{A}')$ is infinite. Consequently, the language of $\mathcal{A}'$ is finite iff that of $\mathcal{A}$ is empty. This, combined with Prp. 3[p332], shows that TAGE₂-Finiteness is ExpTime-hard; since the general problem for TAGE is ExpTime [6, Thm. 14], TAGE₂-Finiteness is ExpTime-complete.   □

4   The Membership Problem

Let us begin with some general observations and notations. We will need to reason about the relation $\approx$; unfortunately, it is not an equivalence relation. For instance, given the constraints $p \approx r$ and $r \approx q$ it is tempting, but in general wrong, to infer $p \approx q$ by transitivity. The crux of the matter here is whether the state r actually appears in the run: if it does, $p \approx q$ is effectively implied, but if it does not, then both constraints $p \approx r$ and $r \approx q$ are moot. Lemma 5 shows that, given the knowledge (or the assumption) of a set $P \subseteq \text{dom} \approx$ of the constrained states which are actually present in runs, the constraints of $\approx$ are interchangeable with an equivalence relation, which we call the *togetherness* relation.

Lemma 5 (*Togetherness*). *Let $\mathcal{A}$ be a TAGE and $P \subseteq \text{dom} \approx$. Then any run ρ such that $(\text{ran}\, \rho) \cap (\text{dom} \approx) = P$ is accepting for $\mathcal{A}$ if and only if it is so for $\mathcal{A}_P = \left\{ \mathcal{A} \mid \approx := \left(\approx \cap P^2 \right)^{\equiv} \right\}$, where the closure is meant under $\text{dom}(\approx \cap P^2)$.*

Given a P, we denote by $\asymp_P = \left(\approx \cap P^2 \right)^{\equiv}$ this equivalence relation, and say that "*p and q are together wrt. P*" if $p \asymp_P q$. Its equivalence classes are denoted by $\mathbb{G}_P = \text{dom}(\approx \cap P^2)/_{\asymp_P}$, and called *groups*. If t is a tree, we write $\sim$ for the *similarity relation on t*, defined on $Pos(t)^2$ such that $\alpha \sim \beta \iff t|_\alpha = t|_\beta$. We denote by $\mathbb{S}_t$ the quotient set $Pos(t)/_\sim$ of the *similarity classes of t*.

Lemma 6 (*Housing Groups*). *Let $\mathcal{A}$ be a TAGE, $P \subseteq \text{dom} \approx$ and ρ a run of $\mathcal{A}$ on a tree t, such that $(\text{ran}\, \rho) \cap (\text{dom} \approx) = P$. Then ρ satisfies the constraints of $\approx$ if and only if $\forall G \in \mathbb{G}_P, \exists C_G \in \mathbb{S}_t : \rho^{-1}(G) \subseteq C_G$.*

Given the hypothesis of P and given a successful run ρ on t, we call the map $G \mapsto C_G$ a *P-housing of ρ in t*, which is said to be *compatible with ρ*, and we denote by $\mathbb{H}_P^t = \mathbb{G}_P \to \mathbb{S}_t$ the set of all possible P-housings on t.

Proposition 7 (*Membership*). *Given an arbitrary but fixed $n \in \mathbb{N}$, the Membership problem for TAGE$_n$ is in PTime — albeit with an overhead exponential in n.*

Proof. Let $\mathcal{A}$ be a TAGE$_n$, and t a tree. The Housing Lemma ($6_{[p334]}$) has already established that a run ρ of $\mathcal{A}$ on t satisfies $\approx$ iff there exists a housing $h \in \mathbb{H}_P^t$ which is compatible with ρ, where $P = (\mathrm{dom} \approx) \cap (\mathrm{ran}\,\rho)$ is the set of constrained states which actually appear in the run. Our strategy to check the membership of t will simply be to try each possible $P \subseteq \mathrm{dom} \approx$ successively, by attempting, for each possible housing $h \in \mathbb{H}_P^t$, to craft an accepting run ρ of ta$(\mathcal{A})$ compatible with h. There are at most 2^{2n} possible P, and given a choice of P, there are $|\mathbb{S}_t|^{|\mathbb{G}_P|} \leqslant \|t\|^{2n}$ P-housings on t, which gives at most $4^n \cdot \|t\|^{2n}$ tests in total. Note that since n is a constant, this remains polynomial. There only remains to show that given a choice of P and $h \in \mathbb{H}_P^t$, the existence of a compatible run can be tested in polynomial time. To do so, we use a variant of the standard reachability algorithm, where only the states of P may appear, and the states of a given group $G \in \mathbb{G}_P$ may only appear at the positions assigned to them by the chosen housing h. Formally, given a choice of P and a housing $h \in \mathbb{H}_P^t$, there exists such a run iff $\Phi_t^{P,h}(\varepsilon) \cap F \neq \varnothing$, where

$$
\Phi_t^{P,h}(\alpha) = \left\{ q \in Q \left| \begin{array}{l} t(\alpha)(p_1, \ldots, p_n) \to q \in \Delta \\ \forall i \in [\![1, n]\!],\ p_i \in \Phi_t^{P,h}(\alpha.i) \\ q \in \bigcup \mathbb{G}_P \implies \alpha \in h\left([q]_{\asymp_P}\right) \\ q \notin \mathrm{dom}(\approx) \setminus P \end{array} \right. \right\}.
$$

The reader will notice that, were the last two conditions removed, $\Phi_t^{P,h}(\alpha)$ would simply be the set of reachable states at position α. The additional two constraints are $O(1)$ operations, thus $\Phi_t^{P,h}(\cdot)$ does run in polynomial time; there only remains to show that our algorithm does what is expected of it. There are two points to this: (**1**) no false negative: every successful run is subsumed by some $\Phi_t^{P,h}(\cdot)$ (**2**) no false positive: every run subsumed by some $\Phi_t^{P,h}(\cdot)$ is accepting.

(**1**) Let ρ a successful run for $\mathcal{A}$, and $P = (\mathrm{ran}\,\rho) \cap (\mathrm{dom} \approx)$; then by the Housing Lemma, it satisfies $\asymp_P$, and there exists a housing $h \in \mathbb{H}_P^t$ with which it is compatible. We propose that ρ is subsumed by $\Phi_t^{P,h}(\cdot)$, which is to say that for each position $\alpha \in \mathcal{P}os(t)$, we must have $\rho(\alpha) \in \Phi_t^{P,h}(\alpha)$. Indeed, let α any position, and $q = \rho(\alpha)$; we check that q satisfies all four conditions for belonging to $\Phi_t^{P,h}(\alpha)$. The first condition is trivially satisfied since ρ is a run. The second one will be the hypothesis of our recursion which, quite conveniently, evaluates to true vacuously if α is a leaf. The third condition is taken care of by the Housing Lemma: suppose $q \in \bigcup \mathbb{G}_P$; then there is a group $G \in \mathbb{G}_P$ such that $q \in G$ (in fact $G = [q]_{\asymp_P}$), and $\rho^{-1}(G) \subseteq h(G)$. Thus we have the chain $\alpha \in \rho^{-1}(\{q\}) \subseteq \rho^{-1}(G) \subseteq h(G)$, and in particular $\alpha \in h([q]_{\asymp_P})$. The fourth and last condition is trivial given our choice of P: Assume its negation $q \in \mathrm{dom}(\approx) \setminus P$, then you have $q \notin \mathrm{ran}\,\rho$, which is absurd.

(**2**) Let ρ be a run subsumed by $\Phi_t^{P,h}(\cdot)$, for some P and h. By the fourth condition, $(\mathrm{ran}\,\rho) \cap (\mathrm{dom}(\approx) \setminus P) = \varnothing$, and thus $(\mathrm{ran}\,\rho) \cap (\mathrm{dom} \approx) \subseteq P$. Let

$\alpha \in \mathcal{P}os(t)$; by the third condition, if $\rho(\alpha) \in G \in \mathbb{G}_P$, then $\alpha \in h(G)$; in other words, $\rho^{-1}(G) \subseteq h(G)$, thus by the Housing Lemma[2], ρ is successful. $\square$

5   Conclusions

In the case of Emptiness and Finiteness we have shown that, perhaps somewhat counter-intuitively, while the limitation to a single equality constraint does lead to tremendously easier complexities (from ExpTime-hardness to quadratic decision procedures), the addition of a second constraint suffices to reintroduce the full complexity of the general, unbounded problem.

This stands in contrast to the behaviour of the Membership problem which, while NP-complete in general, becomes polynomial once the number of constraints is bounded by a constant, regardless of the size of that constant — though admittedly "polynomial" is in that case quite unlikely to mean "efficient" for anything but the smallest constants. Nevertheless, this suggests a potentially more scalable alternative to the existing SAT encoding approach [7].

Acknowledgements. Our thanks go to the reviewers, especially for the suggestion of a simpler approach to the TAGE₁ parts of Propositions 3 and 4.

References

1. Barguñó, L., Creus, C., Godoy, G., Jacquemard, F., Vacher, C.: The emptiness problem for tree automata with global constraints. In: LICS, pp. 263–272. IEEE Computer Society (2010)
2. Bogaert, B., Tison, S.: Equality and Disequality Constraints on Direct Subterms in Tree Automata. In: Finkel, A., Jantzen, M. (eds.) STACS 1992. LNCS, vol. 577, pp. 161–171. Springer, Heidelberg (1992)
3. Comon-Lundh, H., Jacquemard, F., Perrin, N.: Visibly tree automata with memory and constraints. CoRR abs/0804.3065 (2008)
4. Dauchet, M., Mongy, J.: Transformations de noyaux reconnaissables d'arbres, Forêts RATEG. Ph.D. thesis, LIFL (France) (1981)
5. Filiot, E., Talbot, J.-M., Tison, S.: Tree Automata with Global Constraints. In: Ito, M., Toyama, M. (eds.) DLT 2008. LNCS, vol. 5257, pp. 314–326. Springer, Heidelberg (2008)
6. Filiot, E., Talbot, J.-M., Tison, S.: Tree automata with global constraints. Int. J. Found. Comput. Sci. 21(4), 571–596 (2010)
7. Héam, P.-C., Hugot, V., Kouchnarenko, O.: SAT solvers for queries over tree automata with constraints. In: ICST (CSTVA ws.), pp. 343–348. IEEE (2010)
8. Hosoya, H.: Foundations of XML Processing: The Tree-Automata Approach. Cambridge University Press (2010), http://books.google.fr/books?id=xGlH3ADxwn4C
9. Jacquemard, F., Klay, F., Vacher, C.: Rigid tree automata and applications. Inf. Comput. 209(3), 486–512 (2011)

---

[2] The watchful reader will notice that we are slightly cheating here, because Lem. 6[p334] as written requires $(\operatorname{ran}\rho) \cap (\operatorname{dom} \approx) = P$. The inclusion is enough for the "if" part, as shown by the relevant halves of the proofs of Lem. 6[p334] and Lem. 5[p334]. Alternatively, one could replace P and h by adequate $P' \subseteq P$ and $h' \in \mathbb{H}_{P'}^t$ such that we have equality and preserve subsumption. Either way this is an easy technicality with no bearing on any other part of this paper.

SDFA: Series DFA for Memory-Efficient Regular Expression Matching[*]

Tingwen Liu[1,2], Yong Sun[1,3], Li Guo[1,3], and Binxing Fang[3]

[1] Institute of Computing Technology, Chinese Academy of Sciences, Beijing, China
[2] Graduate University of Chinese Academy of Sciences, Beijing, China
[3] National Engineering Laboratory for Information Security Technologies, Beijing
{liutingwen,suny}@software.ict.ac.cn

Abstract. Regular expression (RegEx) matching plays an important role in various network, security and database applications. Deterministic finite automata (DFA) is the preferred representation to achieve online RegEx matching in backbone networks, because of its one single pass over inputs for multiple RegExes and guaranteed performance of $O(1)$ memory bandwidth per symbol. However, DFA may occupy prohibitive amounts of memory due to the explosive growth in its state size. In this work, we propose Series DFA (SDFA) to address the problem. The main idea is to cut a complex RegEx into several ordered and small RegExes carefully, and then concatenate their compact DFAs in series to match. Experimental results show that SDFA can achieve significant reduction in memory size at the cost of limited number of memory bandwidth.

1   Introduction

Deep Packet Inspection (DPI), which searches for predefined signatures over the content of packet payloads, is considered as a powerful and important method in network and security applications. Recently regular expressions (RegExes) are replacing exact strings as the de facto standard to specify signatures in most open-source tools [9,6] and commercial devices. The primary reason is the expressive power, simplicity and flexibility of RegExes. Deterministic Finite Automata (DFA) is an ideal representation for high-speed RegEx matching, because multiple RegExes can be compiled into a composite DFA that performs matching over inputs in a single pass with a guaranteed robust performance of $O(1)$ memory bandwidth per byte. However, the composite DFA constructed for real-world RegEx sets may experience state explosion, as a result it usually consumes prohibitive amounts of memory.

In this paper, we focus on state reduction by cutting complex RegExes into well-designed and ordered RegEx fragments that can be compiled into compact DFAs. To match equivalently as uncutted RegExes, we propose Series DFA

---

[*] Supported by the National High-Tech Research and Development Plan of China under Grant No. 2011AA010703; the National Natural Science Foundation of China under Grant No. 61070026 and No. 61003295.

N. Moreira and R. Reis (Eds.): CIAA 2012, LNCS 7381, pp. 337–344, 2012.

(SDFA) that concatenates the compact DFAs with epsilon transitions in the order of their appearance. We further introduce some optimizations to improve the memory consumption and memory bandwidth of SDFA. Different from prior work [1], SDFA works over RegExes directly to achieve the reduction of states, which makes it being constructed easily and quickly even for large-scale RegEx sets. We perform a systematic experimental study on real RegEx sets and our synthetic RegEx set. The results show that SDFA achieves significant memory reduction, and shows almost the same matching speed comparing with the composite DFA.

2   Related Work

With the widespread use of RegExes in various applications, research interests focus on designing data structures, algorithms and architectures to support fast and memory-efficient RegEx matching. In this context, how to reduce the huge memory consumption is the hotspot of related researches for those matching solutions based on DFAs. In general, prior work can be classified into three categories: *DFA compression, partial determinization* and *history auxiliary.*

DFA compression solutions try to achieve memory reduction by compressing the transition table for a given DFA [5, 3, 8, 7]. They are based on the observation of many common values in the table. However, the memory usage, which have been reduced by 95% after compressing, are still very huge as the composite DFA for real RegEx sets usually costs multiple terabytes. These solutions are orthogonal to our work and can be used to compress the compact DFAs in SDFA.

Partial determinization solutions address the problem by constructing hybrid automata [1] or multiple parallel DFAs [13, 10] at the cost of determinacy by allowing multiple states active during the matching process. Our work improves upon these solutions because our DFAs constructed for the cutting RegEx fragments are compact enough, and are activated when necessary.

History auxiliary solutions introduce counters, queues and other data structures as auxiliary memory to avoid duplication of states by recording matching history [4, 11]. However, the benefit of state reduction does not come for free. They either experience an exponential growth in the size of auxiliary memory, or require much time to update auxiliary memory after processing each symbol.

3   Technical Overview of Series DFA

3.1   State Complexity for RegExes

An analysis of state complexity for DFA of individual RegEx that does not have OR relationship (|) is represented in [13]. Here we consider RegExes in the combination of ^, one unconstrained repetition * and one constrained repetition (three types: fixed repetition {j}, range repetition {j,i} and at-least repetition {j,}) of wildcards, and the detail is shown in Table 1.

Table 1. State complexity for individual RegEx with k characters

RegEx Feature	Example	State Complexity
without constrained repetitions of wildcards	`^abcd, abcd` `^ab.*cd, ab.*cd`	$O(k)$
with `^`, one fixed or one at-least repetition	`^ab.{j}cd` `^ab.{j,}cd`	$O(k+j)$
with `^` and one range repetition	`^ab.{j,i}cd`	$O(k(i-j))$
with only one fixed or one at-least repetition	`ab.{j}cd` `ab.{j,}cd`	$O(k+2^j)$
with only one range repetition	`ab.{j,i}cd`	$O(k(i-j)+2^i)$

From the table, we can find that unconstrained repetitions do not cause state explosion when individual RegEx is compiled into a DFA in isolation (case 1). Constrained repetitions of wildcards lead to exponential growth in DFA state size for individual RegEx not starting with `^` (case 4 and case 5). Because the DFA needs to record the prefix part within each wildcard. The situation becomes even worse when multiple RegExes with constrained repetitions are compiled together into a composite DFA. Because there are more combinations of prefixes and more wildcards in these RegExes.

By comparison, RegExes of the former three cases do not result in a large DFA. Therefore, if we cut RegExes of the latter two cases into multiple RegEx fragments of the former three cases, we can construct a compact DFA for each fragment. In this paper, we investigate its feasibility to reduce DFA state size.

3.2 Main Idea of SDFA

In order to facilitate description, we call a RegEx as its fragments' *father*, each fragment as its *son*. For a given RegEx, the first (last) fragment is called its *eldestson* (*youngestson*), correspondingly other fragments are *non-eldestsons* (*non-youngestsons*). To match multiple RegExes together in a single pass, all the eldestsons are compiled into a composite DFA, and each non-eldestson is compiled into an individual DFA. SDFA organizes all the DFAs in series and perform matching in the follow way: at the beginning only the initial state of the composite DFA is active, all the individual DFAs are sleep; SDFA will add a new instance of the initial state of one individual DFA when its preceding DFA matches successfully, and delete an instance when it moves to the dead-state.

We use an example of two RegExes `ba[^a]*bad.{2}cd` and `de[^e]{3}` to show how SDFA works in detail. It first locates all unconstrained and constrained repetitions in the two RegExes, and then cut them into five fragments: `ba, ^[^a]*bad, ^.{2}cd, de, ^[^e]{3}` at these positions. Note that all the non-eldestsons begin with `^`, because a fragment begins to match from position $j+1$ of input string only when its preceding fragment matches successfully at position j. Fragments `ba` and `de`, which are the eldestsons of the two RegExes, are compiled into a composite DFA. Now we describe how to construct a SDFA with the

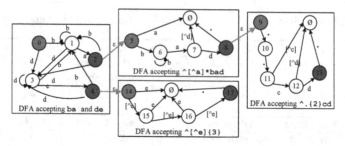

Fig. 1. SDFA accepting `ba[^a]*bad.{2}cd` and `de[^e]{3}`. For each DFA, the state in green (red) is its initial (accepting) state. Transitions to the initial states are omitted.

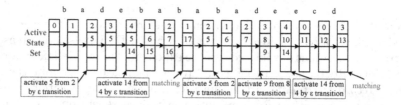

Fig. 2. SDFA traversal with input `badebababadeecd`

four DFAs, as shown in Fig. 1. The initial state of the composite DFA (state 0) is the initial state of the SDFA, and the accepting states of DFAs constructed for the youngestsons (state 13 and 17) are the accepting state of the SDFA. For the accepting states of the other DFAs, adding an epsilon (ϵ) transition that does not consume any symbol to the initial state of the DFA constructed for its following brother. As shown in Fig. 1, the SDFA accepting `ba[^a]*bad.{2}cd` and `de[^e]{3}` has 21 states, while the state-minimized DFA has 58 states (omitted here for readability).

In Fig. 2, we show the matching process of the SDFA in Fig. 1 over input string `badebababadeecd`. For example, fragment `de` is matched two times at the fourth and the twelfth symbol, and then SDFA activates state 14 along an epsilon transition. The first activation reports a successful match of `ba[^a]*bad.{2}cd` after processing the seventh symbol, while the second de-activates immediately because state 14 moves to the dead-state along the next symbol `e`.

4   Optimization for Series DFA

Essentially SDFA trades memory size (size of states) with memory bandwidth (size of *active state set*). In this section, we propose some techniques to optimize the two metrics by improving the cutting process and matching process of SDFA.

4.1   Optimization in Cutting Process

Determining the cutting positions is the main challenge for the construction of good SDFA. Cutting at the repetitions of any character range will have low memory size but high memory bandwidth as each fragment is too short. In contrast, cutting only at the repetitions of wildcards will have low memory bandwidth but high memory bandwidth. Here we give a simple but striking way to finish the determination quantitatively. We define the number of characters allowed in a character range as its size. Then we introduce a threshold μ: if the size of a character range is more than μ, we think the range is large enough to be cut at the positions of its repetitions. When μ is set to 256, the SDFA is essentially a composite DFA for the complete RegEx set because no RegEx is cut.

Furthermore, to obtain good SDFA, the cutting process should comply with the following three rules. In fact, we can also consider these rules trying to combine several adjacent fragments into one.

Rule 1: No constrained repetitions or unconstrained repetitions in any eldest-son. Because repetitions of large character ranges need to duplicate states to record all possible prefixes when multiple RegExes are compiled together as mentioned before. Therefore the composite DFA constructed for eldestsons that violate this rule will experience state explosion.

Rule 2: No constrained repetitions after unconstrained repetitions in each fragment. Obviously eldestsons that satisfy rule 1 also follow this rule. For each non-eldestson, if it violates this rule, it may belong to case 4 or case 5 in Table 1, and cause exponential growth of state size in the worst case.

Rule 3: No constrained repetitions or unconstrained repetitions after range repetitions in each fragment. All the eldestsons also follow this rule just as described above. Any non-eldestson failing to comply with this rule falls into case 3 in Table 1, whose complexity is product.

These rules allow non-eldestsons to have more than one unconstrained repetitions. One vivid example is RegEx `Cookie\s+Monster\s+server\s+engine` in Snort system. It can be cut into fragments `Cookie` and `^s+\s+Monster\s+server\s+engine` if set μ no less than the size of `\s`.

The point that need to be made is that these rules are sufficient conditions but not necessary conditions to combine adjacent fragments. An example is RegEx `ba[^a]*bad.{2}cd` in Fig. 1. The fragment `^[^a]*bad.{2}cd` obviously violate Rule 2, however its DFA does not experience exponential growth in state size. Because the occurrence of a makes `^[^a]*` fails to consume `bad`, as a result the DFA needn't to take into consideration that `bad` may appear in the constrained repetition `.{2}`. Snort and other intrusion detection systems have many RegExes of this type, for example `\/OvCgi\/[^\.]*\.exe[^\x20]{2000,}`.

4.2   Optimization in Matching Process

Most DPI applications such as Snort and L7-Filter are only interested in knowing the set of patterns to be fired by a packet. We call this type of matching as left-most matching, which is formally defined as below.

Left-Most Matching: Consider the matching process M as a function from a pattern P and a string S to a power set of S, such that, $M(P, S) = \{$substring S' of $S | S'$ is the left-most substring which is accepted by the DFA of $P\}$.

Table 2. Primary information of experimental RegEx sets ($\mu = 1$)

RegEx set	# of RegExes	% of * repetitions	% of {} repetitions	min-length range	# of NFA states	# of 7-DFA states
l7filter	107	46.7	21.5	1–76	3325	29047
backdoor	158	36.1	1.3	2–77	3580	6164
synset	300	59	18.7	11-225	19751	$> 10^6$

This specialty can be exploited to decrease memory bandwidth. As left-most matching is enough to know the fired RegExes, once a RegEx is reported it is safe to set its all non-eldestson DFAs inactive forever. To our knowledge, SDFA is the first automata that uses left-most matching to improve matching process. Because all kinds of previous methods must go through the step of constructing a sort of composite finite automata for the complete RegEx set. When a RegEx is matched, they cannot guarantee that the states that have been traversed by the RegEx will not be accessed by other RegExes. On the contrary, SDFA is able to ensure that the fragment DFAs of one RegEx will never be accessed by other RegExes. For the same reason, the composite DFA in SDFA needs to have an always active instance.

5   Experimental Results

We design three representative RegEx sets, as shown in Table 2. Column 3 (4) is the percent of RegExes containing constrained (unconstrained) repetitions of character ranges in each set. The first RegEx set is extracted from L7-Filter [6] system, and the second set is from *backdoor* rule file in Snort [9] system. The third RegEx set is generated by open-source RegEx generator [2]. As shown in column 7, the three RegEx sets can be compiled into 7 DFAs of 29047, 6164 and more than 10^6 states respectively with multiple parallel DFAs [13].

We make experiments using two real traffic traces from different links: one trace named *download* is downloaded from [12], the size is 254 MB; the other trace named *capture* is captured in the interface of a backbone network, the size is 1,538 MB. We also generate some synthetic traces of 50 MB with open-source trace generator [2] under $p_m = \{0, 0.15, 0.3, 0.45, 0.6, 0.75, 0.9\}$. Value p_m is used to model the likelihood of experiencing malicious traffic.

5.1   Evaluation of Memory Consumption

In this section, we use the size of DFA states to evaluate memory consumption of SDFA for the three RegEx sets. Table 3 shows the summary of state size for different values of μ. We can draw the following conclusions from Table 3.

First, DFA-based solutions are infeasible to perform matching for large RegEx sets containing constrained repetitions and unconstrained repetitions. As mentioned before, SDFA is in fact a composite DFA when μ is 256. However, the state size is more than 10^7 (inf) in this case for all experimental RegEx sets.

Table 3. State size of SDFA on varying μ

value of μ	state size of composite DFA / state sums of individual DFAs / # of DFAs		
	l7filter	backdoor	synset
1	5689 / 3293 / 103	2034 / 3009/144	9507 / 20322 / 321
64	6438 / 3246 / 93	45072 / 1451/58	inf / inf / 173
128	inf / 2618 / 56	45072 / 1451/58	inf / inf / 173
256	inf / 0 / 1	inf / 0 / 1	inf / 0 / 1

Second, the number of DFAs decreases as the increase of μ while the state size of the composite DFA grows with μ. The primary reason is that high μ makes some character ranges become small, and SDFA does not cut RegExes at the occurrence of unconstrained repetitions and constrained repetitions of small character ranges. As a result, the eldestsons have more symbols especially more repetitions, which lead to the rapid increase in the state size of the composite DFA. However, the sum of states in individual DFAs appears complexly. The primary reason is that constrained repetitions may appear in the middle of non-eldestsons for large μ, which results in exponential growth in state size even for an individual DFA.

Third, SDFA can greatly reduce memory consumption. When μ is 1, the three SDFAs have 8982, 5043 and 29379 states respectively in all, which can be encoded in on-chip memory directly even without compression. The result is closed to that of NFA, and better than that of multiple parallel DFAs (7-DFA).

5.2   Evaluation of Matching Performance

In this section, we evaluate matching performance of SDFA, which is measured by the size of active state sets. We construct a SDFA for each given RegEx set with $\mu = 1$, and observe its active set size on real traces and synthetic traces in average case and maximum case. In fact, both the average size and the maximum size of active sets increase with μ. When $\mu = 1$, SDFA has the worst average size and maximum size, because it cuts RegExes at the occurrence of repetitions of any character range, as a result fragments are matched frequently.

The result of backdoor set on its synthetic traces is shown in Fig. 3. As l7filter and synset have the similar behavior, we omit them here due to page limitation. We can find that: First, left-most matching can really improve the matching performance of SDFA, especially in average size. Second, the average size grows slowly with the increase of p_m, while the change of maximum size is uncertain.

Fig. 4 shows the results of active set size on real traffic traces for each RegEx set. Each connection carries an application protocol, so almost every packets can be matched by RegExes in l7filter set. As a result, we can regard the behavior of l7filter set as the performance of SDFA under an attack. From Fig. 4 we can find that SDFA works well under attacks, although the maximum size is a little big. The average active set size of SDFA is very close to that of a composite DFA. As each RegEx set can be constructed into 7 DFAs, its average size and

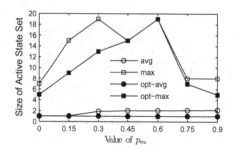 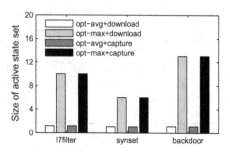

Fig. 3. Size of active state sets for back-door set on its synthetic traces

Fig. 4. Size of active state sets for three experimental RegEx sets on real traces

maximum size are both 7. Obviously SDFA is suitable to perform large-scale RegEx matching in different high-speed network environments.

References

1. Becchi, M., Crowley, P.: A Hybrid Finite Automaton for Practical Deep Packet Inspection. In: Proc. ACM CoNEXT Conference, pp. 1–12 (2007)
2. Becchi, M.: Regular Expression Processor, http://regex.wustl.edu/
3. Ficara, D., Giordano, S., Procissi, G., Vitucci, F., Antichi, G., Di Pietro, A.: An Improved DFA for Fast Regular Expression Matching. ACM SIGCOMM Computer Communication Review 38(5), 29–40 (2008)
4. Kumar, S., Chandrasekaran, B., Turner, J., Varghese, G.: Curing Regular Expressions Matching Algorithms from Insomnia, Amnesia, and Acalculia. In: Proc. ACM/IEEE ANCS, pp. 155–164 (2007)
5. Kumar, S., Dharmapurikar, S., Yu, F., Crowley, P., Turner, J.: Algorithms to Accelerate Multiple Regular Expressions Matching for Deep Packet Inspection. In: Proc. ACM SIGCOMM, pp. 339–350 (2006)
6. Levandoski, J., Sommer, E., Strait, M.: Application Layer Packet Classifier for Linux, http://l7-filter.sourceforge.net/
7. Liu, T., Yang, Y., Liu, Y., Sun, Y., Guo, L.: An Efficient Regular Expressions Compression Algorithm From A New Perspective. In: Proc. IEEE INFOCOM, pp. 2129–2137 (2011)
8. Liu, Y., Guo, L., Liu, P., Tan, J.: Compressing Regular Expressions' DFA Table by Matrix Decomposition. In: Domaratzki, M., Salomaa, K. (eds.) CIAA 2010. LNCS, vol. 6482, pp. 282–289. Springer, Heidelberg (2011)
9. Roesch, M.: Snort - Lightweight Intrusion Detection for Networks. In: Proc. USENIX LISA, pp. 229–238 (1999)
10. Rohrer, J., Atasu, K., van Lunteren, J., Hagleitner, C.: Memory-Efficient Distribution of Regular Expressions for Fast Deep Packet Inspection. In: Proc. IEEE/ACM CODES+ISSS, pp. 147–154 (2009)
11. Smith, R., Estan, C., Jha, S.: XFA: Faster Signature Matching with Extended Automata. In: Proc. IEEE S&P, pp. 187–201 (2008)
12. The Shmoo Group: Internet Traffic Traces, http://cctf.shmoo.com/
13. Yu, F., Chen, Z., Diao, Y., Lakshman, T.V., Katz, R.H.: Fast and Memory-Efficient Regular Expression Matching for Deep Packet Inspection. In: Proc. ACM/IEEE ANCS, pp. 93–102 (2006)

The Removal of Weighted ε-Transitions[*]

Sylvain Lombardy[1] and Jacques Sakarovitch[2]

[1] LIGM, Université Paris-Est Marne-la-Vallée
[2] LTCI, CNRS / Telecom ParisTech

Abstract. The removal of ε-transitions in weighted automata leads to infinite summation when cycles of such transitions are allowed. This paper presents both an algorithm for that purpose, and a framework in which the algorithm is correct.

Introduction

This work addresses the problems raised by the writing of an algorithm for the removal of the ε-transitions in automata with weights in $\mathbb{Q}$ or $\mathbb{R}$. The solution consists in both an algorithm and the setting of a consistent and sensible framework in which the algorithm can be used.

Such an algorithm for Boolean automata belongs to basic automata theory and amounts to the computation of the *transitive closure* of the graph of ε-transitions, more or less intertwined with the construction of the resulting proper automaton itself. The same algorithm for weighted automata also corresponds to a transitive closure computation but is far more complex. Above all, the closure may not exist: if the automaton contains a cycle of ε-transitions then the sum of the weights along the paths following this cycle may well be not defined, as in $\mathcal{Q}_1$ in Fig. 1. For that reason, mathematically oriented works on automata, such as [1] or [3], have ruled out the possibility of having cycles of ε-transitions in automata, either explicitly with the hypothesis of *cycle-free* automata, or implicitly by considering the *discrete topology* on the weight semiring. As a result, $\mathcal{Q}_2$ in Fig. 1 is considered in the quoted works as an incorrect object. Nevertheless Probabilistic automata, with weights in $\mathbb{Q}$ or $\mathbb{R}$ (equipped with the natural topology), or distance automata, with weights in $\mathbb{Z}\mathrm{min}$, are computational models that one should be apt to deal with.

Fig. 1. A non-valid automaton $\mathcal{Q}_1$, and $\mathcal{Q}_2$ that should be considered valid

Given a $\mathbb{K}$-automaton $\mathcal{A}$, the first question is then:

Q 1.— How to compute a *proper* automaton equivalent to $\mathcal{A}$?

---

[*] Work Supported by ANR Project 10-INTB-0203 VAUCANSON 2.

N. Moreira and R. Reis (Eds.): CIAA 2012, LNCS 7381, pp. 345–352, 2012.

A possible answer consists in a generalisation of the algorithm in the Boolean case, that takes the weights into account. Fig. 2 shows a $\mathbb{Q}$-automaton with ε-transitions and an equivalent proper $\mathbb{Q}$-automaton. The generalisation raises correctness and termination issues, far less trivial than in the Boolean case. But above all, computations involved in the algorithm may lead to undefined summations. When this occurs, the behaviour of the automaton is not defined, and the automaton is said to be non-valid. Of course, we want the validity of an automaton be a property defined intrinsically and not depending upon a certain algorithm, nor upon a certain execution of an algorithm. A question that comes thus even before Q1 is:

Q0.— When is a $\mathbb{K}$-automaton $\mathcal{A}$ a *valid* $\mathbb{K}$-automaton?

This paper puts forward a new definition of valid weighted automata, that allows to give a consistent answer to both Q0 and Q1. This definition is a strengthening of the one proposed in [7,8]. Both definitions coincide for all usual weight structures. They give rise to a consistent theory and allow for instance to speak of the behaviour of $\mathcal{Q}_2$ in Fig. 1 or to state that $\mathcal{Q}_3$ in Fig. 2 is valid and equivalent to $\mathcal{Q}_4$. It turns out that *the older definition is not accurate* enough to analyse the situations that occur in the computation of the behaviour of automata with weights in an arbitrary semiring. The new definition, presented at Sect. 1, *does not change* the one of behaviour of automata, leaving thus unchanged all other results of the theory (Kleene–Schützenberger's theorem, *etc.*).

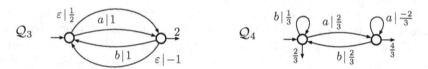

Fig. 2. A transformation which needs theoretical foundation

The ε-transition removal algorithm, described at Sect. 2, is then guaranteed to work properly on a *valid* automaton. As the algorithm implies block summations, it may happens that it successfully terminates on non-valid automata such as those shown at Fig. 3.

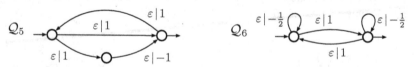

Fig. 3. Two non-valid $\mathbb{Q}$-automata on which the closure algorithm succeeds

Which raises the next question:

Q2.— Is it decidable whether a given $\mathbb{K}$-automaton $\mathcal{A}$ is valid or not?

At Sect. 3, a positive answer to Q2 is given when $\mathbb{K}$ is what we call a *star congruous topological ordered positive semiring*. In a second step, the cases of $\mathbb{R}$

and $\mathbb{Q}$ (which were our primary aim) are settled by considering automata in which every weight is replaced by its absolute value, bringing back to the former case.

In conclusion, this paper fills the *effectively* gap that was left open by our definition of valid automata in [7,8]. The algorithms described in the paper have been *implemented* in VAUCANSON [9].

Due to severe space constraints, classical definitions and notation are understood and to be found in [7], proofs are omitted and the resulting paper is rather an extended abstract. A complete version of the paper is accessible in the ArXiV repository [4].

1   Valid Weighted Automata

Topological Semirings. We deal with *weight semirings* such as $\mathbb{B}$, $\mathbb{N}$, $\mathbb{Z}$, $\mathbb{Q}$ or $\mathbb{R}$, and also the 'tropical' semiring $\mathbb{Z}\min = \langle \mathbb{Z} \cup +\infty, \min, + \rangle$ or $\operatorname{Rat} A^*$. But the semiring structure is weak and the general definitions have to take all cases into account (*cf.* Exm. 1).

We consider *topological* semirings (TS) in order to define the *star* as an infinite sum. The *discrete* topology makes any semiring a TS, and it is the natural one on $\mathbb{B}$, $\mathbb{Z}$, $\mathbb{Z}\min$. On $\mathbb{Q}$ or $\mathbb{R}$, the natural topology is induced by the Euclidean distance, on $\operatorname{Rat} A^*$, by the inclusion order, and on $\mathcal{N} = \mathbb{N} \cup +\infty$, it is the one for which every strictly increasing sequence converges to $+\infty$; we denote by $\mathbb{N}_\infty$ the *same* semiring equipped with the discrete topology, in which the stationary sequences only converge.

An element k in $\mathbb{K}$ is *starable* if the family $\{k^n\}_{n\in\mathbb{N}}$ is summable (*cf.* [7]) and its sum k^* is the 'star of k'. A k in $\mathbb{Q}$ is starable if, and only if, it is smaller than 1. In $\mathbb{Z}\min$, only the non negative integers and $+\infty$ are starable. In $\mathcal{N}$, every element is starable, whereas in $\mathbb{N}_\infty$ only 0 and $+\infty$ are starable. A TS $\mathbb{K}$ is *starable* (resp. non-starable) if so is every element of $\mathbb{K}$ (resp. if no element different from $0_\mathbb{K}$ is starable): $\mathbb{B}$, $\mathcal{N}$, or $\operatorname{Rat} A^*$ are starable semirings, $\mathbb{N}$, $\mathbb{Z}$, are non-starable semirings. The semirings $\mathbb{N}_\infty$, $\mathbb{Q}$, or $\mathbb{S}$ below, are neither starable nor non-starable semirings.

Example 1. Let $\mathbb{S}$ be the semiring of 2×2-matrices of non negative integers, generated by $0_\mathbb{S} = \begin{pmatrix} 0 & 0 \\ 0 & 0 \end{pmatrix}$, $x = \begin{pmatrix} 1 & 0 \\ 0 & 1 \end{pmatrix} = 1_\mathbb{S}$, and $y = \begin{pmatrix} 0 & 1 \\ 1 & 0 \end{pmatrix}$, and quotiented by $x + y = \infty_\mathbb{S} = \infty_\mathbb{S} + x = \infty_\mathbb{S} + y$. In $\mathbb{S}$ equipped with the discrete topology, $0_\mathbb{S}$, $\infty_\mathbb{S}$ and y are starable, whereas $x = y^2$ is not.

Series We denote by A the *alphabet*, by A^* the *free monoid* it generates, and by ε the *empty word*. The semiring $\mathbb{K}\langle\!\langle A^* \rangle\!\rangle$ of *series* over A^* with coefficients in $\mathbb{K}$ is equipped with the *simple convergence topology*. Any *proper* series is then starable. An arbitrary series s is written $s = s_0 + s_\mathsf{p}$ where $s_0 = \langle s, \varepsilon \rangle \varepsilon$ is the *constant term* of s and s_p its *proper part*.

Under the assumption that $\mathbb{K}$ is a *strong semiring* (which holds[1] silently from now on), the following holds:

---

[1] Strongness is not so restrictive: in particular, all semirings quoted above are strong.

Proposition 1 ([5,7,8]). *A series s of $\mathbb{K}\langle\!\langle A^*\rangle\!\rangle$ is starable if, and only if, s_0 is starable and, in this case, $s^* = (s_0^* s_p)^* s_0^* = s_0^* (s_p s_0^*)^*$ holds.*

Weighted Automata. In the sequel, $\mathcal{A} = \langle Q, A, E, I, T \rangle$ is a $\mathbb{K}$-automaton ($\mathbb{K}$ is understood) where $E \subseteq Q \times (A \cup \varepsilon) \times \mathbb{K} \times Q$ is the set of *weighted transitions*. The *label* of the transition $e = (p, x, k, q)$ is x, its *weight* is k, and its *weighted label*, *w-label* for short, written $\mathsf{w}(e)$, is the *monomial* $k\,x$. For consistency with the definitions to come, the set E cannot contain two distinct transitions with the same source, destination, and label. Likewise, the weight of a transition is never equal to $0_\mathbb{K}$. The automaton $\mathcal{A}$ is *finite* if Q is finite, and this condition holds from now on.

The *w-label* of a *path* c in $\mathcal{A}$, written $\mathsf{w}(c)$, is the *product* of the w-labels of the transitions of c. We denote by $\mathsf{W}(P)$ the family of w-labels of paths in any family P of paths of $\mathcal{A}$. A *computation* in $\mathcal{A}$ is a path together with the initial and final functions taken into account.

The set of transitions E may be seen as a $Q \times Q$-matrix, the *transition matrix* of $\mathcal{A}$, and $\mathcal{A}$ is then written as $\mathcal{A} = \langle I, E, T \rangle$. With $\mathcal{A}$, we associate the *automaton of ε-transitions* $\mathcal{A}_0 = \langle I, E_0, T \rangle$, where E_0 is the restriction of E to ε-transitions.

The Behaviour of a $\mathbb{K}$-automaton $\mathcal{A}$, denoted by $|\!|\mathcal{A}|\!|$, is, by definition, the sum of the w-labels of the computations of $\mathcal{A}$. For the behaviour be well-defined, it is then *necessary* that this family of w-labels be summable. It should not be a *sufficient condition* if we want the definition of validity be consistent with natural computations on automata.

Let E^* be the free monoid generated by the set of transitions E of $\mathcal{A}$. The *set of paths* of $\mathcal{A}$ is a (local) rational subset $\mathsf{P}_{\mathcal{A}}$ of E^*. A *rational set of paths* of $\mathcal{A}$ is any rational subset of E^* contained in $\mathsf{P}_{\mathcal{A}}$.

Definition 1. *A $\mathbb{K}$-automaton $\mathcal{A}$ is* valid *if, and only if, the family of w-labels of any rational set of paths of $\mathcal{A}$ is summable.*

We denote by $\mathsf{P}_{\mathcal{A}}(p, q)$ the family of paths with source p and destination q; $\mathcal{A}$ is said (with a slight abuse) to have *summable co-terminal paths* if, for every p and q in Q, $\mathsf{W}(\mathsf{P}_{\mathcal{A}}(p,q))$ is summable. A valid automaton has summable co-terminal paths (and its behaviour is thus well-defined), but the converse is not true, as shown by the three examples of Fig. 4. They have only ε-transitions, as explained by Proposition 2.

Proposition 2. *A $\mathbb{K}$-automaton $\mathcal{A}$ is valid if, and only if, $\mathcal{A}_0$ is.*

Example 2.   (a) Let e be the loop of the $\mathbb{S}$-automaton $\mathcal{T}_1$: $\mathsf{W}\big((e^2)^*\big)$ is not summable (*cf.* Exm. 1).   (b) Let e be the loop of the $\mathbb{N}_\infty$-automaton $\mathcal{T}_2$: $\mathsf{W}((e)^*)$ is not summable.   (c) Let e and f be the two transitions of the $\mathbb{N}_\infty$-automaton $\mathcal{T}_3$ that form a cycle: $\mathsf{W}((e\,f)^*)$ is not summable.

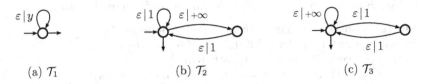

(a) $\mathcal{T}_1$ (b) $\mathcal{T}_2$ (c) $\mathcal{T}_3$

Fig. 4. Three non-valid automata

Coverings. A covering (*cf.* [7]) is an automata *morphism* which induces a *bijection* between the outgoing transitions of a state and its image, and then a bijection between *computations*. The following implies in particular that any covering of a valid automaton is valid, a property that strongly speaks for Definition 1.

Theorem 1. *A $\mathbb{K}$-automaton $\mathcal{A}$ is valid if, and only if, any covering of $\mathcal{A}$ has summable co-terminal paths.*

Transition matrix. Formation of paths corresponds to matrix multiplication, and, for every integer n, the (p,q)-entry of the *matrix* E^n is the sums of the w-labels of paths of length n from p to q in $\mathcal{A}$.

Proposition 3. *If $\mathcal{A} = \langle I, E, T \rangle$ is valid, then E is starable and $|\mathcal{A}| = I \cdot E^* \cdot T$ holds.*

The converse of Proposition 3 does not hold (*cf.* [7]). As $\mathbb{K}\langle\!\langle A^* \rangle\!\rangle^{Q \times Q}$ is homeomorphic to $\mathbb{K}^{Q \times Q}\langle\!\langle A^* \rangle\!\rangle$, Propositions 1, 2, and 3 yield:

Proposition 4. *If $\mathcal{A} = \langle I, E, T \rangle$ is valid, then the matrix E_0 is starable and $|\mathcal{A}| = I \cdot (E_0^* \cdot E_p)^* \cdot E_0^* \cdot T$ holds.*

Algorithm 1 below does not compute the matrix E_0^* itself (what would be called the ε-*closure*), but directly the matrix $E_0^* \cdot E_p$ and the vector $E_0^* \cdot T$ that form the proper automaton $\mathcal{B} = \langle I, E_0^* \cdot E_p, E_0^* \cdot T \rangle$ equivalent to $\mathcal{A}$.

2 The Weighted Closure Algorithm

The ε-transitions of $\mathcal{A}$ will be eliminated one after the other. The elimination process is different when the transition is a loop, the *loop blowing* and when the transition links two distinct states, the *transition killing*.

Loop Blowing Let $e = (q, \varepsilon, k, q)$ be an ε-loop of $\mathcal{A}$ around state q.

- If k is *not starable*, then BLOW(e) is impossible, and *fails*.
- If k is *starable*, then BLOW(e) yields $\mathrm{LB}_e(\mathcal{A})$ by the following:
  (i) every $f = (q, x, h, r)$ is changed into $f' = (q, x, k^*h, r)$;[2]
  (ii) $T(q) := k^* T(q)$;   (iii) and finally e is removed.

No ε-transition is created by the loop blowing but the operation *may fail*.

---

[2] In any TS, the product $k^*h = h + k k^* h$ is never equal to $0_{\mathbb{K}}$ for $h \neq 0_{\mathbb{K}}$.

Transition Killing. Let $e = (p, \varepsilon, k, q)$ be an ε-transition of $\mathcal{A}$, with $p \neq q$.

- KILL(e) is authorized only if *there is no ε-loop around q*.
- In this case, KILL(e) yields $\mathsf{TK}_e(\mathcal{A})$ by the following:
  - (i) for every $f = (q, x, h, r)$, with x in A_ε,
    - (a) if there exists $g = (p, x, l, r)$, then g is *changed* into $g' = (p, x, l + k\,h, r)$ or suppressed if $l + k\,h = 0_{\mathbb{K}}$;
    - (b) otherwise, $g = (p, x, k\,h, r)$ is *created* if $k\,h \neq 0_{\mathbb{K}}$.
  - (ii) $T(p) := T(p) + k\,T(q)$;   (iii) and finally e is removed.

A transition killing, if allowed, *never fails* but may create *new ε-transitions*.

An ε-R emoval Algorithm is made with these two atomic operations:

Algorithm 1. Epsilon Transition Removal Algorithm

---
for EVERY state q in $\mathcal{A}$ **do**
    if EXISTS ε-loop e on q **then**
        BLOW(e) and $\mathcal{A} := \mathsf{LB}_e(\mathcal{A})$
    end if
    for EVERY ε-transition e incoming in q **do**
        KILL(e) and $\mathcal{A} := \mathsf{TK}_e(\mathcal{A})$
    end for
end for

Algorithm 1 applied to $\mathcal{Q}_3$ yields $\mathcal{Q}_4$ at Fig. 2; Fig. 5 shows the intermediate steps: a transition killing from (a) to (b), and a transition blowing from (b) to (c); killing the last ε-transition in (c) yields $\mathcal{Q}_4$.

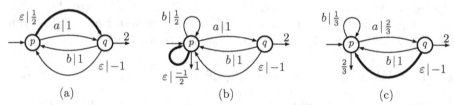

(a)   (b)   (c)

Fig. 5. A transition killing and a transition blowing

Theorem 2. *Algorithm 1 applied to a valid $\mathbb{K}$-automaton computes an equivalent proper $\mathbb{K}$-automaton.*

Algorithm 1 is correct as both blowing and killing operations preserve the behaviour of a valid $\mathbb{K}$-automaton. After a state q is processed in Algorithm 1, no ε-transition ends in q anymore and thus none will be created in the sequel of the execution: the algorithm terminates.

In contrast with the Boolean case, termination is not so obvious in the weighted case. For instance, the 'LIFO policy' which is commonly used in algorithms for transitive closure of graphs may lead to endless procedures in the weighted case.

3   Decision Algorithm

We give here *sufficient conditions* on $\mathbb{K}$ such that the answer to Q2 is positive. All semirings we want to be able to deal with fulfill one of them.

Two situations yield trivial answer to Q2: when the star does not occur in the computation of the behaviour of $\mathcal{A}_0$, or when the star does not induce infinite summation in $\mathbb{K}$. A $\mathbb{K}$-automaton $\mathcal{A}$ is said to be *cycle-free* ([3,2]) if the automaton $\mathcal{A}_0$ of ε-transitions of $\mathcal{A}$ does not contain any cycle. From Proposition 2, follow then the next two statements.

Proposition 5. *Every cycle-free $\mathbb{K}$-automaton is valid, whatever $\mathbb{K}$.*

For non-starable semirings, this sufficient condition is also necessary.

Proposition 6. *If $\mathbb{K}$ is a non-starable semiring, then a $\mathbb{K}$-automaton is valid if, and only if, it is cycle-free.*

The star reduces to a finite sum in *k-closed semirings* ([6]).

Proposition 7. *Every $\mathbb{K}$-automaton is valid if $\mathbb{K}$ is a k-closed semiring.*

3.1   The Case of Topological Ordered Positive Semirings

A semiring $\mathbb{K}$ is an *ordered positive semiring* (OPS, for short) if it is equipped with an ordering $\leqslant$ such that, for every x, y, z in $\mathbb{K}$, it holds:
$$0_{\mathbb{K}} \leqslant x \quad \text{and} \quad x \leqslant y \implies \{x + z \leqslant y + z \quad \text{and} \quad xz \leqslant yz\}.$$
A topology and an ordering on $\mathbb{K}$ are *consistent* if whenever $u = \{u_n\}_{n\in\mathbb{N}}$ and $v = \{v_n\}_{n\in\mathbb{N}}$ are two sequences in $\mathbb{K}$ such that u converges to a limit ℓ, and, for every n, $u_n \leqslant v_n \leqslant \ell$, then v converges to ℓ.

Definition 2. *A topological ordered positive semiring (TOPS, for short) is both a TS and an OPS where topology and ordering are consistent. We call* star congruous *a TOPS where the domain of star is closed downward.*

The semirings $\mathbb{B}$, Rat A^*, $\mathbb{R}_+$, $\mathbb{Q}_+$, $\mathbb{Z}\min^3$, $\mathcal{N}$ or $\mathbb{N}_\infty$ are TOPS; all are star congruous but $\mathbb{N}_\infty$.

Theorem 3. *If Algorithm 1 succeeds on a $\mathbb{K}$-automaton $\mathcal{A}$, where $\mathbb{K}$ is a star congruous TOPS, then $\mathcal{A}$ is valid.*

Algorithm 1 succeeds on $\mathcal{T}_3$ (*cf.* Fig. 4(c)) although it is not valid.

3.2   The Case of $\mathbb{R}$ and $\mathbb{Q}$

In $\mathbb{R}$, topological completeness implies that summability is equivalent to *absolute summability*. For an $\mathbb{R}$-automaton $\mathcal{A}$, let abs($\mathcal{A}$) be the $\mathbb{R}_+$-automaton obtained by replacing every weight k with the absolute value $|k|$.

---

[3] In min-plus semirings, the ordering is the inverse of the usual ordering.

Theorem 4. *An $\mathbb{R}$-automaton $\mathcal{A}$ is valid if, and only if,* $\mathsf{abs}(\mathcal{A})$ *is valid.*

As $\mathbb{Q}$ is not complete, summability is not equivalent to *absolute summability* anymore. As $\mathbb{Q}$ is a subsemiring of $\mathbb{R}$, we may apply however the same procedure to $\mathbb{Q}$-automata as to $\mathbb{R}$-automata, and since $\mathbb{Q}$ is closed for every operation performed in Algorithm 1, the result is a $\mathbb{Q}$-automaton.

References

1. Berstel, J., Reutenauer, C.: Les séries rationnelles et leurs langages. Masson (1984); translation: Rational Series and Their Languages. Springer (1988)
2. Ésik, Z., Kuich, W.: Finite automata. In: Droste, M., et al. (eds.) Handbook of Weighted Automata, pp. 69–104. Springer (2009)
3. Kuich, W., Salomaa, A.: Semirings, Automata, Languages. Springer (1986)
4. Lombardy, S., Sakarovitch, J.: On the weighted closure problem. Available on ArXiV (to appear, 2012)
5. Lombardy, S., Sakarovitch, J.: Derivation of rational expressions with multiplicity. Theoret. Computer Sci. 332, 141–177 (2005)
6. Mohri, M.: Generic ε-removal and input ε-normalization algorithms for weighted transducers. Int. J. Foundations Computer Sci. 13, 129–143 (2002)
7. Sakarovitch, J.: Elements of Automata Theory. Cambridge University Press (2009); corrected English translation of Éléments de théorie des automates, Vuibert (2003)
8. Sakarovitch, J.: Rational and recognisable power series. In: Droste, M., et al. (eds.) Handbook of Weighted Automata, pp. 105–174. Springer (2009)
9. VAUCANSON, www.vaucanson-project.org

Weighted LTL with Discounting*

Eleni Mandrali

Department of Mathematics, Aristotle University of Thessaloniki
54124 Thessaloniki, Greece
elemandr@math.auth.gr

Abstract. We introduce a weighted linear temporal logic over infinite words with weights and discounting parameters over $\mathbb{R}_{\max}$. We translate the formulas of a syntactically defined fragment of our logic to weighted Büchi automata with discounting. We prove that every ω-recognizable series with discounting is the image of a series definable by a formula of this syntactic fragment through a strict alphabetic epimorphism.

1 Introduction

Linear temporal logic (*LTL* for short) was introduced by Pnueli in 1977 [8]. The translation of *LTL* formulas to automata is one of the ways to develop model checking techniques, successfully investigated in [9]. Hence, the weighting of *LTL* and the investigation of its relation to weighted ω-automata is a first step for the development of quantitative model checking theories. In [6] and [4] the authors considered a weighted *LTL* over finite distributive De Morgan lattices, and over strong bimonoids, respectively. *LTL* has been extended to a quantitative setting incorporating discounted methods in [1,5]. In those cases the authors considered as underlying structure the interval $[0,1]$ with operations max and min. Weighted model checking tools were recently introduced in [5].

In this work, we introduce a weighted linear temporal logic for infinite words with discounting parameters d ($0 \leq d < 1$) over $\mathbb{R}_{\max}$. Infinite words are over the powerset of a finite set of atomic propositions AP. The semantics of our weighted *LTL* formulas are infinitary series and our goal is to investigate their relation to ω-recognizable series with discounting. After presenting some preliminary notions in the second section, in Section 3 we introduce weighted generalized Büchi automata with discounting and ε-transitions, and we prove that they are equivalent to weighted Büchi automata with discounting. In Section 4, we define the syntax of formulas of the weighted *LTL* by the grammar: $\varphi ::= k \mid a \mid \neg a \mid \varphi \vee \varphi \mid \varphi \wedge \varphi \mid \bigcirc\varphi \mid \varphi U \varphi \mid \Box\varphi$ where $k \in \mathbb{R}_{\max}$ and $a \in AP$. The discounting parameters d are being involved in the semantics of the *next* $\bigcirc$, *until* U, and *always* $\Box$ operators. Using weights from $\mathbb{R}_{\max}$ is the reason why negation is

* This research has been co-financed by the European Union (European Social Fund - ESF) and Greek national funds through the Operational Program "Education and Lifelong Learning" of the National Strategic Reference Framework (NSRF) - Research Funding Program: Heracleitus II.

N. Moreira and R. Reis (Eds.): CIAA 2012, LNCS 7381, pp. 353–360, 2012.

restricted to atomic propositions, in contrast to [6], [4] where negation is applied to every formula of the proposed logics. Nevertheless, in [6] the values obtained by logical operations are finite. Given any formula of a syntactic fragment of our logic we construct a weighted generalized Büchi automaton with discounting and ε-transitions recognizing its semantics. For this translation we follow the approach investigated in [2,10]. More precisely, the states of the automaton are sets of formulas satisfying discrete conditions of consistency, and the final subsets are defined with respect to the until operators. As in [2] we shall use ε-transitions to reduce formulas. In that paper the goal of the reduction is the production of sets of formulas whose elements are atomic formulas, or their negations, or formulas with outermost connective the *next* operator. In our case, that approach does not allow the effective computation of the weights of the transitions. In order to achieve this, we consider in each state a unique maximal formula (according to subformula relation) that will indicate the induction (and thus the operations) connecting the formulas $k \in \mathbb{R}_{\max}$ of each state. The ε-transitions will be used to reduce the maximal formula of a set, and to ensure that the state set of the automaton is finite.

Finally, we prove that every ω-recognizable series with discounting is the image of a series definable by a formula of the syntactic fragment of our logic, through a strict alphabetic epimorphism.

2   Preliminaries

Let A be an alphabet, i.e., a finite non-empty set. As usual we denote by A^+ (resp. A^ω) the set of finite (resp. infinite) non-empty words over A and by ε the empty word. An infinite word $w = a_0 a_1 \ldots \in A^\omega$ (where $a_i \in A$ for all $i \geq 0$) will be also written as $w = w(0)w(1) \ldots$ with $w(i) = a_i$ for every $i \geq 0$. For every $i \geq 0$ we shall denote by $w_{\geq i}$ the infinite suffix of w starting at position i, i.e., $w_{\geq i} = w(i)w(i+1)\ldots$. Moreover, we shall write $w_{<i}$ for the prefix of w of length i, i.e., $w_{<i} = w(0)\ldots w(i-1)$. Obviously $w_{<0} = \varepsilon$.

In this paper, we consider the *max-plus* (or *arctic*) *semiring* $\mathbb{R}_{\max} = (\mathbb{R}_+ \cup \{-\infty\}, \sup, +, -\infty, 0)^1$ where $\mathbb{R}_+ = \{r \in \mathbb{R} \mid r \geq 0\}$ and $-\infty + x = -\infty$ for every $x \in \mathbb{R}_+ \cup \{-\infty\}$. We extend the multiplication $\cdot$ over $\mathbb{R}_+$ by letting $x \cdot (-\infty) = (-\infty) \cdot x = -\infty$ for every $x \in \mathbb{R}_+ \cup \{-\infty\}$. Then, for every $p \in \mathbb{R}_+ \cup \{-\infty\}$ the mapping $\overline{p} : \mathbb{R}_{\max} \to \mathbb{R}_{\max}$ given by $x \longmapsto p \cdot x$ is an endomorphism of $\mathbb{R}_{\max}$. Conversely, every endomorphism of $\mathbb{R}_{\max}$ is of this form (see Lemma 15 in [3]). We shall alternatively denote the multiplication of $\mathbb{R}_{\max}$ and the composition operation of endomorphisms of $\mathbb{R}_{\max}$ also by concatenation.

A family $d = \left(\overline{d_a}\right)_{a \in A}$ of endomorphisms of $\mathbb{R}_{\max}$ with $0 \leq d_a < 1$ for every $a \in A$ is called a *d-discounting over A and $\mathbb{R}_{\max}$*. For every finite word $w = a_0 a_1 \ldots a_{n-1} \in A^+$ we shall denote by $\overline{d_w}$ the morphism $\overline{d_{a_0} d_{a_1} \ldots d_{a_{n-1}}}$ and by $\overline{d_\varepsilon}$ the identity mapping id on $\mathbb{R}_{\max}$. We put $d_w = \prod_{a \in A} d_a^{|w|_a}$ where $|w|_a$ denotes the number of a's in w.

---

[1] Here we use sup instead of max since we need to compute over infinite words.

An *infinitary formal power series* (*series* for short) *over the alphabet A and the semiring* $\mathbb{R}_{max}$ is a mapping $s : A^\omega \to \mathbb{R}_{max}$. We write (s, w) instead of $s(w)$ for $w \in A^\omega$ and call it the coefficient of s at w. The class of all infinitary series over A and $\mathbb{R}_{max}$ is denoted by $\mathbb{R}_{max} \langle\langle A^\omega \rangle\rangle$. For $s, r \in \mathbb{R}_{max} \langle\langle A^\omega \rangle\rangle$ and $k \in \mathbb{R}_{max}$ the maximum $\max(s, r)$, the scalar sum $k + s$ and the sum $s + r$ are defined elementwise, i.e., $(\max(s, r), w) = \max((s, w), (r, w))$, $(k + s, w) = k + (s, w)$, $(s + r, w) = (s, w) + (r, w)$ for every $w \in A^\omega$.

Let $s \in \mathbb{R}_{max} \langle\langle A^\omega \rangle\rangle$. The *image* $\mathrm{Im}(s)$ *of* s is the set $\mathrm{Im}(s) = \{k \in \mathbb{R}_{max} \mid \exists w \in A^\omega \text{ with } (s, w) = k\}$. We say that s has *bounded image* if there is an $m \in \mathbb{R}_+$ such that $k \leq m$ for every $k \in \mathrm{Im}(s)$. Consider two alphabets A, B and a strict alphabetic epimorphism $h : A^* \to B^*$, i.e., $h(a) \in B$ for every $a \in A$. Then h can be extended to a mapping $h : A^\omega \to B^\omega$ by letting $h(w) = (h(w(i)))_{i \geq 0}$ for every $w \in A^\omega$. For every series $s \in \mathbb{R}_{max} \langle\langle A^\omega \rangle\rangle$ with bounded image we define the series $h(s) \in \mathbb{R}_{max} \langle\langle B^\omega \rangle\rangle$ by $(h(s), u) = \sup_{w \in h^{-1}(u)} ((s, w))$ for every $u \in B^\omega$.

For a finite set C we denote by $\mathcal{P}(C)$ the powerset of C.

3 Weighted Generalized Büchi Automata with Discounting and ε-Transitions

In this section we introduce weighted generalized Büchi automata with discounting and ε-transitions, and weighted Büchi automata with discounting and ε-transitions. We prove that these two models are equivalent and also equivalent to weighted Büchi automata with discounting introduced in [3]. Throughout this section, A will denote an alphabet.

Definition 1. *(i) A weighted generalized Büchi automaton with ε-transitions (ε-WGBA for short) over A and $\mathbb{R}_{max}$ is a quadruple $\mathcal{M} = (Q, wt, in, \mathcal{F})$, where Q is the* finite state set, $wt : Q \times (A \cup \{\varepsilon\}) \times Q \to \mathbb{R}_{max}$ *is a mapping assigning* weights *to the transitions of the automaton,* $in : Q \to \mathbb{R}_{max}$ *is the* initial distribution *and* $\mathcal{F} = \{F_1, \ldots, F_k\}$ *is the set of final subsets, with $F_i \in \mathcal{P}(Q)$, for every $1 \leq i \leq k$. Moreover, for every $t \in Q \times \{\varepsilon\} \times Q$ we require $wt(t) = 0$, or $wt(t) = -\infty$. Finally, for every $(q, \varepsilon, q') \in Q \times \{\varepsilon\} \times Q$ with $wt((q, \varepsilon, q')) = 0$, and every $i \in \{1, \ldots, k\}$, $q \in F_i$ if-f $q' \in F_i$.*

(ii) An ε-WGBA is a weighted Büchi automaton with ε-transitions *(ε-WBA for short) if $k = 1$, i.e., there is only one final subset.*

(iii) An ε-WBA is a weighted Büchi automaton *(WBA for short) if $wt(t) = -\infty$ for every $t \in Q \times \{\varepsilon\} \times Q$. In this case it suffices to define wt over $Q \times A \times Q$.*

Let $w = a_0 a_1 \ldots \in A^\omega$ with $a_i \in A$ ($i \geq 0$). A *path* P_w of $\mathcal{M}$ over w is an infinite sequence of transitions $P_w = ((q_j, b_j, q_{j+1}))_{j \geq 0}$, $b_j \in A \cup \{\varepsilon\}$ ($j \geq 0$), such that $w = b_0 b_1 \ldots$. The set of states that appear infinitely often along P_w is denoted by $In^Q(P_w)$. Given a d-discounting, the *d-weight* of P_w (or simply *weight*) is the value $weight_{\mathcal{M}}(P_w) = in(q_0) + \sum_{j \geq 0} d_{w_{<i_j}} \cdot wt((q_j, b_j, q_{j+1}))$, where for $j = 0$ we set $i_j = 0$, and for $j \geq 1$ we let $i_j \geq 0$ be the unique position of w with

$w_{<i_j} = b_0 b_1 \ldots b_{j-1}$. Observe that the infinite sum converges; its value is bounded by $M \cdot \sum_{i \geq 0} m^i = M \cdot 1/(1-m)$, where $M = \max\{wt(t) \mid t \in Q \times (A \cup \{\varepsilon\}) \times Q\}$ and $m = \max\{d_a \mid a \in A\}$.

The path P_w is called *successful* if $In^Q(P_w) \cap F_i \neq \emptyset$, for every $i \in \{1, \ldots, k\}$. The set of all successful paths of $\mathcal{M}$ is denoted by $succ(\mathcal{M})$. The *d-behavior* (or simply *behavior*) of $\mathcal{M}$ is the infinitary series $\|\mathcal{M}\| : A^\omega \to \mathbb{R}_{\max}$ with coefficients specified, for every $w \in A^\omega$, by $(\|\mathcal{M}\|, w) = \sup_{P_w \in succ(\mathcal{M})} (weight_{\mathcal{M}}(P_w))$. This supremum exists in $\mathbb{R}_{\max}$ since the values $weight_{\mathcal{M}}(P_w)$ are uniformly bounded.

Two ε-WGBA are called *equivalent* if they have the same behavior. Following the construction used in [10] to convert generalized Büchi automata to Büchi automata we can prove that for every ε-WGBA with d-discounting over A and $\mathbb{R}_{\max}$ there exists an equivalent ε-WBA with d-discounting over A and $\mathbb{R}_{\max}$. Moreover, the two conditions imposed on the ε-transitions in Definition 1, i.e., the restriction of their weights to the set $\{0, -\infty\}$ and the condition concerning the final subsets, allow us to produce from every ε-WBA with d-discounting over A and $\mathbb{R}_{\max}$ an equivalent WBA with d-discounting over A and $\mathbb{R}_{\max}$. A series $s : A^\omega \to \mathbb{R}_{\max}$ is called *d-Büchi recognizable* or *(d, ω)-recognizable* if there is a WBA $\mathcal{M}$ such that $s = \|\mathcal{M}\|$.

4   Results

In this section, we first introduce a weighted linear temporal logic (weighted *LTL* for short) with discounting over the max-plus semiring. Let AP be a finite set of atomic propositions. We shall use the symbols a, b, c to denote elements in AP. Moreover, d will denote a discounting over $\mathcal{P}(AP)$ and $\mathbb{R}_{\max}$.

Definition 2. *The syntax of the formulas of the* weighted LTL *over AP and $\mathbb{R}_{\max}$ is given by the grammar*

$$\varphi ::= k \mid a \mid \neg a \mid \varphi \vee \varphi \mid \varphi \wedge \varphi \mid \bigcirc \varphi \mid \varphi U \varphi \mid \Box \varphi$$

where $k \in \mathbb{R}_{\max}$ and $a \in AP$.

We denote by $LTL(\mathbb{R}_{\max}, AP)$ the class of all weighted *LTL*-formulas over AP and $\mathbb{R}_{\max}$. Next, we represent the semantics of formulas $\varphi \in LTL(\mathbb{R}_{\max}, AP)$ as infinitary series $\|\varphi\|_d$ in $\mathbb{R}_{\max}\langle\langle(\mathcal{P}(AP))^\omega\rangle\rangle$.

Definition 3. *Let $\varphi \in LTL(\mathbb{R}_{\max}, AP)$. The d-semantics of φ is a series $\|\varphi\|_d \in \mathbb{R}_{\max}\langle\langle(\mathcal{P}(AP))^\omega\rangle\rangle$ defined inductively as follows. We let for every $w \in (\mathcal{P}(AP))^\omega$*

- $(\|k\|_d, w) = k$
- $(\|a\|_d, w) = \begin{cases} 0 & \text{if } a \in w(0) \\ -\infty & \text{otherwise} \end{cases}$ $(\|\neg a\|_d, w) = \begin{cases} 0 & \text{if } a \notin w(0) \\ -\infty & \text{otherwise} \end{cases}$
- $(\|\bigcirc\varphi\|_d, w) = d_{w(0)} \cdot (\|\varphi\|_d, w_{\geq 1})$ $(\|\Box\varphi\|_d, w) = \sum_{i \geq 0} d_{w_{<i}} \cdot (\|\varphi\|_d, w_{\geq i})$

- $(\|\varphi \vee \psi\|_d , w) = \max ((\|\varphi\|_d , w) , (\|\psi\|_d , w))$
- $(\|\varphi \wedge \psi\|_d , w) = (\|\varphi\|_d , w) + (\|\psi\|_d , w)$

- $(\|\varphi U \psi\|_d , w) = \sup\limits_{i \geq 0} \left(\sum\limits_{0 \leq j < i} d_{w_{<j}} \cdot (\|\varphi\|_d , w_{\geq j}) + d_{w_{<i}} \cdot (\|\psi\|_d , w_{\geq i}) \right) .$

Inductively we can prove that for every $\varphi \in LTL (\mathbb{R}_{\max}, AP)$, $\|\varphi\|_d$ has a bounded image and the supremum and infinite sum appearing in the definitions of U and $\square$, respectively are well defined. Alternatively we shall denote by *true* the formula $0 \in \mathbb{R}_{\max}$.

A formula $\varphi \in LTL (\mathbb{R}_{\max}, AP)$ is *boolean* if it does not contain any constants $k \in \mathbb{R}_{\max} \setminus \{0, -\infty\}$. Moreover, two formulas $\varphi, \psi \in LTL(\mathbb{R}_{\max}, AP)$ are *equivalent*, denoted by $\varphi \equiv \psi$, if for every $w \in (\mathcal{P}(AP))^\omega$ we have $(\|\varphi\|_d , w) = (\|\psi\|_d , w)$. For $\varphi, \psi, \xi \in LTL (\mathbb{R}_{\max}, AP)$, where ξ is boolean, it is straightforward to show the following equivalences.

$$\varphi U \psi \equiv \psi \vee (\varphi \wedge \bigcirc (\varphi U \psi)) \qquad \bigcirc (\varphi U \psi) \equiv (\bigcirc \varphi) U (\bigcirc \psi) \qquad \varphi \wedge true \equiv \varphi$$
$$\square \varphi \equiv \varphi \wedge \bigcirc (\square \varphi) \qquad \bigcirc (\varphi \wedge \psi) \equiv (\bigcirc \varphi) \wedge (\bigcirc \psi) \qquad \xi \wedge \xi \equiv \xi$$
$$\bigcirc (\square \varphi) \equiv \square (\bigcirc \varphi) \qquad \bigcirc (\varphi \vee \psi) \equiv (\bigcirc \varphi) \vee (\bigcirc \psi) .$$

A formula $\varphi \in LTL (\mathbb{R}_{\max}, AP)$ is called *reduced* if (a) for every subformula of the form $\varphi_1 \wedge \ldots \wedge \varphi_k$ with $k \geq 2$ it holds: $\varphi_i \neq true$ for every $1 \leq i \leq k$, and $\varphi_i \neq \varphi_j$ for every boolean φ_i, φ_j with $1 \leq i < j \leq k$, and (b) no *until* operator is in the scope of any *next* operator. For every $\varphi \in LTL (\mathbb{R}_{\max}, AP)$ we can effectively construct an equivalent reduced formula by applying the previous equivalences. We shall denote this formula by φ_{re}. Next, we define a syntactic fragment of our logic.

Definition 4. *A formula $\varphi \in LTL (\mathbb{R}_{\max}, AP)$ will be called* weakly operated *if whenever it contains subformulas of the form $\square \psi_1$, or $\psi_2 U \psi_3$, then ψ_1, ψ_2, ψ_3 may contain the operators U and $\square$ applied to boolean formulas only.*

We denote by $WOLTL (\mathbb{R}_{\max}, AP)$ the class of all weakly operated formulas of $LTL (\mathbb{R}_{\max}, AP)$. An infinitary series $s \in \mathbb{R}_{\max} \langle\langle (\mathcal{P}(AP))^\omega \rangle\rangle$ is called *woLTL-d-definable* if there is a formula $\varphi \in WOLTL (\mathbb{R}_{\max}, AP)$ such that $s = \|\varphi\|_d$.

Our aim is to construct for every reduced weakly operated formula φ an ε-WGBA with d-discounting accepting its semantics. We shall need the subsequent concepts. First we recall the notion of the closure of a formula. So, let $\varphi \in LTL (\mathbb{R}_{\max}, AP)$. The *closure* $cl(\varphi)$ of φ is the smallest set C such that (i) $\varphi \in C$, (ii) if $\psi \wedge \xi \in C$ or $\psi \vee \xi \in C$ or $\psi U \xi \in C$, then $\psi, \xi \in C$, (iii) if $\bigcirc \psi \in C$ or $\square \psi \in C$, then $\psi \in C$.

Furthermore, a subset B of $cl(\varphi)$ will be called φ-*consistent* if whenever $B \neq \emptyset$ it holds: (i) For every $a \in AP : a \in B \Rightarrow \neg a \notin B$, and $\neg a \in B \Rightarrow a \notin B$, (ii) $\varphi \in B$, (ii) $\psi \wedge \xi \in B \Rightarrow \psi \in B$ and $\xi \in B$, (iii) $\psi \vee \xi \in B \Rightarrow \psi \in B$ or $\xi \in B$, (iv) $\psi U \xi \in B \Rightarrow \psi \in B$ or $\xi \in B$, (v) $\square \psi \in B \Rightarrow \psi \in B$.

For every φ-consistent set $B \neq \emptyset$, and every $\psi \in cl(\varphi)$, there exists a maximal (according to subset relation) ψ-consistent subset of B. We shall denote this set by $M_{B,\psi}$. From now on we shall denote a φ-consistent set B by B_φ.

Definition 5. *Let $\varphi \in LTL\,(\mathbb{R}_{max}, AP)$ and B_φ be a φ-consistent set. The finite set $next\,(B_\varphi)$ of formulas is defined in the following way. We set $next\,(\emptyset) = \{-\infty\}$ and for $B_\varphi \neq \emptyset$*

- *if $\varphi = a$ or $\varphi = \neg a$, $a \in AP$, or $\varphi = k$, $k \in \mathbb{R}_{max}$, then $next\,(B_\varphi) = \{true\}$,*
- *if $\varphi = \psi \wedge \xi$, then*
  $next\,(B_\varphi) = \{\psi' \wedge \xi' \mid \psi' \in next\,(M_{B_\varphi,\psi})\,, \xi' \in next\,(M_{B_\varphi,\xi})\}$,
- *if $\varphi = \psi \vee \xi$, then $next\,(B_\varphi) = next\,(M_{B_\varphi,\psi}) \cup next\,(M_{B_\varphi,\xi})$,*
- *if $\varphi = \bigcirc\psi$, then $next\,(B_\varphi) = \{\psi\}$,*
- *if $\varphi = \psi U \xi$, then*
  $next\,(B_\varphi) = \{\varphi \wedge \psi' \mid \psi' \in next\,(M_{B_\varphi,\psi})\} \cup \{\xi' \mid \xi' \in next\,(M_{B_\varphi,\xi})\}$,
- *if $\varphi = \square\psi$, then $next\,(B_\varphi) = \{\varphi \wedge \psi' \mid \psi' \in next\,(M_{B_\varphi,\psi})\}$.*

Clearly, every formula in $next\,(B_\varphi)$ is a finite conjunction of the form $\bigwedge\limits_{1 \leq i \leq k} \psi_i$
where for every $1 \leq i \leq k$, $\psi_i \in cl\,(\varphi)$, or $\psi_i = true$, or $\psi_i = -\infty$.

Next, we define inductively for every formula φ and every B_φ a mapping $v_{B_\varphi} : next\,(B_\varphi) \to \mathbb{R}_{max}$. We let $v_\emptyset\,(-\infty) = -\infty$. Now, assume that $B_\varphi \neq \emptyset$.
For $\varphi = a$, or $\varphi = \neg a$ with $a \in AP$ we set $v_{\{\varphi\}}\,(true) = 0$, and for $\varphi = k \in \mathbb{R}_{max}$, $v_{\{\varphi\}}\,(true) = k$.
For $\varphi = \psi \wedge \xi$, we let $v_{B_\varphi}\,(\psi' \wedge \xi') = v_{M_{B_\varphi,\psi}}\,(\psi') + v_{M_{B_\varphi,\xi}}\,(\xi')$ where $\xi' \in next\,(M_{B_\varphi,\xi})$ and $\psi' \in next\,(M_{B_\varphi,\psi})$. Next, let $\varphi = \psi \vee \xi$. Then, for every $\varphi' \in next\,(M_{B_\varphi,\psi}) \cup next\,(M_{B_\varphi,\xi})$, we let $v_{B_\varphi}\,(\varphi') = \max\left(v_{M_{B_\varphi,\psi}}\,(\varphi'), v_{M_{B_\varphi,\xi}}\,(\varphi')\right)$ where abusing the notations $v_{M_{B_\varphi,\psi}}\,(\varphi')$ (resp. $v_{M_{B_\varphi,\xi}}\,(\varphi')$) will stand for $-\infty$ whenever $\varphi' \notin next\,(M_{B_\varphi,\psi})$ (resp. $next\,(M_{B_\varphi,\xi})$).
Assume that $\varphi = \bigcirc\psi$. Then, for the unique element ψ of $next\,(B_\varphi)$ we set $v_{B_\varphi}\,(\psi) = 0$. For $\varphi = \square\psi$, we set $v_{B_\varphi}\,(\varphi \wedge \psi') = v_{M_{B_\varphi,\psi}}\,(\psi')$ where $\psi' \in next\,(M_{B_\varphi,\psi})$. Finally, for $\varphi = \psi U \xi$, we let $v_{B_\varphi}\,(\varphi \wedge \psi') = v_{M_{B_\varphi,\psi}}\,(\psi')$ where $\psi' \in next\,(M_{B_\varphi,\psi})$, and $v_{B_\varphi}\,(\xi') = v_{M_{B_\varphi,\xi}}\,(\xi')$ with $\xi' \in next\,(M_{B_\varphi,\xi})$.

Now, we consider the formulas $\varphi_1, \ldots, \varphi_j \in WOLTL\,(\mathbb{R}_{max}, AP)$ with φ_1 being reduced, and the sequence $B_{\varphi_1}, \ldots, B_{\varphi_j}$ with the following properties. For every $1 \leq l \leq j-1$, if φ_l is reduced, then $\varphi_{l+1} \in next\,(B_{\varphi_l})$, otherwise $\varphi_{l+1} = (\varphi_l)_{re}$. Then we will say that B_{φ_j} is *reachable* by B_{φ_1}. The set $reach\,(B_{\varphi_1})$ of all reachable by B_{φ_1} sets of formulas is finite and effectively computable. Observe that, since φ_1 is reduced, the formulas φ_l $(1 \leq l \leq j)$ satisfy the second condition in the definition of reduced formulas. This implies that reduction when applied only reduces conjunctions. We give an example to show that restricting φ_1 to the weakly operated fragment is the key to ensure that $reach\,(B_{\varphi_1})$ is finite. We let $\varphi = \square\,(\square\,(a \wedge 2))$ and $B_\varphi = \{\varphi, \square\,(a \wedge 2), a \wedge 2, a, 2\}$. Then, for every

$j \geq 1$, every consistent set of the formula $\varphi \wedge \left(\bigwedge\limits_{1 \leq i \leq j} \psi_i\right)$ with $\psi_i = \square\,(a \wedge 2)$

$(1 \leq i \leq j)$ belongs to the set $reach\,(B_\varphi)$, and hence $reach\,(B_\varphi)$ is not finite.

Let $\varphi, \psi \in WOLTL\,(\mathbb{R}_{max}, AP)$. For every $\pi \in \mathcal{P}\,(AP)$ the triple (B_φ, π, B_ψ) is called a *next transition* if the following two conditions hold: (i) For every

$a \in AP$ we have $a \in B_\varphi \Rightarrow a \in \pi$ and $\neg a \in B_\varphi \Rightarrow a \notin \pi$, and (ii) φ is reduced and $\psi \in next(B_\varphi)$. Moreover, for every $B_\varphi, B_{\varphi_{re}}$ with $B_\varphi \neq \emptyset$ and $B_{\varphi_{re}} \neq \emptyset$ the triple $(B_\varphi, \varepsilon, B_{\varphi_{re}})$ is called an ε-*reduction transition*. In the subsequent definition φ is assumed to be reduced.

Definition 6. *Let* $\varphi \in WOLTL(\mathbb{R}_{max}, AP)$. *We define the* ε-*WGBA* $\mathcal{A}_\varphi = (Q, wt, in, \mathcal{F})$ *with d-discounting over* $\mathcal{P}(AP)$ *and* $\mathbb{R}_{max}$ *as follows. We set*

- $Q = \bigcup_{B_\varphi} (\{B_\varphi\} \cup reach(B_\varphi))$,

- $in(B_\psi) = \begin{cases} 0 \text{ if } \varphi = \psi \\ -\infty \text{ otherwise} \end{cases}$

  for every $B_\psi \in Q$,

- $wt((B_\psi, \pi, B_\xi)) = \begin{cases} v_{B_\psi}(\xi) \text{ if } (B_\psi, \pi, B_\xi) \text{ is a next transition} \\ 0 \text{ if } (B_\psi, \pi, B_\xi) \text{ is an } \varepsilon\text{-reduction transition} \\ -\infty \text{ otherwise} \end{cases}$

  for every $(B_\psi, \pi, B_\xi) \in Q \times (\mathcal{P}(AP) \cup \{\varepsilon\}) \times Q$, *and*
- $\mathcal{F} = \{F_{\varphi'U\varphi''} \mid \varphi'U\varphi'' \in cl(\varphi)\}$ *where*

$$F_{\varphi'U\varphi''} = \left\{ B_{\overline{\varphi}} \in Q \mid B_{\overline{\varphi}} \neq \emptyset, \overline{\varphi} = \bigwedge_{1 \leq i \leq k} \varphi_i \text{ with } \varphi_i \neq \varphi'U\varphi'', 1 \leq i \leq k \right\}, \text{ for}$$

every $\varphi'U\varphi'' \in cl(\varphi)$.

Observe that for every $\varphi'U\varphi'' \in cl(\varphi)$ and every non-empty $B_\psi, B_{\psi_{re}} \in Q$ the relation $B_\psi \in F_{\varphi'U\varphi''}$ implies that $B_{\psi_{re}} \in F_{\varphi'U\varphi''}$, and vice-versa. Thus, the ε-transitions of the automaton are well defined. We note that if φ contains no U operators, then we have no acceptance conditions, which means that all infinite paths are successful.

We present an example. Let $\varphi = \Box(a \wedge 2)$, and $AP = \{a, b\}$. Then, $\mathcal{A}_\varphi = (Q, wt, in, \mathcal{F})$ with $Q = \{q_1, \ldots, q_5\}$ where $q_1 = \emptyset$, $q_2 = \{-\infty\}$, $q_3 = \{true\}$, $q_4 = \{\varphi, a \wedge 2, a, 2\}$, $q_5 = \{\varphi \wedge (true \wedge true), \varphi, a \wedge 2, 2, a, true \wedge true, true\}$. The states with initial weight 0 are the sets q_1, q_4. The transitions with weight different from $-\infty$ are the following: $wt((q_4, \pi, q_j)) = 2$, where $\pi \in \mathcal{P}(AP)$ with $a \in \pi$, and $j = 1, 5$, and $wt((q_3, \pi, q_i)) = wt((q_5, \varepsilon, q_4)) = wt((q_k, \varepsilon, q_k)) = 0$ where $\pi \in \mathcal{P}(AP)$, and $i = 1, 3, k = 2, 3, 4$. The automaton has no final subsets.

For every reduced $\varphi \in WOLTL(\mathbb{R}_{max}, AP)$ we can prove by induction on the structure of φ that $\|\mathcal{A}_\varphi\| = \|\varphi\|_d$. By this, and the fact that for every ε-WGBA with d-discounting there is an equivalent WBA with d-discounting we conclude Theorem 1. Finally, the last result of our paper, stated in Theorem 2, assigns to every WBA over $\mathcal{P}(AP)$ and $\mathbb{R}_{max}$ a weighted weakly operated LTL-formula φ whose semantics is projected to the behavior of the automaton via a strict alphabetic epimorphism (i.e., a letter-to-letter morphism).

Theorem 1. *Every woLTL-d-definable series over* $\mathcal{P}(AP)$ *and* $\mathbb{R}_{max}$ *is* (d, ω)-*recognizable.*

Theorem 2. *Let* $s \in \mathbb{R}_{max} \langle\langle (\mathcal{P}(AP))^\omega \rangle\rangle$ *be a* (d, ω)-*recognizable series over* $\mathbb{R}_{max}$ *and* $\mathcal{P}(AP)$. *Then there is a finite set of atomic propositions* $\widetilde{AP} \supset AP$, a strict alphabetic epimorphism h from $\mathcal{P}\left(\widetilde{AP}\right)$ to $\mathcal{P}(AP)$ and a woLTL-$\tilde{d}$-definable series $r \in \mathbb{R}_{\max}\left\langle\left\langle\left(\mathcal{P}\left(\widetilde{AP}\right)\right)^{\omega}\right\rangle\right\rangle$ such that $h(r) = s$, where $\tilde{d} = \left(\widetilde{d_{\pi'}}\right)_{\pi' \in \mathcal{P}(\widetilde{AP})}$ is a discounting over $\mathcal{P}\left(\widetilde{AP}\right)$ and $\mathbb{R}_{\max}$ determined for every $\pi' \in \mathcal{P}\left(\widetilde{AP}\right)$ by $\widetilde{d}_{\pi'} = d_{h(\pi')}$.

5 Conclusion

We have defined a weighted *LTL* over infinite words with weights and discounting parameters over $\mathbb{R}_{\max}$, and have connected weakly operated formulas to WBA with discounting. Whether the full fragment of our *LTL* exceeds or not (d, ω)-recognizability is an open and important question to study. In [7] the authors connect the fragment of weakly operated formulas of the weighted *LTL* with discounting to weighted first order logic with discounting, ω-star-free series with discounting, and weighted counter-free Büchi automata with discounting. Finally, another perspective is to investigate complexity fragments, as well as decidability results for our construction.

Acknowledgement. The author would like to thank the three anonymous referees, and especially one of them, for useful remarks.

References

1. De Alfaro, L., Henzinger, T.A., Majumdar, R.: Discounting the Future in Systems Theory. In: Baeten, J.C.M., Lenstra, J.K., Parrow, J., Woeginger, G.J. (eds.) ICALP 2003. LNCS, vol. 2719, pp. 1022–1037. Springer, Heidelberg (2003)
2. Dermi, S., Gastin, P.: Specification and verification using temporal logics. In: D'Souza, D., Shankar, P. (eds.) Modern Applications of Automata Theory, IISc, Research monographs, vol. 2. World Scientific (2011)
3. Droste, M., Kuske, D.: Skew and infinitary formal power series. Theoret. Comput. Sci. 366, 189–227 (2006)
4. Droste, M., Vogler, H.: Weighted automata and multi-valued logics over arbitrary bounded lattices. Theoret. Comput. Sci. 418, 14–36 (2012)
5. Faella, M., Legay, A., Stoelinga, M.: Model checking quantitative linear time logic. Electron. Notes Theor. Comput. Sci. 220, 61–77 (2008)
6. Kupferman, O., Lustig, Y.: Lattice Automata. In: Cook, B., Podelski, A. (eds.) VMCAI 2007. LNCS, vol. 4349, pp. 199–213. Springer, Heidelberg (2007)
7. Mandrali, E., Rahonis, G.: Weighted first order logic with discounting (in preparation)
8. Pnueli, A.: The temporal logics of programs. In: 18th IEEE Symposium on Foundations of Computer Science (FOCS), pp. 46–67 (1977)
9. Vardi, M., Wopler, P.: Reasoning about infinite computations. Inform. and Comput. 115, 1–37 (1994)
10. Wolper, P.: Constructing Automata from Temporal Logic Formulas: A Tutorial. In: Brinksma, E., Hermanns, H., Katoen, J.-P. (eds.) FMPA 2000. LNCS, vol. 2090, pp. 261–277. Springer, Heidelberg (2001)

Automata with Modulo Counters
and Nondeterministic Counter Bounds

Daniel Reidenbach and Markus L. Schmid*

Department of Computer Science, Loughborough University,
Loughborough, Leicestershire, LE11 3TU, United Kingdom
{D.Reidenbach,M.Schmid}@lboro.ac.uk

Abstract. We introduce and investigate Nondeterministically Bounded
Modulo Counter Automata (NBMCA), which are two-way one-head automata that comprise a constant number of modulo counters, where the
counter bounds are nondeterministically guessed, and this is the only
element of nondeterminism. NBMCA are tailored to recognising those
languages that are characterised by the existence of a specific factorisation of their words, e. g., pattern languages. In this work, we subject
NBMCA to a theoretically sound analysis.

Keywords: Multi-head automata, Counter automata, Modulo counters, Stateless automata, Restricted nondeterminism.

1 Introduction

In the present paper we introduce and study a novel automata model, the Nondeterministically Bounded Modulo Counter Automata (NBMCA for short), which
comprise several two-way input heads and a number of counters. These NBMCA
are suitable algorithmic tools for recognising those languages that are characterised by the existence of a specific factorisation of their words, e. g., pattern
languages, and are a generalisation of the Janus automata that have been introduced and applied in [11] in order to investigate the membership problem
for pattern languages. In [11], NBMCA with exactly two input heads are used.
In the present work we focus on NBMCA with only one head, since we can
easily simulate several input heads by just a single one. For every counter of
an NBMCA an individual counter bound is provided, and every counter can
only be incremented and counts modulo its counter bound. The current counter
values and counter bounds are hidden from the transition function, which can
only check whether a counter has reached its bound. By performing a reset on
a counter, the automaton nondeterministically guesses a new counter bound be-
tween 0 and $|w|$, where w is the input word. This guessing of counter bounds is
the only possible nondeterministic step of NBMCA, and the transition function
is defined completely deterministically. We can interpret the counter bounds as

---

* Corresponding author.

N. Moreira and R. Reis (Eds.): CIAA 2012, LNCS 7381, pp. 361–368, 2012.
© Springer-Verlag Berlin Heidelberg 2012

positions of the input and, by means of the counter values, the input head can be moved to these positions.

Two aspects of this approach seem to be particularly worth studying. Firstly, all additional resources the automaton is equipped with, namely the counters, are tailored to storing positions in the input word. We can observe that this aspect is not really new; in fact, the idea of separating the mechanisms of storing positions from the functionality of actually processing the input is formalised in the models of partially blind multi-head automata (see, e. g., Ibarra and Ravikumar [7]), Pebble Automata (see, e. g., Chang et al. [1]) and automata with sensing heads (see, e. g., Petersen [10]). Given this similarity between NBMCA and established automata models regarding their emphasis on storing positions in the input word, there is still one difference: the counters of NBMCA are quite limited in their ability to change the positions they represent, since their values can merely be incremented, and their bounds are guessed. The question arises whether or not, for automata using counters as additional resources, their ability to count in both directions is essential with respect to the expressive power.

The second aspect is that the nondeterminism of NBMCA, which merely allows positions in the input word to be guessed, differs quite substantially from the common nondeterminism of automata, which provides explicit computational alternatives. Nevertheless, automata often use their nondeterminism to actually guess a certain position of the input. For example, a pushdown automaton that recognises $\{ww^R \mid w \in \Sigma^*\}$ needs to perform an unbounded number of guesses even though only one specific position, namely the middle one, of the input needs to be found. Despite this observation, the nondeterminism of NBMCA might be weaker, as it seems to *solely* refer to positions in the input. Hence, we also investigate the question of whether or not it is essential that the nondeterminism is explicitly provided by a nondeterministic transition function in order to exploit it to the full extent, in terms of expressive power.

In order to understand the character of these novel, and seemingly limited, resources NBMCA can use, the present paper compares the expressive power of these automata to that of the well-established, and seemingly less restricted, models of multi-head and counter automata. Furthermore, we study some basic decision problems for NBMCA as well as stateless versions of NBMCA, with and without restricted nondeterminism.

Note that, due to space constraints, all proofs have been omitted.

2   Definitions

Let $\mathbb{N}$ denote the set of all positive integers and let $\mathbb{N}_0 := \mathbb{N} \cup \{0\}$. The symbols $\subseteq$ and $\subset$ refer to subset and proper subset relation, respectively. For an arbitrary alphabet Σ, a *word* (*over* Σ) is a finite sequence of symbols from Σ, and ε stands for the *empty word*. The symbol Σ^+ denotes the set of all nonempty words over Σ, and $\Sigma^* := \Sigma^+ \cup \{\varepsilon\}$. For the *concatenation* of two words u, v we write $u \cdot v$ or simply uv, and u^k denotes the k-fold concatenation of u. The notation $|K|$ stands for the size of a set K or the length of a word K.

For an arbitrary class of automata, such as the set DFA of deterministic finite automata, the expression "a DFA" refers to any automaton from DFA. For an arbitrary automaton M, $L(M)$ denotes the set of all words accepted by M and, for an arbitrary class A of automata, let $\mathcal{L}(A) := \{L(M) \mid M \in A\}$. For every $k \in \mathbb{N}$ let $1DFA(k)$, $2DFA(k)$, $1NFA(k)$ and $2NFA(k)$ denote the class of *deterministic one-way, deterministic two-way, nondeterministic one-way* and *nondeterministic two-way automata* with k input heads, respectively. For a comprehensive survey on multi-head automata the reader is referred to Holzer et al. [3] and to the references therein.

Next, we define the central automata model of this paper. A *Nondeterministically Bounded Modulo Counter Automaton*, $NBMCA(k)$ for short, is a two-way one-head automaton with k counters. More precisely, it is a tuple $M := (k, Q, \Sigma, \delta, q_0, F)$, where $k \in \mathbb{N}$ is the number of *counters*, Q is a finite nonempty set of *states*, Σ is a finite nonempty alphabet of *input symbols*, $q_0 \in Q$ is the *initial state* and $F \subseteq Q$ is the set of *accepting states*. The mapping $\delta : Q \times \Sigma \times \{t_0, t_1\}^k \rightarrow Q \times \{-1, 0, 1\} \times \{0, 1, r\}^k$ is called the *transition function*. Instead of writing transitions in the form $\delta(C) = S$, we use the notation $C \rightarrow_\delta S$. If δ is obvious from the context, we simply write $C \rightarrow S$. An input to M is any word of the form $\text{¢}w\$$, where $w \in \Sigma^*$ and the symbols ¢, \$ (referred to as *left* and *right endmarker*, respectively) are not in Σ. Let $(p, b, s_1, \ldots, s_k) \rightarrow_\delta (q, r, d_1, \ldots, d_k)$. We call the element b the *scanned input symbol* and r the *input head movement*. For each $j \in \{1, 2, \ldots, k\}$, the element $s_j \in \{t_0, t_1\}$ is the *counter message of counter j*, and d_j is called the *counter instruction for counter j*. The transition function δ of an $NBMCA(k)$ determines whether the input heads are moved to the left ($r_i = -1$), to the right ($r_i = 1$) or left unchanged ($r_i = 0$), and whether the counters are incremented ($d_j = 1$), left unchanged ($d_j = 0$) or reset ($d_j = r$). In case of a reset, the counter value is set to 0 and a new counter bound is nondeterministically guessed between 0 and the current input length. Hence, every counter is bounded, but these bounds are chosen in a nondeterministic way. In order to define the language accepted by an NBMCA, we need to define the concept of an NBMCA computation.

Let M be an NBMCA and $w := a_1 \cdot a_2 \cdots \cdot a_n$, $a_i \in \Sigma$, $1 \leq i \leq n$. A *configuration of M (on input w)* is an element of $\widehat{C}_M := \{[q, h, (c_1, C_1), \ldots, (c_k, C_k)] \mid q \in Q, 0 \leq h \leq n+1, 0 \leq c_i \leq C_i \leq n, 1 \leq i \leq k\}$. The pair (c_i, C_i), $1 \leq i \leq k$, describes the current configuration of the ith counter, where c_i is the *counter value* and C_i the *counter bound*. The element h is called the *input head position*.

An *atomic move* of M is denoted by the relation $\vdash_{M,w}$ over the set of configurations. Let $(p, b, s_1, \ldots, s_k) \rightarrow_\delta (q, r, d_1, \ldots, d_k)$. Then, for all c_i, C_i, $1 \leq i \leq k$, where $c_i < C_i$ if $s_i = t_0$ and $c_i = C_i$ if $s_i = t_1$, and for every h, $0 \leq h \leq n+1$, with $a_h = b$, we define $[p, h, (c_1, C_1), \ldots, (c_k, C_k)] \vdash_{M,w} [q, h', (c_1', C_1'), \ldots, (c_k', C_k')]$. Here, the elements h' and c_j', C_j', $1 \leq j \leq k$, are defined in the following way. $h' := h + r$ if $0 \leq h + r \leq n+1$ and $h' := h$ otherwise. For each $j \in \{1, \ldots, k\}$, if $d_j = r$, then $c_j' := 0$ and, for some $m \in \{0, 1, \ldots, n\}$, $C_j' := m$. If, on the other hand, $d_j \neq r$, then $C_j' := C_j$ and $c_j' := c_j + d_j$ mod $(C_j + 1)$.

In order to describe a *sequence of (atomic) moves of M (on input w)* we use the reflexive and transitive closure of the relation $\vdash_{M,w}$, denoted by $\vdash^*_{M,w}$. M accepts the word w if and only if $\widehat{c}_0 \vdash^*_{M,w} \widehat{c}_f$, where $\widehat{c}_0 := [q_0, 0, (0, C_1), \ldots, (0, C_k)]$ for some $C_i \in \{0, 1, \ldots, |w|\}$, $1 \leq i \leq k$, is an *initial configuration*, and $\widehat{c}_f := [q_f, h, (c_1, C_1), \ldots (c_k, C_k)]$ for some $q_f \in F$, $0 \leq h \leq n + 1$ and $0 \leq c_i \leq C_i \leq n$, $1 \leq j \leq k$, is a *final configuration*. In every computation of an NBMCA, the counter bounds are nondeterministically initialised, and the only nondeterministic step an NBMCA is able to perform during the computation consists in guessing a new counter bound for some counter.

3 Expressive Power, Hierarchy and Decidability

An NBMCA can be regarded as a finite state control with additional resources. Thus, it is quite similar to classical nondeterministic multi-head automata. The essential differences between the models are those addressed in Section 1. Hence, in order to gain insights with respect to the question of whether these differences affect the expressive power, we study the problem of simulating classical nondeterministic multi-head automata by NBMCA and vice versa. It is almost obvious that NBMCA can be simulated by nondeterministic multi-head automata as NBMCA can be interpreted as just a further restricted version of them. So multi-head automata intuitively seem to be more powerful.

Theorem 1. *For every $k \in \mathbb{N}$, $\mathcal{L}(\text{NBMCA}(k)) \subseteq \mathcal{L}(2\text{NFA}(2k + 1))$.*

The converse question, i.e., whether arbitrary multi-head automata, and particularly their unrestricted nondeterminism, can be simulated by NBMCA, is more interesting. It can be done by using a modulo counter of the NBMCA in order to simulate an input head of the 2NFA(k) in the following way. The modulo counter first guesses $|w|$ as counter bound, which is done by reseting it and checking, by means of the input head, whether or not the guessed bound equals $|w|$, and then the counter value can be used in order to store the position of the input head. Since the counter value cannot be decremented, a decrement has to be performed by $|w| - 1$ increments.

However, for reasons that shall be explained later, we aim for a simulation that is more economic with respect to the usage of modulo counters. More presicely, we want to use a single modulo counter in order to store the positions of two input heads of a 2NFA(k), i.e., the counter value and the counter bound each represent a distinct input head position. A step of the 2NFA(k) is then simulated by first moving the input head of the NBMCA successively to all these positions stored by the counters and record the scanned input symbols in the finite state control. After that, all these positions stored by the counters must be updated according to the transition function of the 2NFA(k). It turns out that this is possible, but, since counter values cannot be decremented and counter bounds cannot be changed directly, the constructions are rather involved and require some technical finesse. Furthermore, we need an additional counter which is also used in order to simulate the possible nondeterministic choices of the 2NFA(k).

Theorem 2. *For every $k \in \mathbb{N}$, $\mathcal{L}(2\mathrm{NFA}(k)) \subseteq \mathcal{L}(\mathrm{NBMCA}(\lceil \frac{k}{2} \rceil + 1))$.*

The above results show that neither the restrictions on the counters of NBMCA nor the special nondeterminism constitute a restriction on the expressive power. Thus, NBMCA can be used whenever classical multi-head automata can be applied, but due to their specific counters and nondeterminism they are particularly suitable algorithmic tools for recognising those languages that are characterised by the existence of a certain factorisation for their words, such as pattern languages (see [11]).

The tight use of the modulo counters in the previous simulation turns out to be worth the effort, as it allows us to prove a hierarchy result on the class NBMCA by applying a classical hierarchy result concerning multi-head automata (Monien [9]).

Corollary 1. *For every $k \in \mathbb{N}$, $\mathcal{L}(\mathrm{NBMCA}(k)) \subset \mathcal{L}(\mathrm{NBMCA}(k + 2))$.*

Next, we investigate the decidability of the emptiness, infiniteness, universe, equivalence, inclusion and disjointness problem with respect to languages given by NBMCA. From the fact that all these problems are undecidable even for 1DFA(2) (cf., Holzer et al. [3]) and Theorem 2, it follows that all these problems are also undecidable for NBMCA. However, it is a common approach to further restrict automata models with undecidable problems in order to obtain subclasses with decidable problems. One respective option is to require the automata to be reversal bounded (see, e.g., Ibarra [4]). Hence, for all $m_1, m_2, l, k \in \mathbb{N}$, let (m_1, m_2, l)-REV-NBMCA(k) denote the class of NBMCA(k) that perform at most m_1 input head reversals, at most m_2 counter reversals and resets every counter at most l times in every accepting computation (here, input head reversals are defined in the same way as by Ibarra [4], whereas a counter reversal is an increment of the counter in case that it has already reached its counter bound). We can directly apply a result by Ibarra [4] about reversal-bounded counter machines in order to obtain the following:

Theorem 3. *The emptiness, infiniteness and disjointness problem for the class (m_1, m_2, l)-REV-NBMCA are decidable.*

Next, we investigate the decidability properties of (m, ∞, l)-REV-NBMCA, i. e., the number of counter reversals is not bounded anymore. This question is motivated as follows. Ibarra [4] shows for counter machines that if only the reversals of the input head are bounded and counter reversals are unrestricted, then the typical decision problems remain undecidable. However, while a counter reversal of a counter machine can happen anytime in the computation and for any possible counter value, a counter reversal of an NBMCA strongly depends on the current counter bound, i. e., as long as a counter is not reset, all the counter reversals of that counter happen at exactly the same counter value. Hence, the modulo counters of (m, ∞, l)-REV-NBMCA are still restricted, since the number of resets is bounded, and the question arises whether or not this restriction is strong enough to maintain positive decidability results. The following answers this question in the negative, even for small m and k, and no counter resets:

Theorem 4. *The emptiness, infiniteness, universe, equivalence, inclusion and disjointness problems are undecidable for* $(3, \infty, 0)$-REV-NBMCA(3).

4   NBMCA without States

In this section, we consider NBMCA without states. Stateless versions of automata have recently been introduced by Yang et al. [12], where they are compared to P-Systems. Ibarra et al. [6] and Frisco and Ibarra [2] investigate stateless multi-head automata, whereas Ibarra and Eğecioğlu [5] consider stateless counter machines. Kutrib et al. [8] study stateless restarting automata.

A *stateless* NBMCA (SL-NBMCA for short) can be regarded as an NBMCA with only one internal state that is never changed. Hence, the component referring to the state is removed from the transition function and transitions do not depend anymore on the state. As a result, the acceptance of inputs by accepting state is not possible anymore. So for stateless NBMCA we define the input to be accepted by a special accepting transition, i.e., the transition that does not change the configuration of the automaton anymore. On the other hand, if the automaton enters a configuration for which no transition is defined, then the input is rejected and the same happens if an infinite loop is entered. For example, $(b, s_1, \ldots, s_k) \to (r, d_1, \ldots, d_k)$ is a possible transition for an SL-NBMCA(k) and $(b, s_1, \ldots, s_k) \to (0, 0, 0, \ldots, 0)$ is an accepting transition. An SL-NBMCA(k) can be given as a tuple (k, Σ, δ) comprising the number of counters, the input alphabet and the transition function. We now consider an example for the languages $S_k := \{a^k, \varepsilon\}$, $k \in \mathbb{N}$. The following SL-NBMCA(5) recognises exactly S_3.

Definition 1. *Let* $M_{S_3} := (5, \{a\}, \delta) \in$ SL-NBMCA(5), *where* δ *is defined by*
$(\text{¢}, \mathsf{t}_0, \mathsf{t}_0, \mathsf{t}_0, \mathsf{t}_0, \mathsf{t}_0) \to_\delta (1, 1, 1, 1, 1, \mathsf{r})$, $(\mathsf{a}, \mathsf{t}_1, \mathsf{t}_1, \mathsf{t}_1, \mathsf{t}_1, \mathsf{t}_0) \to_\delta (-1, 1, 1, 1, 1, 1)$,
$(\text{¢}, \mathsf{t}_0, \mathsf{t}_0, \mathsf{t}_0, \mathsf{t}_0, \mathsf{t}_1) \to_\delta (1, 1, 0, 0, 0, 0)$, $(\mathsf{a}, \mathsf{t}_1, \mathsf{t}_0, \mathsf{t}_0, \mathsf{t}_0, \mathsf{t}_1) \to_\delta (1, 0, 1, 0, 0, 0)$,
$(\mathsf{a}, \mathsf{t}_1, \mathsf{t}_1, \mathsf{t}_0, \mathsf{t}_0, \mathsf{t}_1) \to_\delta (1, 0, 0, 1, 0, 0)$, $(\mathsf{a}, \mathsf{t}_1, \mathsf{t}_1, \mathsf{t}_1, \mathsf{t}_0, \mathsf{t}_1) \to_\delta (1, 0, 0, 0, 1, 1)$,
$(\$, \mathsf{t}_1, \mathsf{t}_1, \mathsf{t}_1, \mathsf{t}_1, \mathsf{t}_0) \to_\delta (0, 0, 0, 0, 0, 0)$.

The question of whether or not states are really necessary for a model, i.e., whether it is possible to simulate automata by their stateless counterparts, is probably the most fundamental question about stateless automata. Regarding SL-NBMCA, we can observe that every NBMCA with states can be turned into an equivalent one without states. Hence, the loss of the finite state control does not lead to a reduced expressive power of the model.

Theorem 5. *For every* $M \in$ NBMCA(k), $k \in \mathbb{N}$, *with a set of states* Q, *there exists an* $M' \in$ SL-NBMCA$(k + \lceil \log(|Q| + 1) \rceil + 2)$ *with* $L(M) = L(M')$.

For the simulation of NBMCA by SL-NBMCA as well as for the automaton M_{S_3} (see Definition 1), it is vital that certain counters have a counter bound of 1. Due to the lack of states, this need for counters to be initialised with a counter bound of 1 involves considerable technical challenges.

Next, we use the model of stateless NBMCA in order to investigate a more general question in automata theory regarding limited nondeterminism. Usually, the nondeterminism is mainly controlled by the finite state control, i. e., certain states allow nondeterministic steps whereas others enforce a deterministic transition. Hence, nondeterminism can be switched on and off and, thus, it is a resource the automaton may use, but it is not forced to. These considerations suggest that the finite state control plays an important role regarding restricted nondeterminism and it is not obvious what consequences, in this regard, an abolishment of the finite state control may have. In the following we try to answer this question by employing SL-NBMCA. As shown in the previous section, if we allow an unbounded number of modulo counters, a finite state control can be simulated and, thus, nondeterminism can be controlled in the usual way. Therefore we consider SL-NBMCA(1) and, furthermore, we assume the input head to operate in a one-way manner. In order to restrict the nondeterminism of the model, we simply limit the number of possible resets for the modulo counter. More precisely, in any computation the first k applications of a reset operation reset the counter in accordance with the definition, whereas every further application of a reset is simply ignored. We shall refer to this model by 1SL-NBMCA$_k$(1), where k stands for the number of possible resets.

This way of restricting automata is unusual compared to the common restrictions that are found in the literature. This can be illustrated by considering input head reversal bounded automata as an example (see, e. g., Ibarra [4]). An input head reversal bounded automaton is an automaton that can recognise each word of a language in such a way that the number of input head reversals is bounded. There is no need to require the input head reversals to be bounded in the non-accepting computations as well, as this does not constitute a further restriction. This is due to the fact that we can always use the finite state control to count the number of input head reversals in order to interrupt a computation in a non-accepting state as soon as the bound of input head reversals is exceeded. However, regarding stateless automata this is not necessarily possible anymore, and it seems that it is a difference whether a restriction is defined for all possible computations or only for the accepting ones. Our definition of bounded resets introduced above avoids these problems by slightly changing the model itself, i. e., in every computation it loses the ability to reset the counter after a number of resets. For every $k \in \mathbb{N}$, there are languages that require at least k resets:

Theorem 6. *For every $k \in \mathbb{N}$, there exists a language $L \in \mathcal{L}(\text{1SL-NBMCA}_k(1))$ with $L \notin \mathcal{L}(\text{1SL-NBMCA}_{k'}(1))$ for every $k' \in \mathbb{N}$, $k' < k$.*

Moreover, by applying Theorem 6 and a simple set-theoretic reasoning, we can show that there are languages that can be recognised by a 1SL-NBMCA$_k$(1), but cannot be recognised by any 1SL-NBMCA$_{k+1}$(1).

Theorem 7. *There exist infinitely many $k \in \mathbb{N}$ such that $\mathcal{L}(\text{1SL-NBMCA}_k(1))$ and $\mathcal{L}(\text{1SL-NBMCA}_{k+1}(1))$ are incomparable.*

The above results yield the following conclusions: For every $k \in \mathbb{N}$, there is a language that can be recognised by a 1SL-NBMCA(1) with k, but not with $k-1$

resets. This meets our expectation of nondeterminism being a useful resource enhancing the expressive power of automata. Theorem 7, on the other hand, does not fit with the usual results on restricted nondeterminism, as it shows that expressive power is lost by increasing the nondeterminism, i. e., for infinitely many $k \in \mathbb{N}$, there is a language that can be recognised by a 1SL-NBMCA(1) with k, but not with $k+1$ resets. Considering the strong restrictions of 1SL-NBMCA$_k$(1), it is maybe not surprising that without any states the nondeterminism cannot be controlled anymore and, thus, a result of the sort mentioned above can be obtained. However, proving this behaviour is quite involved and, to the knowledge of the authors, it is the first result in the literature that formally establishes such a connection between finite state control and nondeterminism.

References

1. Chang, J.H., Ibarra, O.H., Palis, M.A., Ravikumar, B.: On pebble automata. Theoretical Computer Science 44, 111–121 (1986)
2. Frisco, P., Ibarra, O.H.: On Stateless Multihead Finite Automata and Multihead Pushdown Automata. In: Diekert, V., Nowotka, D. (eds.) DLT 2009. LNCS, vol. 5583, pp. 240–251. Springer, Heidelberg (2009)
3. Holzer, M., Kutrib, M., Malcher, A.: Complexity of multi-head finite automata: Origins and directions. Theoretical Computer Science 412, 83–96 (2011)
4. Ibarra, O.H.: Reversal-bounded multicounter machines and their decision problems. Journal of the ACM 25, 116–133 (1978)
5. Ibarra, O.H., Eğecioğlu, Ö.: Hierarchies and Characterizations of Stateless Multicounter Machines. In: Ngo, H.Q. (ed.) COCOON 2009. LNCS, vol. 5609, pp. 408–417. Springer, Heidelberg (2009)
6. Ibarra, O.H., Karhumäki, J., Okhotin, A.: On stateless multihead automata: Hierarchies and the emptiness problem. Theoretical Computer Science 411, 581–593 (2010)
7. Ibarra, O.H., Ravikumar, B.: On partially blind multihead finite automata. Theoretical Computer Science 356, 190–199 (2006)
8. Kutrib, M., Messerschmidt, H., Otto, F.: On stateless two-pushdown automata and restarting automata. International Journal of Foundations of Computer Science 21, 781–798 (2010)
9. Monien, B.: Two-way multihead automata over a one-letter alphabet. RAIRO Informatique Théorique 14, 67–82 (1980)
10. Petersen, H.: Automata with sensing heads. In: Proc. 3rd Israel Symposium on Theory of Computing and Systems, pp. 150–157 (1995)
11. Reidenbach, D., Schmid, M.L.: A Polynomial Time Match Test for Large Classes of Extended Regular Expressions. In: Domaratzki, M., Salomaa, K. (eds.) CIAA 2010. LNCS, vol. 6482, pp. 241–250. Springer, Heidelberg (2011)
12. Yang, L., Dang, Z., Ibarra, O.H.: On stateless automata and p systems. International Journal of Foundations of Computer Science 19, 1259–1276 (2008)

Author Index